Geochemical Processes at Mineral Surfaces

ACS SYMPOSIUM SERIES **323**

Geochemical Processes at Mineral Surfaces

James A. Davis, EDITOR
U.S. Geological Survey

Kim F. Hayes, EDITOR
Stanford University

Developed from a symposium sponsored by
the Division of Environmental Chemistry
and the Division of Geochemistry
at the 190th Meeting
of the American Chemical Society,
Chicago, Illinois,
September 8–13, 1985

American Chemical Society, Washington, DC 1986

Library of Congress Cataloging-in-Publication Data

Geochemical processes at mineral surfaces.
  (ACS symposium series, ISSN 0097-6156; 323)

  "Developed from a symposium sponsored by the
Division of Environmental Chemistry and the Division
of Geochemistry at the 190th Meeting of the American
Chemical Society, Chicago, Illinois, September 8–13,
1985."

  Bibliography: p.
  Includes indexes.

  1. Geochemistry—Congresses. 2. Mineralogical
chemistry—Congresses.

  I. Davis, James A., 1950–   . II. Hayes, Kim F.,
1953–   . III. American Chemical Society. Division of
Environmental Chemistry. IV. American Chemical
Society. Division of Geochemistry. V. American
Chemical Society. Meeting (190th: 1985: Chicago, Ill.)
VI. Series.

QE515.G3724   1986   551.9   86–22173
ISBN 0–8412–1004–7

ACS Symposium Series

M. Joan Comstock, *Series Editor*

Advisory Board

FOREWORD

The ACS SYMPOSIUM SERIES was founded in 1974 to provide a medium for publishing symposia quickly in book form. The format of the Series parallels that of the continuing ADVANCES IN CHEMISTRY SERIES except that, in order to save time, the papers are not typeset but are reproduced as they are submitted by the authors in camera-ready form. Papers are reviewed under the supervision of the Editors with the assistance of the Series Advisory Board and are selected to maintain the integrity of the symposia; however, verbatim reproductions of previously published papers are not accepted. Both reviews and reports of research are acceptable, because symposia may embrace both types of presentation.

CONTENTS

PREFACE

THIS BOOK deals only with the chemistry of the mineral–water interface, and so at first glance, the book might appear to have a relatively narrow focus. However, the range of chemical and physical processes considered is actually quite broad, and the general and comprehensive nature of the topics makes this volume unique. The technical papers are organized into physical properties of the mineral–water interface; adsorption; ion exchange; surface spectroscopy; dissolution, precipitation, and solid solution formation; and transformation reactions at the mineral–water interface. The introductory chapter presents an overview of recent research advances in each of these six areas and discusses important features of each technical paper. Several papers address the complex ways in which some processes are interrelated, for example, the effect of adsorption reactions on the catalysis of electron transfer reactions by mineral surfaces.

Papers in the symposium upon which this book is based were contributed by a diverse group of scientists from the fields of geochemistry, hydrogeology, water chemistry, chemical engineering, soil chemistry, electrical engineering, dental science, and mathematics. This diversity was facilitated by joint sponsorship of the symposium by the Environmental Chemistry and Geochemistry Divisions of ACS. The participants shared a common interest in the chemistry of mineral–water interfaces; however, each of these scientific disciplines has a different perspective on the theories and experimental methods that are used to describe interfacial chemistry. Research publications on the topic are widely dispersed in the literature, and as a result, many investigators are unaware of recent advances in disciplines other than their own. Thus, it appeared important and timely to compile recent developments in these related fields in a single volume. It is hoped that this effort will contribute to a greater understanding of the significant role of interfacial reactions in geochemical processes.

The inspiration for this symposium emerged from stimulating discussions with colleagues at the Gordon Conference on Environmental Sciences: Water held at New Hampton, N.H., in June 1984 (chaired by C. R. O'Melia). Several people assisted with the selection of speakers, including J. O. Leckie, L. N. Plummer, G. A. Parks, D. D. Eberl, A. F. White, A. T. Stone, and J. C. Westall. Financial support for foreign speakers was provided by the ACS Petroleum Research Fund and each of the sponsoring ACS divisions.

The quality of this volume is due in part to the careful work of numerous technical reviewers who submitted detailed comments and criticisms, and we are greatly indebted to these reviewers. We also thank Robin Giroux of the ACS Books Department for her guidance throughout the editorial process.

JAMES A. DAVIS
U.S. Geological Survey
Menlo Park, CA 94025

KIM F. HAYES
Stanford University
Stanford, CA 94305-4020

June 1, 1986

INTRODUCTION

1

Geochemical Processes at Mineral Surfaces: An Overview

James A. Davis[1] and Kim F. Hayes[2]

[1]Water Resources Division, U.S. Geological Survey, Menlo Park, CA 94025
[2]Environmental Engineering and Science Group, Department of Civil Engineering,
Stanford University, Stanford, CA 94305-4020

The phase discontinuity that occurs at the mineral-water interface greatly influences the geochemical cycles of many elements. The composition of natural waters and the flux of material through the hydrosphere are largely controlled by the weathering of minerals and the precipitation of new phases -- processes in which the mineral-water interface plays a fundamental role. In addition, mineral surfaces may act as catalysts for chemical or biological transformations that occur within the hydrosphere. Reactions at the mineral-water interface are of interest in the study of ore genesis, geochemical exploration, mineral separation processes such as flotation and sedimentation, transport of adsorbed nutrients or pollutants in rivers and lakes, scavenging of trace elements in the oceans, and the transport of nuclear or other hazardous waste materials in groundwaters.

Because these processes are so complex, we rely to a great extent on models to understand our observations of the geochemical behavior of solutes in water. The models typically contain several components; for example, solute transport models may be composed of a hydrologic model coupled with a chemical or biological submodel. The chemical submodel can be as simple as a distribution coefficient to represent the partitioning of an element between solid and aqueous phases, or if the aqueous chemical composition is expected to be controlled by the dissolution of a mineral, a solubility product or rate constant for dissolution would be included in the model. Ideally, the chemical model would describe geochemical behavior in terms of a series or combination of elementary processes, e.g. adsorption, ion exchange, precipitation, dissolution, or electron transfer reactions. In reality, the behavior of geochemical systems is often so complex that the actual mechanisms of the processes observed are not well understood.

In this overview we discuss recent advances in the study of chemical reactions at the mineral-water interface as we introduce the

0097-6156/86/0323-0002$06.00/0

chapters contained in this volume. Our objective is to indicate important features of each chapter to the field of aqueous geochemistry and ways in which the chapters relate to each other. The paper is divided into sections on physical properties of the interface; adsorption and ion exchange; surface spectroscopy; dissolution, precipitation, and solid solution formation; and transformation reactions at the mineral-water interface. Each section touches on the importance of that topic in geochemical processes and interdependent relationships among the topics covered.

Physical Properties of the Mineral-Water Interface

The chemical reactivity of the mineral-water interface is influenced by some important properties which distinguish the interfacial environment from that of bulk water, e.g. 1) water is more structured in the interfacial region, 2) ions and water molecules are less mobile, 3) the dielectric constant of water is decreased, and 4) electrical charge and potential may develop at the surface, leading to the formation of an electrical double layer. Although experimental studies of mineral-water interface have been extensive, the effects of the perturbed layer of water and the electrical double layer on chemical reactions at the interface are still unresolved issues (1-11). Mulla (Chapter 2) compares simulations of interfacial water structure from various statistical-mechanical models, including Monte Carlo and Molecular Dynamics models. The results predict the existence of molecular layering with ordered dipole orientations at the interface, density oscillations which extend many Angstroms away from the surface, and fewer hydrogen bonds between water molecules in the interfacial region. The existence of density oscillations at the interface has recently been confirmed experimentally (12). Reduced dipole relaxation times are also predicted, which suggests that interfacial water experiences hindered rotation. Unfortunately the dielectric properties cannot be effectively modeled, but the results do suggest that interfacial water is not a uniform dielectric continuum. The development of improved models in the future appears promising, and these models should increase our understanding of properties which are not easily quantified at present, e.g. hydration forces, hydrophobic effects, and double layer forces.

Giese and Constanzo (Chapter 3) present the results of an infrared study of the bonding of intercalated water in synthetic hydrated kaolinites. Two types of water were identified: 1) water molecules which were bonded to the ditrigonal holes of the silicate layer, and 2) associated water molecules which were hydrogen bonded to those of the first group. The mobility of water molecules bound to the ditrigonal holes was greatly reduced in comparison to the associated water. Although hydrated kaolinites are not found in nature, Giese and Constanzo suggest that the surface-water interactions observed in this study are representative of interactions with the surfaces of other silicate minerals.

The properties of the electrical double layer (EDL) have been the subject of considerable research (1,3,5,8,10). Unlike reversible electrodes, where surface potential is controlled and charge develops in response to changes in electrode potential, mineral surfaces develop potential in response to the formation of surface charge (8). On the surface of hydrous oxides, for example, hydroxyl groups

(Bronsted acid sites) or metal atoms with unsatisfied coordination (Lewis acid sites) react with water to form surface charge (13). Isomorphic substitution in the interlayer region of layered silicates results in a negative surface charge. In each case chemical "exchange" of ions between phases results in the formation of surface charge and the development of an electrical potential.

It is important to establish the origin and magnitude of the acidity (and hence, the charge) of mineral surfaces, because the reactivity of the surface is directly related to its acidity. Several microscopic-mechanistic models have been proposed to describe the acidity of hydroxyl groups on oxide surfaces; most describe the surface in terms of amphoteric weak acid groups (14-17), but recently a monoprotic weak acid model for the surface was proposed (18). The models differ primarily in their description of the EDL and the assumptions used to describe interfacial structure. "Intrinsic" acidity constants that are derived from these models can have substantially different values because of the different assumptions employed in each model for the structure of the EDL (5). Westall (Chapter 4) reviews several different amphoteric models which describe the acidity of oxide surfaces and compares the applicability of these models with the monoprotic weak acid model. The assumptions employed by each of the models to estimate values of thermodynamic constants are critically examined.

The difficulty in characterizing the interface arises from the fact that the electrostatic interactions are closely coupled to the chemical interactions. An independent measurement of electrostatic energy would be useful for probing the separation of coulombic and chemical components in the EDL models. Bousse and Meindl (Chapter 5) describe a technique for measuring the electrical potential at oxide surfaces using ion-sensitive field effect transitors (ISFETs). In this method one may regulate total electrical potential of the interface, and this allows estimates of intrinsic acidity constants that are independent of proton adsorption data. The measurement of the total electrical potential is preferable to that of zeta potential, since the exact location of the latter measurement is indeterminate. Furthermore, it has been shown that the dependence of proton adsorption as a function of pH may depend to a great degree on the extent of complex formation with adsorbed counterions (19-21), whereas the total potential as a function of pH is relatively insensitive to complexation (21).

Chan (Chapter 6) presents a simple graphical method for estimating the free energy of EDL formation at the oxide-water interface with an amphoteric model for the acidity of surface groups. Subject to the assumptions of the EDL model, the graphical method allows a comparison of the magnitudes of the chemical and coulombic components of surface reactions. The analysis also illustrates the relationship between model parameter values and the deviation of surface potential from the Nernst equation.

The relative importance of the EDL for reactions other than adsorption is not well understood. Surface complexation models have recently been applied to processes in which adsorption represents the first step in a sequence of reactions. For example, Stumm et al. (22) have applied a model with an EDL component in their studies of the role of adsorption in dissolution and precipitation reactions. The effect of surface charge and potential on precipitation and the

formation of metastable solid phases is discussed by Zawacki et al.
(Chapter 32). Waite (Chapter 20) reviews the important role of the
interfacial environment in catalyzing light-induced redox reactions.
Voudrias and Reinhard (Chapter 22) discuss the effects of surface
acidity on transformations of organic compounds. The rates of
chemical reactions are also influenced by the EDL. As with
equilibrium models, the relative contribution of the EDL in
determining the overall reaction rate is dependent on the
interfacial model chosen (see Chapters 7 and 12).

Adsorption

Sorption processes have received much study because of their
fundamental importance in geochemistry, analytical chemistry, and in
industrial applications. The three principal sorption processes are
adsorption, absorption, and surface precipitation; the differences
among these processes are discussed by Sposito (Chapter 11). If the
specific process leading to the loss of a solute from aqueous
solution is not known, then the general term, sorption, may be used.
The abundance of literature on experimental studies of ion adsorption
has been reviewed by Kinniburgh and Jackson (23) and Hingston (24).
Recent laboratory studies of sorption processes have been directed
more toward a better understanding of the mechanisms of reactions, a
characterization of the bonding of adsorbed species, and improvements
in adsorption models. It is increasingly clear, however, that the
macroscopic approaches that have been commonly used to study sorption
processes, e.g. adsorption isotherms, solubility calculations, and
kinetic methods, cannot unequivocably distinguish between processes
such as adsorption and surface precipitation (see Chapter 11). It is
likely that future developments in this field will come from studies
utilizing molecular techniques such as in-situ surface spectroscopy.
     Mechanisms of Sorption Processes. Kinetic studies are valuable
for hypothesizing mechanisms of reactions in homogeneous solution,
but the interpretation of kinetic data for sorption processes is more
difficult. Recently it has been shown that the mechanisms of very
fast adsorption reactions may be interpreted from the results of
chemical relaxation studies (25-27). Yasunaga and Ikeda (Chapter 12)
summarize recent studies that have utilized relaxation techniques to
examine the adsorption of cations and anions on hydrous oxide and
aluminosilicate surfaces. Hayes and Leckie (Chapter 7) present new
interpretations for the mechanism of lead ion adsorption by goethite.
In both papers it is concluded that the kinetic and equilibrium
adsorption data are consistent with the rate relationships derived
from an interfacial model in which metal ions are located nearer to
the surface than adsorbed counterions.
     The surfaces of minerals are generally not homogeneous; kinks,
steps, edges, dislocations, or point defects may provide reactive
zones. Microcrystalline preparations of solids, which are frequently
used in laboratory sorption experiments, may have several cleavage
planes with different site energies. The importance of the high
energy sites that result from these imperfections is well recognized
for the processes of dissolution and crystal growth (28). Several
equilibrium studies of ion adsorption have suggested that hydrous
oxide surfaces are composed of heterogeneous sites (29-32). Some
preliminary results from a kinetic study that suggest the existence

of heterogeneous sites on the surface of goethite are given in Chapter 7. Despite the growing evidence of surface site heterogeneity, most surface complexation models are based on the concept of a homogeneous surface with averaged EDL properties (5,8,18,33-38). In addition to heterogeneous sites, mineral surfaces may contain either poorly crystallized or well hydrated material. Thus, multiple sorption mechanisms may operate at the beginning of many laboratory studies of sorption kinetics. Sposito (Chapter 11) argues that this multiplicity prevents a simple interpretation of kinetic or equilibrium data in terms of a simple first order rate law or an adsorption mechanism.

Relaxation studies have shown that the attachment of an ion to a surface is very fast, but the establishment of equilibrium in well-dispersed suspensions of colloidal particles is much slower. Adsorption of cations by hydrous oxides may approach equilibrium within a matter of minutes in some systems (39-40). However, cation and anion sorption processes often exhibit a rapid initial stage of adsorption that is followed by a much slower rate of uptake (24,41-43). Several studies of short-term isotopic exchange of phosphate ions between aqueous solutions and oxide surfaces have demonstrated that the kinetics of phosphate desorption are very slow (43-45). Numerous hypotheses have been suggested for this slow attainment of equilibrium including 1) the formation of binuclear complexes on the surface (44); 2) dynamic particle-particle interactions in which an adsorbing ion enhances contact adhesion between particles (43,45-46); 3) diffusion of ions into adsorbents (47); and 4) surface precipitation (48-50).

Bleam and McBride (51-52) recently presented evidence that the arrangement of groups of sites on mineral surfaces may influence adsorption. These authors argued that, under certain conditions, the formation of a monolayer of adsorbing ions may be less favorable than the formation of a multilayer cluster of polymerized or precipitated material. Several studies have indicated that adsorption may be described by complexation reactions at discrete surface sites at low surface coverage, but that polymerization and hydroxide precipitation may occur at high surface coverage (53-55). Farley et al. (56) recently proposed a model for sorption of cations on hydrous oxides that allows for a continuum between adsorption and surface precipitation as the sorption density increases. Surface coprecipitates (solid solutions) may form when an adsorbing cation is capable of occupying structural sites in the adsorbent lattice. Experimental evidence of this type of process has been given by McBride (57) for alumina and by Davis et al. (49) for calcite.

Spectroscopic techniques may provide the least ambiguous methods for verification of actual sorption mechanisms. Zeltner et al. (Chapter 8) have applied FTIR (Fourier Transform Infrared) spectroscopy and microcalorimetric titrations in a study of the adsorption of salicylic acid by goethite; these techniques provide new information on the structure of organic acid complexes formed at the goethite-water interface. Ambe et al. (Chapter 19) present the results of an emission Mossbauer spectroscopic study of sorbed Co(II) and Sb(V). Although Mossbauer spectroscopy can only be used for a few chemical elements, the technique provides detailed information about the molecular bonding of sorbed species and may be used to differentiate between adsorption and surface precipitation.

Empirical models. Natural systems contain a wide variety of mineral surfaces that may be involved in sorption processes. Application of surface complexation models requires a detailed characterization of each adsorbent present, and the amount of information required generally exceeds our knowledge of the properties of natural materials (58). While the models have been moderately successful in describing the results from laboratory studies with pure mineral phases (e.g. 59), they have not yet been applied in field studies. Some authors (58,60) have made calculations for hypothetical sediments for predictive purposes. In these calculations the conventional approach has been to assume that the overall adsorption of an ion for a mixture of minerals can be described as the sum of the adsorptive contribution of each mineral. However, Honeyman (61) has demonstrated that this concept of "adsorptive additivity" does not hold, even in simple experiments with binary mixtures of oxide phases. In the absence of a unified theoretical model, geochemists have often formulated empirical approaches that utilize macroscopic parameters to describe the adsorption process (60,62-63). Honeyman and Leckie (Chapter 9) show that these macroscopic parameters are the net result of numerous microscopic subreactions occurring in the system. In particular, Honeyman and Leckie review the use of macroscopic proton coefficients (e.g. Kurbatov coefficients) in practical sorption models. The authors show that macroscopic proton coefficients are rarely observed to have integral values, despite the fact that proton coefficients of the microscopic adsorption reactions of interest may have integral values. A mathematical derivation supporting this conclusion is presented in Chapter 7.

Sorption of organic compounds. Adsorption may play an important role in the transformations of organic compounds in the environment. For example, adsorption of organic pollutants at the mineral-water interface may catalyze the conversion of these compounds to less harmful products. The chemical factors controlling the sorption of hydrophobic compounds in sediments with moderate to high organic content have received considerable study (64), but sorption processes for sediments containing low organic content are poorly understood. Curtis et al. (Chapter 10) discuss current problems in understanding the sorption behavior of hydrophobic organic compounds, including reaction kinetics, hysteresis effects, and the influence of dissolved macromolecular organic material.

Ion Exchange

The distribution of major elements between soils and soil solutions is known to be governed primarily by ion exchange processes (65). These processes are important because they greatly influence the uptake of nutrients by plants and other living organisms. Even though an ion exchange reaction could be classified as a type of adsorption reaction, it is usually treated as a separate sorption process as a matter of convenience and tradition. Describing ion exchange reactions as those taking place at "constant charge" surfaces, e.g. in the interlayer regions of clay minerals, distinguishes them from adsorption reactions that occur at "constant potential" surfaces, such as those of hydrous oxides (8).

While the theory and experimental measurements of the equilibria of ion exchange are well established (66,67), some models employ unverified assumptions concerning the structure of exchanged ions. For example, the extent to which cations are dehydrated upon entering an ion exchanger is not known unequivocably (68). Sposito (10) discusses the experimental evidence for formation of both inner- and outer-sphere complexes for various exchanging ions. Ca-montmorillonite suspensions exhibit a d(001) spacing that is consistent with the formation of an outer-sphere surface complex between exchangeable Ca^{2+} ions and a pair of opposing siloxane ditrigonal cavities. Additional evidence for this structure includes quasielastic neutron scattering experiments which suggest the existence of a rigid octahedral solvation shell for Ca^{2+} in Ca-montmorillonite (69). Electron spin resonance spectra for exchangeable Cu^{2+} and Mn^{2+} in montmorillonite also suggest the formation of outer-sphere complexes (70, and see McBride, Chapter 17). Goodman (Chapter 16) reviews the current state of knowledge concerning the sorption of metal ions by aluminosilicate minerals and the various spectroscopic techniques which have been employed to characterize the bonding environment.

When isomorphic substitution of Al^{3+} for Si^{4+} occurs in the tetrahedral sheet of a phyllosilicate, the excess negative charge will distribute itself primarily over the three surface oxygens of one tetrahedron. This allows for the formation of strong complexes with cations (10). In particular, the formation of inner-sphere complexes with K^+ is likely, because the ionic diameter of K^+ is almost equal to the size of the ditrigonal cavity in the basal planes of vermiculite and illitic micas. The influence of such structures and bonding forces on ion exchange processes involving K^+ and Ca^{2+} are discussed in detail by Goulding (Chapter 15) and more generally by Maes and Cremers (Chapter 13). In both papers, evidence for the existence of highly selective and specific surface sites in layered minerals is reviewed. Goulding presents a characterization of several selected layered silicate minerals in terms of specific site types which can be identified by the enthalpies of ion exchange. The author also discusses reasons for the slow exchange or "fixation" of K^+ in soils. In Chapter 13, Maes and Cremers present a comprehensive review of the influence ion exchanger charge density and the relative polarizability of exchanging ions on ion exchange processes. The factors that lead to highly selective exchange behavior in montmorillonites and zeolites are emphasized. The paper by Yasunaga and Ikeda (Chapter 12) examines the mechanisms of intercalation and deintercalation in layered, channeled, and cage-structured minerals.

Eberl et al. (Chapter 14) present evidence for the dynamic role of wetting and drying cycles on the weathering of smectites and K-feldspars. The exchange of K^+ with other cations leads to the formation of illite-like layers in smectites, but this reaction is completely reversible when exchanged again with cations of high hydration energy such as Ca^{2+} (71). However, Eberl et al. show that K-smectite may fix K^+ irreversibly when subjected to wetting and drying cycles. The moisture content of soils can dramatically affect the reactivity of soil minerals. Under very low moisture conditions both the surface and interlayer acidity can be greatly increased, leading to an increase in the rates of acid-catalyzed reactions. Thus, chemical transformations may occur at quite different rates in

unsaturated vs. saturated soil zones. In addition to the paper by
Eberl et al., this topic is addressed in the papers by Voudrias and
Reinhard (Chapter 22) and Velbel (Chapter 30).

Surface Spectroscopy

Surface spectroscopy offers the best opportunity to elucidate the
structures of chemical species at the mineral-water interface (see
Sposito, Chapter 11). The application of spectroscopic methods to
probe the molecular environment of the interface is still a
relatively new field. Chapters 16-19 present reviews and some recent
advances in investigations of molecular structure at the mineral-
water interface. A recent review of spectroscopic methods applied to
soil and clay mineral systems is given in Stucki and Banwart (72).
    Spectroscopic techniques can be classified according to the type
of interfacial environment being investigated. While the terms
"surface" and "interface" are often interchanged, each has a distinct
meaning: Surface refers to the face of a solid which is exposed to a
gas or liquid phase; an interface is a narrow region of finite width
between the surface and a liquid or gas phase. Surface spectroscopic
techniques often require vacuum or ultra high vacuum conditions and
are sometimes referred to as ex-situ techniques. Spectroscopic
techniques applied to mineral-aqueous systems are referred to as in-
situ techniques in that direct investigation of aqueous suspensions
is possible.
    Spectroscopic techniques which analyze the composition and
structure of mineral surfaces include Auger Electron Spectroscopy
(AES), X-ray Photoelectron Spectroscopy (XPS, or the older name,
ESCA), and Secondary Ion Mass Spectroscopy (SIMS). Perry (Chapter
18) discusses the application of XPS, AES, and SIMS to studies of
natural materials. Each of these techniques can yield detailed
information about the structure and bonding of minerals and of
chemical species present on the surfaces of minerals, but Perry
illustrates the fact that a much greater knowledge can be gained by
combining data from two or more methods. The author also discusses
the important spectral parameters that yield structural information,
the application of depth profiling methods, and problems created by
the high vacuum necessary for analysis. The use of XPS and SIMS in an
investigation of the oxidation state of cobalt sorbed by birnessite
is reported in Dillard and Schenck (Chapter 24).
    A variety of in-situ spectroscopic techniques have been used to
investigate the mineral-water interface, including Raman (73),
Fourier Transform Infrared (FTIR)(74-75), Nuclear Magnetic Resonance
(NMR)(76), Electron Paramagnetic Resonance (EPR)(77-80), Electron
Nuclear Double Resonance (ENDOR)(81), Mossbauer (82), and Extended X-
ray Absorption Fine Structure (EXAFS)(83) spectroscopies. In-situ
spectroscopic investigations of the mineral-water interface can be
found in the papers by Zeltner et al. (FTIR, Chapter 8), McBride
(EPR, Chapter 17) and Ambe et al. (Mossbauer, Chapter 19). McBride
(Chapter 17) studied the orientation and mobility of Cu^{2+} ions sorbed
at exchange sites of layer silicate minerals. The EPR spectral data
revealed that the rotational motion of Cu^{2+} in these minerals was
highly dependent on the size of the interlayer region, and hence, the
degree of interlayer expansion. Ambe et al. (Chapter 19) discuss the
structures of sorbed Co(II) and Sb(V) ions on the surface of

hematite. An in-situ emission Mossbauer spectroscopic study of these
surfaces revealed two chemical forms of adsorbed Co(II): one
attributable to coordinative bonds to surface sites and the other to
weakly bound (possibly hydrogen bonded) ions. The relative
proportions of the two forms varied as a function of pH. Two forms of
adsorbed Sb(V) were also found in this study. Goodman (Chapter 16)
discusses the use of EPR and other spectroscopic techniques in
studies of the adsorption of metal ions and complexes by
aluminosilicate minerals.

Many other methods can be used to obtain information about
mineral surfaces including Low Energy Electron Diffraction (LEED),
Scanning Electron Microscopy (SEM) and Transmission Electron
Microscopy (TEM) (72). The diffraction methods are well established
for studying structure and the microscopic techniques can reveal the
morphology of surfaces. Giese and Constanzo (Chapter 3) studied the
bonding of intercalated water molecules in synthetic hydrated
kaolinite by infrared (IR) absorption. Voudrias and Reinhard (Chapter
22) review the application of a variety of spectroscopic techniques
for identifying sorbed reactants, products, and intermediates in
organic reactions at the mineral-water interface.

Dissolution, Precipitation, and Solid Solution Formation

Dissolution. Chemical weathering is certainly among the most
important geochemical processes that occur on the earth's surface. In
the last decade, research in this field has focused on the kinetics
and mechanisms of mineral dissolution reactions, since a departure
from equilibrium in natural systems is apparent (84-85). Hypotheses
for dissolution mechanisms and rate-limiting steps during the
weathering process can be grouped into two schools of thought (86).
One school proposes that the dissolution rate is controlled by the
formation of a residual layer at the surface of the reacting mineral,
through which the reactants and products of weathering must diffuse
(87,88). The second group proposes that the dissolution rate is
controlled by the rate of a surface reaction (89,90). The proponents
of the diffusion-control mechanism base their conclusions on the
temporal evolution of aqueous solution composition, which suggests
that the kinetics of mineral dissolution usually obeys a parabolic
rate law. Evidence for the latter hypothesis includes the results of
surface spectroscopic studies which have failed to detect any leached
layer (89-91). Dibble and Tiller (28) noted that the consistency of
the kinetic data with parabolic rate law may indicate that the rate-
determining step involves diffusion of a reaction product or impurity
ion to or from the interface. While mass transfer through the liquid
phase at the interface is a relatively fast step, diffusion could
become rate-determining if adsorption retards molecular detachments
at kinks or layer edges (92). Velbel (Chapter 30) reviews this area
of research and discusses current applications of the conclusions to
natural systems.

A common phenomenon in the dissolution of silicate minerals is
the formation of etch pits at the surface (90-91,93-94). When this
occurs, the overall rate of mineral dissolution is non-uniform, and
dissolution occurs preferentially at dislocations or defects that
intercept the crystal surface. Preferential dissolution of the
mineral could explain why surface spectroscopic studies have failed

to detect a leached layer at the surface (93). Brantley et al. (Chapter 31) examine the formation of etch pits on quartz in laboratory experiments at 300°C and on quartz particles from a natural soil profile. Their results suggest that the theory of etch pit formation at dislocations may be useful in describing mineral weathering behavior at low temperatures as well as during hydrothermal alteration.

Numerous geochemical studies have attempted to interpret the composition of waters in terms of chemical reactions between parent minerals and weathering products in near-surface weathering environments. The studies suggest, or in some cases assume, that the water is in equilibrium with either observed or inferred weathering products or hypothetical metastable phases (95). However, chemical equilibrium is not necessarily expected in an open system through which water is fluxing rapidly. A more realistic approach is to relate the water chemistry to the kinetics of dissolution and precipitation of primary and secondary phases (87,96). Weathering rates in nature are usually estimated from geochemical mass balance equations (97). Velbel (Chapter 30) compares the few estimates of mineral weathering rates measured in nature which can be normalized on the basis of surface area with those measured in the laboratory (28,98). The results indicate that the rates of dissolution in nature are much slower than predicted from laboratory experiments. Velbel outlines possible reasons for this discrepancy and presents areas in which further research are needed.

Precipitation. An important element of any geochemical analysis of natural waters is an evaluation of which minerals are present and the extent to which the system can be represented by equilibrium models. Typical questions that need to be answered are: 1) Is the water supersaturated, undersaturated, or at equilibrium with a given mineral? and 2) If more than one solid phase can form for a given element, which phase is more stable in that particular environment?

Precipitation can occur if a water is supersaturated with respect to a solid phase; however, if the growth of a thermodynamically stable phase is slow, a metastable phase may form. Disordered, amorphous phases such as ferric hydroxide, aluminum hydroxide, and allophane are thermodynamically unstable with respect to crystalline phases; nonetheless, these disordered phases are frequently found in nature. The rates of crystallization of these phases are strongly controlled by the presence of adsorbed ions on the surfaces of precipitates (99). Zawacki et al. (Chapter 32) present evidence that adsorption of alkaline earth ions greatly influences the formation and growth of calcium phosphates. While hydroxyapatite was the thermodynamically stable phase under the conditions studied by these authors, it is shown that several different metastable phases may form, depending upon the degree of supersaturation and the initiating surface phase.

Precipitation must begin with the formation of nuclei; nucleation can be homogeneous (formed in the aqueous solution by the spontaneous association of ions) or heterogeneous (originating on the surface of an impurity or via seed particles which act as crystallization catalysts). In nature, it is thought that heterogeneous nucleation is the predominant process which begins precipitation (100). The factors determining growth kinetics may be divided into two main groups: 1) transport processes -- the transport of reactants up to a crystal surface and the transport of reaction products away from the surface, and 2) surface processes, which may include adsorption of molecules and ions on a crystal surface, migration of ions along the surface, a change of the degree of hydration of ions, formation of two dimensional surface nuclei, and the fitting of ions into the crystal lattice (101). Neilsen (Chapter 29) reviews some of the important aspects of crystal growth processes and the factors which influence the rate-determining mechanism for crystal growth.

Adsorption may influence precipitation by means other than the processes mentioned above. Davies (Chapter 23) discusses the role of the surface as a catalyst for oxidation of adsorbed Mn^{2+}. Redox reactions may contribute substantially to the formation of manganese oxide coatings on mineral surfaces in soils and sediments.

Solid Solutions. The aqueous concentrations of trace elements in natural waters are frequently much lower than would be expected on the basis of equilibrium solubility calculations or of supply to the water from various sources. It is often assumed that adsorption of the element on mineral surfaces is the cause for the depleted aqueous concentration of the trace element (97). However, Sposito (Chapter 11) shows that the methods commonly used to distinguish between solubility or adsorption controls are conceptually flawed. One of the important problems illustrated in Chapter 11 is the evaluation of the state of saturation of natural waters with respect to solid phases. Generally, the conclusion that a trace element is undersaturated is based on a comparison of ion activity products with known pure solid phases that contain the trace element. If a solid phase is pure, then its activity is equal to one by thermodynamic convention. However, when a trace cation is coprecipitated with another cation, the activity of the solid phase end member containing the trace cation in the coprecipitate will be less than one. If the aqueous phase is at equilibrium with the coprecipitate, then the ion activity product will be less than the solubility constant of the pure solid phase containing the trace element. This condition could then lead to the conclusion that a natural water was undersaturated with respect to the pure solid phase and that the aqueous concentration of the trace cation was controlled by adsorption on mineral surfaces. While this might be true, Sposito points out that the ion activity product comparison with the solubility product does not provide any conclusive evidence as to whether an adsorption or coprecipitation process controls the aqueous concentration.

There is considerable evidence that coprecipitation and the formation of solid solutions is significant in soils and sediments (10). While ideal solution models have been widely proposed for various mineral solid solutions, experimental investigations and studies of natural mineral assemblages have shown that miscibility gaps are common in almost every major mineral group (102). The existence of such gaps requires nonideal solution models to describe the distribution of components in the solid and aqueous phases. Driessens (Chapter 25) reviews the important literature concerning laboratory investigations of solid solutions and presents the theory of ideal solid solutions and nonideal solid solutions, including the more general models of regular solid solutions with and without ordering. Many of the calculated activity coefficients for end-member solid phases in solid solutions are based on an assumption that a system has reached equilibrium when the solid phase composition is measured. Plummer (Chapter 26) examines the criteria for deciding when a system has reached equilibrium; his analysis shows that previous calculations for the KCl-KBr system at $25^{o}C$ may be in error because the system analyzed was only near equilibrium.

Non-lattice sites may play an important role in the incorporation of large foreign ions in crystal structures during coprecipitation; Pingitore (Chapter 27) discusses the importance of these sites in the formation of coprecipitates of calcium carbonate containing Sr^{2+} or Ba^{2+}. White and Yee (Chapter 28) discuss the diffusion of alkali ions into defect structures in the surfaces of glasses and crystalline feldspars.

Transformation Reactions at the Mineral/Water Interface

Certain properties of the mineral-water interface or of reactive sites on mineral surfaces may lower the activation energy of various transformation reactions, e.g. electron transfer reactions or hydrolysis reactions of organic compounds. The presence of Bronsted or Lewis acid sites on mineral surfaces is a primary factor in catalyzing such reactions at the surface. Other factors are the structure and charge of the mineral surface (including any interlayer spacing) and the size and charge of reacting solutes. It is generally believed that the mechanisms of these transformation reactions are similar to those that occur in homogeneous aqueous solutions, but the conditions at the mineral-water interface may accelerate the rates of certain reactions. Although many important reactions of organic compounds are catalyzed by mineral surfaces under dessicated conditions at elevated temperatures (103), little information is available on catalysis by surfaces in aqueous environments.

Electron Transfer Reactions. Important redox reactions involving aqueous solutes and mineral surfaces include oxidative or reductive dissolution, oxidation or reduction of solutes by reaction with surface sites, and polymerization reactions of organic compounds. The theory of electron transfer reactions is well established in homogeneous solution; however, the mechanisms of electron transfer reactions which occur at the mineral-water interface are more difficult to establish because of the difficulty in identifying the reacting species. Two types of electron transfer mechanisms have been found from kinetic studies in homogeneous solution: 1) inner-sphere and 2) outer-sphere. Stone (Chapter 21) discusses the analogy between electron transfer reactions in homogeneous and heterogeneous systems. Waite (Chapter 20) reviews the literature on the ability of light to initiate or enhance the rates of redox reactions which occur on mineral surfaces. A study of accelerated oxidation of Mn^{2+} in the presence of hydrous iron oxide is presented by Davies (Chapter 23). Dillard and Schenck (Chapter 24) report on redox reactions of Co(II) and Co(III)-complexes on the surface of birnessite.

The surfaces of clay minerals can catalyze the polymerization of organic compounds through a free radical-cationic initiation process. This type of reaction is believed to be initiated by the abstraction of an electron by Lewis acid sites on mineral surfaces; however, Bronsted acidity has also been shown to be important in certain cases (see Chapter 22).

Additional Transformation Reactions. Other reactions that can be catalyzed by mineral surfaces are substitution, elimination, and addition reactions of organic molecules. Substitution and elimination are two general types of reactions that occur at saturated carbon atoms of organic molecules. Both types are initiated by nucleophilic attack; however, in elimination reactions it is the basicity of the nucleophile that determine its reactivity rather than its nucleophilicity. Since mineral surfaces are expected to have both nucleophilic and basic properties, these types of reactions should also occur at mineral-water interfaces (see Chapter 22). It remains to be shown whether or not these reactions are catalyzed under environmental conditions.

Hydrolysis reactions occur by nucleophilic attack at a carbon single bond, involving either the water molecule directly or the hydronium or hydroxyl ion. The most favorable conditions for hydrolysis, e.g. acidic or alkaline solutions, depend on the nature of the bond which is to be cleaved. Mineral surfaces that have Bronsted acidity have been shown to catalyze hydrolysis reactions. Examples of hydrolysis reactions which may be catalyzed by the surfaces of minerals in soils include peptide bond formation by amino acids which are adsorbed on clay mineral surfaces and the degradation of pesticides (see Chapter 22).

Concluding Remarks

Our knowledge of the physical and chemical nature of the mineral-water interface is still advancing. The significance of the interface in processes such as sorption, ion exchange, precipitation, and dissolution has been recognized for some time, but new studies are demonstrating that these processes are interrelated in complex and interesting ways. A full appreciation of the fundamental importance of interfacial reactions in geochemical processes is still emerging, and it is increasingly clear that the interface may play a critical role in accelerating the rates of redox reactions, polymerization, hydrolysis, and other transformations. This volume presents a compilation of state-of-the-art theoretical and experimental approaches which are being applied in studies of the mineral-water interface. These new concepts must be integrated into geochemical models if a comprehensive chemical description of natural systems is to be achieved.

Acknowledgments

The authors would like to thank A. Maest, C. Chisholm, and C. Fuller for their criticial reviews of the manuscript.

Literature Cited

1. Weise, G.R.; James, R.O.; Yates, D.E.; Healy, T.W. MPT Int. Rev. Sci. 1975, 6, 53-103.
2. Klier, K.; Zettlemoyer, A.C. J. Colloid Interface Sci. 1977; 58, 216-229.
3. Lyklema, J. J. Colloid Interface Sci. 1977; 58, 242.

4. Arnold, P.W. In "The Surface Chemistry of Soil Constituents"; Greenland, D.J.; Hayes, M.H.B., Eds.; John Wiley and Sons, New York, 1978.
5. Westall, J.; Hohl, H. Adv. Colloid Interf. Sci. 1980, 12, 265-294.
6. Derjaguin, B.V.; Churaev, N.V. In "Progress in Surface and Membrane Science"; Cadenhead, D.A.; Danielli, J.F., Eds.; Academic Press: New York, 1981; Vol. 14, pp. 69-130.
7. Sposito, G.; Prost, R. Chem. Rev. 1982, 82, 553-573.
8. James, R.O.; Parks, G.A. Surface and Colloid Sci. 1982, 12, 119-217.
9. Vold, R.D.; Vold, M.J. "Colloid and Interface Chemistry"; Addison-Wesley: Reading, Mass., 1983; Chaps. 7,9.
10. Sposito, G. "The Surface Chemistry of Soils"; Oxford: New York, 1984.
11. Ruckenstein, E.; Schiby, D. Langmuir 1985, 1, 612-615.
12. Pashley, R.M.; Israelachvili, J.N. J. Colloid Interface Sci. 1984, 101, 511-523.
13. Schindler, P.W. In "Adsorption of Inorganics at the Solid Liquid Interface"; Anderson, M.A.; Rubin, A.J., Eds.; Ann Arbor Science, Ann Arbor, 1981, Chap. 1.
14. Huang, C.P., Stumm, W.J. Colloid Interf. Sci. 1973, 43, 409.
15. Stumm, W.; Hohl, H.; Dalang, F. Croatica Chem. Acta 1976, 48, 491.
16. Bowden, J.W.; Posner, A.M.; Quirk, J.P.Aust. J. Soil Res. 1977, 15, 121.
17. Davis, J.A.; James, R.O.; Leckie, J.O.J. Colloid Interf. Sci. 1978, 63, 480-499.
18. van Riemsdijk, W.H.; Bolt, G.H.; Koopal, L.K.; Blaakmeer, J. J. Colloid Interf. Sci. 1986, 109, 219-228.
19. Sprycha, R. J. Colloid Interf. Sci. 1984, 102, 173-185.
20. Smit, W.; Holten, C.L.M. J. Colloid Interf. Sci. 1980, 78, 1.
21. Bousse, L.; de Rooij, N.F.; Bergveld, P. Surface Sci. 1983, 135, 479.
22. Stumm, W.; Furrer, G.; Kunz, B. Croatica Chem. Acta 1983, 56, 593-611.
23 Kinniburgh, D.G.; Jackson, M.L. In "Adsorption of Inorganics at Solid-Liquid Interfaces"; Anderson, M.A.; Rubin, A.J., Eds.; Ann Arbor Science: Ann Arbor, MI, 1981; Chap. 3.
24. Hingston, F.J. In "Adsorption of Inorganics at Solid-Liquid Interfaces"; Anderson, M.A.; Rubin, A.J., Eds.; Ann Arbor Science: Ann Arbor, MI, 1981; Chap. 2.
25. Mikami, N.; Sasaki, M.; Hachiya, K.; Astumian, R.D.; Ikeda, T.; Yasunaga, T. J. Phys. Chem. 1983, 87, 1454-1458.
26. Hachiya, K.; Sasaki, M.; Saruta, Y.; Mikama, N.; Yasunaga, T. J. Phys. Chem. 1984, 88, 23-27.
27. Hachiya, K.; Sasaki, M.; Ikeda, T.; Mikami, N.; Yasunaga, T. J. Phys. Chem. 1984, 88, 27-31.
28. Dibble, W.E. Jr.; Tiller, W.A. Geochem. Cosmochim. Acta 1981, 45, 79-92.
29. Benjamin, M.M.; Leckie, J.O. J. Colloid Interf. Sci. 1981, 79, 209-221.
30. Kinniburgh, D.G.; Barker, J.A.; Whitfield, M. J. Colloid Interf. Sci. 1983, 95, 370.

31. Benjamin, M.M.; Leckie, J.O. J. Colloid Interface Sci. 1981, 83, 410-419.
32. Hingston, F.J.; Posner, A.M.; Quirk, J.P. Discuss. Faraday Soc. 1972, 52, 334-342.
33. Chan, D.; Perram, J.W.; White, L.R. J. Chem. Soc. Faraday Trans. I 1975, 71, 1046-1057.
34. Sposito, G. J. Colloid Interf. Sci. 1983, 91, 329.
35. Hayes, K.F.; Leckie, J.O. J. Colloid Interf. Sci. 1986, in press.
36. Hohl, H.; Stumm, W. J. Colloid Interface Sci. 1976, 55, 281-288.
37. Davis, J.A.; Leckie, J.O. J. Colloid Interface Sci. 1978, 67, 90-107.
38. Davis, J.A.; Leckie, J.O. J. Colloid Interface Sci. 1978, 74, 32-43.
39. Ahrland, S.; Grenthe, I.; Noren, B. Acta Chem. Scand. 1960, 14, 1059-1076.
40. Zazoski, R.J.; Burau, R.G. Soil Sci. Soc. Am. J. 1978, 42, 372-374.
41. Theis, T.L. "Seminar on Adsorption"; EPA Report 600/X-85/122; U.S. Environmental Protection Agency: Washington, D.C., 1985.
42. Kurbatov, M.H.; Wood, G.B. J. Phys. Chem. 1952, 56, 698.
43. Anderson, M.A.; Tejedor-Tejedor, M.I.; Stanforth, R.R. Environ. Sci. Tech. 1985, 19, 632-637.
44. Atkinson, R.J.; Posner, A.M.; Quirk, J.P. J. Inorg. Nucl. Chem. 1972, 34, 2201-2211.
45. Hansmann, D.D.; Anderson, M.A. Environ. Sci. Tech. 1985, 19, 544-551.
46. Pashley, R.M. J. Colloid Interface Sci. 1984, 102, 23-35.
47. Barrow, N.J. J. Soil Sci. 1983, 34, 751-758.
48. Leckie, J.O.; Stumm, W. In "Advances in Water Quality Improvement--Physical and Chemical Processes"; Golyna E.; Eckenfelder, W., Eds.; Univ. of Texas Press: Austin, 1970, 237-249.
49. Davis, J.A.; Fuller, C.C.; Cook, A.D. Geochim. Cosmochim. Acta, in press.
50. Dzombak, D.A.,; Morel, F. M.M., J. Colloid Interface Sci. 1986, in press.
51. Bleam, W.F.; McBride, M.B. J. Colloid Interface Sci. 1985, 103, 124-132.
52. Bleam, W.F.; McBride, M.B. J. Colloid Interface Sci. 1986, 110, 335-346.
53. Harvey, D.T.; Linton, R.W. Colloids and Surfaces 1984, 11, 81-96.
54. McBride, M.B.; Fraser, A.R.; McHardy, W.J. Clays and Clay Min. 1984, 32, 12-18.
55. Anderson, M.A.; Palm-Gennen, M.H.; Renard, P.N.; Defosse, C.; Rouxhet, P.G. J. Colloid Interface Sci. 1984, 102, 328-336.
56. Farley, K.J.; Dzombak, D.A.; Morel, F.M.M. J. Colloid Interface Sci. 1985, 106, 226-242.
57. McBride, M.B. Soil Sci. Soc. Am. J. 1978, 42, 27-31.
58. Luoma, S.N.; Davis, J.A. Marine Chem. 1983, 12, 159-181.
59. Hsi, C.K.D.; Langmuir, D. Geochimica Cosmochimica Acta 1985, 49, 1931-1942.
60. Davis-Colley, R.J.; Nelson, P.O.; Williamson, K.J. Environ. Sci. Tech. 1984, 18, 491-499.

61. Honeyman, B.D. Ph.D. Thesis, Stanford University, Stanford, Calif., 1984.
62. Tessier, A.; Rapin, F.; Carignan, R. Geochim. Cosmochim. Acta 1985, 49, 183-194.
63. Balistrieri, L.S.; Murray, J.W. Geochim. Cosmochim. Acta 1983, 47, 1091-1098.
64. Karickhoff, S.W. J. Hydraulic Eng. 1984, 110, 707-735.
65. Sposito, G.; Mattigod, S.V. Soil Sci. Soc. Am. J. 1977, 41, 323-329.
66. Sposito, G. "Thermodynamics of Soil Solutions"; Oxford Clarendon Press: Oxford, 1981.
67. Bruggenwert, M.G.M.; Kamphorst, A. In "Soil Chemistry, B. Physico-Chemical Models" Bolt, G.H. Ed.; Elsevier Sci. Publ.: Amsterdam, 1979, Chap. 5.
68. Maes, A.; Cremers, A. In "Soil Chemistry B. Physico-Chemical Models" Bolt, G.H. Ed.; Elsevier Sci. Publ.: Amsterdam, 1979, Chap. 6.
69. Ross, D.K.; Hall, P.L. In "Advanced Chemical Methods for Soil and Clay Mineralogy Research"; Stucki, J.W.; Banwart, W.L., Eds.; Reidel: Boston, 1980, p. 93.
70. McBride, M.B.; Pinnavaia, T.J.; Mortland, M.M. J. Phys. Chem. 1973, 77, 196-200.
71. Goulding, K.W.T.; Talibudeen, O. J. Colloid Interface Sci. 1980, 78, 15-24.
72. Stucki, J.W.; Banwart, W.L., Eds. "Advanced Chemical Methods for Soil and Clay Minerals Research"; Reidel: Boston, 1980.
73. Johnston, C.T.; Sposito, G.; Birge, R.R. Clays and Clay Min. 1985, 33, 483-489.
74. Tejedor-Tejedor, M.I.; Anderson, M.A. Langmuir 1986, 2, 203-210.
75. Foley, J.K.; Pons, S. Anal. Chem. 1985, 57, 945A-956A.
76. Young, J.R. PhD. Thesis, California Institute of Technology, Pasadena, Calif., 1981.
77. Motschi, H. Colloids and Surfaces 1984, 9, 337-347.
78. Fransesca, M.O.; Ceresa, E.M.; Visca, M. J. Colloid Interface Sci. 1985, 108, 114-122.
79. Bassetti, V.; Burlamacchi, L.; Martini, G. J. Amer. Chem. Soc. 1979, 101, 5471-5477.
80. Clementz, D.M.; Pinnavaia, T.J.; Mortland, M.M. J. Phys. Chem. 1973, 77, 196-200.
81. Rudin, M.; Motschi, H. J. Colloid Interface Sci. 1984, 98, 385-393.
82. Ambe, F.; Okada, T.; Ambe, S.; Sekizawa, H. J. Phys. Chem. 1984, 88, 3015.
83. Waychunas, G.A.; Brown, G.E. In EXAFS and Near Edge Structure III; Hodgson, K.O.; Hedman, B.; Penner-Hahn, J.E., Eds.; Springer-Verlag, New York, pp. 336-342.
84. Berner, R.A. Am. J. Sci. 1978, 278, 1235-1252.
85. Lerman, A., "Geochemical Processes"; Wiley: New York, 1979.
86. Wollast, R.; Chou, L. In "The Chemistry of Weathering"; Drever, J.I., Ed.; Reidel: Boston, 1985, pp. 75-96.
87. Paces, T. Geochim. Cosmochim. Acta 1973, 37, 2641-2663.
88. Wollast, R. Geochim. Cosmochim. Acta 1967, 31, 635-648.
89. Holdren, G.R.; Berner, R.A. Geochim. Cosmochim. Acta 1979, 43, 1161-1171.

90. Berner, R.A.; Holdren, G.R. Geochim. Cosmochim. Acta 1979, 43, 1173-1186.
91. Schott, J.; Berner, R.A.; Sjoberg, E.L. Geochim. Cosmochim. Acta 1981, 45, 2123-2135.
92. Zutic, V.; Stumm, W. Geochim. Cosmochim. Acta 1984, 48, 1493-1504.
93. Berner, R.A.; Holdren, G.R.; Schott, J. Geochim. Cosmochim. Acta 1985, 49, 1657-1658.
94. Berner, R.A.; Schott, J. Am. J. Sci. 1982, 282, 1214-1231.
95. Velbel, M.A. In "The Chemistry of Weathering"; Drever, J.I., Ed.; Reidel: Boston, 1985, pp. 231-247.
96. Paces, T. In "Interpretation of Environmental Isotope and Hydrochemical Data in Groundwater Hydrology", International Atomic Energy Agency: Vienna, pp. 85-108.
97. Drever, J.I. "The Geochemistry of Natural Waters"; Prentice-Hall: Englewood Cliffs, NJ, 1982.
98. Paces, T. Geochim. Cosmochim. Acta 1983, 47, 1855-1863.
99. Schwertmann, U. In "The Chemistry of Weathering"; Drever, J.I., Ed.; Reidel: Boston, 1985, pp. 119-120.
100. Stumm, W.; Furrer, G.; Wieland, E.; Zinder, B. In "The Chemistry of Weathering"; Drever, J.I., Ed.; Reidel: Boston, 1985, pp. 55-74.
101. Neilsen, A.E. In "Treatise on Analytical Chemistry", 2nd Ed.; Part I; Vol. 3; Kolthoff, I.M.; Elving, P.J., Eds.; John Wiley: New York, 1983, Chap. 27.
102. Nordstrom, D.K.; Munoz, J.L. "Geochemical Thermodynamics"; Benjamin-Cummings Publ.: Menlo Park, Calif., 1985, pp.152-162.
103. Solomon, D.H.; Hawthorne, D.G. "Chemistry of Pigments and Fillers" Wiley-Interscience: New York, 1983.

RECEIVED August 4, 1986

PHYSICAL PROPERTIES
OF THE MINERAL–WATER
INTERFACE

2

Simulating Liquid Water near Mineral Surfaces: Current Methods and Limitations

David J. Mulla

Department of Agronomy and Soils, Washington State University, Pullman, WA 99164-6420

It is important to propose molecular and theoretical models to describe the forces, energy, structure and dynamics of water near mineral surfaces. Our understanding of experimental results concerning hydration forces, the hydrophobic effect, swelling, reaction kinetics and adsorption mechanisms in aqueous colloidal systems is rapidly advancing as a result of recent Monte Carlo (MC) and molecular dynamics (MD) models for water properties near model surfaces. This paper reviews the basic MC and MD simulation techniques, compares and contrasts the merits and limitations of various models for water-water interactions and surface-water interactions, and proposes an interaction potential model which would be useful in simulating water near hydrophilic surfaces. In addition, results from selected MC and MD simulations of water near hydrophobic surfaces are discussed in relation to experimental results, to theories of the double layer, and to structural forces in interfacial systems.

Recent evidence (1) suggests that reactions at the mineral/liquid interface were involved in the beginnings of life on Earth. Not surprisingly, the nature and properties of mineral/water interfaces are of interest to physicists, chemists, physical chemists, applied mathematicians, colloid scientists, geochemists, soil scientists and civil engineers. Of particular interest is an increased understanding of the role of water in colloidal swelling, solute hydration, reaction kinetics, adsorption mechanisms, and ion exchange.

The theoretical study (2,3) of this interface is made inherently difficult by virtue of the complex, many-body nature of the interaction potentials and forces involving surfaces, counterions, and water. Hence, many models of the interfacial region explicitly specify the forces between colloidal particles or between solutes, but few account for the many-body interaction forces of the solvent.

0097–6156/86/0323–0020$06.00/0
© 1986 American Chemical Society

Experimental studies of the thermodynamic, spectroscopic and transport properties of mineral/water interfaces have been extensive, albeit conflicting at times (4-10). Ambiguous terms such as "hydration forces", "hydrophobic interactions", and "structured water" have arisen to describe interfacial properties which have been difficult to quantify and explain. A detailed statistical-mechanical description of the forces, energies and properties of water at mineral surfaces is clearly desirable.

Molecular predictions of the properties of interfacial systems are now becoming possible as a result of rapid advances in liquid state chemical physics and computer technology. The objectives of this paper are 1) to review the general approaches and models used in Monte Carlo (MC) and molecular dynamics (MD) simulations of interfacial systems, 2) to describe and discuss results from selected simulation studies of interfacial water, and 3) to discuss the major limitations of these techniques and to offer suggestions for overcoming them.

General Simulation Approaches

In most MC (11,12) and MD (12,13) studies, a small number (N) of particles are placed in a cell of fixed volume (V) and the total interaction potential energy (U_N) from all pairwise interaction potentials (U_{ij}) between particles i and j is calculated:

$$U_N = \Sigma_i \ \Sigma_j \ U_{ij} \qquad (1)$$

Particle interactions are not computed beyond a cutoff radius of from four to eight Angstroms to improve computational efficiency by neglecting long-range interactions which contribute little to the overall structure of the fluid. Periodic boundary conditions are imposed by filling the space around the basic cell with image cells translated by multiples of the unit length of the basic cell. Thus, particles near the cell boundaries of the basic cell interact with image particles, not with empty cavities. This technique prevents spurious edge effects from affecting the results of the simulation. Periodic cell boundaries also allow a small sample of particles to exhibit properties characteristic of a much larger sample size.

Although the method of moving particles to new locations and of obtaining equilibrium configurations for the particles differs for the MC and MD methods, in both techniques the positions and orientations of thousands of configurations are generated and used to calculate average properties of the system. In the MC method, ensemble average properties can be computed. These may include structural and thermodynamic properties such as density, dipole direction cosine, hydrogen bond energies and number, radial distribution functions, internal energy, heat capacity and internal pressure. In the MD method, time average properties which are either structural or dynamic can be computed. For instance, these properties may include density, dipole direction cosine, hydrogen bond energies and number, radial distribution functions, dipole relaxation time, and self-diffusion coefficients. Thus, the key difference between the two techniques is that the MC method generally allows only static properties to be

evaluated, while the MD method generally allows both static and time
dependent phenomena to be studied.

Monte Carlo Methods. Although several statistical mechanical
ensembles may be studied using MC methods (2,12,14), the canonical
ensemble has been the most frequently used ensemble for studies of
interfacial systems. In the canonical ensemble, the number of
molecules (N), cell volume (V) and temperature (T) are fixed. Hence,
the canonical ensemble is denoted by the symbols NVT. The choice of
ensemble determines which thermodynamic properties can be computed.
In the NVT ensemble one cannot compute the chemical potential or
entropy of the system; two properties which are of critical importance
for interfacial systems. The choice of an ensemble also determines the
sampling algorithm used to generate molecular configurations from
random moves of the molecules.

In MC methods the ultimate objective is to evaluate macroscopic
properties from information about molecular positions generated over
phase space. To evaluate average macroscopic properties, $\bar{p}$, in the
canonical ensemble from statistical mechanics, the following
expression is used:

$$\bar{p} = \Sigma_q \ [p_q \exp(-U_N(q)/kT)]/Q(N,V,T) \qquad (2)$$

where p_q is the value of the macroscopic property $\bar{p}$ in the qth
configuration, k is Boltzmann's constant, $Q(N,V,T)$ is the canonical
ensemble partition function, and the summation runs over all q
equilibrium molecular configurations. Thus, Equation 2 suggests that
to evaluate properties in the canonical ensemble MC method, the
probability with which any molecular configuration occurs should be
proportional to $\exp(-U_N(q)/kT)$. The specific algorithm for generating
new configurations that satisfy this requirement involves i) selecting
a molecule at random, ii) selecting Cartesian center-of-mass
displacement coordinates randomly over an interval which is not
greater than half the cell length, iii) selecting a rotation angle at
random, and iv) calculating the new energy, $U_N(q+1)$, of the new
configuration generated by moves i)-iii). The final step involves
deciding whether to accept or reject the new, but random,
configuration. This is done by generating a random number between
zero and unity and comparing the random number to the quantity:

$$\exp[-(U_N(q+1) - U_N(q))/kT] \qquad (3)$$

If the random number is less than or equal to the quantity in Equation
3, then the new move is accepted; otherwise the move is rejected.
This acceptance criteria is often made even more stringent by
requiring that as few as 50% of the moves satisfying Equation 3 are
actually selected. Note that these procedures always favor moves
which lead to reduced total interaction energies.

Molecular Dynamics Methods. In contrast to the MC method, both
kinetic and structural properties of a molecular system can be
evaluated from MD studies. These properties are evaluated as averages
over configurations generated during time. In microcanonical ensemble
studies with the MD method, the properties which are controlled

include N, V and total energy, E. Total energy is computed from the sum of total kinetic energy and total potential energy. The potential energy is evaluated from an expression involving a summation over the interaction potentials between individual particles i and j, U_{ij}, as given in Equation 1. Temperature is not fixed in the microcanonical MD method, since it varies with fluctuations in total kinetic energy.

To determine the movement of molecules, the following algorithm (15) is often used. The force acting on the ith atom in a molecule ($\bar{F}_i$) is determined from the spatial derivative of the total interaction potential energy of that particle:

$$\bar{F}_i = -\bar{\nabla} \Sigma_j U_{ij} \tag{4}$$

A centered finite difference scheme is used to calculate the position of the ith atom, $\bar{x}_i$, a short time Δt in the future:

$$\bar{x}_i(t + \Delta t) = -\bar{x}_i(t - \Delta t) + 2\bar{x}_i(t) + (\Delta t^2/m_i)\bar{F}_i \tag{5}$$

where all that is required to determine this new position are the present and past locations, and the force from Equation 4. Typically, to ensure numerical stability of the algorithm, the magnitude of Δt is on the order of 10^{-15} seconds or less. The velocity of the atom, $\bar{v}_i$, is determined by the expression:

$$\bar{v}_i(t) = [1/(2\Delta t)][\bar{x}_i(t + \Delta t) - \bar{x}_i(t - \Delta t)] \tag{6}$$

Hence, the MD method involves numerical integration of the equations of motion for all particles each time a new configuration is generated, while the MC method only involves movement of one random particle for each new configuration. It should be noted that many other numerical algorithms are available for the MD method (13), and that maximum accuracy results from the use of algorithms that include higher powers of Δt than are given in Equation 5.

Interaction Potentials

Equations 3-4 show that the form of the interaction potentials used in simulating interfacial water is critical. Of interest for interfacial systems are both the interaction potential between water molecules and that between the surface and a water molecule.

The first MC (16) and MD (17) studies were used to simulate the properties of single particle fluids. Although the basic MC (11,12) and MD (12,13) methods have changed little since the earliest simulations, the systems simulated have continually increased in complexity. The ability to simulate complex interfacial systems has resulted partly from improvements in simulation algorithms (15,18) or in the interaction potentials used to model solid surfaces (19). The major reason, however, for this ability has resulted from the increasing sophistication of the interaction potentials used to model liquid-liquid interactions. These advances have involved the use of the following potentials: Lennard-Jones 12-6 (20), Rowlinson (21), BNS

(22), ST2 (23), MCY (24), CF (25), PE (26), TIP4P (27), and MCY+CC+DC (28).

Water Potentials. The ST2 (23), MCY (24), and CF (25) potentials are computationally tractable and accurate models for two-body water-water interaction potentials. The ST2, MCY and CF models have five, four, and three interaction sites and have four, three and three charge centers, respectively. Neither the ST2 nor the MCY potentials allow OH or HH distances to vary, whereas bond lengths are flexible with the CF model. While both the ST2 and CF potentials are empirical models, the MCY potential is derived from ab initio configuration interaction molecular orbital methods (24) using many geometrical arrangements of water dimers. The MCY+CC+DC water-water potential (28) is a recent modification of the MCY potential which allows four body interactions to be evaluated. In comparison to the two-body potentials described above, the MCY+CC+DC potential requires a supercomputer or array processor in order to be computationally feasible. Therefore, the ST2, MCY and CF potentials are generally more economical to use than the MCY+CC+DC potential.

A comparison of the bulk water properties predicted by the ST2, MCY, and CF models in simulations is given in Table I. These data were obtained from (2), unless otherwise noted.

Table I. Comparison of water properties for the ST2, MCY and CF simulation models and bulk water at approximately 298 K.

Property	ST2	MCY	CF	bulk water
-U (kJ/mol)	34	28.5	33	34
C_V (J/K/mol)	71	79	--	75
μ (Debye units)	2.35	2.26	1.86	1.86
PV/NkT (V/N=1)	0.09	8.5 (29)	0.1	0.05
D (10^{-9} m^2/sec)	3.1 (30)	2.3 (29)	1.10 (31)	2.85 (32)

Results in Table I illustrate some of the strengths and weaknesses of the ST2, MCY and CF models. All models, except the MCY model, accurately predict the internal energy, -U. Constant volume heat capacity, C_V, is accurately predicted by each model for which data is available. The ST2 and MCY models overpredict the dipole moment, μ, while the CF model prediction is identical with the value for bulk water. The ratio PV/NkT at a liquid density of unity is tremendously in error for the MCY model, while both the ST2 and CF models predictions are reasonable. This large error using the MCY model suggests that it will not, in general, simulate thermodynamic properties of water accurately (29). Values of the self-diffusion coefficient, D, for each of the water models except the CF model agree fairly well with the value for bulk water.

In simulating interfacial water, it is important to use a model for water-water interactions which yields accurate results in simulations of bulk water. Each of the models discussed here have obvious advantages and disadvantages. The CF model is generally more accurate in predicting bulk water properties than the other models in Table I. Two drawbacks of the ST2 model are its rigidness and overly tetrahedral geometry. The MCY potential may lead to spurious results for interfacial water, since it generates excessive internal pressures.

Surface Potentials. Consider the form of the surface-water interaction potential for an interfacial system with a hydrophobic surface. The oxygen atom of any water molecule is acted upon by an explicitly uncharged surface directly below it via the Lennard-Jones potential (U_{LJ}):

$$U_{LJ}(R_{ij}) = A[(\sigma/R_{ij})^a - (\sigma/R_{ij})^b] \qquad (7)$$

where R_{ij} is the distance between the jth surface atom and the oxygen atom on the ith water molecule, and A and σ are parameters which specify the depth of the potential energy well and the distance at which its value first equals zero, respectively. The exponents a and b specify the power law for the repulsive and attractive components of the Lennard-Jones potential, respectively. Three commonly used pairs of values for a and b are 12 and 6, 9 and 3, or 4 and 2, which produce the Lennard-Jones 12-6 (33,34), 9-3 (35), and 4-2 (36) potentials, respectively. Typical values for parameters of the latter Lennard-Jones potentials are reported in Table II. In general, the depth of the potential well for these potentials (about -0.5 kcal/mole) is typical of the energy between hydrophobic surfaces and physisorbed noble gases.

Table II. Typical values for parameters of the Lennard-Jones 12-6, 9-3, and 4-2 potentials.

Lennard-Jones potential	A (kcal/mole)	a	b	σ (Angstroms)
12-6	0.303	12	6	3.1
9-3	1.202	9	3	2.5
4-2	1.728	4	2	2.0

The above forms for the Lennard-Jones surface-water interaction potential have been used as models of hydrophobic surfaces such as pyrophyllite, graphite, or paraffin. If the intention of the study, however, is to understand interfacial processes at mineral surfaces representative of smectites or mica, explicit electrostatic interactions betweeen water molecules and localized charges at the surface become important.

Two methods for including explicit electrostatic interactions are proposed. In the first, and more difficult approach, one would need to conduct extensive quantum mechanical calculations of the potential energy variation between a model surface and one adjacent water molecule using thousands of different geometrical orientations. This approach has been used in a limited fashion to study the interaction potential between water and surface Si-OH groups on aluminosilicates, silicates and zeolites (37-39).

A simpler approach is to use an empirical model consisting of a Lennard-Jones potential plus a Coulombic term to explicitly account for charges on the surface. To calculate the magnitude of this Coulombic charge, consider the physical properties of smectite and mica aluminosilicates. It is known that the unit cell of smectite and mica surfaces has dimensions of about 46 square Angstroms (34). Using this value for the area of a unit cell and the data in Table III, the charge per unit cell and delocalized charge on surface oxygens of typical smectite and mica minerals can be computed.

Table III. Calculation of the number of charges per unit cell on typical smectite and mica surfaces.

Physical Property	Smectite	Mica
surface area (m^2/g)	750	100
charge density (e.s.u./m^2)	4×10^8	10×10^8
cation exchange capacity (meq/g)	1.038	0.346
charge density (number/unit cell)	0.38	0.96
delocalized charge per surface oxygen	0.06	0.16

To simulate smectite or mica minerals, a total of about 0.4 and 1 explicit negative charges, respectively, need to be assigned to each unit cell on the surface. This charge should be delocalized over about six oxygen atoms surrounding the ditrigonal cavities of the smectite and mica surfaces, since these charges originate from octahedral or tetrahedral sites within the crystal and not from the surface atoms. A proposed form for the water-surface interaction potential, U_{WS}, suitable for simulations of smectite or mica surfaces interacting with the ST2 model of water is:

$$U_{WS}(R_{ij},d_{\alpha j}) = A[(\sigma/R_{ij})^a - (\sigma/R_{ij})^b] + S(R_{ij}) \sum_{\alpha=1}^{4} (q_\alpha q_j) /d_{\alpha j} \quad (8)$$

where R_{ij}, A, a, b and σ are as defined in Equation 7 and Table II, $d_{\alpha j}$ is the distance between the jth surface atom and the αth charge (having charge q_α) on the water molecule, q_j is the delocalized charge on the jth surface atom from Table III, and $S(R_{ij})$ is the switching function of the ST2 water potential (23). The magnitude of charge on each of the four point charges for ST2 water in Equation 8 is 0.2357.

A plot of the Lennard-Jones 9-3 form of Equations 7 and 8 for ST2 water interacting with smectite and mica surfaces is shown in Figure 1. Values for the parameters used in Figure 1 are given in Tables II and III, and in reference (23). The water molecule is oriented so that its protons face the surface and its lone pair electrons face away from the surface, and the protons are equidistant from the surface. Note that the depth of the potential well in Figure 1 for interactions with the smectite surface and mica surface are

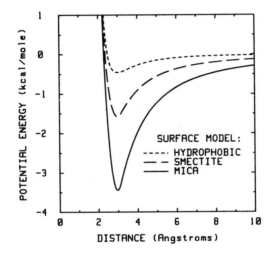

Figure 1. Comparison of ST2 water-surface interactions computed from Equations 7 or 8 using parameters for the Lennard-Jones 9-3 potential in Table II and the delocalized charge magnitude for smectite and mica surfaces in Table III.

about -1.5 and -3.5 kcal/mole, respectively. These interaction
energies are similar in magnitude to weak hydrogen bond energies.

An interaction potential between the surface and ions may also be
needed in simulating counterion diffusion for the smectite and mica
surface models. The form of such an interaction potential remains to
be determined. This may not pose a significant problem, since recent
evidence (40) suggests that over 98% of the cations near smectite
surfaces lie within the shear plane. For specifically adsorbed
cations such as potassium or calcium, the surface-ion interactions can
also be neglected if it is assumed that cation diffusion contributes
little to the water structure. In simulating the interaction
potential between counterions and interfacial water, a water-ion
interaction potential similar to those already developed for MD
simulations (41-43) could be specified.

Simulations of Interfacial Water

Several MC and MD studies of interfacial water near hydrophobic
surfaces have been reported (33-36,44-48). Both of the MC studies
(35,45), as well as the four MD studies (33,34,36,47) reporting
detailed observations of interfacial water are discussed here. This
comparison will show that choice of the water-water potential is
critical for such studies. It will also illustrate the wide range of
interfacial properties which can be studied using computer
simulations. Results from the early pioneering MC studies for
interfacial water are summarized in Table IV.

Table IV.   Results from Monte Carlo Simulations of Water Near
Hydrophobic Surfaces.

	Reference 35	Reference 45
# molecules	216	150
cell dimensions (nm^3)	7.127	4.5
cell density (g/cc)	0.906	0.997
temperature (K)	298	300
# configurations	2.5×10^6	1×10^4
water potential	Rowlinson	MCY
surface potential	L-J 9-3	hard wall
range of density oscillations (g/cc)	1.1 to 1.5	0.5 to 2.3
density trend towards surfaces	decreases	increases
hydrogen bonding trend towards surfaces	decreases	increases
internal energy trend towards surfaces	decreases	increases
preferred dipole orientation		
near surfaces	yes	yes

These results indicate that, compared to bulk water, interfacial
water exhibits unique oscillations in density with distance from the
surface and preferential dipolar orientations. Both simulations
report density values which are unreasonable. Part of this problem
arises from attempting to fix the water density based on the average
cell volume and the number of water molecules; an approach which

overlooks the fact that the cell volume near the surfaces is generally free of water molecules due to repulsive forces arising from the surfaces. For the study using the MCY potential (45), a more serious problem involves the excessive internal pressures generated by the MCY potential, which lead to excessive density oscillations and increasing density near the surfaces. Consistency between the two simulations is poor; whereas the first (35) predicts decreased density, hydrogen bonding and internal energy near the surfaces, the second (45) reports exactly the opposite trends. These differences are all probably related to the use of different water-water potentials.

   Results of selected MD studies of interfacial water are reported in Table V.

Table V.   Results from Selected Molecular Dynamics Studies of Water Near Hydrophobic Surfaces.

	Reference (47)	Reference (33)	Reference (36)	Reference (34)
# molecules	150	216	150	256
cell volume (nm^3)	4.5	8.787	5.198	9.156
cell density (g/cc)	1.0	0.74	0.87	0.84
temperature (K)	301	287	304	286
trajectory time (ps)	25	20	14	0.75
water potential	ST2	ST2	MCY	ST2
surface potential	hard wall	L-J 12-6	L-J 4-2	L-J 12-6
range of density oscillations (g/cc)	0.9 to 1.0	0.9 to 1.0	0.5 to 3.2	0.8 to 1.1
density trend towards surfaces	decreases	decreases	increases	decreases
hydrogen bonding trend towards surfaces	---	---	---	decreases
preferred dipolar orientations	yes	yes	yes	yes
self-diffusion coeff.				
near surfaces (m^2/s)	3.3×10^{-9}	4.8×10^{-9}	3.1×10^{-9}	2.1×10^{-9}
near midplane (m^2/s)	4.2×10^{-9}	3.3×10^{-9}	3.7×10^{-9}	2.7×10^{-9}
dipole relaxation time				
near surfaces (ps)	3.1	---	2.3	---
near midplane (ps)	2.1	---	2.0	---

   The results in Table V illustrate that MD studies, compared to the MC results in Table IV, facilitate the investigation of transport and time-dependent properties. Also, they show that use of the MCY potential leads to very large density oscillations and increasing water density near the surfaces. This appears to be a serious drawback to the use of the MCY potential in simulations of interfacial water. Results from the investigations using the ST2 potential show that interfacial water density is approximately 1.0 g/cc, with a tendency for decreased density and hydrogen bonding near the surfaces. As in the MC simulations, orientations of the water dipole moment are affected by the presence of a solid/liquid interface, and an appreciable decrease in dipole relaxation and the water self-diffusion coefficient are usually observed near the surfaces.

Comparison with Experiment. How do the results of these simulations compare with experimental results on corresponding systems, and what do they infer about interpreting the "hydrophobic effect", "hydration forces", and the "structure" of interfacial water? Structurally, the interfacial water exhibits clear density oscillations which extend to at least 15 Angstroms from the surfaces (34). Since this simulation involved a hydrophobic, neutral surface, these effects are directly attributable to the presence of the surfaces, and not to an effect of charged counterions. Furthermore, since no long-range changes in hydrogen bonding patterns were observed due to this structural reordering (34), the MD results suggest that the hydrophobic effect is due to entropy changes in the interfacial liquid rather than to long-range bonding effects.

The presence of density oscillations in simulation results of interfacial water has stimulated experimental studies which were explicitly designed to detect their presence. Structural forces associated with density oscillations in water next to mica surfaces have recently been measured experimentally (49). Although, mica is not a hydrophobic surface, it should be pointed out that there is theoretical basis for suggesting that molecular layering near surfaces and the accompanying oscillatory forces are responsible for both the "hydration" and "hydrophobic" effects (2). Whether this force is attractive (hydrophobic effect) or repulsive (hydration effect) depends upon how the density oscillations fit into the region between the surfaces. Much MC work is needed with both hydrophobic and hydrophilic surfaces using the grand canonical ensemble to determine the chemical potential and entropy of the interfacial water at various surface separations in order to better understand the magnitude of these effects.

Theoretical explanations for the "hydration force" (50) and the "hydrophobic effect" (51) often involve an analysis of the forces emanating from oriented molecular dipole or quadrupole moments. All of the computer simulations for interfacial water discussed above found significant evidence for preferred dipolar orientations near the surfaces. In most of the studies, all of which used non-polarizable, pairwise interacting water models, the tendency for preferred dipolar orientations diminishes in a continuous fashion with increasing distance from the surfaces and was negligible at a distance of from ten to fifteen Angstroms from the surfaces. When realistic water potentials incorporating cooperative effects become available, this effect can be expected to become even more significant.

According to the Kirkwood theory of polar dielectrics, simple relations (23) between molecular dipole moment vectors and the mean-square total dipole moment of water clusters can be used to compute the static dielectric constant of water. As the normalized mean-square total dipole moment increases towards unity, theory predicts decreases in the static dielectric constant. Since MD results indicate that the mean-square total dipole moment of interfacial water is greater than that for bulk water (48), the static dielectric constant of interfacial water should be lower than that in bulk water. This conclusion is meant to be only qualitative, since calculations of the static dielectric constant using the Kirkwood theory may be in considerable error if the effects of the reaction field are not accounted for (52). The reaction field is an electrical field within the cutoff radius which contributes to the dielectric constant. It results from the polarizing effects of the dipole moments outside the cutoff radius, and its effect can be included (with considerable effort) if accurate computations of the dielectric constant are desired (53).

The heterogeneous dielectric properties in the liquid medium near surfaces have important implications for theories of the structure of the interfacial region. A recent theoretical study of the effect of decreased medium dielectric constant near surfaces (54) shows that it is associated with significant reductions in surface potential computed from double layer theory. A note of caution to theoreticians is in order. The molecular simulations indicate that the static dielectric properties of the interfacial water decay gradually with increasing distance from the surface. Hence, the use of discrete mixture models (54,55) in which water near the surfaces is divided into two zones; one having properties characteristic of water in the first adsorbed layer and the second having bulk properties are not likely to represent actual surfacial conditions. Similar caution should be used in adsorption models that assume discrete molecular layers of surface complexes in the interfacial region.

Reaction kinetics and many transport properties in liquids are controlled by rates of diffusion. Processes that may be controlled by diffusion include, for example, rates of ligand exchange from transition metal ions (56), reactions involving proton transfer (57), and exchange reactions near mineral surfaces (58). Results in Table V (from studies 34,36,47) indicate that the value for the self-diffusion coefficient of water near mineral surfaces is consistently about 80% lower than the value near the midplane. Quantitatively, the values for D near the midplane are higher than values reported for bulk water at comparable temperatures (55). For instance, the values of D in bulk water at 285 and 300 K are about 1.8 and 2.9 x 10^{-9} m^2/s, respectively. The results from one study (33), appear to be excessively high, considering the temperature of the simulation, and are also inconsistent with experimental results in that they predict diffusion rates which are greater near the surfaces than near the midplane. Excluding the latter results, values for D from the MD simulations appear to qualitatively obey the expected trend for increasing values of D with increasing temperature. Furthermore, the decreases in values for self-diffusion coefficient near the surfaces are qualitatively consistent with experimental measurements of decreased water mobility near neutral silicate surfaces (59-62).

Another transport property of interfacial water which can be studied by MD techniques is the dipole relaxation time. This property is computed from the dipole moment correlation function, which measures the rate at which dipole moment autocorrelation is lost due to rotational motions in time (63). Larger values for the dipole relaxation time indicate slower rotational motions of the dipole

moment. Both of the MD studies reporting values for the dipole
relaxation time in Table V indicate that relaxation times are larger
for water near the surfaces than for water near the midplane. The MCY
potential, however, is not as sensitive to changes in relaxation time
as is the ST2 potential. These results can be interpreted to mean
that molecules near the surfaces experience hindered rotational
movement as compared to water near the midplane. Experimental
evidence from near infrared (6), electron spin resonance (59), and
nuclear magnetic resonance (60) studies support the MD simulation
results, in that they indicate hindered rotational motion of water
near uncharged silicate surfaces.

Summary

Monte Carlo and Molecular Dynamics simulations of water near
hydrophobic surfaces have yielded a wealth of information about the
structure, thermodynamics and transport properties of interfacial
water. In particular, they have demonstrated the presence of
molecular layering and density oscillations which extend many
Angstroms away from the surfaces. These oscillations have recently
been verified experimentally. Ordered dipolar orientations and
reduced dipole relaxation times are observed in most of the
simulations, indicating that interfacial water is not a uniform
dielectric continuum. Reduced dipole relaxation times near the
surfaces indicate that interfacial water experiences hindered
rotation. The majority of simulation results indicate that water near
hydrophobic surfaces exhibits fewer hydrogen bonds than water near the
midplane.

Several merits and strengths of molecular simulations of
interfacial water are apparent. Since these methods yield structural,
thermodynamic and transport properties for bulk water which are in
good agreement with many experimentally measured properties of bulk
water over a wide temperature range, they seem to offer a promising
approach for studying interfacial water. Interfacial systems are
generally composed of several components and are difficult to
characterize. The nature of molecular simulations allows the system
being analyzed to be exactly specified in terms of the types of
components, their interaction potentials, the initial atomic or
molecular locations and the types of boundary conditions imposed.
Thus, effects of the surfaces can be studied in detail, separately
from effects of counterions or solutes. In addition, individual
layers of interfacial water can be analyzed as a function of distance
from the surface and directional anisotropy in various properties can
be studied. Finally, one computer experiment can often yield
information on several water properties, some of which would be time-
consuming or even impossible to obtain by experimentation. Examples
of interfacial water properties which can be computed via the MD
simulations but not via experiment include the number of hydrogen
bonds per molecule, velocity autocorrelation functions, and radial
distribution functions.

Several weaknesses and disadvantages of the computer simulation
methods can also be mentioned. Foremost among these limitations is
the fact that none of the commonly used models for water interactions

account for three- or four-body interactions. Yet, the complex behaviour of interfacial water is apparently dictated by these cooperative effects (26). Significant differences exist in the internal energy, hydrogen bonding and diffusion rates of interfacial water as predicted by the ST2 and MCY potentials, with results for the former model being more consistent with experimental results than those for the latter. These differences probably relate to excess internal pressures generated by the MCY potential. A second problem involves uncertainty about what distance of separation should be imposed between surfaces for a particular choice of the number of water molecules. If excluded volume near the surfaces is not accounted for, the resulting average water density and its oscillations will be excessive. Finally, dielectric properties of interfacial water are not easily quantified when rigid, pairwise interacting water potentials are used. This limitation can be partially overcome by accounting for reaction field effects, but even then, water interaction potentials in current use are not effective in modeling the dielectric properties of water (53).

On the whole, the advantages and strengths of MC and MD simulations of interfacial water outweigh their disadvantages and weaknesses. Even if quantitative prediction of interfacial water properties is not possible in some cases, a knowledge of qualitative trends as a function of distance from the surfaces or relative to results from simulations of bulk water are often extremely illuminating.

What is the likely future use of MC and MD techniques for studying interfacial systems? Several promising approaches are possible. Continued investigation of double layer properties, "hydration forces", "hydrophobic effects", and "structured water" are clearly awaiting the development of improved models for water-water, solute-water, surface-water, and surface-solute potentials. Simulations of organics near surfaces are clearly possible since potentials describing water-organic interactions are presently available (64,65). Studies of reaction kinetics and ligand exchange processes near surfaces are clearly possible using the molecular time-scale generalized Langevin equation approach (66). Gaseous adsorption on metal (67), hydrophobic (68) and other simple surfaces have been extensively studied (2), but similar approaches using models for clay and zeolite surfaces are also possible. Mechanisms of crystal growth and defect structures in vitreous silica (69) and glass (70) have been studied, and similar studies on aluminosilicate minerals (even under conditions of high temperature and pressure) are possible. Finally, new theoretical developments are allowing thermodynamic properties (71,72) and non-equilibrium conditions (73) to be studied with MD methods. In short, MC and MD studies of interfacial systems are still in their infancy.

List of Symbols

a	parameter for the Lennard-Jones potential
A	parameter for the Lennard-Jones potential
b	parameter for the Lennard-Jones potential
C_V	constant volume heat capacity
$d_{\alpha j}$	distance between surface and αth charge on water molecule
D	self-diffusion coefficient
E	total energy
$\bar{F}_i$	force acting on ith atom
k	Boltzmann's constant
m_i	mass of ith atom
N	number of molecules
p_q	value of pth property of qth configuration
$\bar{p}$	ensemble average value of property p
P	internal pressure
q_j	charge on jth surface atom
q_α	charge on αth charge center of water molecule
Q	canonical ensemble partition function
R_{ij}	distance between ith and jth atoms
$S(R_{ij})$	switching function of ST2 water potential
T	absolute temperature
U	internal energy
U_{ij}	pairwise interaction potential
U_{LJ}	Lennard-Jones interaction potential
U_N	total interaction potential energy
U_{WS}	potential energy for water interacting with a charged surface
$\bar{v}_i$	velocity of ith atom
V	volume
$\bar{x}_i$	position of ith atom
μ	dipole moment
σ	parameter for the Lennard-Jones potential
Δt	time step for MD algorithm

Literature Cited

1. Cairns-Smith, A. G. Scientific American 1985; 252(6), 90-100.
2. Nicholson, D.; Parsonage, N. G. "Computer Simulation and the Statistical Mechanics of Adsorption"; Academic: New York, 1982; Chap. 4,6.
3. van Megen, W.; Snook, I. Adv. Colloid Interface Sci. 1984; 21, 119-194.
4. Low, P. F. Adv. Agronomy 1961, 13, 269-327.
5. Graham, J. Rev. Pure Appl. Chem. 1964; 14, 81-89.
6. Klier, K.; Zettlemoyer, A. C. J. Colloid Interface Sci. 1977; 58(2), 216-229.
7. Low, P. F. Soil Sci. Soc. Am. J. 1979; 43(5), 651-658.
8. Derjaguin, B. V.; Churaev, N. V. In "Progress in Surface and Membrane Science"; Cadenhead, D. A.; Danielli, J. F., Eds.; Academic: New York, 1981; Vol. 14, pp 69-130.
9. Sposito, G.; Prost. R. Chem. Rev. 1982; 82(6), 553-573.
10. Vold, R. D.; Vold, M. J. "Colloid and Interface Chemistry"; Addison-Wesley: Reading, Mass., 1983; Chap. 7,9.

11. Valleau, J. P.; Whittington, S. G. In "Modern Theoretical Chemistry"; Berne, B. J., Ed.; Plenum: New York, 1977; Vol. 5, Chap. 4.

12. Barker, J. A.; Henderson, D. Rev. Mod. Phys. 1976; 48(4), 587-671.

13. Wood, D. W. In "Water: A Comprehensive Treatise"; Franks, F., Ed.; Plenum: New York, 1979; Vol. 6, Chap. 6.

14. Snook, I. K.; van Megen, W. J. Chem. Phys. 1980; 72(5), 2907-2913.

15. Verlet, L. Phys. Rev. 1967; 159(1), 98-103.

16. Metropolis, N.; Rosenbluth, A. W.; Rosenbluth, M. N.; Teller, A. H. J. Chem. Phys. 1953; 21(6), 1087-1092.

17. Alder, B. J.; Wainwright, T. E. J. Chem. Phys. 1957; 27, 1207-1209.

18. Ryckaert, J. P.; Ciccotti, G.; Berendsen, H. J. C. J. Comp. Phys. 1977; 23, 327-341.

19. Steele, W. A. Surface Sci. 1973; 36, 317-352.

20. Rahman, A. Phys. Rev. 1964; 136(2A), A405-A411.

21. Barker, J. A.; Watts, R. O. Chem. Phys. Lett. 1969; 3(3), 144-145.

22. Rahman, A.; Stillinger, F. H. J. Chem. Phys. 1971; 55(7), 3336-3359.

23. Stillinger, F. H., Rahman, A. J. Chem. Phys. 1974; 60(4), 1545-1557.

24. Matsuoka, O.; Clementi, E.; Yoshimine, M. J. Chem. Phys. 1976; 64, 1351-1361.

25. Stillinger, F. H.; Rahman, A. J. Chem. Phys. 1978; 68(2), 666-670.

26. Barnes, P.; Finney, J. L.; Nicholas, J.; Quinn, J. E. Nature 1979; 282, 459-464.

27. Jorgensen, W. L.; Chandrasekhar, J.; Madura, J. D.; Impey, R. W.; Klein, M. L. J. Chem. Phys. 1983; 79, 926-935.

28. Detrich, J.; Corongiu, G.; Clementi, E. IBM Res. Rept. 1984; KGN-3, 1-11.

29. Impey, R. W.; Madden, P. A.; McDonald, I. R. Mol. Phys. 1982; 46(3), 513-539.

30. van Gunsteren, W. F.; Berendsen, H. J. C.; Rullman, J. A. C. Faraday Disc. 1978; 66, 58-70.

31. Stillinger, F. H.; Rahman, A. J. Chem. Phys. 1978; 68(2), 666-670.

32. Rahman, A.; Stillinger, F. H. J. Chem. Phys. 1971; 55(7), 3336-3359.

33. Sonnenschein, R.; Heinzinger, K. Chem. Phys. Lett. 1983; 102(6), 550-554.

34. Mulla, D. J.; Cushman, J. H.; Low, P. F. Water Resour. Res. 1984; 20(5), 619-628.

35. Christou, N. I.; Whitehouse, J. S.; Nicholson, D.; Parsonage, N. G. Faraday Symp. Chem. Soc. 1981; 16, 139-149.

36. Barabino, G.; Gavotti, C.; Marchesi, M. Chem. Phys. Lett. 1984; 104(5), 478-484.

37. Hobza, P.; Sauer, J.; Morgeneyer, C.; Hurych, J.; Zahradnik, R. J. Phys. Chem. 1981; 85, 4061-4067.

38. Geerlings, P.; Tariel, N.; Botrel, A.; Lissillour, R.; Mortier, W. J. J. Phys. Chem. 1984; 88, 5752-5759.

39. Sauer, J.; Zahradnik, R. Intl. J. Quant. Chem. 1984; 26, 793-822.

40. Low, P. F. Soil Sci. Soc. Am. J. 1981; 45(6), 1074-1078.

41. Probst, M. M.; Radnai, T.; Heinzinger, K.; Bopp, P.; Rode, B. M.
     J. Phys. Chem. 1985; 89, 753-759.
42. Heinzinger, K. Pure Appl. Chem. 1985; 57(8), 1031-1042.
43. Bounds, D. G. Mol. Phys. 1985; 54(6), 1335-1355.
44. Gruen, D. W. R.; Marcelja, S.; Pailthorpe, B. A. Chem. Phys. Lett.
     1981; 82(2), 315-320.
45. Jonsson, B. Chem. Phys. Lett. 1981; 82(3), 520-525.
46. Lee, C. Y.; McCammon, J. A.; Rossky, P. J. J. Chem. Phys. 1984;
     80(9), 4448-4445.
47. Marchesi, M. Chem. Phys. Lett. 1983; 97(2), 224-230.
48. Mulla, D. J.; Low, P. F.; Cushman, J. H.; Diestler, D. J. J.
     Colloid Interface Sci. 1984; 100(2), 576-580.
49. Pashley, R. M.; Israelachvili, J. N. J. Colloid Interface Sci.
     1984; 101(2), 511-523.
50. Gruen, D. W. R.; Marcelja, S. J. Chem. Soc. Faraday Trans. 2
     1983; 79, 225-242.
51. Goldman, S. J. Chem. Phys. 1981; 75(8), 4064-4076.
52. Steinhauser, O. Mol. Phys. 1982; 45(2), 335-348.
53. Neumann, M. J. Chem. Phys. 1985; 82(12), 5663-5672.
54. Helmy, A. K.; Natale, I. M. Clays Clay Min. 1985; 33(4), 329-332.
55. Fripiat, J. J.; Letellier, M.; Levitz, P. Phil. Trans. R. Soc.
     London 1984; A311, 287-299.
56. Margerum, D. W.; Cayley, G. R.; Weatherburn, D. C.; Pagenkopf, G.
     K. In "Coordination Chemistry"; Martell, A. E., Ed.; ACS Monograph
     174, American Chemical Society: Washington, D. C., 1978; Vol. 2,
     Chap. 1.
57. Eigen, M. Angew. Chemie 1964; 3(1), 1-72.
58. Breen, C.; Adams, J.; Riekel, C. Clays Clay Min. 1985; 33(4), 275-
     284.
59. Bassetti, V.; Burlamacchi, L.; Martini, G. J. Am. Chem. Soc. 1979;
     101(19), 5471-5477.
60. Cruz, M. I.; Letellier, M.; Fripiat, J. J. J. Chem. Phys. 1978;
     69(5), 2018-2027.
61. Martini, G. J. Colloid Interface Sci. 1981; 80(1), 39-48.
62. Roberts, N. K.; Zundel, G. J. Phys. Chem. 1980, 84, 3655-3660.
63. Mulla, D. J.; Low, P. F. J. Colloid Interface Sci. 1983; 95(1),
     51-60.
64. Remerie, K.; van Gunsteren, W. F.; Postma, J. P. M.; Berendsen, H.
     J. C.; Engberts, J. B. F. N. Mol. Phys. 1984; 53(6), 1517-1526.
65. Kuharski, R. A.; Rossky, P. J. J. Am. Chem. Soc. 1984; 106, 5786-
     5793.
66. Adelman, S. A. J. Phys. Chem. 1985; 89, 2213-2221.
67. Broughton, J. Q. Surface Sci. 1980; 91, 91-112.
68. Talbot, J.; Tildesley, D. J.; Steele, W. A. Mol. Phys. 1984;
     51(6), 1331-1356.
69. Garofalini, S. H. J. Non-Crystalline Solids 1984; 63, 337-345.
70. Soules, T. F. J. Non-Crystalline Solids 1982; 49, 29-52.
71. Nose', S. Mol. Phys. 1984; 52(2), 255-268.
72. Ray, J. R.; Rahman, A. J. Chem. Phys. 1984; 80(9), 4423-4428.
73. Hoover, W. G.; Moran, B.; Haile, J. M. J. Stat. Phys. 1984;
     37(1/2), 109-121.

RECEIVED June 18, 1986

Behavior of Water on the Surface of Kaolin Minerals

R. F. Giese, Jr., and P. M. Costanzo

Department of Geological Sciences, State University of New York at Buffalo, 4240 Ridge Lea Road, Amherst, NY 14226

Study of hydrated kaolinites shows that water molecules
adsorbed on a phyllosilicate surface occupy two different structural sites. One type of water, "hole" water,
is keyed into the ditrigonal holes of the silicate layer, while the other type of water, "associated" water,
is situated between and is hydrogen bonded to the hole
water molecules. In contrast, hole water is hydrogen
bonded to the silicate layer and is less mobile than
associated water. At low temperatures, all water molecules form an ordered structure reminiscent of ice; as
the temperature increases, the associated water disor-
ders progressively, culminating in a rapid change in
heat capacity near 270 K. To the extent that the kaolinite surfaces resemble other silicate surfaces, hydrated kaolinites are useful models for water adsorbed on
silicate minerals.

To a large extent, the study of terrestrial geology is the study of
the interaction of water and rock materials. Much of the modification of the earth's surface, involving chemical weathering, transport, and deposition of sediment, results from the contact of water,
often containing reactive chemical species, with the surfaces of
mineral grains. The majority of the chemical activity, as far as we
presently know, takes place on a microscopic scale; at the interface
between a mineral and the first, or perhaps the first few, layers of
adsorbed water molecules. Such interfacial regions have complex
physical chemical properties, often very different from the phases
which they separate. This is compounded, in the case of the water-
silicate interface, by the fact that the structure of bulk water is
very complex itself, and, while we know in general terms the crystal
structures of the major silicate minerals, we often do not have a
clear picture of the structure of the mineral surface, nor do we
know in detail the structure of disordered minerals, at least not on
an atomic scale.
    The amount of water in the interfacial region is very small
compared to the bulk water in the system. For many experimental

0097–6156/86/0323–0037$06.00/0

techniques, this means that the signal from the surface water is too
weak to be easily separated from that of the bulk water. One solution to this problem is to study materials with large specific surface areas. Traditionally, these have been clay minerals, zeolites,
and gels. Of these, clay minerals are conceptually the simplest
because they present two dimensional planar surfaces to the external
environment. This paper is divided into two parts; an introduction
to the properties of water at the mineral-water interface, followed
by a description of the work done in our laboratory on the structure
of water in the interlayer region of kaolinite with particular reference to the structure and properties of water at the interface.

Clay Mineral Structures

The basic structures of the clay minerals were described by Pauling
(1, 2) and illustrations of each type can be found in the text of
Grim (3). These structure models are based on regular tetrahedra
and octahedra formed by oxygen or hydroxyl groups, symmetrically
disposed in planar layers. The models are esthetically pleasing, but
actual minerals are far more complex (see (4) for a recent summary).

The two types of clay mineral structures which are of interest
in the present discussion are the expanding 2:1 structures (the
smectites and vermiculites) and the 1:1 structures (the kaolins).
The smectites and vermiculites have a fundamental layer made up of
two sheets of tetrahedra which incorporate small, highly charged
cations and one sheet of octahedra coordinating larger cations. The
octahedra share edges, the tetrahedra share corners, and the three
sheets share oxygens in common planes to form the 2:1 layer (Figure
1A). A similar scheme, but involving only one sheet of tetrahedra
and one of octahedra, produces the 1:1 layer silicates (Figure 1B)
of which kaolinite is perhaps the most important mineral.

In the smectites and vermiculites, substitution of differently
charged ions is common. These may involve aluminum for silicon in
the tetrahedral sites, and, in the octahedral sites, ferrous or
ferric iron for aluminum, magnesium for aluminum, or lithium for
magnesium; in addition, vacant sites are commonly found. The substitutions create a charge imbalance which is neutralized by the
adsorption of cations on the external and internal surfaces of the
crystals. These compensating cations are not firmly attached to the
clay surfaces and they can be exchanged by treatment with dilute
solutions of appropriate salts. When exposed to water and many
organic molecules, the layers of these minerals separate allowing
the guest molecules to enter between the layers. Thus, both surfaces of every layer of the clay crystal become equivalent to external surfaces and the total surface area increases to as much as 800
m^2/g in the case of the smectites.

The 1:1 kaolin structures are chemically simpler; the tetrahedral sites are occupied by silicon and the octahedral sites by
aluminum. There is a minor amount of substitution, largely of
ferric iron for aluminum, but the amounts are generally only a few
tenths of a percent by weight of oxide. The kaolin minerals do not
expand in the presence of water and their surface area, approximately 10 to 15 m^2/g, represents the external area of the crystals.

Because of the difference between the 2:1 and 1:1 structures,
their external and internal surfaces are fundamentally different.

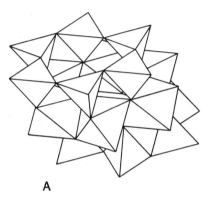 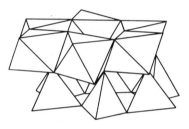

A   B

Figure 1. Polyhedral representations of the layers in 2:1 smectites (A) and 1:1 kaolinites (B). Only a single layer is shown in
each. The view is down the [041] axis of the unit cell.

All surfaces of the 2:1 clays (excluding adsorbed species) consist
of oxygen, while the 1:1 minerals have one surface of each layer
formed by oxygen and the other surface formed by hydroxyl groups.

Hydration of Clays. There is general agreement, and ample experimental and theoretical evidence to support the point of view, that
water molecules in contact with or in the close vicinity of clay
surfaces are perturbed. The distance over which these perturbations
produce measurable changes in the properties of the adsorbed water
molecules is much more controversial. Two conflicting models for
water adsorbed on clay mineral surfaces have been proposed; one
states that the perturbation by the surface is limited to between 3
and 4 water layers (roughly 10A) (5), while the other holds that the
influence of the clay surfaces extends much further, up to 30 or
more water layers (roughly 100A) (6). Evaluating the evidence for
each model is complicated because they are based on different experimental approaches which probe the clay-water interactions at different time scales (7). In addition, it is difficult to separate
the influence of the clay layer itself from the structuring effect
of the exchangeable cations adsorbed on or near the surface. The
following discussion is not meant to be an exhaustive evaluation of
the two models, but rather a selective discussion of what is known
about the structure of water in close contact with a clay mineral
surface and what the general arguments are in favor of or against
the short-range and long-range interaction models. More detailed
discussion of the problem can be found in (7) and (8).

Water on Halloysite. Central to the controversy is the observation
that clay crystals present a planar array of oxygens (and hydroxyls
in the case of kaolinite) which have hexagonal (or nearly) symmetry
with a periodicity similar to that found in the crystal structure of
ice. Because of this geometric similarity, it has frequently been
assumed that water adsorbed on a clay surface will preferentially
adopt an ice-like configuration. When looked at in detail, it is
difficult to find unequivocal evidence to support this.
    Halloysite-10A is frequently referenced as an example of water
molecules with an ice-like structure, epitaxially adsorbed on a clay
surface (9). In fact, this model has been so appealing that the
original figure illustrating the structural model of Hendricks and
Jefferson (9) has been reproduced innumerable times by others (see
(7) and (10) for example). It is often stated in the literature
that the Hendricks and Jefferson model is based on X-ray diffraction
data, implying that a structure was proposed and tested by comparison of observed and calculated intensities. What seems to have
been overlooked is the fact that the original paper of Hendricks and
Jefferson presented no substantial experimental evidence to support
their model other than the observed increase in the thickness of the
clay structure resulting from hydration. That distance, roughly 2.9
A, does impose some geometric restriction on the possible arrangement of the interlayer water molecules, but hardly a definitive one.
Much subsequent work has shown that it is very difficult to obtain
good agreement between observed and calculated intensities even for
the simple case of the one-dimensional structure along the c-axis
which involves only the z coordinate of the atoms (see (11) for
discussion).

Halloysite-10A represents a structure with few if any interlayer cations, allowing one to investigate the relatively simple
case of water interacting with a clay surface. Similarly, ice-like
models have been proposed for water adsorbed on smectite and vermiculite surfaces (9, 12, 13). These represent cases of charged clay
layers with adsorbed exchangeable cations.

Water on Vermiculite. For low water contents (that is, one or two
water layers), the evidence for highly structured water in the
interlayer spaces of smectites and vermiculites is most easily seen
in X-ray diffraction structure determinations of ordered hydrate
structures such as the two-water layer hydrate of Ca-vermiculite
(14, 15) and Na-vermiculite (15, 16).

In the Ca-vermiculite, the interlayer calcium ions are of two
types; one with six water molecules in an octahedral arrangement,
the other with eight water molecules arranged in a distorted cube.
The vermiculite layers are stacked so that the ditrigonal holes in
the tetrahedral surfaces are opposite each other, as are the tetrahedral sites. The six-coordinated calcium ions lie between adjacent
tetrahedra where they can most effectively compensate the charge
deficiency resulting from the aluminum for silicon substitution. In
contrast, the eight coordinated calcium ions are positioned between
the ditrigonal holes, a position where they are also very close to
the tetrahedral charge deficits. the cations in the Na-vermiculite
are all octahedrally coordinated by water molecules and lie exclusively between tetrahedral oxygens of the adjacent layers.

Water on Smectites. Compared to vermiculites, smectites present a
more difficult experimental system because of the lack of stacking
order of the layers. For these materials, the traditional technique
of X-ray diffraction, either using the Bragg or non-Bragg intensities, is of little use. Spectroscopic techniques, especially nuclear magnetic resonance and infrared, as well as neutron and X-ray
scattering have provided detailed information about the position of
the water molecules, the dynamics of the water molecule motions, and
the coordination about the interlayer cations.

As an example, infrared spectroscopy has shown that the lowest
stable hydration state for a Li-hectorite has a structure in which
the lithium cation is partially keyed into the ditrigonal hole of
the hectorite and has 3 water molecules coordinating the exposed
part of the cation in a triangular arrangement (17), as proposed in
the model of Mamy (13). The water molecules exhibit two kinds of
motion; a slow rotation of the whole hydration sphere about an axis
through the triangle of the water molecules, and a faster rotation
of each water molecule about its own C_2 axis (18). A similar
structure for adsorbed water at low water contents has been observed
for Cu-hectorite, Ca-bentonite, and Ca-vermiculite (17).

Multilayer Adsorption of Water. As the amount of water in the clay
increases over that needed for a one- or two-layer hydrate, the
study of the properties of the water becomes experimentally more
difficult. This is important because it is only at water contents
in excess of the two-layer hydrate that a conflict arises between
the short-range and long-range interaction models. In support of
the short-range model, two studies are noteworthy. A small angle

X-ray diffraction study of clay-water gels ([19]) has shown that there are strong interactions between the clay surfaces and their adsorbed ions and water molecules, but these interactions exist only over short distances. An entirely different approach has been taken by Fripiat et al. ([5]). They examined clay-water mixtures over a wide range of clay concentrations using kaolinite and smectites. For microscopic water properties, the nuclear magnetic resonance relaxation time for hydrogen and deuterium were measured. The macroscopic properties were determined by measuring the heats of wetting of previously equilibrated clay-water mixtures. Both experimental techniques, operating on very different time scales, showed that the number of water layers influenced by the clay was less than 5, with an average for all samples of 3.4 layers. This gives an average thickness of about 10 Å for the perturbed water.

The view that the clay surface perturbs water molecules at distances well in excess of 10 Å has been largely based on measurements of thermodynamic properties of the adsorbed water as a function of the water content of the clay-water mixture. There is an extensive literature on this subject which has been summarized by Low ([6]). The properties examined are, among others, the apparent specific heat capacity, the partial specific volume, and the apparent specific expansibility ([6]). These measurements were made on samples prepared by mixing predetermined amounts of water and smectite to achieve the desired number of adsorbed water layers. The number of water layers adsorbed on the clay is derived from the amount of water added to the clay and the surface area of the clay.

The value of the thermodynamic property in question is the difference in values for the clay-water sample and the same measurement on an equivalent amount of pure, anhydrous clay ([6]). This procedure involves two assumptions: 1) the added water is uniformly adsorbed on all clay layers, and 2) the thermodynamic properties of the clay itself do not change when the clay expands and is intercalated by water molecules.

The assumption that the water is adsorbed in uniform layers on all the clay surfaces for a wide range of mixtures has been criticized ([5], [20]). The argument is that the individual clay particles in the clay-water mixture do not expand beyond a certain distance regardless of the quantity of water which is added. The clay layers group themselves into tactoids resulting in two populations of water; those molecules which are found between the tactoids and those directly perturbed by the clay layers. If true, this would invalidate the procedure used to calculate the thermodynamic properties of the adsorbed water. However, other workers have reported complete delamination of certain smectites ([21], [22]). It is not clear under what conditions tactoids will form, or not, and this uncertainty is underlined in ([21]) (see remarks by Nadeau and Fripiat, pages 146-147).

The validity of the assumption that the various thermodynamic properties of the smectite remain invariant, regardless of the state of hydration, has been addressed in detail by Sposito and Prost ([7]). They point out that one would, for example, expect hydrolysis of the clay to occur at high water contents, and also, it is likely that the exchangeable cations will change their spatial relationship with the clay layers. Thus, the derived thermodynamic properties of the adsorbed water would not represent correct values.

Finally, the whole concept of using macroscopic (i.e. thermodynamic) properties to derive a microscopic picture of the adsorbed
water is open to question (7, 8).

It is difficult to reconcile these very different views of the
interaction of water and clay surfaces. Sposito (8) has attempted
this. He points out that the thermodynamic properties have an
essentially infinite time scale, whereas the spectroscopic measurements look at some variant of the vibrational or a predecessor of
the diffusional structure of water. It is possible that the thermodynamic properties reflect a number of cooperative interactions
which can be seen only on a very long time scale. Still, the X-ray
diffraction studies seemingly also operate on as long a time scale
as the thermodynamic properties. There is still not a clear choice
between the short-range and long-range interaction models.

Experimental Studies of Water on Kaolin Minerals

In general, the 2:1 clays are not very simple systems in which to
study the interaction of water and surfaces. They have complex and
variable compositions and their structures are poorly understood.
Water occurs in several different environments: zeolitic water in
the interlayer regions, water adsorbed on the external surfaces of
the crystallites, water coordinating the exchangeable cations, and,
often, as pore water filling voids between the crystallites. Thus,
there are many variables and the effects of each on the properties
of water are difficult to separate.

In view of the problems associated with the expanding 2:1 clays,
the smectites and vermiculites, it seemed desirable to use a different clay mineral system, one in which the interactions of surface
adsorbed water are more easily studied. An obvious candidate is the
hydrated form of halloysite, but studies of this mineral have shown
that halloysites also suffer from an equally intractable set of
difficulties (10). These are principally the poor crystallinity,
the necessity to maintain the clay in liquid water in order to
prevent loss of the surface adsorbed (intercalated) water, and the
highly variable morphology of the crystallites. It seemed to us
preferable to start with a chemically pure, well-crystallized, and
well-known clay mineral (kaolinite) and to increase the normally
small surface area by inserting water molecules between the layers
through chemical treatment. Thus, the water would be in contact
with both surfaces of every clay layer in the crystallites resulting
in an effective surface area for water adsorption of approximately
$1000 \ m^2 \ g^{-1}$. The synthetic kaolinite hydrates that resulted from
this work are nearly ideal materials for studies of water adsorbed
on silicate surfaces.

Kaolin Minerals. The 1:1 structures include a group of aluminosilicate minerals which are termed collectively the kaolin minerals;
specifically these are kaolinite, dickite, nacrite, and halloysite.
The basic 1:1 layer for all of these minerals has the composition
$Al_2Si_2O_5(OH)_4$; there is a small amount of substitution of iron for
aluminum, and fluoride for hydroxyl ion. All, except halloysite,
are normally anhydrous and do not expand (as do the smectites) upon
exposure to water and most organic molecules. As a result, they
generally have a rather small surface area, on the order of $10 \ m^2$

g^{-1}. In spite of their relatively simple composition and the
abundance of well characterized samples, the small surface area of
kaolinite and dickite has, until recently, effectively eliminated
them as candidates for surface chemistry studies. Assuming little
or no isomorphic substitution, the bonding between layers is largely
due to hydrogen bonds from the hydroxyls of one surface to adjacent
oxygens of the next surface (23, 24). Hydrogen bonds are not
normally thought of as being very strong relative to ionic and covalent bonds, and one might, therefore, expect that the layers of the
kaolin minerals would be easily separated, at least by small molecules and water. Such is not the case for water, and intercalation is found to be possible only for a relatively small number of
organic molecules and salts.

The types of organic molecules that are able to intercalate the
kaolin minerals are generally small with large dipole moments.
These include hydrazine, dimethylsulfoxide (DMSO), formamide and
some derivatives (N-methylformamide and dimethylformamide), acet-
amide and some derivatives, and pyridine N-oxide. Some salts such
as potassium acetate also intercalate kaolinites. Once intercalated
by one of these small molecules or salts, other molecules which
normally do not directly intercalate kaolins can be introduced by
replacement. Further, the exposure of the inner surfaces by intercalation gives one the opportunity to alter the interlayer bonding
of the kaolin layers by chemical modification of the inner surfaces.

Synthesis of Kaolinite Hydrates. Our work in synthesizing a water-
silicate system that has a simpler chemistry than the smectites or
vermiculites is based on the concept of reducing the total number of
interlayer hydrogen bonds by a chemical replacement of some inner
surface hydroxyls by fluorine (25). In principle, one need only
expand the kaolinite by intercalation with an appropriate organic
molecule and then expose the intercalated clay to an environment
containing fluoride ions. In practice, the organic molecule, the
type of fluoride salt added, as well as the time and temperature at
each stage in the development of a hydrated kaolinite, play an
important role in determining the success of the synthesis and the
yield of hydrated clay.

Early work showed that a 10A hydrate, similar to naturally
hydrated halloysite, could be synthesized from a well-crystallized
kaolinite from Cornwall, England (26). The procedure was to intercalate the clay with DMSO which contained about 8% by weight of
water. The clay expanded from 7.2 A to 11 A upon intercalation by
DMSO. Ammonium fluoride was dissolved in the DMSO solution and,
presumably, the fluoride diffused into the interlayer region where
it replaced some hydroxyl groups. This suspension was continuously
stirred at 60° C for periods varying from a few hours to as much as
12 hours or more. The clay was then separated from the water-DMSO
by centrifugation and redispersed in distilled water. This cycle of
water washing was repeated several times to remove intercalated DMSO
molecules. Instead of collapsing back to 7 A, the kaolinite exhibited a 10 A spacing similar to that of a fully hydrated halloy-
site-10A. Infrared spectra showed a broad absorption in the 3400-
3500 cm^{-1} region and a single band near 1650 cm^{-1} indicating that
the kaolinite was intercalated primarily by water. As is the case
with hydrated halloysites, this 10A product was not stable under

ambient humidity and thus was of limited use in examining the properties of surface adsorbed water.

Subsequent work showed that a modification of the synthesis procedure produced a 10A hydrate which, if dried carefully, would maintain the interlayer water in the absence of excess water (27). This material is optimal for adsorbed water studies for a number of reasons: the parent clay is a well-crystallized kaolinite with a negligible layer charge, there are few if any interlayer cations, there is no interference from pore water since the amount is minimal, and the interlayer water molecules lie between uniform layers of known structure. Thus, the hydrate provides a useful model for studying the effects of a silicate surface on interlayer water.

Characterization of Interlayer Water. X-ray diffraction studies of the 10A hydrate show no hkl reflections indicating a lack of regularity in the stacking of the kaolin layers. In addition to the 10A hydrate, two other less hydrated kaolinites were synthesized. Both have one molecule of water for each formula unit in contrast to the 10A hydrate which has two. These less hydrated clays consequently have smaller d(001) spacings of 8.4 and 8.6 A. The synthesis conditions for these two hydrates are described in (27). By studying the interlayer water in the 8.4 and 8.6A hydrates, it was possible to formulate a model of the water in the more complicated 10A hydrate.

An isolated hydroxyl group, such as is found in the 2:1 micas, absorbs infrared radiation at approximately 3700 cm^{-1} (28). Kaolinite is more complex since there are two distinct types of hydroxyl groups, the single hydroxyl at the interface between the tetrahedral and octahedral sheets (the inner hydroxyl), and three hydroxyls (the inner surface hydroxyls) which form one of the external surfaces of the kaolinite layer. Because the inner hydroxyl is isolated it produces a single vibration at approximately 3620 cm^{-1}. The three external hydroxyls do not vibrate independently; their stretching vibrations are coupled to produce several bands, none of which can be related to a specific hydroxyl group (29, 30). There are three of these bands for kaolinite; a strong absorption at 3695 cm^{-1}, and two much weaker absorptions at 3665 and 3650 cm^{-1}. The four bands may vary in frequency and intensity of absorption from sample to sample. In contrast, the vibrational bands of water molecules in the solid phase occur at 3350 and 3250 cm^{-1} with a bending mode at 1640 cm^{-1} (31, 32). Hence the infrared bands from the structural hydroxyls and intercalated water do not overlap. Of importance to the present subject is the fact that the three bands corresponding to the inner surface hydroxyls can be used to infer the bonding state of molecules in the interlayer region of the kaolinite. Insertion of guest molecules between the kaolinite layers disrupts the hydrogen bonds of the original kaolinite and results in the formation of new bonds with the guest molecules.

To record the infrared spectra, samples of the parent kaolinite and the three hydrates were dispersed in a fluorinated hydrocarbon. The mulls were squeezed between calcium fluoride plates and the sample was placed directly in the beam of a Perkin-Elmer 683 spectrometer. This mounting technique results in a tendency for the clay layers to align themselves perpendicular to the beam of the spectrometer. Infrared spectra of these materials have been published elsewhere (27) and there are slight differences in frequencies (2-3 cm^{-1} or less) and band intensities compared to the results reported here. There are several reasons for these differences: for this work, we used freshly prepared hydrates which seem to give a better signal, the sample preparation for the infrared spectrometer has improved over our original work, and the present spectra were run in the absorbance rather than in the transmission mode.

The spectra of the hydrated kaolinites (Figure 2A, 2B, 2C) differ substantially from each other and also from the parent kaolinite (Figure 2D). The 8.4A hydrate (Figure 2A) has an ordered layer stacking, as shown by X-ray diffraction (33). The crystal structure shows that the water molecules are associated with and are keyed into the ditrigonal holes of the tetrahedra. The conditions of the 8.4A synthesis result in up to 20% of the inner surface hydroxyls being replaced by fluoride ions. Because it seems likely that some of the interlayer water molecules are hydrogen bonded to fluoride ions of the adjacent layer, at least two hydrogen bond schemes can be envisioned: one where water bonds to the oxygens of the adjacent tetrahedral sheet; the other where the water bonds to an oxygen of one layer and a fluoride of the other layer (or, possibly, to two fluorides of the same layer). Hydrogen bonds to fluoride are normally stronger than similar bonds to oxygen so one would expect to see separate absorption bands in the infrared spectrum of the 8.4A hydrate if there are substantial numbers of water molecules bonding to fluoride (27).

In contrast to the 8.4A hydrate, the 8.6A hydrate (Figure 2B) has only a minor amount of replacement of fluoride for hydroxyl ions and a non-crystalline character, like the 10A hydrate. Comparison of the infrared spectra of the 8.4 and 8.6A hydrates (Figures 2A, 2B) does indeed show more bands for the former in the region between 3600 and 3200 cm^{-1}, in agreement with the argument stated above. In the bending mode region, the 8.4A hydrate has two clearly separated bands, as one would expect for water hydrogen bonded to oxygen (1645 cm^{-1}) and also hydrogen bonded to fluoride (1590 cm^{-1}). In contrast, the 8.6A hydrate has a single band at 1653 cm^{-1}.

Because the inner hydroxyl is buried in the kaolinite layer and is not perturbed substantially by intercalation, its vibration, at 3620 cm^{-1}, is common to all four clays in Figure 2. The 8.4A hydrate has two high frequency bands above 3620 cm^{-1}; 3690 and 3650 cm^{-1}. Similar bands are evident in the 8.6A hydrate although their intensities are somewhat different. Below 3620 cm^{-1}, the 8.4A hydrate has small bands at 3584 and 3538 cm^{-1} and larger bands at 3443 and 3340 cm^{-1}. The latter two do not appear in the spectrum of the 8.6A hydrate and have been assigned to hydrogen bonding from water molecules to fluoride. Bands similar to the 3584 and 3538 cm^{-1} bands of the 8.4A hydrate appear in the 8.6A hydrate spectrum, but shifted slightly to 3595 and 3547 cm^{-1}. The close match between these comparable bands at frequencies above 3500 cm^{-1} in the two hydrates is good evidence that these are due to hydroxyls and water molecules in similar environments. The water molecules attached to the ditrigonal holes have been termed "hole water" (27). These represent a discontinuous monolayer of water adsorbed onto a silicate surface (33). Weight loss measurements (27) show that there are as many water molecules as there are ditrigional holes.

Comparison of the bands at frequencies above 3500 cm^{-1} for the

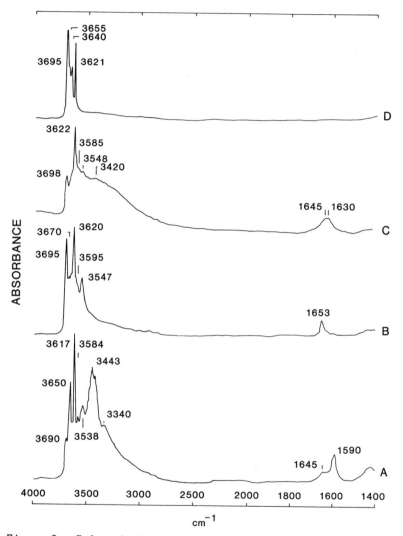

Figure 2.   Infrared absorption spectra of the 8.4A hydrate (A), the 8.6A hydrate (B), the 10A hydrate (C), and the original kaolinite used to synthesize the three hydrates (D).

10A hydrate spectrum with the other two hydrates shows that the
general features are similar; the 10A hydrate has two bands at 3585
and 3548 cm^{-1} from hole water vibration modes, and the high frequency bands above 3620 cm^{-1} have features which are grossly similar to
those in the 8.6A hydrate. The similarity in bands between 3600 and
3500 cm^{-1} for the 10A and 8.6A hydrates indicates the existence of
hole water in the former. Below 3500 cm^{-1}, the 10A hydrate, in
common with hydrated halloysites, shows a very broad absorption centered approximately at 3250 cm^{-1}. This feature is absent from the
two other hydrates and from the kaolinite. This absorption falls in
the range of liquid water and has been assigned to the extra water
introduced to expand the hydrated clay from 8.4/8.6A to 10A. Thus,
the infrared spectra along with the observed increase in layer
thickness from roughly 8.4 to 10 A, and the dehydration data (9,
27) indicate that when a dihydrate is formed, half the water (hole
water) is attached to the ditrigonal holes of the silicate surface
while the other half occupies other sites at a greater distance from
the silicate surface. Lacking a three dimensionally ordered 10A
hydrate, we cannot directly determine the location of the second
layer of water molecules. We do know that they must occupy sites
other than the ditrigonal holes because these are completely filled
by hole water (9). To distinguish the two types of water, the
non-hole water is termed "associated water".

Heat Capacity Measurements and Interlayer Water Structure. The heat
capacity of the interlayer water has been measured for the 10A,
8.6A, and 8.4A hydrates between 110 and 275 K (34) and provides an
important clue in determining the structure of the associated water
molecules. At all temperatures, the Cp for the interlayer water in
the 8.4A and 8.6A hydrates was not significantly different from
published values for ice. The water in the 10A hydrate deviated
from the ice values at roughly 160 K and the deviation increased as
the temperature rose. A sharp peak in the Cp began at roughly 240 K
and ended near the melting point of ice (Figure 3). The initial
departure of the Cp from ice values coincided with an increase in
the proton NMR signal (35). This behavior suggests that at very low
temperatures, the water molecules occupy relatively static positions
in the interlayer region. The ice-like Cp of the hole water
throughout the temperature range investigated is in agreement with
the supposition described earlier that the hole water is relatively
strongly bonded to the ditrigonal holes and remains fixed. The peak
in the Cp then must involve primarily the associated water molecules.

Examination of a projection of the tetrahedral surface of the
kaolin layer shows a pseudo-hexagonal arrangement of oxygen atoms
(Figure 4). When all the ditrigonal sites are occupied by hole
water (open circles in the figure), it can be seen that they are
approximately 5 A apart (Figure 4A). Hence, there is no possibility
of hydrogen bonding between hole water molecules in either the 8.4A
and 8.6A hydrates. The associated water molecules (filled circles
in the figure) can be added in two different ordered arrangements,
one of which is shown in Figure 4. In either arrangement, the distance between associated and hole water molecules is on the order of
3 A, a reasonable distance for hydrogen bonding to occur between
hole water and associated water. These hydrogen bonds from associ-

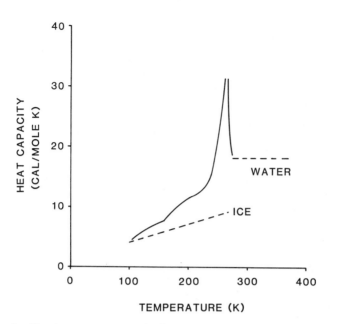

Figure 3. The heat capacity (Cp) for the water intercalated
between the layers of kaolinite in the 10Å hydrate. Standard
values for ice and liquid water are also shown. The heat capacity
of the intercalated water was measured using the procedure
described in Reference 2.

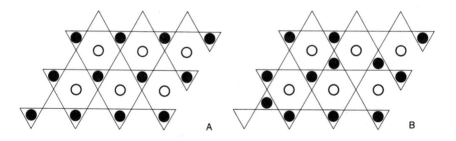

Figure 4. A schematic representation of the tetrahedral surface
of kaolinite (triangles) showing the position of the hole water
molecules (open circles) keying into the ditrigonal holes. The
associated water (filled circles in A) molecules are arranged in
an ordered pattern which exists at low temperatures. Disorder in
the associated water (filled circles in B) is created by increasing the temperature.

ated water to hole water are stronger than the bonds from the hole
water to the ditrigonal oxygens as shown by the lower infrared
absorption frequencies of the former. As the temperature rises, the
model suggests that a few associated water molecules become mobile.
Those which have sufficient kinetic energy can jump to the vacant
sites of the alternate configuration (as in Figure 4B). This disrupts the orderly hydrogen bond scheme which exists at lower temperatures and makes it easier for other associated molecules to jump.
As the temperature continues to rise, this jumping between the two
configurations, similar to melting, leads to the peak in Cp with a
maximum at about 270 K (Figure 3). In summary, the model suggests
that the water in direct contact with the mineral surface (hole
water) is strongly bonded to the silicate layer. The second layer
of water (associated water) behaves very differently because it has
few if any hydrogen bonds directly to the silicate layer.

Application of Results to other Silicate Minerals

Our model for the adsorption of water on silicates was developed for
a system with few if any interlayer cations. However, it strongly
resembles the model proposed by Mamy (13) for smectites with
monovalent interlayer cations. The presence of divalent interlayer cations, as shown by studies of smectites and vermiculites,
should result in a strong structuring of their primary hydration
sphere and probably the next nearest neighbor water molecules as
well. If the concentration of the divalent cations is low, then the
water in interlayer space between the divalent cations will correspond to the present model. On the other hand, if the concentration
of divalent cations approaches the number of ditrigonal sites, this
model will not be applicable. Such a situation would only be found
in concentrated electrolyte solutions.
      In discussing the applicability of the present model to sili-
cate minerals in general, there are two considerations: to what
extent do the exposed surfaces of a given silicate mineral mimic the
ditrigonal holes of the clay minerals, and, during chemical weathering, is the silicate mineral directly exposed to the aqueous phase
or is there an intervening phase of different structure and composition? The first point is fairly easily determined by inspection of
the crystal structures of the major silicate groups. Nesosilicates
(isolated tetrahedra) and the sorosilicates (double tetrahedra) have
little resemblance to the structure of the clay minerals, the ino-
silicates (single and double chains) which have corner shared tetrahedra are similar, particularly the double chains, phyllosilicates
(sheet structures) clearly are similar, and the tektosilicates may
or may not be similar, depending on the external surfaces exposed to
the aqueous phase. To the extent that the minerals are similar, one
would expect the model to apply. It is well known that dissolution
of silicates, particularly the compositionally complex ones such as
the feldspars, is often incongruent and that precipitated product
may coat the mineral surfaces. These coatings may be (compositionally or structurally) similar to clay minerals or related layer
structures (kaolinite, smectite, boehmite), and under these conditions, the model for adsorbed water may also be very good. Little
is known in detail about the amorphous phases which form at the
interface between silicate minerals and the aqueous phase, and the

model must be used with caution for minerals other than phyllosilicates.

Summary

An understanding of much of aqueous geochemistry requires an accurate description of the water-mineral interface. Water molecules in
contact with, or close to, the silicate surface are in a different
environment than molecules in bulk water, and it is generally agreed
that these adsorbed water molecules have different properties than
bulk water. Because this interfacial contact is so important, the
adsorbed water has been extensively studied. Specifically, two
major questions have been examined: 1) how do the properties of
surface adsorbed water differ from bulk water, and 2) to what distance is water perturbed by the silicate surface? These are difficult questions to answer because the interfacial region normally is
a very small portion of the water-mineral system. To increase the
proportion of surface to bulk, the expanding clay minerals, with
their large specific surface areas, have proved to be useful experimental materials.
   Based on the study of expanding clay minerals, two models of
water adsorbed on silicate surfaces have been proposed. One states
that only a few layers (<5) of water are perturbed by the silicate
surface, the other concludes that many layers (perhaps 10 times that
number) are involved. The complexity of the interactions which
occur between water molecules, surface adsorbed ions, and the atoms
of the silicate mineral make it very difficult to unequivocally
determine which is the correct view. Both models agree that the
first few water layers are most perturbed, yet neither has presented
a clear picture of the structure of the adsorbed water, nor is much
known about the bonding of the water molecules to the silicate
surface and to each other.
   Our approach has been to study a very simple clay-water system
in which the majority of the water present is adsorbed on the clay
surfaces. By appropriate chemical treatment, the clay mineral kaolinite will expand and incorporate water molecules between the layers, yielding an effective surface area of approximately 1000 m^2
g^{-1}. Synthetic kaolinite hydrates have several advantages compared
to the expanding clays, the smectites and vermiculites: they have
very few impurity ions in their structure, few, if any, interlayer
cations, the structure of the surfaces is reasonably well known, and
the majority of the water present is directly adsorbed on the kaolinite surfaces.
   By combining experimental results from infrared spectroscopy,
X-ray diffraction, thermal analysis, and heat capacity measurements,
a model for the structure of the water molecules adsorbed on the
kaolinite surfaces has been developed. Based on this structure,
reasonable inferences can be made about the bonding of the water
molecules and their behavior at different temperatures. The kaolinite layer is polar, having one surface formed by hydroxyls coordinating aluminum and the other by oxygens bonded tetrahedrally to silicon. The latter surface contains cavities which we refer to as
ditrigonal holes. It is into these cavities that the first intercalated water molecules fit. These water molecules (hole water in
our terminology) hydrogen bond, in a disordered manner, directly to

the oxygens forming the ditrigonal hole. The distances between the
hole water molecules are too great (about 5 A) for there to be any
direct bonding among them. In one of our synthetic hydrates (with
d(001) = 8.6 A), this is essentially the only type of water present.
Once the ditrigonal holes of the kaolinite layers are fully filled,
additional water occupies positions between the hole water molecules. Since the hole water keys into the ditrigonal holes, the
other water molecules (associated water in our terminology) are at a
slightly greater distance from the silicate surface. Complete filling of the interlayer region by hole and associated water occurs in
a synthetic hydrated kaolinite with d(001) = 10 A. The associated
water is not directly bonded to the oxygens of the silica tetrahedra; instead, it hydrogen bonds to other water molecules and to the
hydroxyl surfaces which form the other boundary of the interlayer
region.

At very low temperatures, the two types of water occupy ordered
positions similar to the oxygen positions in ice. Increasing the
temperature results in the associated water molecules occupying a
larger number of possible sites in an increasingly disordered
manner. The change from an ordered to a disordered arrangement is
reflected in a maximum in the heat capacity near 270 K. At room
temperature, the associated water molecules are relatively mobile,
much more mobile than the hole water. Dehydration at temperatures
just above ambient conditions results in nearly total loss of associated water and retention of the hole water. The very different
properties of the hole and associated water, which are at slightly
different distances from the mineral surface, suggest that the
perturbing effect of the silicate surface decreases very rapidly
with distance from the surface. If that is true, then it is difficult to see how the silicate surface could perturb water molecules
which are more than 10 A from the surface.

To the extent that the surfaces of the kaolinite layers resemble the surfaces of other silicate minerals, the structure of the
adsorbed hole and associated water can serve as a useful model. To
determine the applicability of our model to a specific mineral, it
will be necessary to know in some detail the structure of the external surfaces of that mineral.

Acknowledgments

The authors thank the National Science Foundation for generous support of this work through grants EAR8213888 and EAR8307685. Many of
the ideas presented here were developed during collaborative work
with Dr. Max Lipsicas.

Literature Cited

1. Pauling, L. Proc. Natn. Acad. Sci. 1930a, 16, 123-129.
2. Pauling, L. Proc. Natn. Acad. Sci. 1930b, 16, 578-582.
3. Grim, R. E. "Clay Mineralogy"; McGraw-Hill Book Co.: New York,
    1968.
4. Brindley, G. W.; Brown, G. "Crystal Structures of Clay Minerals
    and Their X-ray Identification" Mineralogical Society: London;
    1982, p 495

5. Fripiat, J. J.; Cases, J.; Francois, M.; Letellier, M. J. Colloid Int. Science 1982, 89, 378-400.
6. Low, P. F. Soil Sci. Soc. Am. J. 1979, 43, 651-658.
7. Sposito, G.; Prost, R. Chem. Reviews 1982, 82, 553-573.
8. Sposito, G. "The Surface Chemistry of Soils" Oxford Univ. Press: New York, 1984, p 234.
9. Hendricks, S. B.; Jefferson, M. E. Amer. Mineral. 1938, 23, 863-875.
10. Cruz, M.; Letellier, M.; Fripiat, J. J. J. Chem. Phys. 1978, 69, 2018-2027.
11. Costanzo, P. M.; Giese, R. F. Clays Clay Min. 1985, 33, 415-423.
12. Macey, H. H. Trans. Ceram. Soc. 1942, 41, 73-424.
13. Mamy, J. Ann. Agron. 1968, 19, 175-246.
14. de la Calle, C.; Pezerat, H.; Gasperin, M. J. de Physique 1977, 128-133.
15. Slade, P. G.; Stone, P. A.; Radoslovich, E. W. Clays Clay Min. 1985, 33, 51-61.
16. de la Calle, C.; Suquet, H.; Pezerat, H. Bull. Groupe fr Argiles 1975, 27, 31-49.
17. Poinsignon, C.; Estrade-Szwarckopf, H.; Dianoux, A. J. Int. Clay Conf.; Denver 1985, p 185.
18. Conard, J. Proc. Int. Clay Conf. 1975, 221-230.
19. Pons, C. H.; Tchoubar, C.; Tchoubar, D. Bull. Soc. Fr. Miner. Crist. 1980, 103, 452-456.
20. Fripiat, J. J.; Letellier, M.; Levitz, P. "Clay Minerals: Their Structure, Behaviour and Use" The Royal Society: London; 1984, 67-79.
21. Lubetkin, S. D.; Middleton, S. R.; Ottewill, R. H. "Clay Minerals: Their Structure, Behaviour and Use" The Royal Society: London; 1984, 133-148.
22. Nadeau, P. H.; Wilson, M. J.; McHardy, W. J.; Tait, J. M. Science 1984, 225, 923-925.
23. Giese, R. F. Clays & Clay Minerals 1973, 21, 145-149.
24. Giese, R. F. Bull. Mineral. 1982, 105, 417-424.
25. Wolfe, R.; Giese, R. F. Clays & Clay Minerals 1978, 26, 76-78.
26. Costanzo, P. M.; Clemency, C. V.; Giese, R. F. Clays & Clay Minerals 1980, 128, 155-6.
27. Costanzo, P. M.; Giese, R. F.; Lipsicas, M. Clays & Clay Minerals 1984, 32, 419-428.
28. Serratosa, J. M.; Bradley, W. F. J. Phys. Chem. 1958, 62, 1164-1167.
29. Rouxhet, P. G.; Samudacheata, Ngo; Jacobs, M.; Anton, O. Agronom. 1984, 4, 403-406.
30. Cruz-Complido, M.; Sow, G.; Fripiat, J. J. Bull. Min. 1982, 105, 493-498.
31. Kyogoku, Y. Nippon Kagaku Zasshi 1960, 81, 1648-1652.
32. Brun, G. Rev. Chim. Miner. 1968, 5, 899-934.
33. Giese, R. F.; Costanzo, P. M. Geol. Soc. Amer. Abstracts 1979, 11, 432.
34. Costanzo, P. M.; Giese, R. F.; Lipsicas, M. Nature 1982, 296, 549-51.
35. Lipsicas, M.; Straley, C.; Costanzo, P. M.; Giese, R. F. J. Colloid. Surface Sci. 1985, 107, 221-230.

RECEIVED June 3, 1986

4

Reactions at the Oxide–Solution Interface: Chemical and Electrostatic Models

John C. Westall

Department of Chemistry, Oregon State University, Corvallis, OR 97331

Surface complexation models for the oxide-electrolyte interface are reviewed: two models for surface hydrolysis reactions are considered (diprotic surface groups and monoprotic surface groups) and four models for the electric double layer (Helmholtz, Gouy-Chapman, Stern, and triple layer). Methods which have been used for determining thermodynamic constants from experimental data for surface hydrolysis reactions are examined critically. One method of linear extrapolation of the logarithm of the activity quotient to zero surface charge is shown to bias the values which are obtained for the intrinsic acidity constants of the diprotic surface groups. The advantages of a simple model based on monoprotic surface groups and a Stern model of the electric double layer are discussed. The model is physically plausible, and mathematically consistent with adsorption and surface potential data.

Any complete mechanistic description of chemical reactions at the oxide-aqueous electrolyte interface must include a description of the electrical double layer. While this fact has been recognized for years, a satisfactory description of the double layer at the oxide-electrolyte interface still does not exist.

Part of the difficulty of characterizing this interface stems from the fact that oxide surfaces, particularly those encountered in geochemistry, are very irregular; many different microcrystalline structures, which exhibit quite different chemical properties, are exposed to the solution. Thus examination of the surface by virtually any experimental method yields only averaged characteristics of the surface and the interface. Parsons (1) has discussed the surface chemistry of single crystals of pure metals, and has shown that the potential of zero charge of different crystal faces of the same pure metal can differ by over 400 mV. For an oxide surface, this difference would be energetically equivalent to a variation in the pH of zero protonic charge (pH_{zpc}) of more than six pH units. This example indicates that an observable macroscopic property of a polycrystalline surface might be the result of a combination of widely different microscopic properties, and that characterizations of these surfaces will remain somewhat operational in nature.

Another fundamental problem encountered in characterizing reactions at the oxide-electrolyte interface is the coupling between electrostatic and chemical interactions, which makes it difficult to distinguish the effects of one from the effects of the other. Westall and Hohl (2) have shown that many models for reactions at the oxide-electrolyte interface are indeterminate in this regard.

Many of the studies, from which our current understanding of reactions at the oxide-electrolyte interface has developed, were based on titrations of colloidal suspensions of oxides. The key to resolving questions left open by this work lies in the study of better defined oxide surfaces, the examination of a particular interface by many different experimental methods, and the development of mathematical methods for interpreting the data.

This paper is based primarily on the understanding of reactions at the oxide-electrolyte interface gained through the study of collodial suspensions of oxides. From the point of view of the geochemist, these suspensions of pure oxides are pristine systems, interesting, but perhaps of only marginal relevance to geochemistry. To the pure chemist, these systems are almost too ill-defined to warrant serious scientific consideration.

As an introduction to the discussion of electrical and chemical models for reactions at oxide-electrolyte interfaces, some reflections on the importance of these interfaces in geochemistry are presented.

The Oxide-Electrolyte Interface in Geochemistry

Empirical Models vs. Mechanistic Models. Experimental data on interactions at the oxide-electrolyte interface can be represented mathematically through two different approaches: (i) empirical models and (ii) mechanistic models. An empirical model is defined simply as a mathematical description of the experimental data, without any particular theoretical basis. For example, the general Freundlich isotherm is considered an empirical model by this definition. Mechanistic models refer to models based on thermo-dynamic concepts such as reactions described by mass action laws and material balance equations. The various surface complexation models discussed in this paper are considered mechanistic models.

Empirical models are often mathematically simpler than mechanistic models, and are suitable for characterizing sets of experimental data with a few adjustable parameters, or for interpolating between data points. On the other hand, mechanistic models contribute to an understanding of the chemistry at the interface, and are very often useful for describing data from complex multicomponent systems, for which the mathematical formulation (i.e., functional relationships) for an empirical model might not be obvious. Mechanistic models can also be used for interpolation and characterization of data sets in terms of a few adjustable parameters; however, mechanistic models are often mathematically more complicated than empirical relationships.

Requirements for a Mechanistic Model. A significant problem in the
use of mechanistic models for the description of the
oxide-electrolyte interface is the separation of observed energy of
interaction into electrostatic and chemical components. If the
separation of energy into these components is completely
indeterminate, the apparent mechanistic model may degenerate to an
empirical model, being of the correct mathematical form to represent
the data, but offering no insight into the chemical nature of the
interface.

      If there is so much difficulty in distinguishing electrostatic
and chemical components of energy, one could raise the question,
does it make any difference? In particular, for application to
surface chemical reactions occurring in geochemistry, would not an
empirical model be as good as a mechanistic one? To answer these
questions one must consider the types of data to be interpreted and
the purpose for which the model is to be used.

Electrostatic vs. Chemical Interactions in Surface Phenomena. There
are three phenomena to which these surface equilibrium models are
applied regularly: (i) adsorption reactions, (ii) electrokinetic
phenomena (e.g., colloid stability, electrophoretic mobility), and
(iii) chemical reactions at surfaces (precipitation, dissolution,
heterogeneous catalysis).

      The adsorption reactions considered in geochemical studies
include surface hydrolysis, adsorption of metal ions by surface
complexation, adsorption of anions by ligand exchange, and ion
association, as illustrated in Figure 1. The chemical energy of
such a reaction is often expressed as the equilibrium constant for
the reaction of a species in solution with hydroxyl groups on the
uncharged surface. The electrostatic interactions associated with
adsorption broaden the energy of the adsorption reaction from a
single discrete value to a distribution of values; however, it is
difficult to know how much of the broadening is attributable to the
variation in electrostatic energy associated with variation in
surface charge, and how much to the chemical heterogeneity of the
oxide surface. In principle, the study of adsorption in
electrolytes of different ionic strengths should allow the
electrostatic and chemical components of adsorption energy to be
distinguished. However, in practice, it is difficult to reach an
indisputable conclusion from these experiments, for reasons to be
discussed.

      Surface equilibrium models are also applied in the study of
electrokinetic phenomena, including colloid stability, which is of
great practical importance in geochemistry, and electrophoretic
mobility, which is convenient for experimentation. The
electrophoretic mobility of a particle is related to the
electrostatic potential at a mean "slipping plane". Where this
plane is located within the structural model of the interface is not
clear, although there has been a great deal of thought on the
subject (3,4). Furthermore, electrokinetic data yield no direct
information on surface charge. Thus, while data for electrokinetic
phenomena offer additional information on electrical energy at the
interface, these data alone cannot really resolve the problem of
separating electrical and chemical components of adsorption energy.

A third phenomenon, the kinetics of surface reactions such as dissolution, precipitation, and surface-catalyzed reactions, could be among the most sensitive indicators of chemical and electrostatic energies at the surface. Chemical energies are specific (i.e. short-range and highly dependent on molecular structure), while electrostatic energies are not. However, up to the present, only a very few (5) of the studies of these sorts of reactions at oxide surfaces have been cast explicitly in terms of electrochemical energies at the interface. As a better understanding of electrochemical potentials at the interface evolves, both electrical and chemical energies can be considered in studies of reaction kinetics at oxide surfaces. At some point in the future, it might even be possible to use data on the kinetics of these reactions to aid in separating interfacial energies into electrical and chemical components.

To return to the question of the necessity of distinguishing between chemical and electrostatic energies at interfaces in problems of geochemical interest, we reach the following conclusion: many phenomena, if considered separately, can be interpreted reasonably well without a clear separation of electrostatic and chemical energies. Part of the reason for our moderate satisfaction with this less-than-perfect result is that natural surfaces and interfaces are so heterogeneous that an explicit mathematical description of them is impossible. However, if different phenomena are considered simultaneously, it becomes not only possible, but also necessary, to distinguish between electrostatic and chemical energies.

Conceptual Model of an Adsorption Reaction

The nature of the problem in establishing a mechanistic model of the oxide-electrolyte interface, in which chemical and electrostatic energies are described explicitly, can be appreciated by consideration of the adsorption reaction depicted in Figure 2. The adsorption of a hydrogen ion from the bulk of a monovalent electrolyte is considered. The oxide-solution interface is divided conceptually into four regions: the bulk oxide (not shown in the figure), the oxide surface at which the adsorption reaction takes place, the solution part of the double layer containing the counterions, and the bulk of solution.

The total energy of this adsorption reaction can be found experimentally from the microscopic activity quotient, and separated theoretically into the following components: (1) transfer of the ion to be adsorbed from the bulk of solution to the oxide surface plane, at which the mean electrostatic potential is ψ_0 with respect to the bulk of solution; (2) reaction of the adsorbate in the surface plane with a functional group at the surface; (3) transfer of a fraction of the counter charge from solution into the solution part of the double layer by attraction of counter ions; and (4) transfer of the remainder of the counter charge by expulsion of co-ions from the solution part of the double layer to the solution.

In creating a mechanistic model of this adsorption reaction, it is necessary to separate the observed energy of interaction into components associated with each step. As can be inferred from Figure 2, the separation of one observable energy into four separate

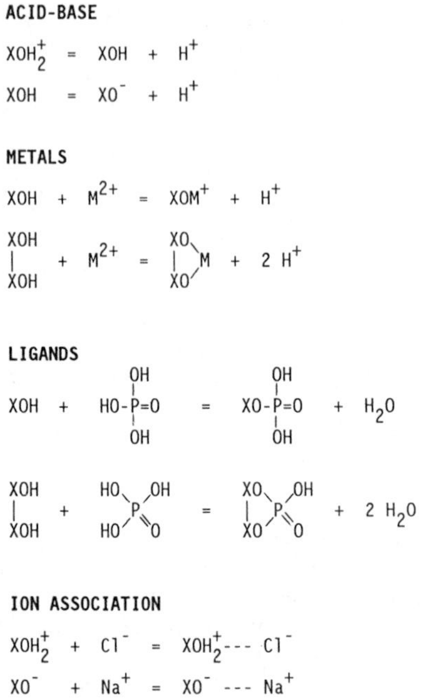

ACID-BASE

$$XOH_2^+ = XOH + H^+$$

$$XOH = XO^- + H^+$$

METALS

$$XOH + M^{2+} = XOM^+ + H^+$$

$$\begin{matrix} XOH \\ | \\ XOH \end{matrix} + M^{2+} = \begin{matrix} XO \\ \diagdown \\ XO \diagup \end{matrix} M + 2\ H^+$$

LIGANDS

$$XOH + \begin{matrix} OH \\ | \\ HO-P=O \\ | \\ OH \end{matrix} = \begin{matrix} OH \\ | \\ XO-P=O \\ | \\ OH \end{matrix} + H_2O$$

$$\begin{matrix} XOH \\ | \\ XOH \end{matrix} + \begin{matrix} HO \diagdown \ \diagup OH \\ P \\ HO \diagup \diagdown O \end{matrix} = \begin{matrix} XO \diagdown \ \diagup OH \\ P \\ XO \diagup \diagdown O \end{matrix} + 2\ H_2O$$

ION ASSOCIATION

$$XOH_2^+ + Cl^- = XOH_2^+ \text{---} Cl^-$$

$$XO^- + Na^+ = XO^- \text{---} Na^+$$

Figure 1. Representative adsorption reactions. The symbol XOH represents surface hydroxyl groups such as =FeOH, =AlOH, =SiOH, etc. Direct evidence for the structure of these surface species is generally lacking; the stoichiometries presented here are based on adsorption data.

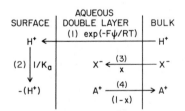

Figure 2. A complete adsorption reaction, including transfer of counter charge to the solution part of the electric double layer. While the reaction is usually divided conceptually into four steps, it is usually only the total energy of reaction that is observed experimentally.

components is bound to be indeterminate to some degree. The energy associated with Steps 1, 3, and 4 is generally found from electrostatic theory, and the chemical energy of Step 2 is found from the difference between the observed net energy and the calculated electrostatic energy. Thus an independent measurement of electrostatic energy is almost essential for a mechanistic characterization of the complete adsorption reaction.

Surface Complexation Models

Many models, which could be classified as "surface complexation models (6-8)," have been used to describe reactions at the oxide-solution interface. Although there are differences in the way these models are formulated, they all have two features in common: (i) the reactions of surface hydroxyl groups are described by conventional mass action and material balance equations, and (ii) the surface potential is related to surface charge by an electrostatic model of the electric double layer.

The approach to the mathematical definition of the interface model is very simple. For every layer in the interface, the charge is defined once as a function of chemical parameters and once as a function of electrostatic parameters. The functions for charge are set equal to each other and solved for the unknown electrochemical potentials. Mathematical techniques for solving the equations have been worked out and described in detail (9).

Following a review of the chemical and electrical fundamentals of these models, the interpretation of data in terms of a combined chemical-electrostatic model will be discussed. Detailed discussion of other aspects of these models can be found in recent reviews (2,6-8).

Although adsorption of many types of species could be considered, this discussion will focus on surface hydrolysis reactions, that is, adsorption of H^+ and OH^-. Virtually all surface hydrolysis experiments are carried out in the presence of a "background electrolyte," many of which appear to exhibit weak specific chemical interactions (e.g., ion-pair formation) with the surface (10-12). While consideration of these interactions is essential to a complete understanding of the interfacial chemistry, the topic is a subject in itself, and will not be considered in detail here. Treatment of these interactions is readily incorporated within the framework that is presented here.

Chemical Reactions at the Oxide Surface. The formation of surface hydroxyl groups on a hydrous oxide or hydroxide is depicted in Figure 3. Water adsorbs on the oxide surface, hydrolysis occurs and surface hydroxyl groups are formed. According to the figure, at least two chemically distinct types of hydroxyl groups are formed, those with interior and those with exterior metal ions. However, if different crystal faces, edges, steps, kinks, and various kinds of dislocations are considered, many more distinct groups could be imagined.

In order to maintain the complexity of the model at a level consistent with the resolution of the experimental data, the reactivity of these surface groups has been described by relatively simple models: (i) as diprotic weak acids, and (ii) as monoprotic

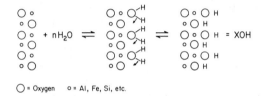

Figure 3. Surface hydrolysis and formation of surface hydroxyl groups.

weak acids. Almost all of the recent work has been interpreted in terms of the diprotic weak acid model; however, there are appealing features to the monoprotic weak acid model which have not yet been appreciated fully.

Diprotic Surface Groups. According to the diprotic model, the surface is represented as an ensemble of identical diprotic weak acid groups, which react according to:

$$XOH_2^+ = XOH + H^+ \qquad\qquad K_{a1} \qquad\qquad I$$

$$XOH = XO^- + H^+ \qquad\qquad K_{a2} \qquad\qquad II$$

The representation of surface groups as diprotic weak acids is appealing because it includes a modest degree of complexity (two acidity constants), allows convenient representation of the condition of zero surface excess of hydrogen ion, and is still quite manageable mathematically. However, it must be borne in mind that this model is still a grossly simplified representation of the actual surface. It remains to be shown that this simplification is significantly better than any other simplification.

Corresponding to Reactions I and II are the following mass action equations:

$$n_{XOH_2^+} \; K_{a1} \;=\; n_{XOH} \; a_{H^+} \; \exp(-e\psi_0/kT) \tag{1}$$

$$n_{XOH} \; K_{a2} \;=\; n_{XO^-} \; a_{H^+} \; \exp(-e\psi_0/kT) \tag{2}$$

where n_i represents the surface concentration (number of groups per square meter) of species i, a_i is the solution activity of species i, K_{a1} and K_{a2} are the surface acidity constants, and ψ_0 is the electrostatic potential, with respect to the bulk of solution, at the mean plane of adsorption. The exponential term represents the electrostatic energy required to bring a charged species from the bulk of solution to the plane of adsorption which is at potential ψ_0 (13). The definitions of all symbols are listed at the end of the text.

Activity coefficients appear in thermodynamic models to account for various energies that are not expressed explicitly in the model of the "ideal" case. The energy expressed by the exponential term in Equations 1 and 2 could be regarded as an activity correction; however, since the calculation of this correction is strongly coupled to the model for surface speciation itself, we do not refer to the exponential term as an activity correction, but as an integral part of the model. (In contrast, the Debye-Hueckel activity coefficient for an ion in solution is only weakly coupled to the solution speciation, and can be treated conveniently as an independent activity coefficient.) Activity corrections associated with the energy of lateral interactions among adsorbed species will be neglected, primarily on account of the practical difficulty of determining what they are. The topic of surface activity coefficients is discussed by Chan et al. (14) and Sposito (15).

The material balance equation for surface hydroxyl groups is of the form

$$N_s = n_{XOH_2^+} + n_{XOH} + n_{XO^-} \tag{3}$$

where N_s is the surface density of reactive surface sites. The value of N_s can be determined either by surface titration experiments, or by independent means such as negative adsorption, tritium exchange, or consideration of surface area and crystal structure ([2]). The mass action and material balance equations used in the surface complexation model are the same as those on which the Langmuir isotherm is based ([16]).

In addition to the material balance equation for the surface hydroxyl groups listed above, a charge balance equation can be written:

$$\sigma_0^H = e \ (n_{XOH_2^+} - n_{XO^-}) \tag{4}$$

The superscript H on the surface charge σ_0 indicates that the surface charge has been calculated exclusively from the surface concentration of H^+ ions; the assignment of adsorbed ions to "mean planes of adsorption" is subject to different interpretations, as will be discussed later.

The surface complexation model described by Equations 1-4 includes five unknowns (n_X, n_{X^+}, n_{X^-}, σ_0, ψ_0) and three parameters (K_{a1}, K_{a2}, N_s), if the experimental pH is known. (The subscripts X, X^+, and X^- are used as abbreviations for XOH, XOH_2^+, and XO^-.) It remains to develop one more equation based on an electrostatic model of the electric double layer to relate σ_0 and ψ_0.

$$\sigma_0 = f(\psi_0, \ldots) \tag{5}$$

Following the presentation of the chemical model with monoprotic surface groups, four electrostatic models will be developed, from which the necessary relationship between ψ_0 and σ_0 can be found.

Monoprotic Surface Groups. The monoprotic surface group model presented here has been discussed elsewhere ([17],[18]), but has not yet been applied widely. With this model it is possible to represent both a positive charge and a negative charge on the surface; in this respect it is different from other monoprotic surface group models, such as one discussed by Healy and White ([19]), with which it is not possible to represent the transition between positive and negative surface charge.

The monoprotic surface group model can be described in terms of the more familiar diprotic surface group model. In the diprotic model, the surface is thought of as an ensemble of N_s diprotic surface groups, which, under the condition of zero protonic charge, are occupied by N_s protons.

In the monoprotic model, the surface is thought of as an ensemble of 2 N_s monoprotic surface groups, which, under the condition of zero protonic charge, are occupied by N_s protons.

The protonation and deprotonation of these monoprotic groups can be represented formally as the reaction:

$$ZOH^{+\frac{1}{2}} = ZO^{-\frac{1}{2}} + H^{+} \qquad K_a \qquad \text{III}$$

with the mass action equation:

$$n_{ZOH^{+\frac{1}{2}}} K_a = n_{ZO^{-\frac{1}{2}}} a_{H^{+}} \exp(-e\psi_0/kT) \qquad (6)$$

with surface site balance:

$$2 N_s = n_{ZOH^{+\frac{1}{2}}} + n_{ZO^{-\frac{1}{2}}} \qquad (7)$$

and with charge density:

$$\sigma_0^H = e \ (n_{ZOH^{+\frac{1}{2}}} - n_{ZO^{-\frac{1}{2}}}) \qquad (8)$$

The association of the formal charge $\frac{1}{2}$ with each surface group is simply a bookkeeping device. No information about the actual distribution of charge among neighboring ions or groups on the surface is implied. The surface is still thought of as an ensemble of $2 N_s$ surface groups occupied by $N_s + n$ protons when the surface charge is $\sigma_0 = en$. An alternative derivation of this model [20] shows how the behavior of the ensemble of diprotic acid groups resembles the behavior of an ensemble of monoprotic acid groups when $K_{a1} \ll K_{a2}$.

The monoprotic model described by Equations 6-8 includes four unknowns ($n_{ZOH^{+\frac{1}{2}}}$, $n_{ZO^{-\frac{1}{2}}}$, σ_0^H, and ψ_0) and two adjustable parameters (K_a and N_s) if the experimental pH is known. To complete the definition of the monoprotic model an equation similar to Equation 5 is required to relate surface charge to surface potential. The electrostatic models used with the monoprotic model are the same as those used with the diprotic model, and are described in the following section.

Other similarities exist between the two models: reactions for adsorption of other species are written in the same way, and the same mathematical techniques [9] can be used to solve the equations.

Westall and Hohl [2], have observed that it is possible to represent experimental data with a diprotic model in which $K_{a1} = K_{a2}$, or even $K_{a1} < K_{a2}$. Such models are mathematically similar to the monoprotic model, since the diprotic model becomes similar to a monoprotic model if the acidity constants in the diprotic acid model are such that the neutral XOH group is insignificant in the material balance equations.

The monoprotic model is appealing since it is very simple, realistic, and based on one less adjustable parameter than the diprotic model. The value of the parameter K_a can be found directly from the H^{+} concentration in the bulk of solution at $\sigma_0^H = 0$, since $K_a = a_{H^{+}}$ at this condition, according to Equation 6. Since the surface complexation models are already recognized as being underdetermined, any physically realistic model with fewer adjustable parameters is welcomed.

Electrostatic Models for the Electric Double Layer

To complement the models for the surface reactions, a model for the electric double layer is needed. Current models for the electric double layer are based on the work of Stern (21), who viewed the interface as a series of planes or layers, into which species were adsorbed by chemical and electrical forces. A detailed discussion of the application of these models to oxide surfaces is given by Westall and Hohl (2).

The structure of the interface according to the Stern model and several limiting-case approximations is presented in Figure 4. The electrostatic models of the interface will be introduced in terms of the most complete one, the triple layer model (Figure 4a). Then the relationship of the triple layer model to the simplified models in Figures 4b-d will be discussed.

The Triple Layer Model and the Stern Model. The ions most intimately associated with the surface are assigned to the innermost plane where they contribute to the charge σ_0 and experience the potential ψ_0. These ions are generally referred to as primary potential determining ions. For oxide surfaces, the ions H^+ and OH^- are usually assigned to this innermost plane. In Stern's original model, the surface of a metal electrode was considered, and the charge σ_0 was due to electrons.

Beyond the surface plane is a layer of ions attracted to the surface by specific chemical interactions. The locus of the center of these ions is known as the inner Helmholtz plane (IHP). The charge in this plane, which results from the specifically adsorbed ions is denoted by σ_1, and the electrostatic potential at the IHP by ψ_1. The species usually assigned to this plane include chemically adsorbed metals and ligands, as well as weakly adsorbed electrolyte ions. In the "site-binding model," Yates et al. (22) considered the weakly bound electrolyte ions to be bound pair-wise with oppositely charged groups in the surface plane.

Beyond the IHP is a layer of charge bound at the surface by electrostatic forces only. This layer is known as the diffuse layer, or the Gouy-Chapman layer. The innermost plane of the diffuse layer is known as the outer Helmholtz plane (OHP). The relationship between the charge in the diffuse layer, σ_2, the electrolyte concentration in the bulk of solution, c_b, and potential at the OHP, ψ_2, can be found from solving the Poisson-Boltzmann equation with appropriate boundary conditions (for 1:1 electrolytes (13))

$$\sigma_2 = - (8\epsilon\epsilon_0 RT\ c_b)^{\frac{1}{2}} \quad \sinh(F\psi_2/2RT) \tag{9}$$

The region between the surface plane and the IHP, and the region between the IHP and the OHP are considered to behave electrostatically as parallel plate capacitors, with charge related to potential by the capacitances C_1 and C_2:

$$\sigma_0 = C_1\ (\psi_0 - \psi_1) \tag{10}$$

$$\sigma_1 = C_2\ (\psi_1 - \psi_2) \tag{11}$$

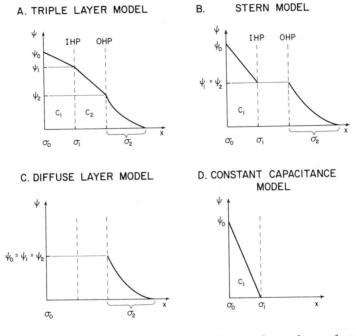

Figure 4. Electrostatic models for the surface-electrolyte
solution interface. These models were conceived for metal
surfaces but have been used for oxide surfaces as well.

Helmholtz had proposed such a parallel plate capacitance model for
the entire interface in 1853.

While Stern recognized that formally one should include the
capacitance between the IHP and OHP in the interface model, he
concluded that the error introduced into the electrical properties
predicted for the interface would usually be small if the second
capacitance were neglected, and ψ_2 were set equal to ψ_1.
Stern used this simplification in his calculations. The simplified
model with only one Helmholtz capacitance is commonly referred to as
the Stern model (Figure 4b), while the "extended" Stern model
(Figure 4a) is designated the triple layer model.

The final equation to complete the electrostatic constraints on
the interface is the electroneutrality equation,

$$\sigma_0 + \sigma_1 + \sigma_2 = 0 \tag{12}$$

Equations 10-12 allow the charges σ_0, σ_1, and σ_2 in the
three regions to be calculated in terms of the three electrostatic
potentials ψ_0, ψ_1, ψ_2.

Incorporation of Chemical Constraints. To complete the model of the
oxide-electrolyte interface, it is necessary to return to the
preceding section on models of chemical reactions to find values of
σ_0, σ_1, and σ_2 calculated from chemical considerations.
Combination of the electrostatic equations with the chemical
equations then completely defines the interface.

The chemical model for charge in the surface plane (σ_0) was
given by Equations 1-4 for the diprotic model or Equations 6-8 for
the monoprotic model. In general, either of these sets of equations
can be represented by

$$\sigma_0 = e \sum_i z_i n_i^0 = g\ (\psi_0,\ a_j,\ \dots) \tag{13}$$

For adsorption at the IHP, mass action and material balance
equations could be set up for specifically adsorbed ions yielding
equations similar to those for the surface plane

$$\sigma_1 = e \sum_i z_i n_i^1 = g\ (\psi_1,\ a_j,\ \dots) \tag{14}$$

In the absence of specific adsorption at the IHP, σ_1 is simply
set equal to zero. The chemical constraint for σ_2 is given
directly by Equation 9.

Equations 9-14 provide the framework for combining either of
the two surface hydrolysis models that were presented with any of
the four electric double layer models to define the interface model
completely and to solve for all unknown potentials, charges, and
surface concentrations. In the following section some specific
limiting cases are considered.

Limiting Cases of the Stern Model. The electrostatic energies
through the interface have been formulated in terms of
capacitances. For the basic Stern model, the total capacitance of

the interface, C_T, is given by the equation for two capacitances in series:

$$C_T^{-1} = C_1^{-1} + C_d^{-1} \qquad (15)$$

where C_1 is the Helmholtz (compact) layer capacitance and C_d is the diffuse layer capacitance found by differentiating Equation 9 with respect to ψ_2. At low ionic strengths and relatively low potentials, the diffuse layer capacitance is much less than the compact layer capacitance; then the total capacitance is approximately equal to the diffuse layer capacitance. At high ionic strengths and high potentials, the diffuse layer capacitance is much larger than the compact layer capacitance and the total capacitance is approximately equal to the compact layer capacitance. In other words, the value of the variable diffuse layer capacitance relative to the constant compact layer capacitance determines whether the total capacitance of the interface appears more like a diffuse layer capacitance (Figure 4c) or more like a Helmholtz constant capacitance (Figure 4d).

<u>Application of Electric Double Layer Models</u>. Now the relationships among the four electrical double layer models described in Figure 4 can be summarized. The Helmholtz constant capacitance model (23-24) may be regarded as the "high ionic strength" limiting case of the basic Stern model described above. In contrast, the diffuse layer model (25-26) may be regarded as the "low ionic strength, low potential" limiting case of the basic Stern model. The basic Stern model (27) is applicable at all ionic strengths and represents a minor simplification with regard to the outer Helmholtz capacitance of the triple layer model (22,28-29).

With regard to the assignment of ions to the mean planes in the interface, the oxide-solution interface can be compared to two ideal interfaces which are more thoroughly characterized: the metal-solution interface and the silver iodide-solution interface.

As mentioned before, Stern had a metal electrode in mind when he described the surface-solution interface; then σ_0 referred to the electronic charge on the surface of the metal itself, σ_1 to the charge formed by electrostatically (or chemically) bound electrolyte ions at the IHP, and σ_2 to the charge in the diffuse layer. In the case of silver iodide, the surface charge σ_0 is assumed to be made up of the adsorbed "potential determining ions" Ag^+ and I^-; σ_1 and σ_2 are as described for the metal-electrolyte interface.

Oxide surfaces have usually been regarded as being similar to the AgI surface: the adsorbed "potential determining ions," H^+ and OH^-, form the charge σ_0, and σ_1 and σ_2 are as described for the other interfaces. However, other interpretations are possible, in which the primary potential determining ions, H^+ and OH^-, and electrostatically adsorbed electrolyte ions occupy the same interfacial layer.

As stated previously, it is difficult to assign ions to a few discrete "mean planes of adsorption" and have these mean planes correspond in every case to the location of ions expected from hypothetical structure and bonding at the hydrous oxide surface.

The "AgI" model for location of ions in the oxide-water interface is probably the best reasonably simple interpretation.

One other issue arises with respect to potential determining ions. In the case of AgI, the potential difference at the surface-solution interface varies with the activity of Ag^+ or I^- in solution according to the Nernst equation,

$$\Delta\psi = \frac{RT}{z_i F} \ln a_i \tag{16}$$

since the activities of Ag^+ and I^- at the surface are fixed by the presence of these ions in the bulk of the solid (20, 30).

At oxide surfaces, the surface activities of H^+ and OH^- are not fixed in a similar way. Then the variation in surface potential with solution activity of H^+ depends on the chemical and electrostatic properties of the interface. For the many oxides that are insulators, it is much more difficult to obtain a measurement of the surface-solution potential differences than it is for conductors such as AgI. Thus there is uncertainty whether the dependence of surface potential on pH is approximately Nernstian or significantly sub-Nernstian.

In the following section, it is shown that mathematical methods which have been used to interpret adsorption data bias the interpretation towards chemical and electrostatic properties which lead to a significantly sub-Nernstian response; this bias arises out of the need for mathematical simplifications, not from physical considerations.

Interpretation of Data

Two models of surface hydrolysis reactions and four models of the electrical double layer have been discussed. In this section two examples will be discussed: the diprotic surface group model with constant capacitance electric double layer model and the monoprotic surface group model with a Stern double layer model. More details on the derivation of equations used in this section are found elsewhere (31).

Diprotic Surface Groups. Most of the recent research on surface hydrolysis reactions has been interpreted in terms of the diprotic surface hydrolysis model with either the triple layer model or the constant capacitance model of the electric double layer. The example presented here is cast in terms of the constant capacitance model, but the conclusions which are drawn apply for the triple layer model as well.

As discussed in the introduction, it is difficult to separate the observed energy of interaction at the interface into electrical and chemical components. In addition, the necessity to consider both components of energy simultaneously complicates the reduction of the model to a simple linear form for determining thermodynamic constants from the data.

Three methods have been used for interpretation of surface hydrolysis data, the first two of which involve approximations and reduction to linear form: Method I: assumption that on the acidic branch of the titration curve the concentration of XO^- groups is negligible, and on the basic branch, the concentration of XOH_2^+ groups is negligible; Method II: linear expansion of the exponential term in the mass action equations, and rearrangement of the equations to a linear form; Method III: nonlinear least squares optimization procedure, in which no assumptions are made for computational convenience. It will be shown that the approximation of Method I introduces an assumption about the separation between electrostatic and chemical energy, and an artificial physical constraint on the system.

Method I. To illustrate the application of these methods, the experimental data for the adsorption of H^+/OH^- on TiO_2 (32) are considered (Figure 5). Use of Method I involves the approximations in the material balance equations that, on the acidic branch of the titration curve,

$$\text{if } n_{XOH}, n_{XOH_2^+} \gg n_{XO^-} \text{ then } n_{XOH_2^+} \approx \sigma_0^H / e \qquad (17)$$

and on the basic branch,

$$\text{if } n_{XOH}, n_{XO^-} \gg n_{XOH_2^+} \text{ then } n_{XO^-} \approx -\sigma_0^H / e \qquad (18)$$

Then the material balance equations can be simplified to allow the reaction quotients Q_{a1} and Q_{a2} to be approximated from the experimentally observed protonic surface charge σ_0^H:

$$Q_{a1} = \frac{(n_{XOH})\,(a_{H^+})}{(n_{XOH_2^+})} \approx \frac{(N_s - n_{XOH_2^+})\,(a_{H^+})}{(n_{XOH_2^+})} \qquad (19)$$

$$Q_{a2} = \frac{(n_{XO^-})\,(a_{H^+})}{(n_{XOH})} \approx \frac{(n_{XO^-})\,(a_{H^+})}{(N_s - n_{XO^-})} \qquad (20)$$

The intrinsic acidity constants K_{a1} and K_{a2} can then be determined from Q_{a1} and Q_{a2} as follows. Combination of Equations 19 and 20 with Equations 1 and 2 show that the acidity constants are related to the acidity quotients by

$$Q_{ai} = K_{ai} \exp(e\psi_0/kT) \qquad (21)$$

or, in logarithmic form, with the constant capacitance constraint $\sigma_0^H = C\,\psi_0$ substituted for ψ_0,

$$\log Q_{ai} = \log K_{ai} + \frac{F}{C\,RT\,\ln 10}\,\sigma_0^H \qquad (22)$$

Then a plot of $\log Q_{ai}$ vs. σ_0^H yields the value of $\log K_{ai}$ as the intercept, and the value of the capacitance is related to the slope (Figure 6).

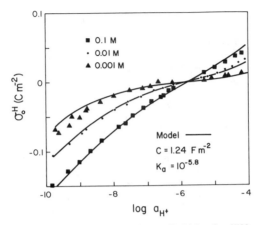

Figure 5. Titration of a suspension of TiO_2 in KNO_3 at three different concentrations. σ_0^H is the protonic surface charge, calculated from the acid-base mole balance equations, the specific surface area of 20 m^2 g^{-1}, and the solids concentration of 12 g L^{-1}. The lines are calculated from the monoprotic surface group model with Stern double layer model: N_s = 12 sites nm^{-2}, C = 1.23 F m^{-2}, log a_H^Z = -5.8. The experimental data is from Yates and Healy (32).

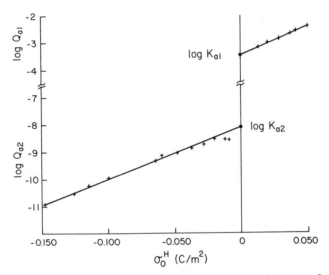

Figure 6. Use of Method I, Equation 22, to calculate surface acidity constants with N_s = 12 sites nm^{-2}: log K_{a1} = -3.5 and log K_{a2} = -8.1 or, in other terms, pH_{zpc} = 5.8 and log K_d = -2.3. Capacitances were determined from the slopes according to Equation 22: acid branch, 0.77 F m^{-2}; base branch, 0.89 F m^{-2}. Data are from Figure 5, TiO_2 in 0.1 M KNO_3 (32).

An extension of this method has been developed for the triple layer model which allows data obtained at several values of ionic strength to be considered simultaneously ($\underline{7}$, $\underline{33}$). However, this "double extrapolation technique" involves the same sort of approximation.

An example of the use of this method with the constant capacitance model on the data for TiO_2 in 0.1 M KNO_3 is illustrated in Figure 6. It appears from the figure that the problem is perfectly well determined, and that unique values of K_{a1} and K_{a2} can be determined. However, as is shown below, the values of K_{a1} and K_{a2} determined by this method are biased to fulfill the approximations made in processing the data: (i) on the acidic branch, n_X^+, $n_X \gg n_{X^-}$, which yields a small value for K_{a2}, and (ii) on the basic branch, n_{X^-}, $n_X \gg n_X^+$, which yields a large value of K_{a1}. Thus the approximation used to find values for Q_{a1} and Q_{a2} leads to values of K_{a1} and K_{a2} consistent with the approximation of a large domain of predominance of the XOH group. This constraint arose out of the need for mathematical simplicity, not out of any physical considerations.

Method II. Some insight into the significance of this constraint can be seen by consideration of other methods of interpreting the data. In Method II the linear region of the σ_0^H vs. log a_H^+ curve near the pH_{zpc} is considered (Figure 5).

Three quantities are defined for convenience. (i) A surface dissociation constant is defined for the reaction:

$$XOH = \frac{1}{2} XOH_2^+ + \frac{1}{2} XO^- \qquad K_d = (K_{a2}/K_{a1})^{\frac{1}{2}} \qquad (23)$$

The value of K_d corresponding to values of K_{a1} and K_{a2} determined from Method I would be relatively small, reflecting the predominance of the neutral XOH group. (ii) The hydrogen ion activity in solution that yields zero protonic surface charge is defined from a combination of Equations 1 and 2 with n_X^+ set equal to n_{X^-}:

$$a_H^z{}^+ = (K_{a1} K_{a2})^{\frac{1}{2}} \qquad (24)$$

(iii) The function α, which is related to the deviation of the surface potential from the value predicted from the Nernst equation, is defined by:

$$\alpha = \ln \frac{a_H^+}{a_H^z{}^+} - \frac{F\psi_0}{RT} \qquad (25)$$

Then Equations 23-25 can be combined with Equations 1-4 to yield the protonic surface charge as a function of α:

$$\sigma_0^H = \frac{2 K_d e N_s \sinh(\alpha)}{1 + 2 K_d \cosh(\alpha)} \qquad (26)$$

Equation 26 was derived directly and contains no mathematical
approximations. A similar equation was derived by Healy et al.
(30). If the hyperbolic functions are approximated by a linear
expansion, Equation 26 can be approximated

$$\sigma_0^H \approx \frac{2 \, K_d \, e \, N_s \, \alpha}{1 + 2 \, K_d} \tag{27}$$

This linear approximation is certainly valid about the pH_{zpc}, and
valid over a much wider domain if the surface potential-pH
relationship is nearly Nernstian, that is, if the value of α
defined in Equation 25 remains small over a wide domain.

The parameters in the chemical and electrostatic models which
appear in Equation 27 can be related to the experimentally
observable slope of the σ_0^H vs. log a_{H^+} curve by combining
Equation 27 with Equation 25 and the constant capacitance constraint
$\sigma_0^H = C \, \psi_0$ to yield:

$$\sigma_0^H \approx \beta \, \ln \frac{a_{H^+}}{a_{H^+}^z} \tag{28}$$

with the factor β, which is the slope of the σ_0 vs. ln a_{H^+}
curve,

$$\beta = \frac{1 + 2 \, K_d}{2 \, K_d \, e \, N_s} + \frac{F}{C \, RT}^{-1} \tag{29}$$

Equation 28 is certainly valid about the pH_{zpc}, and over a wider
range if α remains small.

The value of the proportionality constant β can be determined
experimentally. In the absence of any other experimental data, an
acceptable model would be based on any combination of the adjustable
parameters C, K_d, and N_s that yields the correct value of β,
according to Equation 29. Since three adjustable parameters are
available to define the value of one experimentally observable
quantity, covariability among these parameters is expected. In
reality an independent estimate of N_s might be available, and
curvature of the σ_0 vs. log a_{H^+} plot might reduce some of the
covariability, but Equation 29 provides an initial step in
understanding the relationship between covarying adjustable
parameters.

The value of β determined from Figure 5 is 0.010 C m^{-2}.
Combinations of values of the parameters C and log K_d, (with N_s
set to 12.04 sites nm^{-2}) that yield the observed value of β
through Equation 29 are given by the smooth curve in Figure 7. The
values of C and log K_d are strictly appropriate to describe the
data in a limited domain about pH = pH_{zpc}. The extent of this
domain can be evaluated by determining values of K_d and C that are
appropriate to represent the entire dataset, as can be done with
Method III.

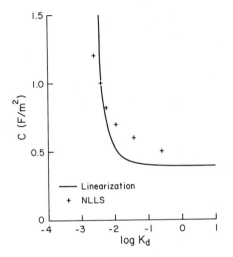

Figure 7. Covariability between values of C and K_d yielding
best fit of diprotic surface hydrolysis model with constant
capacitance model to titration data for TiO_2 in 0.1 M KNO_3
(Figure 5). The line is consistent with Equation 29. The
crosses represent values of C and log K_d found from a nonlinear
least squares (NLLS) fit of the model to the data, with the value
of capacitance imposed; in all cases the fit was quite
acceptable. The values of K_d and C found by Method I (Figure
6) also fall near the line consistent with Equation 29. The
agreement between these results supports the use of the
linearized model (Equation 29) for developing an intuitive feel
for surface reactions.

Method III. The weighted nonlinear least squares parameter
optimization procedure (34, 35) was applied to the entire set of
points shown in Figure 5 for 0.1 M KNO_3. The value of C was fixed
and the optimal values of K_d and $a_H{}^z$ were obtained. The model
with diprotic surface groups and constant capacitance electric
double layer model was used with no mathematical simplifications.

The pairs of log K_d and C that yielded quite acceptable
representations of the data are shown by the symbols on Figure 7.
The agreement between the symbols and the smooth curve in Figure 7
support the validity of the approach of Method II, and the insight
that is gained from this approach.

Comparison of Methods. The use of Method I for interpretation of
data has resulted in the conclusions that, for a great number of
oxides, the value of K_d is small, the relationship between ψ_0
and ln a_{H^+} is significantly sub-Nernstian, and the value of C is
relatively large.

Here it has been shown that the conclusion about K_d is
related to the mathematical approximation used in interpreting the
data by way of Method I; it is easy to show that the second and
third conclusions are also dependent on the initial assumptions in
Method I. Rearrangement of Equation 28 and substitution of the
constant capacitance constraint yields a relationship between ψ_0
and ln a_{H^+}:

$$\psi_0 = \frac{\beta}{C} \ln \frac{a_{H^+}}{a_{H^+}^z} \tag{30}$$

The value of the slope of ψ_0 vs. ln a_H depends on the relative
magnitude of the two terms in the definition of β in Equation 29.
If the second term on the right-hand side of Equation 29 is
dominant, then Equation 30 yields a Nernstian slope; if the first
term on the right-hand side of Equation 29 is significant, then the
slope given by Equation 30 will be sub-Nernstian.

It is clear that small values of K_d, as are obtained from
Method I, favor the predominance of the first term over the second,
and a sub-Nernstian slope; furthermore, if K_d is constrained to be
small, and the proportionality constant β is to be consistent with
the experimental value, then it is necessary that C is large.

Direct measurement of the change in interfacial potential
difference at the oxide-electrolyte interface with change in pH of
solution can be measured with semiconductor or semiconductor-oxide
electrodes. These measurements have shown d ψ_0/d log a_{H^+}
approaching 59 mV for TiO_2 (36, 37). These values are
inconsistent with the highly sub-Nernstian values predicted from the
models with small values of K_d. (Similar studies (38,39) have
been performed with other oxides of geochemical interest. Oxides of
aluminum have yielded a value of d ψ_0/d log a_{H^+} greater than
50 mV, while some oxides of silicon have yielded lower values.)

The discussion above pertains to the diprotic acid chemical
model and the constant capacitance electrostatic model. It is
interesting to note that in some applications of the triple layer
model with site binding of electrolyte ions at the IHP, the

ionization of the surface hydroxyl groups is effectively dominated by the ion pairs. Then the effective value of K_d for this surface is much larger, and the model is more consistent with the surface potential measurements discussed above, than would be expected from the small value of K_d calculated directly from K_{a1} and K_{a2}, without consideration of site-binding of counter ions.

If the constraint that K_d is small is removed in interpretation of the data, one may consider the physical nature of the interface and other forms of experimental data in deciding what combinations of parameters are appropriate for describing the interface. In particular, the relationship of the diprotic surface group model to the monoprotic surface group model can be examined.

Monoprotic Surface Groups. It has been stated that for some oxides the surface potential response to pH is close to the Nernstian value, and that relatively large values of K_d in the model were necessary to be consistent with this observation. For $K_d = 1$, the acidity constants K_{a1} and K_{a2} are equal. If these acidity constants are equal, it could be argued that the groups are indistinguishable, and instead of representing the surface as an ensemble of N_s diprotic acid groups, it would be as appropriate to represent it as an ensemble of $2 N_s$ monoprotic acid groups.

How realistic is such a model? The real difference between the two chemical models lies in the number of nearest-neighbor interactions that influence the acidity of a particular surface hydroxyl group. In the diprotic model, the acidity of a particular surface hydroxyl group depends on the state of one nearest neighbor: if the nearest neighbor is protonated, the energy associated with the loss of a proton is related to K_{a1}, and if the nearest neighbor is deprotonated, the energy is related to K_{a2}. In the monoprotic model, zero nearest neighbor interactions are considered, and the energy associated with the loss of a proton is related to K_a for all groups.

It is apparent that consideration of only one nearest-neighbor interaction is a greatly simplified way to represent a surface; however, it is not apparent that consideration of no nearest-neighbor interactions is a significantly worse way to represent the surface.

The representation of the data for TiO_2 in terms of the monoprotic surface group model of the oxide surface and the basic Stern model of the electric double layer is shown in Figure 5. It is seen that there is good agreement between the model and the adsorption data; furthermore, the computed potential ψ_0 (not shown in the figure) is almost Nernstian, as is observed experimentally.

Concluding Remarks

In addition to the overview of models that are used for adsorption at the oxide-electrolyte interface, examples for the application of these models were discussed. It has been stated that there is a great deal of uncertainty associated with models of the oxide-electrolyte interface, and, in the opinion of the author, it is better to cast uncertainty in terms of a simple model than in terms of a complex model.

It was shown that a method which has been used for determining acidity constants for the diprotic surface groups biases the result in favor of $K_{a1} \gg K_{a2}$; the arguments presented here do not prove the contrary, they simply affirm that the question is still open. If the prejudgment that $K_{a1} \gg K_{a2}$ is dropped, then the model with monoprotic surface groups becomes a logical special case of the model with diprotic surface groups, and, if simpler is better, the model with monoprotic surface groups warrants further consideration.

One other caveat concerning the approach used here must be made. This discussion, and the studies to which it relates, are based on some version of the Stern model for the oxide-electrolyte interface. Oxide surfaces are rough and heterogeneous. Even for the mercury-electrolyte interface, or single crystal metal-electrolyte interfaces, the success of some form of the Stern model has been less than satisfactory. It is important to bear in mind the operational nature of these models and not to attach too much significance to the physical picture of the planar interface.

Acknowledgments

The author acknowledges the Department of Energy (through Battelle Pacific Northwest Laboratories Contract No. B-N8267-A-H) for partial support of this work, and Dr. Gerhard Furrer for review of the manuscript.

Legend of Symbols

a_i	solution activity of species i	-
a_H^z	solution activity of H^+ when $\sigma_0^H = 0$	-
c_b	concentration of monovalent electrolyte in bulk of solution	$mol\ m^{-3}$
C_i	specific capacitance of region i	$F\ m^{-2}$
e	elementary charge	C
F	Faraday constant	$C\ mol^{-1}$
k	Boltzmann constant	$J\ mol^{-1}$
K_a	surface acidity constant	-
K_d	surface dissociation constant	-
n_i	surface concentration of species i	m^{-2}
N_s	surface site concentration	m^{-2}
Q_a	acidity quotient (charge dependent)	-
R	gas constant	$J\ mol^{-1}\ K^{-1}$
T	temperature	K
XOH	surface hydroxyl group (diprotic)	-
z_i	integral charge on species i	-
ZOH	surface hydroxyl group (monoprotic)	-

α	defined in Equation 25	-
β	slope of σ_0 vs. ln a_{H^+} plot	$C\ m^{-2}$
ϵ	dielectric constant	-
ϵ_0	permitivity of free space	$F\ m^{-1}$
σ_i	specific surface charge	$C\ m^{-2}$
σ_0^H	specific surface charge calculated from surface excess of H^+	$C\ m^{-2}$
ψ_i	electric potential at plane i	V

Literature Cited

1. Parsons, R. J. Electroanal. Chem. 1982, 118, 3-18.
2. Westall, J.; Hohl, H. Adv. Coll. Interface Sci. 1980, 12, 265-94.
3. Lyklema, J. J. Coll. Interface Sci. 1977, 58, 242-250.
4. Hunter, R. J. "Zeta Potential in Colloid Science"; Academic Press: London, 1981;
5. Brown, G. T.; Darwent, J. R. J. Chem. Soc. Chem. Commun. 1985, 98-100.
6. Schindler, P. W.; Stumm, W. In "Aquatic Surface Chemistry"; Stumm, W. Ed.; John Wiley: New York, in press.
7. James, R. O.; Parks, G. A. In "Surface and Colloid Science", Vol. 12; Matijevic, E., Ed.; Plenum: New York, 1982.
8. Schindler, P. W. In "Adsorption of Inorganics at Solid-Liquid Interfaces"; Anderson, M. A.; Rubin, A. J., Eds.; Ann Arbor Science: Ann Arbor, 1981; Chap. 1.
9. Westall, J. In "Particles in Water"; Kavanaugh, M. C.; Leckie, J. O., Eds.; ADVANCES IN CHEMISTRY SERIES No. 189, American Chemical Society: Washington, D.C., 1980; Chap. 2.
10. Smit, W.; Holten, C. L. M.; Stein, H. N.; De Goeij, J. J. M.; Theelen, H. M. J. J. Coll. Interface Sci. 1978, 63, 120-128.
11. Foissy, A. M.; Pandou, A.; Lamarche, J. M.; Jafferzic-Renault, N. Colloids Surfaces 1982, 5, 363 .
12. Sprycha, R. J. Coll. Interface Sci. 1984, 102, 173-185.
13. Stumm, W.; Morgan, J. "Aquatic Chemistry"; John Wiley: New York, 1981.
14. Chan, D.; Perram, J. W.; White, L. R.; Healy, T. W. J. Chem. Soc. Faraday Trans. 1 1975, 71, 1046-1057.
15. Sposito, G. "Surface Chemistry of Soils"; Oxford University Press: New York, 1984; Chap. 5.
16. Morel, F.; Yeasted, J.; Westall, J. In "Adsorption of Inorganics at Solid-Liquid Interfaces"; Anderson, M. A.; Rubin, A. J., Eds.; Ann Arbor Science: Ann Arbor, 1981; Chap. 7.
17. Van Riemsdijk, W. H.; Bolt, G. H.; Koopal, L. K.; Blaakmeer, J. J. Coll. Interface Sci. 1986, 109, 219-228.
18. Bolt, G. H.; Van Riemsdijk, W. H. In "Soil Chemistry, Part B, Physicochemical Methods"; Bolt, G. H., Ed.; Elsevier: Amsterdam, 1982; Chap. 13.
19. Healy, T. W.; White, L. R. Adv. Coll. Interface Sci. 1978, 9, 303-45.
20. Westall, J. In "Aquatic Surface Chemistry"; Stumm, W. Ed.; John Wiley: New York, in press.

21. Stern, O. Z. Elektrochem. 1924, 30, 508-16.
22. Yates, D. E.; Levine, S.; Healy, T. W. Trans. Far. Soc. 1974,
     70, 1807.
23. Schindler, P. W.; Gamsjaeger, H. Kolloid Z. u. Z. Polymere
     1972, 250, 759.
24. Stumm, W.; Hohl, H.; Dalang, F. Croat. Chem. Acta 1976, 48,
     491.
25. Stumm, W.; Huang, C. P.; Jenkins, S. R. Croat. Chem. Acta 1970,
     42, 223.
26. Huang, C. P.; Stumm, W. J. Coll. Interface Sci. 1973, 43, 409.
27. Bowden, J. W.; Posner, A. M.; Quirk, J. P. Aust. J. Soil Res.
     1977, 15, 121.
28. Davis, J. A.; James, R. O.; Leckie, J. O. J. Coll. Interface
     Sci. 1978, 63, 480-499.
29. Davis, J. A.; Leckie, J. O. J. Coll. Interface Sci. 1978, 67,
     90-107.
30. Healy, T. W.; Yates, D. E.; White, L. R.; Chan, D.
     J. Electroanal. Chem. 1977, 80, 57-66.
31. Westall, J. Manuscript in preparation.
32. Yates, D. E.; Healy, T. W. J. Chem. Soc., Faraday Trans. 1
     1980, 76, 9-18.
33. James, R. O.; Davis, J. A.; Leckie, J. O. J. Coll. Interface
     Sci. 1978, 65, 331.
34. Westall, J. "FITEQL. A Computer Program for Determination of
     Chemical Equilibrium constants from Experimental Data,"
     Version 1.2, Report 82-01, Department of Chemistry, Oregon
     State University, Corvallis, OR 97331, 1982.
35. Westall, J. "FITEQL. A Computer Program for Determination of
     Chemical Equilibrium constants from Experimental Data,"
     Version 2.0, Report 82-02, Department of Chemistry, Oregon
     State University, Corvallis, OR 97331, 1982.
36. Watanabe, T.; Fujishima, A.; Honda, K. Chem. Letters 1974,
     890-900.
37. Kinoshita, K.; Madou, M. J. J. Electrochem. Soc. 1984, 131,
     1089-1094.
38. Bousse, L; De Rooij, N. F.; Bergveld, P. IEEE Trans. Electron
     Devices, 1983, ED-30, 1263-1270.
39. Abe, H.; Esashi, M.; Matsuo, T. IEEE Trans. Electron Devices
     1979, ED-26, 1939-1944.

RECEIVED June 25, 1986

5

Surface Potential–pH Characteristics in the Theory of the Oxide–Electrolyte Interface

Luc Bousse and J. D. Meindl

Stanford Electronics Laboratories, Stanford University, Stanford, CA 94305

The measurement of the surface potential ψ_0 as a function of pH for an oxide provides valuable information for the determination of the parameters which describe the surface reactions. Ionizable surface site theories of the formation of surface charge and potential at an oxide surface in contact with a liquid electrolyte involve many more parameters than can be directly experimentally determined. Additional assumptions are required to evaluate these parameters, which explains why there is often no agreement in the literature about their value. A mathematical treatment of the amphoteric surface site model is given which exhibits the characteristic quantities which can be experimentally measured. It is shown that the measurement of both the surface potential ψ_0 and the surface charge σ_0 are required to completely determine these characteristic quantities. This approach is applied to SiO_2 and Al_2O_3, two surfaces for which both charge and potential measurements are available.

Quantitative explanations of the formation of charge and potential on oxide surfaces based on chemical reactions of surface sites have become increasingly popular in the past two decades. Among the earliest authors proposing surface complexation models are Levine et al. ([1]), Schindler et al. ([2]), Huang and Stumm ([3]), and Yates et al. ([4]). These last authors also introduced the important idea that the solution counter-ions enter the double layer, and form surface complexes with oppositely charged surface sites. This can explain the high values of σ_0 commonly measured on oxide surfaces. The site-binding theory of Yates et al. was reformulated by Davis et al. ([5]), who gave a convenient procedure for extracting the parameters of the theory from measured σ_0/pH data. Davis' formulation of the site-binding theory has been applied by several authors recently to data on oxide colloids ([6,7,8]), with satisfactory results, although Sprycha ([9]) reports only a limited agreement and suggests some model assumptions are inadequate. Some authors ([10,11]) have presented extensions of the theory which take more possible surface reactions into account than the version by Davis et al. ([5]).

However, providing good fits to data is not sufficient to make a theory convincing. One of the unsatisfactory aspects of the site-binding theories is that the various methods of parameter extraction give widely different results. For instance, values of the parameter ΔpK for Al_2O_3 have been reported ranging from 1.2 ([3]) to 5.8 ([5]). A related phenomenon is that the theory contains too many parameters, so that it is possible to generate almost identical theoretical σ_0/pH curves with many sets of parameters. This problem was first noted by Westall and Hohl ([12]), and has also been pointed out by Sposito ([13]) and Johnson ([11]). This overabundance of parameters to be determined

0097–6156/86/0323–0079$06.00/0

makes the values found by various authors depend on additional assumptions, and explains the wide range of reported values.

One possible solution is to obtain new experimental data, which is independent of σ_0/pH curves. The zeta potential is of course a possibility, but it suffers from the intrinsic indeterminacy of the exact location in the double layer where it occurs. Another possibility is the surface potential, ψ_0, which will be defined below. Variations of ψ_0 can be measured by using electrolyte/insulator/semiconductor structures. It has been shown by Bousse et al. (14) that the ψ_0/pH characteristics are determined mainly by the number of charged but uncomplexed surface sites, and are insensitive to complexation. This means that combined consideration of σ_0/pH and ψ_0/pH characteristics should lead to a more complete and reliable determination of model parameters.

In this paper, a set of approximate solutions to model equations will be presented in a way intended to make clear which parameters can be determined from experimental data. This leads to a methodology of extracting a maximum amount of information from $\psi_0/\sigma_0/pH$ data.

The Measurement of ψ_0/pH Curves

The measurement of changes of the surface potential ψ_0 at the interface between an insulator and a solution is made possible by incorporating a thin film of that insulator in an electrolyte/insulator/silicon (EIS) structure. The surface potential of the silicon can be determined either by measuring the capacitance of the structure, or by fabricating a field effect transistor to measure the lateral current flow. In the latter case, the device is called an ion-sensitive field effect transistor (ISFET). Figure 1 shows a schematic representation of an ISFET structure. The first authors to suggest the application of ISFETs or EIS capacitors as a measurement tool to determine the surface potential of insulators were Schenck (15) and Cichos and Geidel (16).

A complete and quantitative theory of the dependence of the flat-band or threshold voltage of an EIS structure on ψ_0 can be found in (17) and (18), and need not be repeated here. It can be shown (17) that the expression of the flat-band voltage of an EIS structure is given by:

$$V_{FB} = E_{ref} - \frac{\Phi^{Si}}{q} - \psi_0 - Q_{ins}/C_{ins} \tag{1}$$

where E_{ref} is the reference electrode potential on a scale relative to vacuum, Φ^{Si} is the work function of silicon, q is the absolute value of the charge of an electron, C_{ins} is the insulator capacitance, and Q_{ins} is the charge inside the insulator which is assumed to be located at the insulator/silicon interface. ψ_0 is the potential drop inside the electrolyte at the insulator/electrolyte interface. A number of dipole potential terms which should also be present in the equation above are small and can be neglected; the potential due to water dipoles at the insulator/electrolyte interface is not negligible, but is included in ψ_0, as will be made clear below.

Equation (1) is the basis for using EIS capacitors or ISFETs to measure changes in ψ_0. If an ISFET is used, the measured quantity is the threshold voltage, which is equal to the flat-band voltage plus a constant. In both cases only variations of ψ_0 can be measured, because the other terms in (1) are not independently known with sufficient accuracy.

Model for an Amphoteric Surface without Complexation

In this section, we will treat the case of an oxide whose surface OH sites only undergo amphoteric acid/base reactions. We will write the surface site as SOH, following the notation of James and Parks (19). The two surface acid/base reactions and their

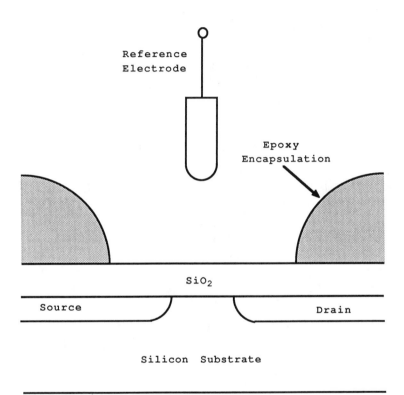

Figure 1.   Schematic representation of the basic pH-sensitive ISFET structure, assuming that the insulator is SiO_2.

associated equilibrium constants can then be written as:

$$SOH_2^+ \xrightarrow{\leftarrow} SOH + H_s^+ \qquad K_{a1} = \frac{\{H^+\}[SOH]}{[SOH_2^+]} \exp\left(-\frac{q\psi_o}{kT}\right) \qquad (2)$$

$$SOH \xrightarrow{\leftarrow} SO^- + H_s^+ \qquad K_{a2} = \frac{\{H^+\}[SO^-]}{[SOH]} \exp\left(-\frac{q\psi_o}{kT}\right) \qquad (3)$$

in which $\{H\}$ stands for the H^+ activity in the bulk solution, and quantities in square brackets are numbers of surface sites per unit area. The activity coefficients for the surface sites are assumed to be constant and are incorporated in the equilibrium coefficients. Healy and White ([20]) have given a derivation of these equations, together with the implied assumptions, which are similar to those of the Langmuir isotherm. Smit and Holten ([21]) discuss the validity of assuming the surface site activity coefficients to be constant, and give experimental support for that assumption.

By obtaining $\{H^+\}$ from the product of Equations 2 and 3, and taking the logarithm, it follows that

$$2.303 \left(pH_{pzc} - pH\right) = \frac{q\psi_0}{kT} + \ln\left(\frac{[SOH_2^+]}{[SO^-]}\right)^{1/2} \qquad (4)$$

where we introduce the definition $pH_{pzc} = -\log_{10}(K_{a1}K_{a2})^{1/2}$. This equation shows the important role played by the ratio of the two charged site forms. The desired relation between ψ_0 and pH can be found if this ratio can be expressed in terms of the net surface charge, which in turn is connected to ψ_0 by the electrostatic laws of the double layer. Thus, it appears that to calculate ψ_0/pH characteristics, two relations are needed: a double layer theory for the connection between ψ_0 and the surface charge, and $\left([SOH_2^+]/[SO^-]\right)^{1/2}$ as a function of this charge. Equations to solve for this quantity were first formulated by Dousma ([24]), and it will be referred to as the F-function after Bousse et al. ([14]) and Harame ([25]).

At this point, the problem has been reduced to solving the surface chemistry equations to calculate the F-function. It is convenient to introduce dimensionless notations by normalizing all surface concentrations to the total density of sites N_S:

$$\alpha_+ = \frac{[SOH_2^+]}{N_S} \qquad \alpha_- = \frac{[SO^-]}{N_S} \qquad \alpha_n = \frac{[SOH]}{N_S} \qquad (5)$$

The normalized conservation-of-site equation is therefore:

$$\alpha_+ + \alpha_- + \alpha_n = 1 \qquad (6)$$

Another relation which follows from the definitions is that the surface charge is given by $q\left([SOH_2^+] - [SO^-]\right)$. In normalized notations, this relation becomes:

$$\chi = \alpha_+ - \alpha_- \qquad (7)$$

From Equation 7 and the definition of the F-function as

$$F = \left(\frac{\alpha_+}{\alpha_-}\right)^{1/2} \qquad (8)$$

it is possible to calculate α_+ and α_- as a function of χ and F:

$$\alpha_+ = \frac{F^2\chi}{F^2 - 1}, \qquad \alpha_- = \frac{\chi}{F^2 - 1} \qquad (9)$$

We need a similar relation for α_n; this can be obtained by substituting these results into the quotient of Equations 2 and 3, which can be written as:

$$\alpha_n = (2/\delta)(\alpha_+ \, \alpha_-)^{1/2} = \frac{(2/\delta)F\chi}{F^2 - 1} \tag{10}$$

where we have introduced $\delta = 2(K_{a2}/K_{a1})^{1/2}$, following the notation of Healy *et al.* (26). This constant is an indication of the strength of the tendency to dissociate of the surface sites. The higher the tendency of the surface sites to become charged through either reaction 2 or 3, the higher the value of δ will be. Another quantity which is often used in the literature for this purpose is ΔpK, which is related to δ as follows: $\Delta pK = pK_{a2} - pK_{a1} = -\log_{10}(\delta^2/4)$. Substitution of the expressions for α_+, α_-, and α_n into Equation 6 yields:

$$F^2(\chi - 1) + 2(\chi/\delta)F + (\chi + 1) = 0 \tag{11}$$

from which it follows that F is given by:

$$F(\chi) = \frac{\chi/\delta + \sqrt{1 + (\chi/\delta)^2(1 - \delta^2)}}{1 - \chi} \tag{12}$$

This method for computing the F-function is due to Harame (25); it has the virtue of being easy to generalize to more complex cases. Note that the substitution of Equation 12 into Equations 9 and 10 gives the exact solution of the initial system of equations for the surface site concentrations. Together with Equation 4 which can now be written as:

$$2.303 \left(pH_{pzc} - pH\right) = \frac{q\psi_o}{kT} + \ln F(\chi) \tag{13}$$

we have the desired relation between pH and ψ_0, provided that the charge $qN_S\chi$ in the double layer can be expressed in terms of the potential ψ_0. All charge is assumed to be evenly spread out in planes separated by a constant Stern layer capacitance. According to the Gouy-Chapman-Stern theory, the relation between potential and voltage can then be written as:

$$\psi_0 = \frac{\sigma_0}{C_{St}} + \frac{2kT}{q}\sinh^{-1}\left\{\frac{\sigma_0}{\sqrt{8RTc\epsilon}}\right\} \tag{14}$$

which assumes that the double layer is globally neutral , i.e. $\sigma_0 + \sigma_d = 0$. The legitimacy of this assumption has been discussed in (18).

At this point, the combination of Equation 12 to 14 represents a solution to the problem. It is possible, however, to simplify this solution further by introducing a few approximations. First, it is well-known in electrochemistry that the relation between charge and potential in the double layer is dominated by the Stern capacitance at high electrolyte concentrations. The relation of Equation 14 is then nearly linear; the linearization of the $\sinh^{-1}$ function around the point $\sigma_0 = 0$ yields:

$$\psi_0 = \frac{\sigma_0}{C_{DL}} \tag{15}$$

in which the equivalent double layer capacitance, C_{DL}, is given by:

$$C_{DL}^{-1} = \frac{2kT}{q}(8RTc\epsilon)^{-1/2} + C_{St}^{-1} \tag{16}$$

where c is the electrolyte concentration.

Another approximation we will make is that $1 - \delta^2 \approx 1$, since δ is usually considered to be small for oxide surfaces. According to the table of values obtained by Davis *et al.* (5), δ is at most equal to 3×10^{-3}. Other authors tend to find somewhat higher values, such as Grauer and Stumm $(\underline{27})$ whose values for various oxides range from 0.07 to 0.1. One of the highest values in the literature is due to Huang and Stumm $(\underline{3})$, who find $\delta = 0.5$. In almost all cases, therefore, the approximation that δ^2 is much smaller than one should be quite accurate. Even when $\delta = 0.5$, its influence on Equation 12 is small compared to that of the variable χ. It is only if one assumes that ΔpK is around zero that it becomes essential to use the exact expression of Equation 12.

Finally, it can be verified that the denominator of Equation 12 only becomes significant when χ is close to one. This never occurs on oxide surfaces in aqueous solutions, where the available pH range is too small to cause all surface sites to be charged. Thus, by substituting Equations 12 and 15 into Equation 13, and applying the approximations justified above, we obtain:

$$2.303 \left(pH_{pzc} - pH \right) = \frac{q\psi_0}{kT} + \sinh^{-1} \left\{ \frac{\sigma_{DL}}{\sigma_S} \frac{q\psi_0}{kT} \right\} \tag{17}$$

and:

$$2.303 \left(pH_{pzc} - pH \right) = \frac{\sigma_0}{\sigma_{DL}} + \sinh^{-1} \left\{ \frac{\sigma_0}{\sigma_S} \right\} \tag{18}$$

In these equations, the following characteristic charges have been defined:

$$\sigma_{DL} = \frac{kTC_{DL}}{q} \tag{19}$$

$$\sigma_S = qN_S\delta \tag{20}$$

σ_{DL} represents the magnitude of the charge on the double layer capacitance C_{DL} when the voltage on that capacitor is the thermal voltage kT/q. σ_S is the charge generated when a fraction δ of the surface sites is ionized; it can be verified that at the point of zero charge the proportion of charged sites is $\delta/(1 + \delta)$. Thus, σ_S is approximately equal to the sum of the absolute value of the charge on all charged sites at pH_{pzc}, and characterizes the tendency of the surface sites to become ionized. It is apparent from the equations above that in this model the behavior of the oxide surface depends only on three constants: pH_{pzc}, σ_S, and σ_{DL}.

Focusing our attention first on ψ_0/pH curves, it can be seen that the ratio σ_S/σ_{DL}, which is called β in $(\underline{14,18})$, determines how close to Nernstian the ψ_0/pH curves are. The significance of the ratio β can be understood as follows: the two processes which must occur to generate charge on an oxide surface are surface site ionization, described by σ_S, and the charging of the double layer capacitance, described by σ_{DL}. If β is much larger than one, surface ionization is easier than the double layer charging, which is then the limiting step in determining the surface charge. In that case Nernst's law gives the potential, and Equation 16 the charge. If β is much smaller than one, surface site ionization is the surface charge determining process, which will then depend strongly on the solution pH. For most oxides, β is larger than one, which ensures reasonably linear curves. In the case of SiO_2, and perhaps also of SnO_2 $(\underline{8})$, β is small and the Nernstian relation does not apply. Equation 17 shows that the ψ_0/pH relation only contains β as a parameter (apart from pH_{pzc}). β and pH_{pzc} are therefore the only two parameters which can be determined from a ψ_0/pH curve; however, if values for C_{DL} and N_S are known or assumed, K_{a1} and K_{a2} can be found. Experimental measurements of ψ_0 are available for many insulator surfaces, including SiO_2 $(\underline{18,28})$, Al_2O_3 $(\underline{14,18})$, and ZrO_2 $(\underline{29})$, which are particularly interesting because these surfaces have also been studied as colloids. Most oxides have roughly linear ψ_0/pH characteristics, with slopes higher than about 50 mV/pH. Since it follows from Equation 17 that the ψ_0/pH slope around pH_{pzc} is $\beta/(\beta + 1)$ times Nernstian, this corresponds to values of β of at least 4 or 5. The main exception is SiO_2 which has a very low slope around its pH_{pzc}, and a β of about 0.14 $(\underline{18})$.

The σ_0/pH relation involves two parameters, σ_S and σ_{DL} in addition to the pH_{pzc}. However, it will be legitimate to linearize Equation 18 whenever the second term on the right hand side is small compared to the first term. This depends on the magnitude of β, as for the potential/pH curves, and the linearization is thus legitimate for most oxides whose β is large. The linear equation is:

$$2.303 \left(\text{pH}_{pzc} - \text{pH} \right) = \frac{\sigma_0}{\sigma_i} \tag{21}$$

where σ_i is given by:

$$\sigma_i = \left(\sigma_S^{-1} + \sigma_{DL}^{-1} \right)^{-1} \tag{22}$$

This constant contains the first order information from a σ_0/pH curve, namely its slope. Only if the slope varies appreciably could an experimental result be used to find σ_S and σ_{DL} separately. Knowledge of both the ψ_0/pH and σ_0/pH curves is also sufficient in principle to obtain σ_S and σ_{DL} separately. The conclusion that an experimental σ_0/pH curve can only yield two parameters, pH_{pzc} and σ_i, implies that many values of the set (δ, C_{DL}, N_S) can be used to interpret a measurement. This has been found by Westall and Hohl (12), who interpreted surface charge data on Al_2O_3 using computer-generated solutions of the site dissociation model. Table (1) below shows that the nearly identical curves these authors calculate with different sets of parameters all correspond to a constant value of $\sigma_i = 1.3 \ \mu\text{C/cm}^2$. Each line corresponds to a different set of parameters; the last three columns contain the corresponding characteristic charges.

Table 1. *Interpretation of the data from computer-generated fitting of the surface charge of Al_2O_3 by Westall and Hohl (12)*

N_S cm^{-2}	C_{DL} F/cm^2	ΔpK	σ_S $\mu\text{C/cm}^2$	σ_{DL} $\mu\text{C/cm}^2$	σ_i $\mu\text{C/cm}^2$
6.40×10^{13}	1.06×10^{-4}	1.84	2.46	2.70	1.29
8.17×10^{13}	0.66×10^{-4}	1.16	6.89	1.68	1.35
1.01×10^{14}	0.57×10^{-4}	0.82	12.63	1.45	1.30

Comparison with the computer simulations in (12) therefore confirms the role of σ_i as the characteristic parameter in this theory. This does not necessarily mean, however, that this theory corresponds to reality. In fact, consideration of the influence of electrolyte concentration shows it probably does not. The ionic strength only influences σ_{DL} indirectly, through C_{DL} whose dependence on concentration is given in Equation 16. This equation implies that at high concentrations $C_{DL} \approx C_{St}$, and that σ_i varies little. Thus, theory without complexation predicts that that for high concentrations (> 0.01M, for instance) σ_0/pH curves should not depend much on electrolyte concentration. As a rule, the opposite is observed. Another contradiction of the present model which is experimentally observed is that while the ψ_0/pH characteristics are linear for most oxides (such as Al_2O_3 or ZrO_2), the σ_0/pH curves are very non-linear. According to Equations 17 and 18, the degree of linearity of ψ_0 and σ_0 should be the same. The measured non-linear σ_0/pH curves can thus only be interpreted as the result of counter-ion adsorption, which has been neglected so far.

Model for an Amphoteric Surface with Complexation

The concentration dependence and non-linearity observed in σ_0/pH curves indirectly show that it is not legitimate to ignore complexation. Direct experimental evidence, obtained with radiotracer methods (21,30,31), confirms the importance of complexation in the generation of surface charge σ_0. Thus, we need to add the following equations to Equations 2 and 3, assuming for simplicity that the electrolyte is NaCl:

$$SO^- + Na_s^+ \overset{\leftarrow}{\rightarrow} SO^-Na^+ \qquad K_{Na^+} = \frac{[SO^-Na^+]}{[SO^-]\{Na^+\}} \exp\left(\frac{q\psi_{Na^+}}{kT}\right) \quad (23)$$

$$SOH_2^+ + Cl_s^- \overset{\leftarrow}{\rightarrow} SOH_2^+Cl^- \qquad K_{Cl^-} = \frac{[SOH_2^+Cl^-]}{[SOH_2^+]\{Cl^-\}} \exp\left(-\frac{q\psi_{Cl^-}}{kT}\right) \quad (24)$$

The equilibrium constants K_{Na^+} and K_{Cl^-} introduced here characterize the extent of counterion complexation that occurs. Two other constants characterize the potential generation that results from this complexation, namely the capacitances C_{Na^+} and C_{Cl^-}. These are the capacitances between the planes of counterion complexation and the surface plane where σ_0 is located. The potentials ψ_{Na^+} and ψ_{Cl^-} are the electrostatic potential at the location in the double layer where the ions adsorb and form a surface complex.

The derivation of Equation 13, which relates ψ_0, pH, and the F-function is unaffected by the presence of more surface site forms in this case. What changes is the calculation of the number of the different site forms, and therefore the expression for the F-function. Using normalized surface concentrations as before, the new normalized conservation-of-site equation is:

$$\alpha_+ + \alpha_- + \alpha_{Cl^-} + \alpha_{Na^+} + \alpha_n = 1 \qquad (25)$$

Note that the net surface charge, defined in Equation 7 is not the same as the charge on the plane of the surface oxygens, which is called σ_0, and is given by:

$$\sigma_0 = q\left([SOH_2^+] + [SOH_2^+Cl^-] - [SO^-] - [SO^-Na^+]\right) \qquad (26)$$

The effect of the counterion binding reactions 23 and 24 on the surface chemistry will be treated in terms of ratios p and n, defined as:

$$p = \frac{\alpha_{Na^+}}{\alpha_-} = cK_{Na^+} \exp\left(-\frac{q\psi_{Na^+}}{kT}\right) \qquad (27)$$

$$n = \frac{\alpha_{Cl^-}}{\alpha_+} = cK_{Cl^-} \exp\left(\frac{q\psi_{Cl^-}}{kT}\right) \qquad (28)$$

where $c = \{Na\} = \{Cl\}$, the bulk electrolyte activity. With these equations, the conservation-of-site relationship can be written as:

$$(1 + n)\alpha_+ + (1 + p)\alpha_- + \alpha_n = 1 \qquad (29)$$

The expressions of Equations 9 and 10 which give the normalized surface site densities as a function of χ and F are still valid; when substituted into Equation 29 an equation is obtained which can be solved as before to give:

$$F(\chi) = \frac{(\chi/\delta) + \sqrt{(\chi/\delta)^2 + [1 - \chi(1+n)][1 + \chi(1+p)]}}{[1 - \chi(1+n)]} \qquad (30)$$

For surfaces for which δ is small, this new expression for the F-function is in fact similar to the old one, since then the dominant term under the square root is $(\chi/\delta)^2$.

To relate the various potentials in the double layer to the charge in the Gouy-Chapman layer σ_d, the model of Figure 2 can be used; again, constant capacitances between the planes of charge are assumed. Some complications arise when $C_{Na^+} \neq C_{Cl^-}$; in Reference (14) a general method is discussed to calculate $\psi_0/\sigma_0/pH$ curves in that case. This method is an iteration which converges very rapidly, and yields results free of any approximations. Figure 3 shows a theoretical calculation of surface charge for a hypothetical surface with $pH_{pzc} = 8$, $C_{Na^+} = 150\ \mu F/cm^2$, and $C_{Cl^-} = 75\ \mu F/cm^2$. However, in addition to the exact results obtained by that method, it is possible to find approximations such as those developed in the previous section, which illustrate the role of characteristic constants.

For simplicity, we will consider the case in which surface charge and potential are positive, and that only anions adsorb. Furthermore, the potential drop in the Gouy-Chapman layer will be assumed to be small enough that its charge/potential relation can be linearized. The $\psi_0/\sigma_0/pH$ relationship can then be derived parametrically, with the charge in the Gouy-Chapman layer σ_d as the parameter. The potential at the plane of anion adsorption can then be calculated and substituted in Equation 28 to give:

$$n = cK_{Cl^-}\ \exp\left\{\sigma_d(\frac{1}{\sigma_{DL}} - \frac{1}{\sigma_n})\right\} \tag{31}$$

where σ_n is defined as $(kT/q)C_{Cl^-}$, the characteristic charge associated with the capacitance of the anion adsorption plane. Using Equation 30, the F-function can be calculated, and then, from Equations 9, 26, and 28 it follows that:

$$\sigma_0 = \sigma_d\left\{1 + \frac{nF^2}{F^2 - 1}\right\} \tag{32}$$

The total potential in the double layer also follows from the double layer model of Figure 2, and can be calculated to be:

$$\frac{q\psi_0}{kT} = \frac{\sigma_d}{\sigma_{DL}} + \frac{\sigma_d}{\sigma_n}\frac{nF^2}{F^2 - 1} \tag{33}$$

and then from Equation 13 it follows that:

$$2.303\left(pH_{pzc} - pH\right) = \frac{\sigma_d}{\sigma_{DL}} + \ln(F) + \frac{\sigma_d}{\sigma_n}\frac{nF^2}{F^2 - 1} \tag{34}$$

Two regions of behavior can be distinguished in the case that cK_{Cl^-} is much smaller than one:

1. When $n << 1$, we have $\sigma_0 \approx \sigma_d$, and the last term in Equations 33 and 34 is negligible compared to the other terms. Then the equations reduce to the case without specific adsorption which was discussed previously.

2. The parameter n increases exponentially with increasing σ_d, and therefore beyond a certain value of σ_d the last term in Equations 32 to 34 becomes dominant. At that point, almost all of the surface charge σ_0 is due to the counterion adsorption reactions. In this case, the last two equations can be further approximated as follows:

$$\frac{q\psi_0}{kT} = \frac{\sigma_d}{\sigma_{DL}} + \frac{\sigma_0}{\sigma_n} \tag{35}$$

$$2.303\left(pH_{pzc} - pH\right) = \frac{\sigma_d}{\sigma_{DL}} + \ln(F) + \frac{\sigma_0}{\sigma_n} \tag{36}$$

In addition, once the point is reached where the surface charge increases exponentially with σ_d, the parameter σ_d will vary much less than the other parameters,

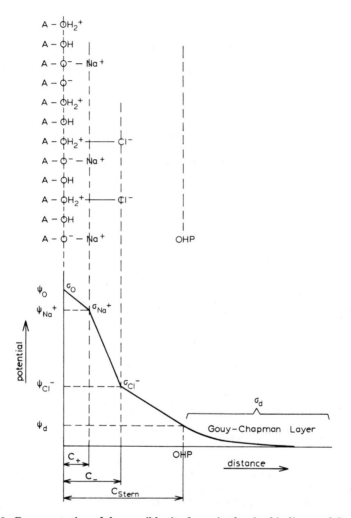

Figure 2. Representation of the possible site forms in the site-binding model which includes the effect of counter-ion adsorption, combined with a diagram of charges and potentials at the insulator/electrolyte interface. Reproduced with permission from Ref. (14). Copyright 1983, North Holland.

and will be approximately constant. In other words, Equation 36 predicts that sufficiently far from the pH_{pzc}, the σ_0/pH curve will tend to become a straight line given by:

$$2.303 \left(pH_{pzc} - pH\right) = f\left(cK_{Cl^-}, \sigma_{DL}, \sigma_S\right) + \frac{\sigma_0}{\sigma_n} \qquad (37)$$

The intercept of this line is a function of cK_{Cl^-}. Thus, this theory predicts that the σ_0/pH curves at various concentrations should all tend to become parallel lines far from the pH_{pzc}. This is indeed the usual experimental observation. The slope of the σ_0/pH curves far from pH_{pzc} is given by $\sigma_n = (kT/q)C_{Cl^-}$, which means that the adsorption capacitance follows directly from surface charge measurements. This had already been noted by Smit ([21]). The intercept of the straight line predicted by Equation 37 can be used to find cK_{Cl^-}, provided that σ_S and σ_{DL} are known.

Between these two limiting cases there is a transition region, the extent of which can be judged from Figure 3. In practice, the limiting slope of the surface charge/pH curves is often not reached at the limit of the pH range of a measurement, and the value of C_{Cl^-} is somewhat higher than the highest measured slope.

The effect of counterion adsorption on the ψ_0/pH characteristics can be seen by considering Equations 35 and 36. Since the last term on the right hand side, which is dominant far from pH_{pzc}, is common to these two equations, the resulting ψ_0/pH relation tends to become Nernstian. However, in this region, the surface potential always becomes Nernstian anyway, even without the effect of counterion adsorption. Therefore, adsorption will not affect the ψ_0/pH curves very much. This is demonstrated in Figure 4, which shows the surface potential in the same conditions as for Figure 3. If cK_{Cl^-} is much larger than one, then there will be an effect on the potential/pH curves, namely an increase in the slope at the pH_{pzc} (see Reference ([14])).

Determination of Surface Equilibrium Constants

Determination of K_{a1} and K_{a2} from ψ_0/pH curves. The results of the previous section show that around the pH_{pzc} the ψ_0/pH characteristics become insensitive to complexation. This fact will be used in this section to obtain estimates of the intrinsic surface reaction constants K_{a1} and K_{a2}. We have shown that surface potential measurements can be used to determine the ratio $\beta = \sigma_S/\sigma_{DL}$ and the point of zero charge. This is done by measuring the pH where the slope of the ψ_0/pH curve is minimal, and the value of that slope. Note that this procedure is only possible when β is not higher than about 10, because otherwise the slope is practically Nernstian for all pH. To deduce the value of δ from the measured value of β, Equations 19 and 20 show that it is necessary to know N_S and C_{St}. The surface site density has been measured by various means, including tritium exchange ([32]) and crystallography ([33]), and for oxides the results lie in the range 5 to $15 \times 10^{14} \, cm^{-2}$. These values are much larger than those obtained by measuring the exchange capacity in aqueous solutions, as was done by several authors ([3,34]); however, it is not certain that the pH range available in aqueous solutions or before dissolution occurs is sufficient to fully charge a surface. Unfortunately, very little is known about the Stern capacitance on oxides, C_{St}. Yates et al. ([4]) have suggested a value of 20 $\mu F/cm^2$, which corresponds roughly to the capacitance of a single layer of adsorbed water molecules with no rotational mobility. There is an experimental study of the capacitance at the SiO_2/water interface which supports this value ([35]), and we will make the assumption here that $C_{St} = 20 \, \mu F/cm^2$. It must be kept in mind, however, that the double layer theory currently used for oxide surfaces contains many approximations, such as neglecting discreteness of charge effects and

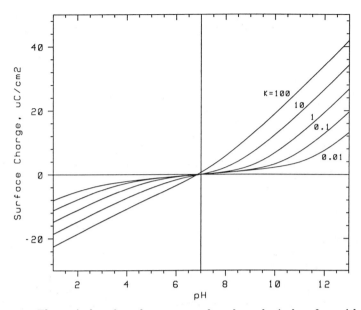

Figure 3. Theoretical surface charge curves for a hypothetical surface with $N_S = 8 \times 10^{14}\,\text{cm}^{-2}$, $\text{pH}_{pzc} = 7$, $C_{cation} = 150\,\mu\text{F}/\text{cm}^2$, $C_{anion} = 75\,\mu\text{F}/\text{cm}^2$, and $c = 0.1$ mol/dm^3. $K_{cation} = K_{anion} = 100$ to 0.01.

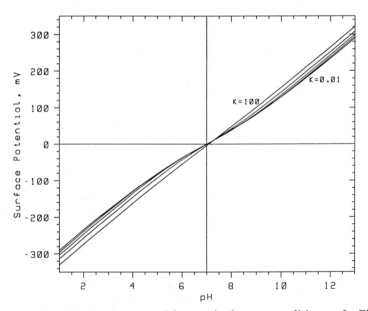

Figure 4. Theoretical surface potential curves in the same conditions as for Figure (3).

lumping the influence of water dipoles into the Stern capacitance. Thus, the possibility exists that the Stern capacitance in this model assumes a value which departs from what would follow from the physical picture of a layer of water dipoles.

Assuming $N_S = 5 \times 10^{14}$ cm^{-2} ([37]), values of $\Delta pK = -\log_{10}(\delta^2/4)$ ranging from 6.4 ([36]) to 8.2 ([38]) have been measured for SiO$_2$ surfaces. Our own measurements are shown in Figure 5, and lead to $\Delta pK = 6.9$ ([18]) for a 0.1M solution; there does exist a slight influence of ionic strength as predicted by Equation 16. The relatively large spread in experimental results for ΔpK measured with electrolyte/SiO$_2$/Si systems may be due to the tendency of SiO$_2$ to hydrate very slowly ([37]), which may cause widely varying surface densities of silanol sites. For both colloidal dispersions and electrolyte/SiO$_2$/Si structures it is found that $pH_{pzc} \approx 2.2$. Taking an average value of ΔpK to be 7.0, it follows that for SiO$_2$:

$$pK_{a1} = pH_{pzc} - \frac{\Delta pK}{2} = 2.2 - 3.5 = -1.3 \qquad (38)$$

$$pK_{a2} = pH_{pzc} + \frac{\Delta pK}{2} = 2.2 + 3.5 = 5.7 \qquad (39)$$

It is to be noted that the ψ_0/pH curves on SiO$_2$ did not depend on the type of cation in solution, as also reported by Fung ([38]), which confirms that these curves are not significantly affected by adsorption of counterions.

In the case of γ-Al$_2$O$_3$, the only ψ_0/pH measurements in which a ΔpK is reported are in ([18]) and ([14]) (see Figure 6); the value observed is $\Delta pK = 4.2$, assuming that $N_S = 8 \times 10^{14}$ cm^{-2}. Again, the type of electrolyte did not affect this result. The pH_{pzc} found with electrolyte/insulator/Si structures was 8, which is close to the value of 8.4 usually reported for colloidal dispersions of the same material. Thus, for γ-Al$_2$O$_3$, the following results are obtained:

$$pK_{a1} = pH_{pzc} - \frac{\Delta pK}{2} = 8 - 2.1 = 5.9 \qquad (40)$$

$$pK_{a2} = pH_{pzc} + \frac{\Delta pK}{2} = 8 + 2.1 = 10.1 \qquad (41)$$

Interpretation of experimental σ_0/pH curves. Since σ_0/pH curves mainly depend on counterion adsorption, it is clear that these data should be used to determine pK_{cation}, pK_{anion}, C_{cation}, and C_{anion}. In fact, the experimental slopes of the linear part of the σ_0/pH curve far from the pH_{pzc} yield the adsorption capacitance directly (provided dissolution effects do not interfere). To determine the equilibrium constants, it is sufficient to generate a family of theoretical plots with different values of pK_{cation} and pK_{anion}, and observe with which curve the experimental data corresponds. This procedure requires the independent knowledge of pK_{a1} and pK_{a2} which was obtained in the previous paragraph. The surface charge data has been obtained from the experiments published in the literature.

For SiO$_2$, we have only considered sources for silica suspensions which were nonporous, such as Ludox ([39]), pyrogenic silica ([40]), heat-treated BDH silica ([32]), or ground quartz ([41]). The data from these sources at 0.1M concentration has been collected in Figure 7. The data of the various researchers is quite consistent, in spite of the differences in origin of the suspensions, and the different electrolytes used. The slope of the points above pH 7 shows that the adsorption capacitance for cations is very large for both sodium and potassium ions, around 200 μF/cm^2. Such a capacitance corresponds to a distance of 0.25Å, when using the dielectric constant of immobilized water molecules. The equilibrium constant for adsorption is low, however, since both K_{Na^+} and K_{K^+} lie between 0.1 and 0.01 dm^3/mol. A possible interpretation of these results is as follows: there is little specific attraction between SiO$_2$ and alkali cations,

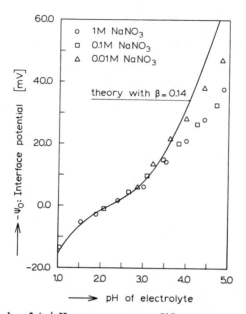

Figure 5. Examples of ψ_0/pH measurements on SiO_2: measurements around pH_{pzc} at various ionic strengths of a $NaNO_3$ background electrolyte. From Reference (18). Reproduced with permission from Ref. (18). Copyright 1983, IEEE.

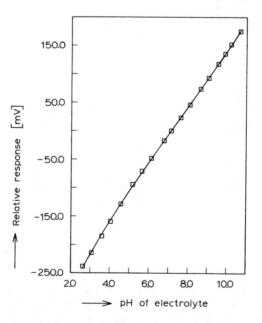

Figure 6. The ψ_0/pH response of Al_2O_3 in 0.1M $NaNO_3$. Note that the curve is not quite linear. Reproduced with permission from Ref. (18). Copyright 1983, IEEE.

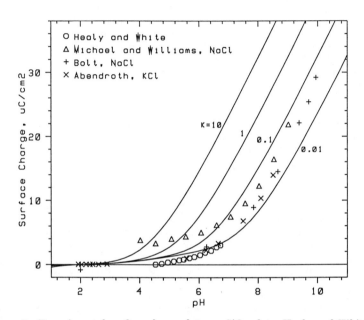

Figure 7. Experimental surface charge data on SiO_2, from Healy and White ($\underline{20}$) (pH_{pzc}= 2 assumed), Micheal and Williams ($\underline{41}$), Bolt ($\underline{39}$), and Abendroth ($\underline{40}$). The theoretical curves are calculated with $pK_{a1} = -1.3$, $pK_{a2} = 5.7$, and $C_{cation} = 200\ \mu F/cm^2$.

and adsorption is only significant when the surface is highly negatively charged; the SiO_2 surface, however, is able to accommodate these adsorbed or complexed cations in such a way that their average plane of charge is nearly identical to the plane of charge on the surface oxygens. This agrees with the idea, due to de Bruyn *et al.* (42,43), and Yates *et al.* (4), that there is space for complexed cations between the surface oxygens.

Similar results for γ-Al_2O_3 at 0.1M are shown in Figure 8. Unfortunately, the data in the literature are much more dispersed in this case, even for γ-Al_2O_3 in the same electrolyte solutions, which precludes quantitative conclusions. Some clear differences from the previous figure can be observed, however. The adsorption capacitances are much smaller on alumina, about 80 $\mu F/cm^2$ for perchlorate ions, and 140 $\mu F/cm^2$ for sodium. These values can be compared to Smit and Holten's results for NaBr adsorption on α-Al_2O_3 (21), who find that the capacitance for adsorption of bromine ions is 348 $\mu F/cm^2$, which is much higher than for perchlorate. Their value for sodium adsorption is 153 $\mu F/cm^2$, which agrees very well with the data in Figure 8. The considerable difference between bromine and perchlorate ions suggests that the adsorption capacitances can depend strongly on the type of ion. The simplest interpretation for the difference between cations and anions is in terms of the larger size of anions such as perchlorate relative to sodium ions. The adsorption for both cations and anions is much stronger than on SiO_2, however, and seems to be driven by chemical in addition to electrostatic forces. $K_{ClO_4^-}$ seems to lie between 1 and 100 dm^3/mol, and K_{Na^+} around 10 dm^3/mol.

Interpretation of Experimental ψ_0/pH Curves.   Close to the pH_{pzc}, ψ_0/pH curves are expected to be independent of counterion adsorption. Far from the pH_{pzc}, however, deviations in the 10 millivolt range due to such adsorption are to be expected. Measurements of the surface potential of SiO_2 are not stable enough in the pH range above 5 to permit such precise measurements (18). Al_2O_3 surfaces are much more stable, though, and it is possible to measure the effects of counterion adsorption (14). Figure 9 shows potential measurements on γ-Al_2O_3 in various electrolytes, together with a number of theoretical curves. The theoretical curves are clustered close together, in spite of the wide range of adsorption equilibrium constants considered. Thus, this figure confirms that counterion adsorption and complexation occur, but it is difficult to estimate the order of magnitude of the equilibrium constants involved. One indication is that around pH_{pzc} the slope $d\psi_0/dpH$ does not depend on the type of anions present, nor does the ionic strength of the electrolyte have much influence at high concentrations. These indications suggest that for anions such as nitrate or chloride ions the value of K_{anion} is not much higher than 1 dm^3/mol, which is the low end of the range deduced from the data in Figure 8.

Conclusions

σ_0/pH and ψ_0/pH results are sensitive to different aspects of the surface chemistry of oxides. Surface charge data allow the determination of the parameters which describe counterion complexation. Surface potential data allow the determination of the ratio $\beta = \sigma_S/\sigma_{DL}$. Given assumptions about the magnitude of the site density N_S and the Stern capacitance C_{St}, this quantity can be combined with the pH_{pzc} to yield values of K_{a1} and K_{a2}. Surface charge/pH data contain direct information about the counterion adsorption capacitances in their slope. To find the equilibrium constants for adsorption, a plot such as those in Figures 7 and 8 can be used, provided that K_{a1} and K_{a2} are independently known from ψ_0/pH curves.

Problems are encountered in practice in the application of this procedure. The data in the literature for both ψ_0/pH and σ_0/pH curves tend to vary considerably, even for the same material and electrolyte. This means the error bounds for the resulting set of equilibrium constants are quite large. Also, it is difficult to evaluate to which extent

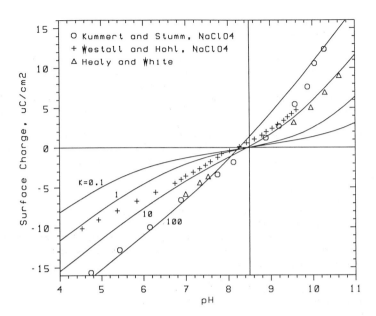

Figure 8. Experimental surface charge data on Al_2O_3, from Healy and White ([20]) (pH_{pzc} = 8.5 assumed), Westall and Hohl ([12]), and Kummert and Stumm ([34]). The theoretical curves are calculated with pK_{a1} = 5.9, pK_{a2} = 10.1, C_{cation} = 140 $\mu F/cm^2$, and C_{anion} = 80 $\mu F/cm^2$.

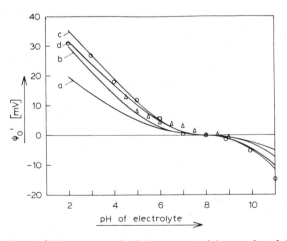

Figure 9. The ψ_0/pH response of Al_2O_3 presented in a reduced fashion, to remove the dominant linear response: $\psi_0' = \psi_0 + 48(pH - 8)$ (mV). The various electrolytes at 0.1M concentration are: ○: NaCl, △: NaH_2PO_4/Na_2HPO_4, □: Merck buffer. Line a is the site-dissociation theory without counterion adsorption, and the other lines include various amounts of counterion adsorption. Reproduced with permission from Ref. ([14]). Copyright 1983, North Holland.

the materials in a colloidal dispersion have the same surface properties as the thin films grown and deposited at high temperatures which are typically used for EIS capacitors or ISFETs.

The ideal situation would be a combined and simultaneous experiment in which an electrolyte/insulator/semiconductor device is used to monitor ψ_0 while a colloid dispersion of the same oxide is titrated to measure σ_0. To our knowledge, such experiments have yet to be carried out. The most complete parameter determination that could result from such an experiment would be the set: σ_S, σ_{DL}, K_{Na^+} and K_{Cl^-}, C_{Na^+} and C_{Cl^-}.

However, even using results from various sources in the literature for the surface charge data, we believe that the combined use of ψ_0 and σ_0 curves leads to more reliable values of the surface equilibrium constants than is obtainable otherwise.

Legend of Symbols

σ_0	The surface charge at the oxide/electrolyte interface;
ψ_0	The surface potential at the oxide/electrolyte interface;
V_{FB}	The flat-band voltage of an electrolyte/insulator/silicon (EIS) capacitor;
E_{ref}	The reference electrode potential on a scale relative to vacuum;
Φ^{Si}	The work function of silicon;
q	The absolute value of the charge of an electron;
Q_{ins}	The insulator charge of an EIS capacitor;
C_{ins}	The insulator capacitance of an EIS capacitor;
k	Boltzmann's constant;
T	The absolute temperature;
K_{a1}	Equilibrium constant of a surface acid/base reaction;
pK_{a1}	A notation for $-\log_{10}(K_{a1})$;
$\{H^+\}$	Bulk activity of an ion;
[SOH]	Surface concentration of a surface species, in numbers per unit area;
pH_{pzc}	The pH of zero charge of the oxide surface, equal to $(pK_{a1} + pK_{a2})/2$;
F	A function equal to $\left([SOH_2^+]/[SO^-]\right)^{1/2}$;
N_S	The total number of sites per unit area on the oxide surface;
α_+	A notation for surface concentrations divided by the site density N_S;
χ	The dimensionless net surface charge;
δ	A quantity indicating surface reactivity, equal to $2(K_{a2}/K_{a1})^{1/2}$ and $2 \times 10^{-(\Delta pK/2)}$;
ΔpK	A quantity indicating surface reactivity, equal to $pK_{a2} - pK_{a1}$ and $-\log_{10}(\delta^2/4)$;
C_{St}	The Stern capacitance;
C_{DL}	The double layer capacitance in a linearized model;
c	The ionic strength of the electrolyte;
R	The universal gas constant;
ϵ	The dielectric constant of bulk water;
σ_{DL}	A characteristic charge indicating the double layer capacitance;
σ_S	A characteristic charge indicating the tendency of the surface to ionize;
σ_i	A charge characterizing the first-order slope of σ_0/pH curves;
β	The ratio σ_S/σ_{DL} which characterizes the pH sensitivity of ψ_0 at an oxide surface;
C_{Na^+}	The capacitance between the plane of cation complexation and the surface plane where σ_0 is located;
C_{Cl^-}	The capacitance between the plane of anion complexation and the surface plane where σ_0 is located;
n	The ratio α_{Na^+}/α_- which characterizes the extent of adsorption of cations;
p	The ratio α_{Cl^-}/α_+ which characterizes the extent of adsorption of anions.

Literature Cited

[1] Levine, S.; Smith, A.L. Disc. Faraday Soc., 1971, **52**, 290.

[2] Schindler, P.W.; Gamsjäger, H. Kolloid Z. und Z. Polymere, 1972, **250**, 759.

[3] Huang, C.P.; Stumm, W. J. Colloid Interface Sci., 1973, **43**, 409.

[4] Yates, D.E.; Levine, S.; Healy, T.W. J. Chem. Soc. Faraday Trans. I, 1974, **70**, 1807.

[5] Davis, J.A.; James, R.O.; Leckie, J.O. J. Colloid Interface Sci. 1978, **63**, 480.

[6] Regazzoni, A.E.; Blesa, M.A.; Maroto, A.J.G. J. Colloid Interface Sci., 1983, **91**, 560.

[7] Milonjić, S.K.; Ilić, Z.E.; Kopecni, M.M. Colloids and Surfaces, 1983, **6**, 167.

[8] Houchin, M.R.; Warren, L.J. J. Colloid Interface Sci., 1984, **100**, 278.

[9] Sprycha, R. J. Colloid Interface Sci., 1984, **102**, 173.

[10] Helmy, A.K.; Ferreiro, E.A.; de Bussetti, S.G. Z. Phys. Chemie (Leipzig), 1980, **261**, 1065.

[11] Johnson, R.E., Jr. J. Colloid Interface Sci., 1984, **100**, 540.

[12] Westall, J.; Hohl, H. Advan. Colloid Interface Sci., 1980, **12**, 265.

[13] Sposito, G. J. Colloid Interface Sci., 1983, **91**, 329.

[14] Bousse, L.; de Rooij, N.F.; Bergveld, P. Surface Science, 1983, **135**, 479.

[15] Schenck, J.F. J. Colloid Interface Sci., 1979, **61**, 569.

[16] Cichos, C.; Geidel, T. Colloid Polym. Sci., 1978, **256**, 1140.

[17] Bousse, L. J. Chem. Phys., 1982, **76**, 5128.

[18] Bousse, L.; de Rooij, N.F.; Bergveld, P. IEEE Trans. Electron Devices, 1983, **ED–30**, 1263.

[19] James, R.O.; Parks, G.A. In "Surface and Colloid Science"; Matijevic, E., Ed.; Plenum: New York, 1982; Vol. 12, p. 119.

[20] Healy, T.W.; White, L.R. Advan. Colloid Interface Sci., 1978, **9**, 303.

[21] Smit, W.; Holten, C.L.M. J. Colloid Interface Sci., 1980, **78**, 1.

[22] Trassatti, S. J. Electroanal. Chem., 1983, **150**, 1.

[23] Parsons, R. In "Modern Aspects of Electrochemistry"; Bockris, J.O'M.; Conway, B.E., Eds.; Butterworths: London, 1954; p. 103.

[24] Dousma, K. Ph.D. Dissertation, Rijksuniversiteit Utrecht, The Netherlands, 1979.

[25] Harame, D.L. Ph.D. Dissertation, Stanford University, California, 1984.

[26] Healy, T.W.; Yates, D.E.; White, L.R.; Chan, D. J. Electroanal. Chem., 1977, **80**, 57.

[27] Grauer, R.; Stumm, W. Coll. Polymer Sci., 1982, **260**, 959.

[28] Siu, W.M.; Cobbold, R.S.C. IEEE Trans. Electron Devices, 1979, **ED-26**, 1805.

[29] Sobczyńska, D.; Torbicz, W. Sensors and Actuators, 1984, **6**, 93.

[30] Smit, W.; Holten, C.L.M.; Stein, H.N.; de Goeij, J.J.M.; Theelen, H.M.J. J. Colloid Interface Sci., 1978, **63**, 120.

[31] Foissy, A.; M'Pandou, A.; Lamarche, J.M.; Jaffrezic-Renault, N. Colloids Surf., 1982, **5**, 363.

[32] Yates, D.E.; Healy, T.W. J. Colloid Interface Sci., 1976, **55**, 9.

[33] Furlong, D.N.; Yates, D.E.; Healy, T.W. In "Electrodes of Conductive Metallic Oxides" Part B; Trasatti, S., Ed., Elsevier: Amsterdam, 1981; p. 367.

[34] Kummert, R.; Stumm, W. J. Colloid Interface Sci., 1980, **75**, 373.

[35] Bousse, L.; Bergveld, P. J. Electroanal. Chem., 1983, **152**, 25.

[36] Amari, A. Thesis, Université Paul Sabatier de Toulouse, France, 1984.

[37] Iler, R.K. "The Chemistry of Silica," Wiley: New-York, 1979.

[38] Fung, C.W.; Cheung, P.W.; Ko, W.H. Proc. of the 1980 International Electron Devices Meeting, p. 689.

[39] Bolt, G.H. J. Phys. Chem., 1957, **61**, 1166.

[40] Abendroth, R.P. J. Colloid Interface Sci., 1970, **34**, 591.

[41] Michael, H.L.; Williams, D.J.A. J. Electroanal. Chem., 1984, **179**, 131.

[42] Bérubé, Y.G.; de Bruyn, P.L. J. Colloid Interface Sci., 1968, **28**, 92.

[43] Blok, L.; de Bruyn, P.L. J. Colloid Interface Sci., 1970, **32**, 533.

RECEIVED June 25, 1986

6

Free Energies of Electrical Double Layers at the Oxide-Solution Interface

Derek Y. C. Chan

Department of Mathematics, University of Melbourne, Parkville, Victoria 3052, Australia

A simple graphical method is used to illustrate the roles of the charge-potential relationship of surfaces bearing ionizable groups and the charge-potential relationship of the diffuse double layer in determining the free energy of formation of a charged oxide/solution interface as well as the double layer interaction free energy involving such surfaces.

In all physical and chemical processes, and in particular those of relevance to geochemistry, that involve the oxide/aqueous solution interface, it is important to understand the general, non-specific characteristics of that interface before focussing on those specific processes or mechanisms of interest. Due to the structure of mineral surfaces, the mineral oxide/aqueous solution interface will invariably acquire a net charge or electrostatic potential relative to the bulk solution. The electrical state of the interface will depend in part on the chemical reactions that can take place on the mineral surface, and in part on the electrolytic composition of the aqueous environment.

From the preceding contributions in this volume it is evident that the techniques of modelling the electrical double layer properties at the oxide/electrolyte interface have been well developed (2, 11). However, the problem still contains a certain amount of 'art form' in the sense that there is more than one school of thought as to how the various modelling techniques should be applied.

The aim of this paper is not to add to the current debate but to present a simple graphical method of analysing the free energy of formation of the electrical double layer at the oxide/solution interface (1). This will provide a simple way of visualizing the complementary roles of chemical reactions or surface properties of

0097-6156/86/0323-0099$06.00/0
© 1986 American Chemical Society

the mineral and the nature of the diffuse layer in determining the thermodynamics of the interface. While an amphoteric surface is being used as a specific example to illustrate the key ideas, the concepts presented here are quite general and can be applied to other types of surfaces such as semiconductors.

The same graphical method can also be used to illustrate the nature of the double layer interaction free energy and to bring out a simple physical result which can be used to check numerical algorithms commonly used to calculate the interaction free energy.

The Equations of State

a. The Oxide Surface

Let us consider the basic physical mechanisms whereby a surface can develop a surface charge or surface potential.

For the familiar dropping mercury electrode, the electrical potential ψ_0 at the metal surface relative to the bulk region of the electrolyte is controlled by an external potential source – a constant voltage source. In this case, ψ_0 can be set to any value (within reasonable physical limits) as the mercury/electrolyte interface does not allow charge transfer or chemical reactions to occur (at least to a good approximation for the case of NaF). Therefore, we can say that the equation of state of the mercury surface is

$$\psi_0 = \text{constant} \tag{1}$$

where the constant can be adjusted by changing the constant voltage source.

In a typical inorganic oxide, the oxide surface acquires a charge by the dissociation or adsorption of potential determining ions at specific amphoteric surface groups or sites. As a consequence the equation of state of such surfaces will involve parameters that characterize surface reactions. In addition, one may also allow for the adsorption of anions and cations of the supporting electrolyte. However, in this paper we shall ignore this possibility to keep the discussion clear. Such embellishments of the model of the surface do not alter the key ideas presented here.

We derive the equation of state of an amphoteric surface by considering the generic dissociation reactions involving potential determining ions:

$$AH_2^+ \rightleftharpoons AH + H^+$$

$$AH \rightleftharpoons A^- + H^+ \tag{2}$$

where we have assumed the potential determining ions to be hydrogen ions. The change in free energy associated with these reactions are characterized by the intrinsic dissociation constants K_+ and K_-:

$$[AH] \cdot H_s / [AH_2^+] = K_+$$

$$[A^-] \cdot H_s / [AH] = K_- \tag{3}$$

Strictly, the intrinsic dissociation constants should be defined in terms of the activities of the various species at the surface. However, we shall neglect corrections due to activity coefficients and use surface concentrations instead. This approximation is consistent with the use of the Gouy–Chapman theory to describe the behaviour of the diffuse double layer in the electrolyte.

If the surface is populated with these amphoteric groups at N_s sites/unit area, the surface charge density of the surface is (e = protonic charge)

$$\sigma_0^s = eN_s \{[AH_2^+] - [A^-]\}/\{[AH] + [AH_2^+] + [A^-]\} \tag{4}$$

$$= eN_s \{H_s/K_+ - K_-/H_s\}/\{1 + H_s/K_+ + K_-/H_s\} \tag{5}$$

We have expressed the surface charge density in terms of the solution concentration of potential determining ions H_s at the surface and the dissociation constants K_+, K_- of the positive and negative groups. [K_+, K_- are sometimes written as K_{a1} and K_{a2} in the literature (3)]. From Equation 5 we see that σ_0 can be varied by changing the solution concentration of potential determining ions at the surface, which is in turn related to the concentration of potential determining ions H, in the bulk solution – we shall assume that this relation is given by the Boltzmann distribution

$$H_s = H \, e^{-e\psi_0/kT} \tag{6}$$

where ψ_0 is the mean electrostatic potential at the interface relative to the bulk solution. Equation 6 is consistent with the Gouy–Chapman model which we shall use to describe the diffuse double layer.

Combining Equations 5 and 6, the expression for the surface charge density can be written as

$$\sigma_0^s = eN_s \{\delta \sinh [e(\psi_N - \psi_0^s)/kT]/$$

$$\{1 + \delta \cosh [e(\psi_N - \psi_0^s)/kT]\} \tag{7}$$

where

$$\delta \equiv 2(K_+/K_-)^{1/2} = 2 \times 10^{\Delta pK/2} \tag{8}$$

$$\Delta pK \equiv pK_+ - \Delta pK_- \tag{9}$$

$$pH_{pzc} \equiv (pK_+ + pK_-)/2 \qquad (10)$$

$$\psi_N = 2.303(kT/e) \, [pH_{pzc} - pH] \qquad (11)$$

and pH_{pzc} is the value of the bulk pH at the point of zero charge. Equation 7 is an equation of state of the surface which is dictated by the equilibrium dissociation of surface groups. That is, the dissociation reaction of the surface groups imposes a relation between the potential ψ_0^s and the charge density σ_0^s which the interface can acquire. Note that this relationship between the surface charge and surface potential is dependent only on the density of surface groups N_s, the dissociation constants K_+, K_- and the bulk pH – the concentration of potential determining ions; but independent of any other solution properties. Thus, if the surface potential is known then the association or dissociation of the surface groups will ensure that the surface charge at equilibrium will be that given by Equation 7. For a particular value of the Nernst potential ψ_N, the surface charge σ_0^s can be calculated as a function of the surface potential ψ_0^s. But since σ_0^s only depends on $(\psi_N - \psi_0^s)$, see Equation 7, the location of this $\sigma_0^s - \psi_0^s$ curve is determined by the sign of the Nernst potential ψ_N or the bulk concentration of potential determining ions (pH) (see Equation 7 and Figure 1) while the overall shape of this curve is controlled by the dissociation constants via the quantity δ (see Equation 8) (2). The superscript s serves as a reminder that we are dealing with a charge-potential relationship that is dictated by chemical reactions of the surface groups.

Although a family of $\sigma_0^s - \psi_0^s$ values are allowed under Equation 7 the actual equilibrium state of the oxide/solution interface will be determined by the dissociation of the surface groups and the properties of the electrolyte or the diffuse double layer near the surface. For surfaces that develop surface charges by different mechanisms such as for semiconductor, there will be an equation of state or charge-potential relationship that is analogous to Equation 7 which characterizes the electrical response of the surface.

b. The Diffuse Double Layer

We shall use the familiar Gouy-Chapman model (3) to describe the behaviour of the diffuse double layer. According to this model the application of a potential ψ_0^d at a planar solid/electrolyte interface will cause an accumulation of counter-ions and a depletion of co-ions in the electrolyte near the interface. The disposition of diffuse double layer implies that if the surface potential of the planar interface at a 1:1 electrolyte is ψ_0^d then its surface charge density σ_0^d will be given by (3)

$$\sigma_0^d = (2\kappa\varepsilon_0\varepsilon_r kT/e) \, \sinh \, (e\psi_0^d/2kT) \qquad (12)$$

where κ is the inverse Debye length of the electrolyte and ε_r is the relative permittivity of the solvent. Equation 12 for σ_0^d as a function of ψ_0^d may be regarded as an equation of state of the interface which is dictated by the diffuse double layer – thermal motions of and coulombic interactions between ions in the electrolyte will control the equilibrium amount of charge accumulated as a result of the applied potential. A more sophisticated treatment of the diffuse layer which includes effects such as finite ion size and fluctuation potential effects will result in a different equation connecting σ_0^d and ψ_0^d (4). The superscript d denotes a charge–potential relationship of the interface that is controlled by the diffuse double layer in the electrolyte.

c. The Equilibrium Point

The properties of the equilibrium state of the oxide/solution interface must satisfy both the equation of state of the amphoteric surface and the diffuse double layer. Graphically, the equilibrium state corresponds to the unique point of intersection between the curves representing Equations 7 and 12. An illustration of this result is given in Figure 2. Given all other parameters being held constant, the equilibrium point may be moved by changing the value of ψ_N or equivalently by adjusting the pH relative to the point of zero charge of the interface (see Equation 11). Detailed numerical calculations and examples of the variations of the surface charge density σ_0 and surface potential ψ_0 as a function of solution pH have been given for a range of parameters that are representative of the oxide/aqueous solution interface (2, 7, 10). A distinctive feature of these calculations is that with an amphoteric model for the development of a charged oxide/solution interface, the surface potential as a function of solution concentration of potential determining ions is non–Nernstian. To illustrate this result, we use Equations 7 and 12 to obtain an expression for the slope of the ψ_0 vs pH curve

$$\frac{d\psi_0}{dpH} = 2.303 \left(\frac{kT}{e}\right)$$

$$\{1 + \frac{\cosh(e\psi_0/2kT)[1 + \delta \cosh[e(\psi_N - \psi_0)/kT]^2}{2 \gamma \delta[\delta + \cosh(e(\psi_N - \psi_0)/kT)]}\}^{-1} \qquad (13)$$

where $\gamma = (10^3 N_s \kappa/4N_A C)$, with N_A being Avogadro's constant and C the bulk 1:1 electrolyte concentration in mol dm^{-3}. For a Nernstian system we have $d\psi_0/dpH = 2.303(kT/e) = 59.2mV$ at 298K. In Figure 3 we have displayed the slope function $d\psi_0/dpH$ given by Equation 13 to illustrate the degree of non–Nernstian behaviour.

As mentioned earlier the possibility of the adsorption of ionic species of the supporting electrolyte will complicate the equation of state of the surface, but the general ideas discussed so far remain valid.

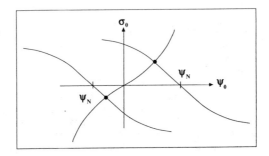

Figure 1. The charge-potential curves for a Gouy-Chapman diffuse layer and an amphoteric surface where $\psi_N > 0$ (< 0) corresponds to pH $<$ pH$_{pzc}$ ($>$ pH$_{pzc}$).

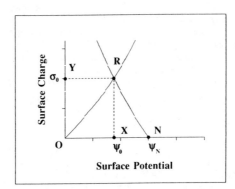

Figure 2. The charge-potential curves for a Gouy-Chapman diffuse layer and an amphoteric surface: 10^{-1}M 1:1 electrolyte, $\Delta pK = 2$, $N_s = 1 \times 10^{18}$ m^{-2}.

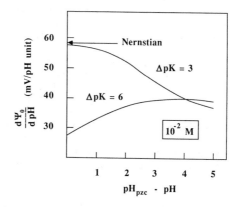

Figure 3. The function $d\psi_0/dpH$, Equation 13, for an amphoteric surface: 10^{-2} M 1:1 electrolyte, $N_s = 5 \times 10^{18}$ m^{-2}.

Since the equations of state of the system are summarized by the curves in Figure 2, all interesting thermodynamic properties of the interface will have a simple representation in such a diagram. We shall consider the free energy of formation of a single charged surface and the interaction free energy due to the overlap of two identical planar double layers.

Free Energy of Formation of One Surface

In order to calculate the free energy change associated with the spontaneous formation of the electrical double layer at the oxide/solution interface we choose as the initial state one in which the amphoteric surface is net neutral and the ionic profile in the electrolyte is uniform up to the surface. Consistent with the Gouy-Chapman model imaging or fluctuation potential effects are neglected. The final state will be a charged interface characterized by surface potential ψ_0 and surface charge σ_0.

The free energy of formation of the charged interface is calculated as follows (3). Starting from the reference state, infinitesimal amounts of potential determining ions are transferred from the bulk solution on to the surface. After each step, ions in the solution are allowed to re-establish equilibrium and the configuration of the surface groups is also allowed to come to the new equilibrium state. The change in free energy per unit area df, in transferring $d\Gamma$ moles per unit area of potential determining ions from the bulk solution to the surface is given by

$$df = [\mu^s - \mu^b]d\Gamma \tag{14}$$

where μ^s is the electrochemical potential of potential determining ions bound to the surface at coverage

$$\Gamma = [AH_2^+] - [A^-] \tag{15}$$

and μ^b is the constant chemical potential in the bulk solution. We formally separate the electrical contribution to μ^s from the chemical part:

$$\mu^s \equiv \tilde{\mu}^s(\Gamma) + e\,\psi_0 \tag{16}$$

with the assumption that the chemical part $\tilde{\mu}^s(\Gamma)$ is only a function of Γ and not of the mean-electrostatic potential ψ_0. This assumption is met in our mass action/Gouy-Chapman model of the interface as well as in all models proposed so far in the literature.

The total change in free energy per unit area in charging up the interface is obtained by integrating Equation 14

$$f = \int_0^{\sigma_0} \psi_0(\sigma)d\sigma + \int_0^{\Gamma_0} [\tilde{\mu}^s(\Gamma) - \mu^b]d\Gamma \tag{17}$$

where $\sigma_0 \equiv e\Gamma_0$ is the equilibrium surface charge density, Γ_0 is the total amount of potential determining ions transferred to the surface at equilibrium and is determined by the condition

$$\mu^b = \mu^s \equiv \tilde{\mu}^s(\Gamma_0) + e\ \psi_0(\sigma_0) \tag{18}$$

The first integral in Equation 17 is identified as the electrical contribution to the change in free energy in forming the charged interface (3) and may be evaluated using Equation 12

$$\psi_0(\sigma) \equiv \psi_0^d(\sigma) = (2kT/e)\ \sinh^{-1}(\sigma/2\kappa\varepsilon_0\varepsilon_r kT) \tag{19}$$

The second integral in Equation 17 is the chemical contribution due to chemical reactions of the potential determining ions with the surface groups. This term may be recast as follows. Chemical equilibrium between potential determining ions bound on the surface and those in the solution adjacent to the surface during the changing process means that the chemical part of the chemical potentials are equal, i.e.

$$\tilde{\mu}^s(\Gamma) = \mu_0 + kT\ \ell n\ H_s \tag{20}$$

but since the chemical potential of potential determining ions in the bulk solution is

$$\mu^b = \mu_0 + kT\ \ell n\ H \tag{21}$$

we have

$$\mu^s(\Gamma) - \mu^b = kT\ \ell n(H_s/H) \tag{22}$$

Thus using Equations 5-7, 15 and 22 the second integral in Equation 17 over $d\Gamma$ may be written as an integral in terms of the charge-potential carve given by Equation 7. The expression for the free energy per unit area in forming the charged interface then has the compact form

$$f = \int_0^{\sigma_0} \psi_0^d(\sigma)d\sigma - \int_0^{\sigma_0} \psi_0^s(\sigma)d\sigma \tag{23a}$$

$$= -\int_0^{\psi_0} \sigma_0^d(\psi)d\psi - \int_{\psi_0}^{\psi_N} \sigma_0^s(\psi)d\psi \tag{23b}$$

Equation 23b follows from 23a using integration by parts – the integrated terms cancel because of the equilibrium condition:

$$\psi_0^d(\sigma_0) = \psi_0^s(\sigma_0) \tag{24}$$

We recall that the first integral in Equation 23a represents the change in electrical free energy in forming the diffuse double layer. This contribution to f, the free energy of formation of the charged interface, is positive and hence represents an unfavourable component which opposes the formation of the charged interface.

This point has been noted in earlier treatment of this problem (3). The value of this contribution depends on the model we adopt to describe the diffuse layer.

The second term in Equation 23a represents the change in free energy due to chemical reactions associated with the formation of an amphoteric surface with a net charge σ_0. This chemical contribution is negative and hence favours the formation of the charged interface. Indeed, if the charged interface is to form spontaneously, the lowering of the free energy of the system due to chemical reactions must outweigh the unfavourable contribution necessary in forming the diffuse double layer. In other words, it is the chemical reactions of the surface groups that provides the main driving mechanism for the formation of the charged interface. The magnitude of this chemical contribution depends specifically on the reactions that can take place at the oxide surface.

Apart from the assumptions made in deriving Equation 23, this expression for the free energy is independent of the details or types of reactions that can take place at the surface.

As it stands, Equation 23a is cumbersome to use directly as we need to solve Equations 7 and 12 to obtain ψ_0^d and ψ_0^s as functions of the surface charge σ. However, graphically the two terms can be represented as areas under the equations of state of the surface and the diffuse layer. The first term in Equation 23a, the positive electrical contribution to the free energy of formation of the charged interface is simply the area in the region denoted by ORYO in Figure 2. The negative chemical contribution, the second term in Equation 23a, is the negative of the area denoted by ONRYO in Figure 2.

We can obtain an explicit expression for the change in free energy per unit area in forming an amphoteric charged surface using Equations 7, 12 and 23a:

$$f \equiv -(4\kappa\varepsilon_0\varepsilon_r)(kT/e)^2[\cosh(e\psi_0/2kT) - 1]$$

$$- N_s kT \, \ell n \, \{[1 + \delta \cosh \, (e(\psi_N - \psi_0)/kT)]/[1 + \delta]\} \qquad (25)$$

The graphical representation for f is the negative of the area designated as ORNO in Figure 2. Note that f is always negative irrespective of the sign of the charge on the surface.

The result for the free energy of formation of a charged interface at an amphoteric surface is a generalization of results previously obtained for surfaces with only one type of ionizable groups (1, 5). Equation 25 also differs from the expression for the free energy of formation of a charged interface given in (3) due to the presence of the second term on the right hand side. The expression given in (3) is pertinent to a 'constant potential' surface. That is, the equation of state of the surface is taken to be of the form (3):

$$\tilde{\mu}^s(\Gamma) - \mu^b = -e\psi_0 \qquad (26)$$

where ψ_0 is assumed to be a constant, <u>independent</u> of the surface charge density. The graphical representation of the free energy of formation of a charged interface under the constant potential assumption is the negative of the area given by ORXO in Figure 2 – this corresponds to the first term on the right hand side of Equation 25. The second term on the right hand side of Equation 25 is represented by the region XRNX in Figure 2.

Interaction Free Energy Between Two Plates

The interaction free energy per unit area between two planar double layers $V(L)$ is just the difference between the change in free energy per unit area in charging up two surfaces at a distance L apart and that at infinite separation

$$V(L) = 2 \ (f_L - f_\infty) \qquad (27)$$

The change in free energy f_L in charging up a surface at a given separation L can be different from the corresponding quantity f_∞ at infinite separation L = ∞, (given by Equation 23) if the equations of state of the surface and that of the diffuse layer varies with separation. In all models proposed so far, the equation of state of the surface (the adsorption isotherm of potential determining ions) does not depend explicitly on the separation between the surfaces. This is physically reasonable since we do not expect the processes that determine the association and dissociation of surface groups to depend on L until the plate separation is of the order of atomic or molecular dimensions. In this regime, other effects such as the molecular nature of the surface, the ions and solvent which had so far been neglected must also be taken into account. On the other hand, the change in free energy in forming the diffuse layer will change with separation when the plates are of the order of a few Debye lengths apart.

Analogous to Equation 22 we can write the change in free energy in forming a single charged surface as

$$(f_L - f_\infty) = \int_0^{\sigma_0} [\psi_0^d(\sigma, L) - \psi_0^s(\sigma)] \, d\sigma$$

$$- \int_0^{\sigma_0} [\psi_0^d(\sigma, \infty) - \psi_0^s(\sigma)] \, d\sigma \qquad (28)$$

where we have assumed that the equation of state of the surface $\psi_0^s(\sigma)$ is independent of the plate separation. The quantity $\psi_{0d}(\sigma, L)$ is the charge–potential relationship for overlapping diffuse layers at a separation L. For the Gouy–Chapman model this can readily be obtained by solving the Poisson-Boltzmann equation (<u>3</u>). In Figure 4 we show the function $\psi_0^d(\sigma, L)$ for various values of the plate separation. Note that at a fixed surface potential the surface charge decreases as the separation decreases. This is an expected result since from Gauss's law the surface charge is proportional to the slope of the potential profile at the surface, and at small separations the potential profile will be nearly constant between the two surfaces.

The graphical representation of the two integrals on the right hand side of Equation 28 is given in Figure 5. The first integral is the negative of the area of the region OSNO, the second integral corresponds to the negative of the area ORNO and the difference is simply the shaded area ORSO. This area is always positive, which corresponds to the well known result that within the Gouy-Chapman model the double layer interaction between identical surfaces is always repulsive.

As the separation L approaches zero, we note that the charge of the diffuse layer is essentially zero for small surface potentials – see the curve for $\kappa L = 0.1$ in Figure 4. In the limit $L \to 0$, it is easy to see from Figure 5 that the area ORSO will approach the area ORNO. In other words

$$V(L \to 0) \to -2\, f_\infty \qquad (29)$$

with f_∞ given by Equation 23. This result states that the work needed to bring two charged surfaces from infinity to zero separation (when the surfaces will become uncharged) is, apart from the sign, twice the energy of formation of two charged surfaces at infinite separation. That is, the energy needed to completely 'discharge' two surfaces is $V(0)$. Furthermore, as the surface charge σ_0 reduces to zero as $L \to 0$, we can see from Figure 5 that the surface potential ψ_0 will approach the Nernst potential ψ_N (8). These observations are useful for checking the numerical accuracy of implementations of algorithms for calculating the double layer interaction (6) because Equations 25 and 29 gives us an exact analytical result on the value of $V(0)$.

We note that the interaction free energy under the assumptions of constant charge or constant potential rather than at chemical equilibrium can also be represented in Figure 5. The interaction free energy at constant potential is given by the area designated by ORPO while the interaction free energy under constant surface charge corresponds to the region ORCSPO. It is clear from Figure 5 that the double layer repulsion under constant charge is higher than that under chemical dissociation equilibrium in our amphoteric model; which in turn is higher than the repulsion under constant potential interaction.

Detailed numerical examples of the behaviour of the surface charge and surface potential when the electrical double layer of two identical amphoteric surfaces overlap and interact are available in the literature (8). Examples of the differences between the form of the interaction free energy under constant

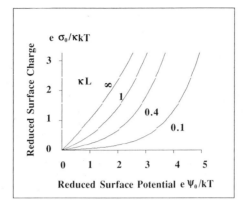

Figure 4. The charge–potential curves for overlapping Gouy–Chapman diffuse layers as a function of separation.

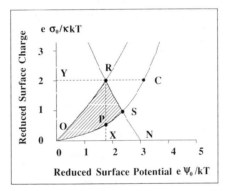

Figure 5. Graphical representation of the interaction free energy for two amphoteric surfaces: 10^{-1} M 1:1 electrolyte, $\Delta pK = 2$, $N_s = 1 \times 10^{18}$ m^{-2}.

charge, constant potential and dissociation equilibrium have also been given (8). The interaction involving dissimilar surfaces bearing ionizable groups is more cumbersome to analyse. A method of tracking the changes in the surface charge, the surface potential of the two surfaces as well as the force between the surfaces has been given earlier (9).

Concluding Remarks

Using a simple amphoteric model for the mineral surface, we have demonstrated the role specific chemical binding reactions of potential determining ions in determining the electrical properties and thermodynamics of the oxide/solution interfaces. A by-product of our study is that under appropriate conditions, an amphoteric surface can show marked deviations from ideal Nernstian behaviour. The graphical method also serves to illustrate the difference between double layer interactions under the assumptions of constant potential, constant charge or dissociation equilibrium.

Returning to our introductory remarks about the existence of various models for the oxide/solution interface, it may be appropriate to point out that the results of very relevant experiments based on electrokinetic measurements are often not used in conjunction with titration data. Granted that there may be additional difficulties in identifying the precise location the slipping plane and hence the significance of the electrokinetic ζ potential may be open to debate, both titration and electrokinetic data ought to be combined where possible to elucidate the behaviour of the oxide/solution interface.

Literature Cited

1. Chan, D.Y.C.; Mitchell, D.J. J. Colloid Interface Sci. 1983, 95, 193.
2. James, R.O.; Parks, G.A. Surface and Colloid Sci. 1982, 12, 110.
3. Verwey, E.J.W.; Overbeek, J.Th.G. "Theory of the Stability of Lyophobic Colloids"; Elsevier: Amsterdam, 1948; Chap. 2.
4. Carnie, S.L.; Torrie, G.M. Adv. Chem. Phys. 1984, 56, 141.
5. Payens, Th.A.J.; Phillips Res. Rep. 1955, 10, 425.
6. Chan, D.Y.C.; Pashley, R.M.; White, L.R. J. Colloid Interface Sci. 1980, 77, 283.
7. Healy, T.W.; White, L.R. Adv. Colloid Interface Sci. 1978, 9, 303.
8. Chan, D.; Perram, J.W.; White, L.R.; Healy, T.W. J. Chem. Soc., Faraday Trans. I. 1976, 71, 1046.
9. Chan, D.Y.C.; Healy, T.W.; White, L.R. J. Chem. Soc. Faraday I 1977, 76, 2844.
10. Healy, T.W.; Yates, D.E.; White, L.R.; Chan, D. J. Electroanal. Chem. 1977, 80, 57.
11. Sposito, G. J. Colloid Interface Sci. 1983, 91, 329.
12. James, R.O. Colloid and Surfaces 1981, 2, 201.

RECEIVED June 18, 1986

ADSORPTION

7

Mechanism of Lead Ion Adsorption at the Goethite-Water Interface

Kim F. Hayes and James O. Leckie

Environmental Engineering and Science Group, Department of Civil Engineering, Stanford University, Stanford, CA 94305-4020

The pressure-jump relaxation technique has been used to determine the mechanism of the reaction of lead ion with the goethite/solution interface. The kinetic study required three types of information to unambiguously specify the detailed reaction mechanism: (1) overall equilibrium, (2) reaction stoichiometry, and (3) reaction rate data. Electrical double layer models must be used in both equilibrium and time-dependent relationships for reactions occurring at charged interfaces. Based on the kinetic results of this study, a bimolecular adsorption/ desorption reaction has been postulated for Pb^{2+} adsorption/desorption at the goethite/solution interface. The desorption step associated with breaking the bond of the inner-sphere lead hydroxyl surface complex is the rate-limiting step. Dependence of the rate constants with pressure-jump magnitude is consistent with adsorption/ desorption from sites with a distribution of bond energies.

Due to the fast kinetics of adsorption/desorption reactions of inorganic ions at the oxide/aqueous interface, few mechanistic studies have been completed that allow a description of the elementary processes occurring (half lives < 1 sec). Over the past five years, relaxation techniques have been utilized in studying fast reactions taking place at electrified interfaces (1-7). In this paper we illustrate the type of information that can be obtained by the pressure-jump method, using as an example a study of Pb^{2+} adsorption/desorption at the goethite/water interface.

Based on the pressure-jump relaxation results reported here, the following mechanism is postulated for the adsorption/desorption of Pb^{2+} ion at the goethite/water interface (8):

$$SOH + Pb^{2+} \underset{k_{-1}}{\overset{k_1}{\rightleftharpoons}} SOPb^+ + H^+ \qquad (1)$$

0097-6156/86/0323-0114$08.00/0

i.e., a bimolecular adsorption/desorption reaction, where k_1 and k_{-1} represent the forward and reverse rate constants. In order to verify the above mechanism, three types of information are required: (1) establish the overall equilibrium partitioning, (2) establish the reaction stoichiometry, e.g., the number of protons released per Pb^{2+} adsorbed, and (3) determine the rate law that correctly relates the observed product and reactant changes to the inverse of the relaxation time constant. In the discussion that follows the mechanism shown in Equation 1 is verified through the development of the required relationships based on (1)-(3) above.

Experimental Materials and Methods

The goethite used in this study was prepared according to the procedure described by Atkinson et al. (9). Adsorption experiments were performed in polypropylene tubes that were mixed by end-over-end rotation at 8 RPM at 25°C. The amount of metal ion adsorbed was determined using the ^{210}Pb isotope as a tracer and analyzing the supernatant for ^{210}Pb activity following solid/liquid separation. Solid/liquid separation was accomplished by centrifugation at 22,000 RCF for 1 hour at 25°C. Equilibration times for partitioning experiments were varied from 2 to 24 hrs with no significant difference in amount adsorbed observed.

All chemicals used were ACS reagent grade quality. Ionic strength, pH, and metal ion concentrations were adjusted using $NaNO_3$, HNO_3 and $NaOH$, and $Pb(NO_3)_2$, respectively. For high ionic strength solutions (I = 1.0 or 0.1M) the background electrolyte concentration was adjusted to give the desired ionic strength value; for the 0.01M system, the ionic strength was only approximately 0.01M. In this case the nitrate concentration was adjusted to be exactly 0.01M (e.g., HNO_3 + $NaNO_3$ + $Pb(NO_3)_2$ adjusted to nitrate = 0.01M). The solids concentration of 30 g/l for each experiment was obtained by dilution of a 60 g/l stock slurry. Each point on a pH versus percent adsorbed curve was obtained from a separate polypropylene tube.

The constants and parameters required for making triple layer model (TLM) computations (e.g., K_{a1}^{int}, K_{a2}^{int}, $K_{NO_3}^{int}$, K_{Na}^{int}, C_1, C_2, and N_s) were determined by the standard procedures.[3] The site density, N_s, of the goethite was estimated from isotherm data and surface area measurements. The surface area was measured by N_2 gas adsorption and the BET model. A site density of 7.0 sites/nm^2 was used for all model calculations; this value falls between the calculated value of 3.0 sites/nm^2 from crystallographic analysis of goethite (10) and tritium exchange results of 11.0 sites/nm^2 measured in this laboratory. As shown in Figure 1, the value of 7.0 sites/nm^2 gave a good model fit to experimental data obtained over a range of surface coverage and solution concentrations of lead. The acidity and electrolyte binding constants were estimated from potentiometric titrations by the extrapolation technique described in Davis et al. (11), assuming a site density of 7.0 sites/nm^2. C_2 was fixed at 20 $\mu F/cm^2$ in accordance with literature estimates and C_1 was obtained from a best fit of model calculations to experimental titration data. The TLM constants and parameters are summarized in Table I.

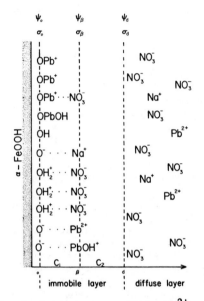

Figure 1.   TLM calculations at variable Pb^{2+} concentration to get best estimate of site density.

Table I. Triple Layer Model Parameters and Values

Surface Area = 52.0 m^2/g

(a)Site Density = 7.0 sites/nm^2

Capacitance: C_1 = 110 $\mu F/cm^2$; C_2 = 20 $\mu F/cm^2$

(b)K_{al}^{int} = -5.80

(b)K_{a2}^{int} = -11.1

(b)K_{Na}^{int} = -8.80

(b)$K_{NO_3}^{int}$ = 7.60

(a)Based on model fit to isotherm data (Figure 1).

(b)For reactions as written in Table II.

Kinetic experiments were conducted using a pressure-jump apparatus with conductivity detection. Details of the apparatus and its
operation can be found in Appendix A. Sample equilibration time can
have an effect on the kinetic results (e.g., slow processes (on the
order of hours-days) occurring concurrently but not monitored in the
time frame of the p-jump technique (milliseconds-seconds)); hence,
it is important to run kinetic experiments on samples with similar
equilibration history. All samples were equilibrated between 3 and
4 hours for the p-jump kinetic studies. The temperature of the
p-jump apparatus, which includes sample and reference solution
cells, was maintained at 25.0°C ± 0.1°C.

Equilibrium Model

At equilibrium the rate of all elementary reaction steps in the
forward and reverse directions are equal; therefore, this condition
provides a check point for studying reaction dynamics. Any postulated mechanism must both satisfy rate data and the overall equilibrium condition. Additionally, for the case of reactions occurring
at charged interfaces, the appropriate model of the interface must
be selected. A variety of surface complexation models have been
used to successfully predict adsorption characteristics when certain
assumptions are made and model input parameters selected to give the
best model fit (12). One impetus for this work was to establish a
self-consistent set of equilibrium and kinetic data in support of a
given modeling approach.
    Surface complexation models attempt to represent on a molecular
level realistic surface complexes; e.g., models attempt to distinguish between inner- or outer-sphere surface complexes, i.e., those
that lose portions of or retain their primary hydration sheath,
respectively, in forming surface complexes. The type of bonding is
also used to characterize different types of surface complexes;
e.g., a distinction between coordinative (sharing of electrons) or
ionic bonding is often made. While surface coordination complexes
are always inner-sphere, ion-pair complexes can be either inner- or
outer-sphere. Representing model analogues to surface complexes has
two parts: stoichiometry and closeness of approach of metal ion to

surface site. Generally, in multi-layer models, a closer distance
of approach of the adsorbing ion reflects a stronger interaction
with the surface. Some models assume the metal ions are bonded
directly to the surface (13, 14), while others place the metal ion
at a plane farther away (11, 15). In the TLM, analogues for either
surface coordination or ion-pair complex can be chosen. Also, background electrolyte surface reactions can be modeled allowing the
effects of changing ionic strength on metal ion adsorption to be
tested. Because of this flexibility, the TLM was chosen to establish the overall equilibrium behavior in this study.

Triple Layer Model

The triple layer model has been described in detail elsewhere (11,
16, 17); however, the model as reported here has been slightly modified from the original versions (11, 15) in two ways: (i) metal
ions are allowed to form surface complexes at either the o- or
β-plane insted of at the β-plane only, and (ii) the thermodynamic
basis of the TLM has been modified leading to a different relationship between activity coefficients and interfacial potentials. The
implementation and basis for these modifications are described below.

TLM Surface Complexes. As the name implies, the TLM is composed of
three planes where charge and potential are evaluated. The charge
at the o- and β-planes depends on the concentration and charge of
species adsorbed at the given plane. The concentration and charge,
in turn, depend on the constants of the mass law expressions and
reaction stoichiometries specified. Assuming protons bond to the
surface plane and background electrolyte (e.g., Na^+ and NO_3^-) form
ion-pairs with protonated or deprotonated surface hydroxyl sites at
the β-plane (11), the following mass action expressions can be
written:

$$SOH_2^+ \xrightleftharpoons{K_{a1}} SOH + H^+ \qquad K_{a1} = \frac{[SOH][H^+]}{[SOH_2^+]} \qquad (2)$$

$$SOH \xrightleftharpoons{K_{a2}} SO^- + H^+ \qquad K_{a2} = \frac{[SO^-][H^+]}{[SOH]} \qquad (3)$$

$$SOH + H^+ + NO_3^- \xrightleftharpoons{K_{NO_3}} SOH_2^+\!-\!NO_3^- \qquad K_{NO_3} = \frac{[SOH_2^+\!-\!NO_3^-]}{[SOH][H^+][NO_3^-]} \qquad (4)$$

$$SOH + Na^+ \xrightleftharpoons{K_{Na}} SO^-\!-\!Na^+ + H^+ \qquad K_{Na} = \frac{[SO^-\!-\!Na^+][H^+]}{[SOH][Na^+]} \qquad (5)$$

The charge balances based on these expressions are

$$\sigma_o = B([SOH_2^+] + [SOH_2^+\!-\!NO_3^-] - [SO^-] - [SO^-\!-\!Na^+]) \qquad (6)$$

$$\sigma_\beta = B([SO^-\!-\!Na^+] - [SOH_2^+\!-\!NO_3^-]) \qquad (7)$$

where B is equal to 10^6 F/A, F is faradays constant, and A is surface area per unit volume of suspension (e.g., m^2/l). From the electro-neutrality condition,

$$\sigma_o + \sigma_\beta + \sigma_d = 0 \tag{8}$$

The charge at the diffuse layer plane is calculated from Gouy-Chapman-Stern-Grahame theory, which for a symmetrical monovalent electrolyte of concentration C_s is given by

$$\sigma_d = -11.74 C_s^{1/2} \sinh\left(\frac{ze\psi_d}{2kT}\right) \text{ in } \mu C/cm^2 \tag{9}$$

The relationship between charge and potential are derived by assuming that the planes can be treated as plates of two parallel plate capacitors in series (18) with

$$\sigma_o = C_1(\psi_o - \psi_\beta) \tag{10}$$

and

$$-\sigma_d = C_2(\psi_\beta - \psi_d) \tag{11}$$

In the original versions of this TLM (16), metal ion reactions were written as ion-pair analogues with the metal ion placed in the β-plane:

$$SOH + Me^{2+} \underset{\longleftarrow}{\overset{K_{Me}}{\longrightarrow}} SO^{-}{-}Me^{2+} + H^+ \qquad K_{Me} = \frac{[SO^{-}{-}Me^{2+}][H^+]}{[Me^{2+}][SOH]} \tag{12}$$

with charge balances, e.g., in the o- and β-plane given by

$$\sigma_o = B([SOH_2^+] + [SOH_2^+{-}NO_3^-] - [SO^-] - [SO^-{-}Na^+] - [SO^-{-}Me^{2+}]) \tag{13}$$

$$\sigma_\beta = B([SO^-{-}Na^+] + 2[SO^-{-}Me^{2+}] - [SOH_2^+{-}NO_3^-]) \tag{14}$$

However, an alternative is to consider the TLM analogue of an inner-sphere surface coordination complex by placing the metal ion in the o-plane (19), e.g.,

$$SOH + Me^{2+} \underset{\longleftarrow}{\overset{K_{Me}}{\longrightarrow}} SOMe^+ + H^+ \tag{15}$$

By allowing metal ions to be placed in the o-plane, the surface charge is now given by (6):

$$\sigma_o = \sigma_H + \sigma_{Me} \tag{16}$$

where σ_H and σ_{Me} represent the contribution from protonated and de-protonated sites and metal bound sites, respectively. This leads to

$$\sigma_o = B([SOH_2^+] + [SOH_2^+{-}NO_3^-] + [SOMe^+] - [SO^-] - [SO^-{-}Na^+]) \tag{17}$$

$$\sigma_\beta = B([SO^-{-}Na^+] - [SOH_2^+{-}NO_3^-]) \tag{18}$$

with

$$\sigma_{Me} = B([SOMe^+]) \tag{19}$$

Figure 2 gives a schematic illustration of the TLM with examples of ion-pair and surface coordination complexes.

In representing TLM surface complex analogues, it is tempting to use interchangeably the terms ion-pair or outer-sphere complex when metal ions are placed in the β-plane and surface coordination or inner-sphere complex when metal ions are placed in the o-plane; however, placement of metal ions in these planes only implies a relative closeness of approach of the ion to the surface. Historically the TLM (i.e., the Grahame-Stern model) (20) was developed with a distinct physical picture of the interface in mind. The closest approaching ions located at the IHP (inner Helmholtz plane) were considered those that lost their primary hydration sheath on the side facing the surface so that the radius of the central ion touched the surface, OHP (outer Helmholtz plane) ions attached to the surface retaining their primary hydration sphere, and diffuse layer ions, which also retained their primary hydration sheath, approached no closer than the outside of the hydration sphere of ions in the OHP. In this historical context, the interpretation of the TLM discussed here would equate the surface plane to the IHP, the β-plane to the OHP, and have an additional diffuse layer d-plane. The need for identifying the IHP with the surface plane arises from considering that "surface" charge results from surface protolysis and coordination reactions. In this model construct, the terms inner- and outer-sphere have their implied structural meaning; however, depending on the values of interfacial capacitance, TLM estimates of the distance of ion approach may differ somewhat from the Grahame-Stern physical model. This serves as a warning that model analogues should not be taken literally to imply a given type of complex. Model results can, however, allow direct conclusions to be drawn about the relative closeness of approach and the best choice of planes in which to place ions regardless of the actual structure of surface complexes.

Final justification for using terms such as inner- or outer-sphere awaits direct spectroscopic confirmation. Electron Spin Resonance, Mossbauer, and Fourier Transform Infrared-Cylindrical Internal Reflection Spectroscopic techniques are being used to establish the structure of surface complexes (see, e.g., McBride, Ambe et al., and Zeltner et al., this volume). The potential for using EXAFS (extended x-ray absorption fine structure) to establish the type of surface complex for Pb^{2+} adsorbing onto goethite is currently being undertaken in our laboratory.

TLM Activity Coefficients. In the version of the TLM as discussed by Davis et al. (11), mass action equations representing surface complexation reactions were written to include "chemical" and "coulombic" contributions to the overall free energy of reaction, e.g., the equilibrium constant for the deprotonation reaction represented by Equation 2 has been given as

$$K_{al}^{int'} = \frac{[SOH][H^+]}{[SOH_2^+]} \frac{\gamma'_{SOH}\gamma'_{H^+}}{\gamma'_{SOH_2^+}} \exp(-e\psi'_o/kT) \tag{20}$$

where $K_{al}^{int'}$ is defined as the intrinsic acidity constant in absence of electrical double layer effects, the γ'_i are activity coefficients,

and the exponential term represents the coulombic contribution to
the overall equilibrium. The thermodynamic basis of the activity
coefficients and exponential term described by Equation 20 has been
attributed to short- and long-range interactions by Chan et al. (21)
and Sposito (10). In the modeling approach used for calculations in
this paper, no distinction is made between short- and long-range
contributions, leading to the result that

$$K_{al}^{int} = \frac{[SOH][H^+]}{[SOH_2^+]} \exp(-e\psi_o/kT) \qquad (21)$$

with

$$\frac{\gamma_{SOH}\gamma_{H^+}}{\gamma_{SOH_2^+}} = \exp(-e\psi_o/kT) \qquad (22)$$

where these γ_i represent activity coefficients due to both short-
and long-range interactions (19). In applying Equation 20 Davis et
al. (11) assume $\gamma'_{SOH}/\gamma'_{SOH_2^+} = 1$, leading to

$$K_{al}^{int'} = \frac{[SOH][H^+]}{[SOH_2^+]} \gamma'_{H^+} \exp(-e\psi_o'/kT) \qquad (23)$$

Comparing Equations 21 and 23 it is apparent that the main difference computationally lies in the appearance of γ'_{H^+} in Equation 23.
The TLM has been modified accordingly to eliminate activity corrections like γ'_{H^+} in Equation 23 for the modeling computations reported
here. A detailed discussion of the thermodynamic basis for making
this change can be found in Hayes (8) and Hayes and Leckie (19).
This change has two attractive features: (1) It eliminates the
separation of activity coefficients into short- and long-range
interactions, which cannot be evaluated separately in practice, and
(2) implicitly incorporates an expected effect of surface potential
on solution activity through the activity coefficient relationship
of Equation 22. Table II summarizes the relevant reaction and
activity coefficient terms based on the above modifications of
the TLM.

Equilibrium Results

Results of a two order-of-magnitude change in ionic strength (I =
1.0M–0.01M) on the adsorption of lead (2.0 × 10^{-3}M total Pb) onto
goethite (30.0 g/l) are shown in Figure 3. As shown, there is relatively little effect of changes in background electrolyte concentration on the position of the adsorption edge. The ionic strength
dependence data for Pb^{2+} ion adsorption at the goethite/water interface was modeled using the TLM as described above. Modeling results
for the inner- and outer-sphere cases are shown in Figures 4 and 5,
respectively. The surface protolysis and background electrolyte
constants and model parameters shown in Table I were used for all
computations. The general approach was to fit 0.1M ionic strength
adsorption data using either the lead inner- or outer-sphere model
analogue assuming a reaction stoichiometry of 1 or 2 protons released per lead ion adsorbed. Using the reaction stoichiometry and

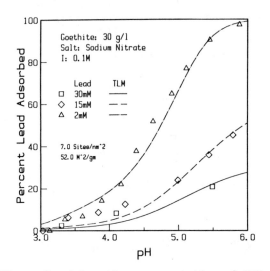

Figure 2.   Schematic representation of TLM.

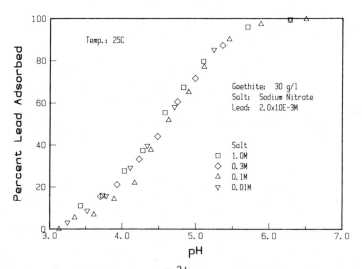

Figure 3.   The dependence of Pb^{2+} adsorption on goethite as a
function of background electrolyte concentration.

Table II. TLM Reactions and Equilibrium Expressions

Reaction	Equilibrium Expression	Activity Coefficients
Surface Protolysis Reactions		
$SOH_2^+ \overset{K_{a1}}{\underset{}{\rightleftarrows}} SOH + H^+$	$K_{a1}^{int} = \dfrac{[SOH][H^+]}{[SOH_2^+]}\exp(-e\psi_o/kT)$	$\dfrac{\gamma_{SOH}\gamma_{H^+}}{\gamma_{SOH_2^+}} = \exp(-e\psi_o/kT)$
$SOH \overset{K_{a1}}{\underset{}{\rightleftarrows}} SO^- + H^+$	$K_{a2}^{int} = \dfrac{[SO^-][H^+]}{[SOH]}\exp(-e\psi_o/kT)$	$\dfrac{\gamma_{SO^-}\gamma_{H^+}}{\gamma_{SOH}} = \exp(-e\psi_o/kT)$
Electrolyte Surface Reactions		
$SOH + Na^+ \overset{K_{Na}}{\underset{}{\rightleftarrows}} SO^- - Na^+ + H^+$	$K_{Na}^{int} = \dfrac{[SO^- - Na^+][H^+]}{[SOH][Na^+]}\exp(-e(\psi_o - \psi_\beta)/kT)$	$\dfrac{\gamma_{H^+}\gamma_{SO^- - Na^+}}{\gamma_{SOH}\gamma_{Na^+}} = \exp(-e(\psi_o - \psi_\beta)/kT)$
$SOH + H^+ + NO_3^- \overset{K_{NO_3}}{\underset{}{\rightleftarrows}} SOH_2^+ - NO_3^-$	$K_{NO_3}^{int} = \dfrac{[SOH_2^+ - NO_3^-]}{[SOH][H^+][NO_3^-]}\exp(e(\psi_o - \psi_\beta)/kT)$	$\dfrac{\gamma_{SOH_2^+ - NO_3^-}}{\gamma_{SOH}\gamma_{H^+}\gamma_{NO_3^-}} = \exp(e(\psi_o - \psi_\beta)/kT)$
Outer-Sphere Surface Reaction		
$SOH + Pb^{2+} \overset{K_{o.s.}}{\underset{}{\rightleftarrows}} SO^- - Pb^{2+} + H^+$	$K_{o.s.}^{int} = \dfrac{[SO^- - Pb^{2+}][H^+]}{[SOH][Pb^{2+}]}\exp(-e(\psi_o - 2\psi_\beta)/kT)$	$\dfrac{\gamma_{H^+}\gamma_{SO^- - Pb^{2+}}}{\gamma_{SOH}\gamma_{Pb^{2+}}} = \exp(-e(\psi_o - 2\psi_\beta)/kT)$
Inner-Sphere Surface Reaction		
$SOH + Pb^{2+} \overset{K_{i.s.}}{\underset{}{\rightleftarrows}} SOPb^+ + H^+$	$K_{i.s.}^{int} = \dfrac{[SOPb^+][H^+]}{[Pb^{2+}][SOH]}\exp(e\psi_o/kT)$	$\dfrac{\gamma_{H^+}\gamma_{SOPb^+}}{\gamma_{Pb^{2+}}\gamma_{SOH}} = \exp(e\psi_o/kT)$

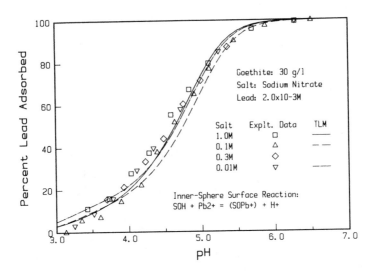

Figure 4.   Inner-sphere TLM calculations as a function of
background electrolyte concentration.

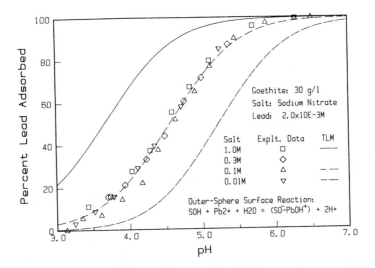

Figure 5.   Outer-sphere TLM calculations as a function of background electrolyte concentration.

constant that fit best at ionic strength of 0.1M, model computations were then made at different ionic strengths and compared to the adsorption data.

As shown in Figure 5, the modeling simulations for the outer-sphere case do not compare well with the observed data. It is also important to note that the best fit at I = 0.1M ($NaNO_3$) required a reaction stoichiometry of two protons released per lead ion adsorbed. On the other hand, the modeling results assuming an inner-sphere surface complex, shown in Figure 4, agree quite well with the observed experimental result that ionic strength changes have little effect on the adsorption behavior of lead ion adsorption at the goethite/water interface. In the inner-sphere case a reaction stoichiometry of one proton released per lead ion adsorbed is required to obtain a good model fit.

Potentiometric titrations of suspensions of 30.0 g/1 goethite in 0.01M ($NaNO_3$) were conducted in the presence and absence of 2.0 mM lead to determine the number of protons released per lead ion adsorbed. The net number of protons released is related to the (stoichiometry of reaction or) ligand number as described in Appendix B. Following the approach of Hohl and Stumm (22), the ligand number was found to be 1 (Figure 6). As discussed in Appendix B, it is not correct to equate the net number of protons released to the ligand number. The α values in the denominator of the term of the ordinate axis in Figure 6 reflect the need to account for the redistribution of surface species in the presence of adsorbed lead. The ligand number of 1 obtained from the slope of the line in Figure 6 agrees with the inner-sphere reaction stoichiometry required to model data shown in Figure 4. This analysis establishes the inner-sphere analogue as the proper choice for modeling the overall equilibrium behavior of lead ion adsorption at the goethite/water interface and establishes the reaction stoichiometry of 1 proton per lead ion adsorbed. What now remains is the description of the kinetic model and mechanism consistent with both the overall equilibrium and kinetic results.

Kinetic Model
_____

Having chosen a particular model for the electrical properties of the interface, e.g., the TLM, it is necessary to incorporate the same model into the kinetic analysis. Just as electrical double layer (EDL) properties influence equilibrium partitioning between solid and liquid phases, they can also be expected to affect the rates of elementary reaction steps. An illustration of the effect of the EDL on adsorption/desorption reaction steps is shown schematically in Figure 7. In the case of lead ion adsorption onto a positively charged surface, the rate of adsorption is diminished and the rate of desorption enhanced relative to the case where there are no EDL effects.

As an example of the manner in which EDL effects are incorporated into the kinetic analysis, consider the following bimolecular adsorption/desorption mechanism:

$$SOH + Me^{2+} \underset{k_{-1}}{\overset{k_1}{\rightleftharpoons}} SOMe^+ + H^+ \qquad (24)$$

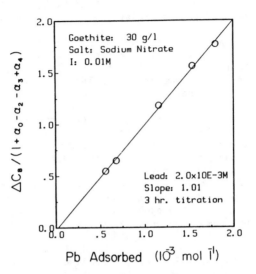

Pb Adsorbed $(10^{-3}$ mol $l^{-1})$

Figure 6. Plot of $\Delta C_B/(1 + \alpha_0 - \alpha_2 - \alpha_3 + \alpha_4)$ versus Pb^{2+} adsorbed (see Appendix B for definition of α_i). Slope of plot gives ligand number.

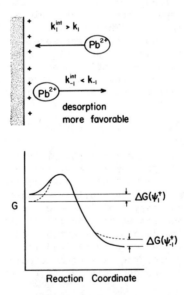

Reaction Coordinate

Figure 7. Change in activation energy of adsorption/desorption due to EDL. The solid and dashed lines in the free energy diagram represent the free energy path in the presence and absence of the EDL activation potential.

$$K_1 = \frac{k_1}{k_{-1}} = \frac{[SOMe^+][H^+]}{[SOH][Me^{2+}]} \qquad (25)$$

Using the TLM, the intrinsic equilibrium constant is defined using the activity coefficients, as described previously, i.e.,

$$K_1^{int} = K_1 \exp(e\psi_o/kT) \qquad (26)$$

Defining intrinsic rate constants as rate constants that would be observed in the absence of an electric field leads to

$$K_1^{int} = \frac{k_1^{int}}{k_{-1}^{int}} = \frac{k_1}{k_{-1}} \exp(e\psi_o/kT) \qquad (27)$$

What remains is to relate the surface potential to activation potentials for the adsorption/desorption reaction steps. Defining the activation potentials as $\psi_1^{\ddagger}$, $\psi_{-1}^{\ddagger}$ for the activation required to overcome the EDL potential for the adsorption, desorption steps, respectively, allows the intrinsic rate constants to be directly related to the rate constants k_1, k_{-1} (4), i.e.,

$$k_1^{int} = k_1 \exp(e\psi_1^{\ddagger}/kT) \qquad (28)$$

$$k_{-1}^{int} = k_{-1} \exp(e\psi_{-1}^{\ddagger}/kT) \qquad (29)$$

Since

$$K_1^{int} = \frac{k_1^{int}}{k_{-1}^{int}} = \frac{k_1 \exp(e\psi_1^{\ddagger}/kT)}{k_{-1} \exp(e\psi_{-1}^{\ddagger}/kT)} = K_1 \exp(e\psi_o/kT) \qquad (30)$$

and

$$\frac{\exp(e\psi_1^{\ddagger}/kT)}{\exp(e\psi_{-1}^{\ddagger}/kT)} = \exp(e\psi_o/kT) \qquad (31)$$

then

$$\psi_1^{\ddagger} = -\psi_{-1}^{\ddagger} = \psi_o/2 \qquad (32)$$

In arriving at Equation 32 it is assumed that the magnitude of the activation potentials for adsorption and desorption are equal and opposite in sign (a reasonable assumption if the surface potential is relatively unchanged during the course of the reaction, which would be the case for small perturbation of the equilibrium). In this manner the EDL properties are developed consistently for both the equilibrium and kinetic analysis based on the TLM.

Kinetic Results

The equilibrium and inner-sphere modeling results discussed above are consistent with the following overall equilibrium

$$-FeOH + Pb^{2+} \xrightleftharpoons{K_1} -FeOPb^+ + H^+ \tag{33}$$

$$K_1 = \frac{[-FeOPb^+][H^+]}{[-FeOH][Pb^{2+}]} \tag{34}$$

Pressure-jump kinetic experiments were conducted over the same pH range as the equilibrium results, at 2.0 mM total lead, I = 0.01M, and 30.0 g/l α-FeOOH (see, e.g., Figure 4). A typical relaxation curve from the pressure-jump kinetic experiments is shown in Figure 8 (for a discussion of the p-jump apparatus with conductivity detection, including the relationship between conductivity and reactant concentrations, see Appendix A). Determination of the relaxation time constants from the semi-log plot of the relaxation curve of Figure 8 (23) shows two relaxation events occur following the pressure perturbation (see Figure 9). In the following discussion only an interpretation of the fast process is given in terms of a mechanism. The analysis of the slow step, complicated by the dependence of τ_F^{-1} on ΔP, is the subject of a future paper (24).

The pH dependence of the inverse of the fast relaxation time constant, τ_F^{-1}, is shown in Figure 10 (error bars represent 95% confidence level) for pressure-jump magnitudes of 70-140 atmospheres. A mechanism is determined from these data by choosing one which is consistent with the overall equilibrium behavior and which correctly matches the rate relationships derived for the postulated mechanism; e.g., assuming the bimolecular adsorption/desorption reaction mechanism, as given in Equation 1, and using the kinetic model described above, the following relationship between τ_F^{-1} and reactant and product concentrations can be derived (see Appendix C):

$$\tau_F^{-1} = k_1^{int}\{(\exp(-e\psi_o/2kT) \times ([-FeOH] + [Pb^{2+}]) +$$

$$(K_1^{int})^{-1} \times \exp(e\psi_o/2kT) \times ([-FeOPb^+] + [H^+])\} \tag{35}$$

If this mechanism is consistent with the experimental relaxation data, then a plot of τ_F^{-1} versus the expression in the brackets of Equation 35 will give a straight line with a slope of k_1^{int} and an intercept at the origin. As shown in Figure 11, the data fit this proposed mechanism quite well. Values for ψ_o, reactant and product concentrations, and K_1^{int} input into Equation 35 are from the equilibrium modeling results calculated at each pH value for which kinetic runs were made. Normally a variety of different mechanisms are tested against the experimental data. Several other more complex mechanisms were tested, including those postulated for metal ion adsorption onto γ-Al$_2$O$_3$ (7); however, only the above mechanism was consistent with the experimental data. Hence it was concluded that the bimolecular adsorption/desorption reaction was the most plausible mechanism for Pb^{2+} ion adsorption onto α-FeOOH.

No mention has yet been made of the variation of τ_F^{-1} with the pressure-jump magnitude as shown in Figures 10 and 11. In fact, this result is quite surprising. For small perturbations, only the

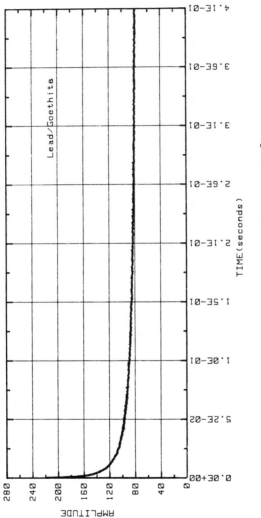

Figure 8. Pressure-jump relaxation curve for Pb^{2+} ion adsorption on goethite. Plot of conductivity change versus time.

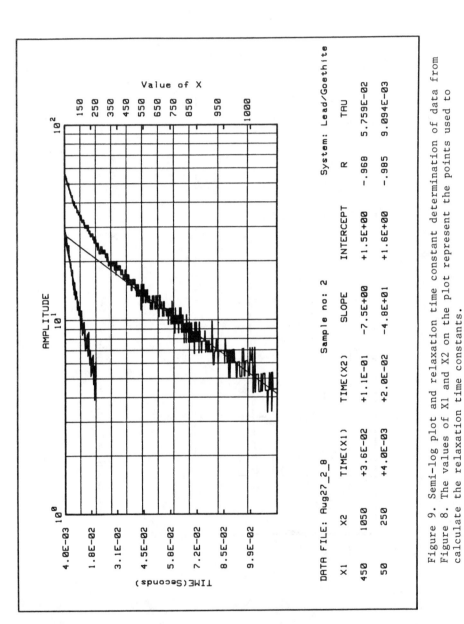

Figure 9. Semi-log plot and relaxation time constant determination of data from Figure 8. The values of X1 and X2 on the plot represent the points used to calculate the relaxation time constants.

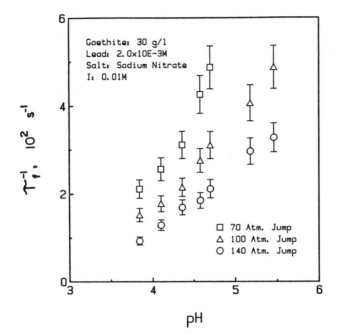

Figure 10. The pH dependence of τ_F^{-1} as function of ΔP.

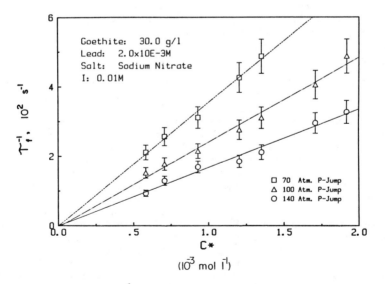

Figure 11. Plot of τ_F^{-1} versus the term in brackets on the right-hand side of Equation 35 (denoted C* in figure) at ΔP = 70, 100, and 140 atmospheres.

amplitude of the relaxation curve should change as the pressure-jump magnitude is varied (23). This variation is interpreted as an indication that a distribution of site types exists for the adsorption of Pb^{2+} onto oxide surfaces (8, 24). The results of the analysis of the data in Figure 11 are summarized in Table III. The trend in the intrinsic rate constants k_1^{int}, k_{-1}^{int} with pressure-jump magnitude are consistent with the notion that at higher pressure, more higher-energy sites are perturbed. This is reflected in the lower value of the rate constants at higher ΔP, which implies a higher activation energy and a slower overall rate. Equilibrium evidence that a distribution of site types exist for metal ion adsorption has been previously reported (25).

Table III. Intrinsic Rate Constants

ΔP (atm)	k_1^{int} ($mol^{-1}dm^3s^{-1}$)	k_{-1}^{int} ($mol^{-1}dm^3s^{-1}$)	K_1^{int}
140	1.7×10^5	4.2×10^2	4.0×10^2
100	2.4×10^5	6.1×10^2	"
70	3.6×10^5	8.9×10^2	"

In an interesting analysis of the effects of reduction of dimensionality on rates of adsorption/desorption reactions (26), the bimolecular rate of $10^5 M^{-1}s^{-1}$ has been reported as the lower limit of diffusion control. Based on this value, the rates given in Table III indicate the desorption step is chemical-reaction-controlled, likely controlled by the chemical activation energy of breaking the surface complex bond. On the other hand, the coupled adsorption step is probably diffusion controlled.

Summary

Relaxation techniques can be used in determining mechanisms of reactions at the mineral/water interface. In this paper the pressure-jump kinetic study of lead ion adsorption at the goethite/water interface was reported. At least three pieces of information are required to establish a plausible mechanism: (1) overall equilibrium, (2) stoichiometry of reaction, and (3) rate data. For reactions occurring at charged interfaces, electrical double layer models must be incorporated in equilibrium and rate relationships. The equilibrium results suggest an inner-sphere surface complex is formed. Based on the kinetic results for lead adsorption on goethite of this study, a bimolecular adsorption/desorption reaction has been postulated for Pb^{2+} adsorption/desorption at the goethite/water interface. The rate-limiting step is the desorption step associated with breaking the surface bond of the inner-sphere lead hydroxyl complex. The variation of the rate constants with pressure-jump magnitude is suggested as evidence of adsorption/desorption from sites with a distribution of bonding energies.

APPENDIX A

Pressure-Jump Apparatus with Conductivity Detection

Apparatus. The principle of the pressure-jump method is based on the pressure dependence of the equilibrium constant, i.e.,

$$\left(\frac{\partial \ln K}{\partial P}\right)_T = -\frac{\Delta V}{RT} \qquad (A-1)$$

where ΔV is the standard molar volume change of the reaction, P, the pressure, R, the universal gas constant, and T, the temperature. A pressure perturbation results in the shifting of the equilibrium; the return of the system to the original equilibrium state (i.e., the relaxation) is related to the rates of all the elementary reaction steps. The relaxation time constant associated with the relaxation can be used to evaluate the mechanism of reaction.

Section views of the pressure-jump apparatus used in this investigation are shown in Figure 12. The main components include the central pressure-jump chamber or autoclave, a pressure pump, sample and reference cells, bayonet socket, and vacuum pump. Pressurization of the autoclave is accomplished with the pressure pump. The autoclave has ports for both a sample and reference cell. The caps of the cells have a plastic membrane which effectively transmits pressure. The confined volume of the autoclave is filled with water which transmits the pressure changes to the cells. A piece of brass shimstock is clamped onto one wall of the autoclave with the bayonet socket. When the pressure in the autoclave gets high enough the shimstock bursts and the pressure returns to ambient pressure "spontaneously," i.e., in less than 100 μs. After the shimstock bursts, the sample solution having equilibrated at the higher pressure is out of equilibrium due to the "instantaneous" pressure-jump (instantaneous relative to chemical relaxation processes occurring slower than 1 millisecond). The chemical relaxation of the sample solution back to the equilibrium condition of ambient pressure is monitored by conductivity detection. The reference cell is filled with a background electrolyte solution which has no relaxation in the time range of the pressure-jump apparatus. Water is circulated around the autoclave and through a constant-temperature bath to maintain the temperature at 25°C ± 0.1°C. In the pressure-jump experiments described in this paper different thicknesses of brass shimstock were used to obtain pressure-jumps of 70, 100, and 140 atmospheres.

Conductivity Detection. The conductivity detector of the pressure-jump apparatus is based on a simple Wheatstone bridge circuit. The sample and reference cell make up two arms of the bridge; the other two arms are made up of variable resistors and capacitors. A nonrelaxing solution (e.g., sodium chloride) with nearly the same conductivity as the sample is placed in the reference cell. The resistances and capacitances are adjusted to balance the bridge. During pressurization of the autoclave, the chemical equilibrium of the sample solution shifts, resulting in a change in conductivity and an unbalancing of the bridge. After the brass shimstock bursts the sample solution is out of equilibrium with respect to ambient pressure and must "relax" to the equilibrium state defined by

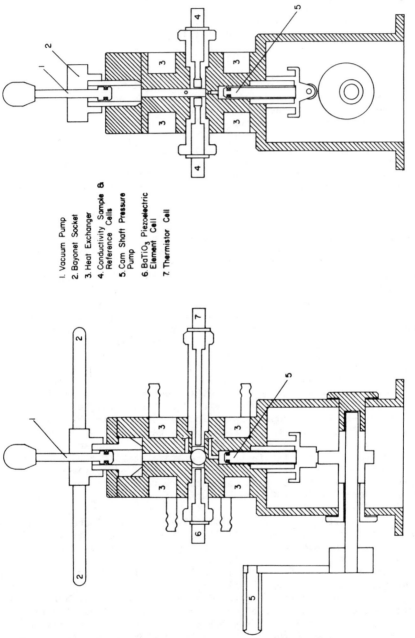

1. Vacuum Pump
2. Bayonet Socket
3. Heat Exchanger
4. Conductivity Sample &
   Reference Cells
5. Cam Shaft Pressure
   Pump
6. BaTiO₃ Piezoelectric
   Element Cell
7. Thermistor Cell

Figure 12. Sectional views of p-jump apparatus with conductivity detection.

ambient pressure; this relaxation process is monitored by the conductivity bridge as a voltage change. The electrical signal is
stored by a Tektronix 7612AD Programmable Digitizer. The digitized
data is sent to an HP9836 computer for analysis.

The conductivity change measured by the Wheatstone bridge
arrangement is directly related to the extent of reaction. The conductivity and change in conductivity following a pressure perturbation are given by

$$\sigma = \frac{F}{1000} \sum c_j |z_j| \mu_j = \frac{F}{1000} \rho \sum m_j |z_j| \mu_j \qquad \text{(A-2)}$$

$$\Delta\sigma = \frac{F}{1000} \left(\rho \sum |z_j| \mu_j \Delta m_j + \rho \sum |z_j| m_j \Delta\mu_j + \sum |z_j| m_j \mu_j \Delta\rho \right) \qquad \text{(A-3)}$$

where σ is the specific conductivity ($\Omega^{-1} cm^{-1}$), F is Faraday's constant, z, the valence of ion j, m and C, the molal and molar concentration of ion j, respectively; μ is the electrical mobility (in
$cm^2 V^{-1} sec^{-1}$), and ρ the density of the solution. The first term of
Equation A-3 is due to chemical relaxation effects while the later
two terms are from physical effects resulting from pressure and temperature changes which occur during a pressure-jump. The bridge
arrangement described above fully compensates for the physical
effects and thus the conductivity change is directly related to
changes from chemical relaxation, i.e., the first term in Equation
A-3. Taking as an example the bimolecular mechanism postulated for
lead, i.e.,

$$-FeOH + Pb^{2+} \rightleftarrows -FeOPb^+ + H^+ \qquad \text{(A-4)}$$

then

$$\Delta\sigma = \frac{F}{1000} \left(2\mu_{Pb}(\Delta[Pb^{2+}]) + \mu_H(\Delta[H^+]) \right) \qquad \text{(A-5)}$$

Since

$$\left(\Delta[Pb^{2+}]\right) = -\left(\Delta[H^+]\right) = X \qquad \text{(A-6)}$$

then Equation A-5 can be rewritten as follows:

$$\Delta\sigma = \frac{F}{1000} (2\mu_{Pb} - \mu_H)X \qquad \text{(A-7)}$$

and so $\Delta\sigma$ is directly related to X, the change in concentration of
the reacting species. In the case of the lead adsorption/desorption
studies, the conductivity decreases as the system relaxes (see Figure 8). This can be explained as resulting from a shift in the
equilibrium to the right for the reaction written above in Equation
A-4. A shift of the equilibrium to the right favors desorption of
protons and adsorption of lead. Since protons have a higher mobility and thus higher conductance than lead, this results in an
increase in the conductivity and subsequent decrease during the
relaxation in accordance with the observed results of Figure 8.

APPENDIX B

Net Proton Release and Ligand Number

The determination of the ligand number (27) for adsorption reactions has been discussed by Hohl and Stumm (22). The following example illustrates the relationship between the net proton release and ligand number.

Consider the sodium nitrate/goethite suspension titrated in the presence and absence of lead ion. The following equations governing the titrations are easily derived from the charge and mass balance equations:

$$(C_B - C_A + [H^+] - [OH^-]) = ([SO^-] - [SOH_2^+] + [SO^--Na^+]$$
$$- [SOH_2^+-NO_3^-]) \tag{B-1}$$

$$(C_B^* - C_A^* + [H^+] - [OH^-]) = ([SO^-]^* - [SOH_2^+]^* + [SO^--Na^+]^*$$
$$- [SOH_2^+-NO_3^-]^* - [SOPb^+] - 2[(SO)_2Pb]) \tag{B-2}$$

where the superscript * represents the concentration of surface species in the presence of lead, and where two types of lead surface complexes are allowed that would release 1 and 2 protons respectively. In order to find the number of protons released in the presence and absence of Pb^{2+}, Equation B-1 is subtracted from B-2. When the following relationships are substituted into B-1 and B-2

$$[SOH_2^+] = S_T\alpha_0 \qquad [SOH_2^+]^* = S_T^*\alpha_0 \tag{B-3}$$

$$[SOH] = S_T\alpha_1 \qquad [SOH]^* = S_T^*\alpha_1 \tag{B-4}$$

$$[SO^-] = S_T\alpha_2 \qquad [SO^-]^* = S_T^*\alpha_2 \tag{B-5}$$

$$[SO^--Na^+] = S_T\alpha_3 \qquad [SO^--Na^+]^* = S_T^*\alpha_3 \tag{B-6}$$

$$[SOH_2^+-NO_3^-] = S_T\alpha_4 \qquad [SOH_2^+-NO_3^-]^* = S_T^*\alpha_4 \tag{B-7}$$

where

$$S_T = ([SOH_2^+] + [SOH] + [SO^-] + [SO^--Na^+] + [SOH_2^+-NO_3^-]) \tag{B-8}$$

$$S_T^* = ([SOH_2^+]^* + [SOH]^* + [SO^-]^* + [SO^--Na^+]^*$$
$$+ [SOH_2^+-NO_3^-]^*) \tag{B-9}$$

$$S_T = S_T^* + [SOPb^+] + 2[(SO)_2Pb] \tag{B-10}$$

and when Equation B-1 is subtracted from B-2

$$\Delta C_B = (\alpha_0 - \alpha_2 - \alpha_3 + \alpha_4 + 1) \times ([SOPb^+] + 2[(SO)_2Pb]) \tag{B-11}$$

where

$$\Delta C_B = (C_B^* - C_A^*) - (C_B - C_A)$$

The α_i are defined as follows:

$$\alpha_0 = \left(1 + \frac{K_{a1}}{[H^+]} + \frac{K_{a1}K_{a2}}{[H^+]^2} + \frac{K_{a1}K_{Na}[Na^+]}{[H^+]^2} + K_{a1}K_{NO_3}[NO_3^-]\right)^{-1} \quad \text{(B-12)}$$

$$\alpha_1 = \left(\frac{[H^+]}{K_{a1}} + 1 + \frac{K_{a2}}{[H^+]} + \frac{K_{Na}[Na^+]}{[H^+]} + K_{NO_3}[H^+][NO_3^-]\right)^{-1} \quad \text{(B-13)}$$

$$\alpha_2 = \left(\frac{[H^+]^2}{K_{a1}K_{a2}} + \frac{[H^+]}{K_{a2}} + 1 + \frac{K_{Na}[Na^+]}{K_{a2}} + \frac{[H^+]^2[NO_3^-]K_{NO_3}}{K_{a2}}\right)^{-1} \quad \text{(B-14)}$$

$$\alpha_3 = \left(\frac{[H^+]^2}{[Na^+]K_{a1}K_{Na}} + \frac{[H^+]}{K_{Na}[Na^+]} + \frac{K_{a2}}{[Na^+]K_{Na}} + 1 + \frac{K_{NO_3}[NO_3^-][H^+]^2}{K_{Na}[Na^+]}\right)^{-1}$$

$$\text{(B-15)}$$

$$\alpha_4 = \left(\frac{1}{K_{a1}K_{NO_3}[NO_3^-]} + \frac{1}{[NO_3^-][H^+]K_{NO_3}} + \frac{K_{a2}}{K_{NO_3}[H^+]^2[NO_3^-]}\right.$$

$$\left. + \frac{[Na^+]K_{Na}}{K_{NO_3}[NO_3^-][H^+]^2} + 1\right)^{-1} \quad \text{(B-16)}$$

Since the ligand number is defined as

$$\bar{n} = \frac{\Sigma_i\left((SO)_iPb^{2-i}\right)}{Pb_T} \quad \text{(B-17)}$$

where Pb_T is total bound lead, then

$$\bar{n} = \frac{\Delta C_B/Pb_T}{(1 + \alpha_0 - \alpha_2 - \alpha_3 + \alpha_4)} \quad \text{(B-18)}$$

As can be seen by Equation B-18, $\bar{n}$ does not simply equal $\Delta C_B/Pb_T$ but is lowered by the factor $(1 + \alpha_0 - \alpha_2 - \alpha_3 + \alpha_4)$, which reflects the redistribution of surface speciation due to Pb^{2+} adsorption.

APPENDIX C

Derivation of the Time Constant Relationship of Equation 35

The relationship between the inverse of the relaxation time
constant, τ, and the concentration of the reacting species shown by
Equation 35 is developed as follows. For the bimolecular reaction

$$-FeOH + Pb^{2+} \underset{k_{-1}}{\overset{k_1}{\rightleftharpoons}} -FeOPb^+ + H^+ \qquad (C-1)$$

the rate is defined as

$$r = -\frac{d[-FeOH]}{dt} = -\frac{d[Pb^{2+}]}{dt} = \frac{d[-FeOPb^+]}{dt} = \frac{d[H^+]}{dt} \qquad (C-2)$$

or

$$r = -k_1[-FeOH][Pb^{2+}] + k_{-1}[-FeOPb^+][H^+] \qquad (C-3)$$

At equilibrium $r = 0$ and Equation C-3 becomes

$$0 = -k_1[\overline{-FeOH}][\overline{Pb^{2+}}] + k_{-1}[\overline{-FeOPb^+}][\overline{H^+}] \qquad (C-4)$$

where the overbar denotes equilibrium concentrations related by the
mass law expression

$$\frac{[\overline{-FeOPb^+}][\overline{H^+}]}{[\overline{-FeOH}][\overline{Pb^{2+}}]} = \frac{k_1}{k_{-1}} = K_1 \qquad (C-5)$$

Following a small perturbation, e.g., a pressure-jump, equilibrium
concentrations are shifted some small amount Δ, and the time-
dependent concentrations are given by

$$[-FeOH] = [\overline{-FeOH}] + (\Delta[-FeOH]) \qquad (C-6)$$

$$[Pb^{2+}] = [\overline{Pb^{2+}}] + (\Delta[Pb^{2+}]) \qquad (C-7)$$

$$[-FeOPb^+] = [\overline{-FeOPb^+}] + (\Delta[FeOPb^+] \qquad (C-8)$$

$$[H^+] = [\overline{H^+}] + (\Delta[H^+]) \qquad (C-9)$$

According to the law of mass conservation

$$(\Delta[-FeOH]) = (\Delta[Pb^{2+}]) = -(\Delta[-FeOPb^+]) = -(\Delta[H^+]) = X \qquad (C-10)$$

or from Equations C-6 through C-9

$$[-FeOH] = [\overline{-FeOH}] + X \qquad (C-11)$$

$$[Pb^{2+}] = [\overline{Pb^{2+}}] + X \qquad (C-12)$$

$$[-\text{FeOPb}^+] = \overline{[-\text{FeOPb}^+]} - X \qquad (C\text{-}13)$$

$$[H^+] = \overline{[H^+]} - X \qquad (C\text{-}14)$$

Substituting C-11 through C-14 into C-3

$$r = \frac{dX}{dt} = -k_1(\overline{[-\text{FeOH}]} + X)(\overline{[Pb^{2+}]} + X) + k_{-1}(\overline{[-\text{FeOPb}^+]} - X)(\overline{[H^+]} - X)$$

$$= -k_1(\overline{[-\text{FeOH}]}\,\overline{[Pb^{2+}]}) + k_{-1}(\overline{[-\text{FeOPb}^+]}\,\overline{[H^+]})$$

$$- [k_1(\overline{[-\text{FeOH}]} + \overline{[Pb^{2+}]}) + k_{-1}(\overline{[-\text{FeOPb}^+]} + \overline{[H^+]})]X - k_1 X^2 + k_{-1} X^2$$

$$(C\text{-}15)$$

The first two terms sum to zero via relation C-4 and for small perturbation (i.e., small X), the last two terms become vanishingly small leading to

$$\frac{dX}{dt} = -\frac{1}{\tau} X \qquad (C\text{-}16)$$

with

$$\tau^{-1} = k_1(\overline{[-\text{FeOH}]} + \overline{[Pb^{2+}]}) + k_{-1}(\overline{[-\text{FeOPb}^+]} + \overline{[H^+]}) \qquad (C\text{-}17)$$

Combining the following relationships based on Equations 28–32, i.e.,

$$k_1 = k_1^{int} \exp(-e\psi_o/2kT) \qquad (C\text{-}18)$$

and

$$k_{-1} = k_{-1}^{int} \exp(e\psi_o/2kT) \qquad (C\text{-}19)$$

and Equation C-17

$$\tau^{-1} = k_1^{int}[(\exp(-e\psi_o/2kT) \times (\overline{[-\text{FeOH}]} + \overline{[Pb^{2+}]})$$

$$+ (K_1^{int})^{-1} \times \exp(e\psi_o/2kT) \times (\overline{[-\text{FeOPb}^+]} + \overline{[H^+]})] \qquad (C\text{-}20)$$

Legend of Symbols

A	$\equiv$	surface area/l
B	$\equiv$	10^6 F/A
C_A	$\equiv$	concentration of acid
C_B	$\equiv$	concentration of base
C_s	$\equiv$	concentration of symmetrical electrolyte
C_1	$\equiv$	inter-layer capacitance
C_2	$\equiv$	outer-layer capacitance
e	$\equiv$	electrostatic charge
F	$\equiv$	Faraday's constant

k $\equiv$ Boltzmann's constant
T $\equiv$ temperature
z $\equiv$ charge of ion
α_i $\equiv$ distribution coefficient
γ_i $\equiv$ activity coefficient of species i
ψ_o $\equiv$ potential at o-plane
ψ_β $\equiv$ potential at β-plane
ψ_d $\equiv$ potential at d-plane
$\psi_1^\ddagger$ $\equiv$ activation potential of adsorption
$\psi_{-1}^\ddagger$ $\equiv$ activation potential of desorption
σ_o $\equiv$ charge density in o-plane
σ_β $\equiv$ charge density in β-plane
σ_d $\equiv$ charge density in d-plane
τ_F $\equiv$ relaxation time constant for fast step
$*$ $\equiv$ superscript to indicate concentration of given species in presence of lead ion

Acknowledgments

This research was supported by the Environmental Protection Agency (Grant No. EPA 809561). K.F.H. gratefully acknowledges the Japanese Ministry of Education for a research scholarship. Also this work would not have been completed without the kindness and support of Professor T. Yasunaga and Drs. M. Sasaki, T. Ikeda, and N. Mikami of Hiroshima University. The authors wish to thank Dr. V. Tripathi for his help in the computer modeling of the equlibrium data.

Literature Cited

1.   Ashida, M.; Sasaki, M.; Hachiya, K.; Yasunaga, T. J. Colloid Interf. Sci. 1980, 74, 572-4.
2.   Astumian, R. D.; Sasaki, M.; Yasunaga, T.; Schelly, Z. A. J. Phys. Chem. 1981, 85, 3832-5.
3.   Sasaki, M.; Moriya, M.; Yasunaga, T.; Astumian, R. D. J. Phys. Chem. 1983, 87, 1449-53.
4.   Mikami, N.; Sasaki, M.; Hachiya, K.; Astumian, R. D.; Ikeda, T.; Yasunaga, T. J. Phys. Chem. 1983, 87, 1454-8.
5.   Mikami, N.; Sasaki, M.; Kikuchi, T.; Yasunaga, T. J. Phys. Chem. 1983, 87, 5245-8.
6.   Hachiya, K.; Sasaki, M.; Ikeda. T.; Mikami, N.; Yasunaga, T. J. Phys. Chem. 1984, 88, 23-7.
7.   Hachiya, K.; Sasaki, M.; Ikeda, T.; Mikami, N.; Yasunaga, T. J. Phys. Chem. 1984, 88, 27-31.
8.   Hayes, K. F. Ph.D. Thesis, Stanford University, CA 1986.
9.   Atkinson, R. J.; Posner, A. M.; Quirk, J.P. J. Inorg. Nucl. Chem. 1972, 34, 2201-11.
10.   Sposito, G. "The Surface Chemistry of Soils," p 40; Oxford University Press, New York, 1984.
11.   Davis, J. A.; James, R. O.; Leckie, J. O. J. Colloid Interf. Sci. 1978, 63, 480-99.
12.   Westall, J.; Hohl, H. Advan. Colloid Interf. Sci. 1980, 12, 265.

13. Stumm, W.; Huang, C. P.; Jenkins, S. R. Croatica Chem. Acta 1970, 42, 223.
14. Stumm, W.; Hohl, H.; Dalang, F. Croatica Chem. Acta 1976, 48, 491.
15. Yates, D. E.; Levine, S.; Healy, T. W. J.C.S. Faraday I 1974, 70, 1807.
16. Davis, J. A.; Leckie, J. O. J. Colloid Interf. Sci. 1978, 67, 90-107.
17. Davis, J. A.; Leckie, J. O. J. Colloid Interf. Sci. 1980, 74, 32-43.
18. Levine, S. J. Colloid Interf. Sci. 1971, 37, 619-34.
19. Hayes, K. F.; Leckie, J. O. J. Colloid Interf. Sci. submitted for publication.
20. Grahame, D. C. Chem. Reviews 1947, 41, 441-501.
21. Chan, D.; Perram, J. W.; White, L. R. J. Chem. Soc. Faraday Trans. I 1975, 71, 1046-1057.
22. Hohl, H.; Stumm, W. J. Colloid Interf. Sci. 1976, 55, 281.
23. Bernasconi, C. F. "Relaxation Kinetics"; Academic Press: New York, 1976.
24. Hayes, K. F.; Leckie, J. O. J. Phys. Chem. in preparation.
25. Benjamin, M. M.; Leckie, J. O. J. Colloid Interf. Sci. 1981, 79, 209-21.
26. Astumian, R. D.; Schelly, Z. A. J. Am. Chem. Soc. 1984, 106, 304-8.
27. Rossotti, F. J. C.; Rosotti, H. "The Determination of Stability Constants"; McGraw-Hill: New York, 1961.

RECEIVED June 18, 1986

8

Characterization of Anion Binding on Goethite Using Titration Calorimetry and Cylindrical Internal Reflection-Fourier Transform Infrared Spectroscopy

W. A. Zeltner, E. C. Yost, M. L. Machesky, M. I. Tejedor-Tejedor, and M. A. Anderson

Water Chemistry Program, University of Wisconsin, Madison, WI 53706

Titration calorimetry and cylindrical internal
reflection-Fourier transform infrared (CIR-FTIR)
spectroscopy are two techniques which have seldom
been applied to study reactions at the solid-liquid
interface. In this paper, we describe these two
techniques and their application to the investigation
of salicylate ion adsorption in aqueous goethite
(α-FeOOH) suspensions from pH 4 to 7. Evidence
suggests that salicylate adsorbs on goethite by
forming a chelate structure in which each salicylate
ion replaces two hydroxyls attached to a single iron
atom at the surface.

Lucio Forni ($\underline{1}$), while discussing catalysis research, commented
that: "Each technique, in fact, taken by itself, allows some
useful information to be collected, but can give rise to
criticisms. The combination of information obtainable by two or
more of them can often be the only way to give a complete picture
of the surface acid properties of the solid."
　　To a certain extent a similar statement could be made
about research on the chemistry of mineral-water interfaces. Some
theoretical models ($\underline{2},\underline{3}$) developed to date have focused primarily
on their ability to fit data collected from one experimental
technique, namely potentiometric titration. While these models
have done much to improve our understanding of the oxide-water
interface, we do not have a complete picture of the interfacial
region at present. Although potentiometric titrations can still
provide new insights, failure to utilize other techniques
may result in the problem mentioned in Forni's statement above.
　　Several alternative methods for examining the chemistry of
interfacial reactions are currently being developed, as evidenced
by the many fine chapters in this volume. While it is certainly
not necessary to utilize all techniques described, many can be
employed in a given system to better understand the reaction
chemistry involved in these complicated interfacial processes.

0097-6156/86/0323-0142$06.00/0
© 1986 American Chemical Society

To put things into perspective, we can broadly classify these analytical methods into bulk, dry surface, and in situ interfacial techniques. This chapter focuses on the last category, illustrating two in situ techniques used to study anion binding at the goethite (α-FeOOH)-water interface: titration calorimetry and cylindrical internal reflection-Fourier transform infrared (CIR-FTIR) spectroscopy. In fact, CIR-FTIR could prove to be extremely powerful, since it allows direct spectroscopic observation of ions adsorbed at the mineral-water interface. Both of these in situ techniques have proven very beneficial in our research.

Background

Several general experimental points will be presented before discussing these two methods in detail. All experiments described in this chapter were performed using goethite prepared in polyethylene vessels by the method of Atkinson et al. (4), with a OH/Fe ratio of 2 and 50 hour aging time. Sample morphology was determined by electron microscopy, which showed that the primary particles were needle shaped, with the 100 plane as the predominant face. The needles averaged 50 nm in length and 20 nm in width, and B.E.T. analysis of nitrogen gas adsorption indicated a surface area of 81 m^2/g.

Anion adsorption densities are directly affected by the net surface proton balance (often referred to as surface "charge" in the theoretical models) of hydrated metal oxide surfaces. Surface proton balances are determined by potentiometric titration (5-7), which also yields the pH value for the zero point of charge (ZPC). Our goethite titrations gave a ZPC of pH 8.1 ± 0.1 using sodium perchlorate as the inert electrolyte. This is higher than other reported goethite ZPC values of 7.5-7.6 (8-10) but in excellent agreement with that of Kavanagh et al. (11), who found a ZPC of 8.2. All anion adsorption measurements were made at pH values at least 1.5 pH units below the ZPC to ensure that initial anion adsorption was occurring on positively charged surfaces.

The primary anion studied in both the titration calorimetry and CIR-FTIR experiments reported here was the salicylate (2-hydroxybenzoate) ion (SAL). Acidity constants for salicylic acid are pK_1 = 3.0 and pK_2 = 13 (12), and the aqueous solubility of salicylic acid is 2.4 g/L, while that of NaSAL is 975 g/L. SAL has been shown to adsorb on both iron oxides (13) and aluminum oxides (14). Several other anions were also studied, and results for these anions are given as needed to illuminate certain features of the salicylate-goethite adsorption process.

Titration Calorimetry

Titration calorimetry involves the measurement of heat evolved while adding a titrant. This technique is well established for determining reaction enthalpies in homogeneous solution (see refs. 15 and 16 for general reviews of the method) but has been used far less often to measure adsorption enthalpies in heterogeneous suspensions. Instead, adsorption studies have relied mainly on the determination and interpretation of adsorption free energies (usually through adsorption isotherm experiments) using one of several available models. While the understanding of the overall adsorption process has increased with this approach, enthalpic and entropic components of adsorption free energies have rarely been determined. Reaction enthalpies can provide important information on site heterogeneity and temperature dependence of adsorption reactions, while entropies reflect increased system ordering or disordering as adsorption occurs.

Most previous adsorption studies employing calorimetry have investigated cation exchange phenomena on zirconium phosphates (17-21), titanium phosphate (22) and clay minerals (23,24). Proton exchange with alkali metal cations on zirconium phosphates is initially exothermic, but subsequently becomes progressively endothermic after reaching some critical exchange fraction. This transition depends on the crystallinity of the exchanger as well as the nature of the alkali cation. Entropies were generally negative when calculated from measured enthalpies and free energies. This was largely attributed to the net water structuring resulting from exchange of the metal cation, which may keep all or part of its hydration sheath intact. During investigations of cation exchange phenomena on clay minerals, regions of relatively discrete exchange enthalpies were observed which were explained as being due to the heterogeneous nature of the aluminosilicate surfaces studied (also see Goulding, this volume).

Calorimetric investigations of hydrous metal oxide suspensions are more scarce. A study of starch adsorption on hematite revealed that the adsorption process became less exothermic as surface coverage increased (25). This was attributed to a rearrangement of starch molecules to less favorable configurations.

Free energy variations with temperature can also be used to estimate reaction enthalpies. However, few studies devoted to the temperature dependence of adsorption phenomena have been published. In one such study of potassium octyl hydroxamate adsorption on barite, calcite and bastnaesite, it was observed that adsorption increased markedly with temperature, which suggested the enthalpies were endothermic (26). The resulting large positive entropies were attributed to loosening of ordered water structure, both at the mineral surface and in the solvent surrounding octyl hydroxamate ions during the adsorption process, as well as hydrophobic chain association effects.

The major conclusions which can be drawn from previous work are:

1) Adsorption enthalpies vary with surface coverage and adsorbent type but are usually exothermic – at least during initial stages of the adsorption process.

2) A large portion of the calculated entropy change (+ or -) results from the relative ordering or disrupting of water structure during adsorption.

3) Heat changes occur over a period of minutes in contrast to most solution phase reactions (seconds).

Methods. The equipment used consists of a TRONAC Model 450 (TRONAC, Inc., Orem, Utah) isoperibol titration calorimeter which

has been modified by adding (1) a high precision buret driven by a
stepping motor and (2) micro glass and reference pH electrodes
(Microelectrodes, Inc., Londonberry, NH). This system is
interfaced to an Apple IIe computer through an ISAAC 41A (Cyborg,
Inc., Newton, MA) data acquisition and control system equipped with
a 2 channel, 16 bit A/D converter to digitize temperature and pH
data, as well as a binary I/O card that permits automated titrant
delivery and heater calibration. Operator control is obtained by
setting required input parameters before starting data acquisition.
These parameters include: (1) a threshold digital value to initiate
data acquisition, (2) the number and length of titrant additions,
(3) time between additions, and (4) a "digital window" that is used
in the pH stat mode to activate a second buret containing standard
acid or base. These inputs allow experimental conditions to be
easily altered and optimized.

In order to adequately resolve the μV level thermistor
signals, 16 bit A/D conversion is necessary, rather than the more
conventional 12 bit. Also, signal to noise ratio is enhanced by
averaging 90 A/D conversions for each channel every 0.5 sec, an
averaging time well within the thermistor ($\sim$3 sec) and pH
electrode ($\sim$10 sec) response times.

Digitized thermistor and pH data (stored on floppy disk) are
converted to temperature and pH values using a previously
determined thermistor constant (°C/digital value) and digitized pH
buffer values. Several corrections must be included, however,
before temperature data can be converted correctly to heat content
values (in Joules) corresponding to the interval including and
following each titrant addition. This data reduction process
must account for such physical effects as heat produced by stirring
the suspension, heat lost to the surroundings, and resistance
heating of the thermistor. The heat of water formation which
results from acid-base neutralization as the system pH is changed
must also be determined. This is particularly important outside
the pH range of 5 to 9. Detailed explanations and equations for
these corrections are published elsewhere (16,27,28).

The standard calorimetric reaction of tris(hydroxymethyl)
aminomethane (THAM) neutralization with HCl was used in several
initial experiments to determine both precision and accuracy for
the data acquisition and reduction process. Three to five minutes
were allowed between acid additions, since this same time frame was
used for all later suspension titrations in order to minimize the
effects of slow surface reactions which occur during a titration
(9,29,30). The amount of acid added in each experiment was varied
to generate heat changes of 40-400 mJ (typical heat changes
observed in our adsorption studies with goethite suspensions).

Precision (% Standard Deviation (SD)) estimates at 400, 160,
80 and 40 mJ are 3.1, 6.4, 11.1 and 19.3% respectively. The
corresponding accuracy values (%) are +0.6, +4.3, +1.5 and -13.0%.
Thus, precision and accuracy decrease as heat evolved decreases,
and the magnitude of this decrease permits error estimates to be
obtained.

Suspensions were prepared in 500 mL polycarbonate bottles by
ultrasonically dispersing freeze-dried goethite in dilute nitric

acid (with enough solid $NaNO_3$ added to give a 0.05 M solution) for
$\sim$24 hours. This gave a suspension concentration of 10 g/L at a pH
between 4.2 and 4.5, following which all suspensions were aged at
least one week (one month maximum). Suspensions kept at pH 4 for
two months showed undetectable (< 1 μg/L) dissolved Fe levels in
the filtered (0.05 μm Nuclepore filters) supernate when analyzed
by graphite furnace atomic absorption. Fifty mL aliquots of the
suspensions were used for all titrations, with the only addition
being nitric acid to adjust the system to the desired starting pH.

Proton Adsorption Results. The capability of measuring pH as well
as temperature change allowed us to perform combined potentiometric-
calorimetric titrations on suspensions. Data representative of
such titrations are presented in Figure 1. All titrations were
begun at pH $\sim$3.8 (0.05 M $NaNO_3$) and equal volumes of standard NaOH
were added every five minutes until pH $\sim$10. Standard HNO_3 was
then similarly dispensed until the pH returned to $\sim$4. Over the
entire pH range, heats were exothermic for proton adsorption and
endothermic for proton desorption, but the absolute values of
these heats were largely identical when partitioned according to
the following scheme:

Adsorption

1) $MOH + H^+ = MOH_2^+$ exothermic

Desorption

1) $MOH_2^+ + OH^- = MOH + H_2O$ exothermic

2) $H_2O = H^+ + OH^-$ endothermic
       $----------------------------$
3) $MOH_2^+ = MOH + H^+$ endothermic

Data from different experiments exhibited considerable
variation at a particular pH value, but the overall trend was for
proton adsorption – desorption heats to increase over the pH range
4-9. Above pH 9, adsorption – desorption heats may decrease, but
the difficulty in obtaining data above pH 10 precluded further
delineation of this trend. At present our calorimetry system is
not equipped for nitrogen purging during titration, so carbonate
interference is possible at pH values above 7. However, carbonate
contamination effects were probably small, since suspensions were
aged at pH 4 and held under basic conditions for less than one hour
during a titration. As mentioned earlier, our goethite suspensions
had a zero point of charge at pH 8.1. Thus, over the pH range 4 to
8, we are likely measuring heats associated with only one general
type of reaction (i.e., $MOH_2^+ = MOH + H^+$), and electrostatic
considerations can justify the data trend. As the goethite surface
becomes less positive (titrating from low to high pH), proton
desorption becomes less favorable, which leads to more endothermic
reaction heats as observed. Surface site heterogeneity could also
explain the data, but "step-like" discontinuities are not
apparent (in contrast to the clay mineral study (23) discussed

earlier), showing that surface site heterogeneity with respect to proton reactions in these systems may not be extreme. There is also an unknown influence of carbonate contamination and particle-particle aggregation phenomena on the data at all pH values.

Anion Adsorption Results. To investigate heat changes associated with the adsorption of various anions on goethite it was necessary to fix all other important system parameters. The conditions chosen were: a goethite suspension concentration of 10 g/L, an ionic strength of 0.05 M $NaNO_3$, and a pH of 4.0 ± 0.1. Since adding an aliquot of anion invariably resulted in a pH increase, it was necessary to add acid to maintain the pH at 4.0. This was done by dispensing standard HNO_3 from a calorimeter buret one minute after the anion addition and subsequent pH rise. Then the system was allowed an additional four minutes for temperature equilibration before adding the next anion aliquot. Separate adsorption isotherm experiments were performed in a similar manner outside of the calorimeter to determine the fraction of any particular aliquot adsorbed. Both calorimetric and adsorption isotherm experiments were necessary to reduce the data to kJ produced/mole adsorbed. Because these experiments were constrained to five minute intervals between anion additions, resulting adsorption isotherms were not true equilibrium isotherms. Since the experimental reaction enthalpies did not necessarily represent equilibrium conditions, these enthalpies were not used to derive any additional thermodynamic information.

Evidence from pressure jump kinetic studies (31,32) has suggested that anion adsorption proceeds according to the following mechanism:

Step 1: $M-OH + H^+ = M-OH_2^+$

Step 2: $M-OH_2^+ + A^- = M-A + H_2O$

Proton abstraction from solution is thought to activate a site for subsequent ligand exchange. Our adsorption isotherm experiments have indicated that the $[H^+]$ abstracted to anion adsorbed ratio measured during an anion adsorption experiment is less than one and can vary with surface coverage (28). This is possible because the suspension has been preequilibrated near pH 4 for at least one week before use, so that many surface sites have already been activated by the Step 1 reaction before anion adsorption experiments are begun. Because of electrostatic attraction, these sites would be expected to react with adsorbing anions by Step 2 above. However, Step 2 does not provide a direct mechanism for proton abstraction from the supernate. This occurs indirectly, since Step 2 produces a less charged surface which then can adsorb more protons from the supernate at pH 4. After this, further anion adsorption is possible. Because our systems have been preequilibrated at pH 4, the number of protons abstracted for initial site activation has not been measured. Consequently, although this postulated mechanism would lead to a proton abstracted to anion adsorbed ratio of one on the microscopic level, our measured ratios of less than one appear reasonable.

The surface proton adsorption which occurs after Step 2, however, complicates the determination of the heat content change resulting from anion adsorption. In order to make this correction, the heat associated with proton adsorption must be determined from the previous potentiometric-calorimetric titrations. Proton adsorption on goethite is exothermic, and Figure 1 provides an average value of -29.6 kJ/mol near pH 4. This value, when multiplied by the moles of protons required to return to pH 4 after anion adsorption, allows correction for the heat associated with proton adsorption. This correction, however, is based on the assumption that the proposed two-step anion adsorption mechanism described above represents the only surface reactions which occur during anion adsorption. As such, the results obtained by this procedure are model dependent and are best used for comparative purposes.

Figure 2 depicts the differential adsorption heats as a function of surface coverage for $H_2PO_4^-$ and salicylate (SAL). The large enthalpy ($\sim$24 kJ/mol) observed for SAL adsorption on goethite at very low surface coverage ($\sim$10%) tentatively argues for bidentate bonding, as this enthalpy is similar to that of phosphate, which is thought to adsorb as a bidentate complex at low surface coverages (33-35). Fluoride ions, on the other hand, which would not be expected to show bidentate bonding, gave a maximum adsorption enthalpy of only $\sim$7 kJ/mol at low surface coverage (28), although this is not illustrated here. However, the conclusion that phosphate undergoes bidentate adsorption on goethite at low coverages, as mentioned above, is based on infrared studies of dry or slightly moist goethite in vacuum, rather than in aqueous suspensions. Further support for the bidentate bonding of phosphate on suspended goethite comes from comparing the maximum adsorption densities of phosphate and fluoride on our goethite at pH 4, which were 220 and 550 μmol/g respectively (28). Since monodentate SAL-goethite bonds could have an energy of 24 kJ/mol, though, the conclusion that SAL-goethite bonding is bidentate must be considered speculative without additional evidence.

A sharp decrease in adsorption enthalpy between 10 and 30% surface coverage of SAL can also be seen in Figure 2. This decrease may indicate that only a small number of surface sites are favorably oriented for SAL-goethite bond formation, although possible SAL-SAL interactions on the surface may also have an effect. Separate measurements of SAL adsorption on goethite, gave relatively small adsorption maxima (when compared to the phosphate and fluoride adsorption maxima discussed above) of 22 and 11 μmol/g at pH 4.8 and 6.3, respectively, in either 0.001 M $NaNO_3$ or 0.001 M KCl (36).

At high surface coverage (>30%), bonding enthalpies for SAL are low and comparable to enthalpies observed for fluoride (28), which suggests that little, if any, SAL-goethite surface bonding occurs. Instead, electrostatic attractions could account for the small exothermic heats found in this region, especially given the large measurement errors shown by the error bars. Such attractions would also account for the higher adsorption maximum of SAL on goethite at pH 4.8, as mentioned above, because the goethite surface would have a larger positive charge at the lower pH.

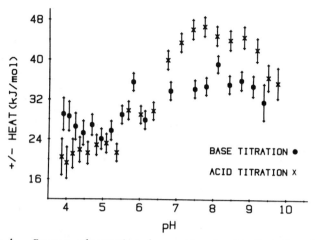

Figure 1. Proton adsorption-desorption heats (± 1 Standard Deviation or SD) as a function of pH for goethite at 10 g/L solid concentration in 0.05 M NaNO$_3$. All titrations were NaOH titrations starting from pH 4 followed by acid titrations back to pH 4.

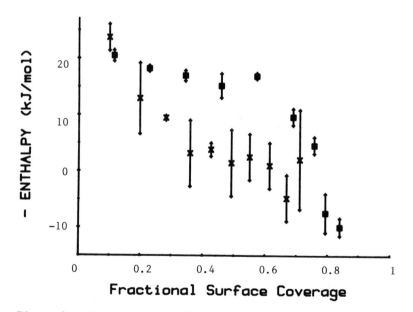

Figure 2. Adsorption heat (± 1 SD) as a function of fractional surface coverage for phosphate (■) and salicylate (✗) at 10 g goethite/L in 0.05 M NaNO$_3$ at pH 4.0.

It may prove possible to apply titration calorimetry data in one further direction. If ΔG can be estimated for SAL-goethite complexation and reaction enthalpies can be obtained under equilibrium conditions, then an entropy change for this reaction can also be derived. This can only be done, however, if the adsorption reaction can be shown to be reversible. Since this has not been proven as yet in our systems, such thermodynamic extensions of titration calorimetry can only be speculative at this time.

We have shown that titration calorimetry, by measuring the heat change resulting from adsorption reactions, can provide useful information on the strength of adsorbate-adsorbent interactions. This information, in turn, provides indirect evidence which can indicate possible adsorption mechanisms. Our titration calorimetry results, when combined with other evidence, suggest that salicylate ions bind to goethite by forming some type of bidentate complex on only relatively few surface sites. However, this is a tentative conclusion which still requires direct support. As discussed below, direct CIR-FTIR spectroscopic investigation of the interfacial region provides even more useful information about the bonding mechanism.

Cylindrical Internal Reflection - Fourier Transform Infrared (CIR-FTIR) Spectroscopy

CIR-FTIR spectroscopy provides a direct technique for studying in situ hydrous metal oxide surfaces and molecules adsorbed on these surfaces ($\underline{37}$). By itself, FTIR spectrometry is a well established technique which offers numerous advantages over dispersive (grating) IR spectrometry: (1) improved accuracy in frequency measurements through the use of a HeNe laser; (2) simultaneous frequency viewing; (3) rapid, repetitive scanning which allows many spectra to be collected in a small time interval; (4) minimal thermal effects from IR beam; and (5) no detection of sample IR emissions ($\underline{38}$).

Cylindrical internal reflection (CIR) is accomplished by using polished crystal rods (ZnSe in this study) with cone-shaped ends ($\underline{39}$). In CIR, the infrared beam enters the crystal obliquely, passes a fraction of a wavelength beyond the crystal into a less dense medium (e.g., water), and then "reflects" back into the crystal. Thus, for a cylindrical rod, the infrared beam "bounces" from edge to edge while traveling in the direction of the rod's diameter, providing multiple reflections through the crystal. (Harrick ($\underline{40}$) provides a more detailed explanation of internal reflection spectroscopy.) Because of the slight beam penetration into a medium like water, only the vibrations of groups that are near the crystal surface ($\sim 0.4 - 2.4$ μm depending on wavelength) can be detected. The advantages of CIR over transmission IR include small infrared beam penetration into water (a strong IR absorber), easy sample addition or withdrawal, and very reproducible sample path lengths. CIR-FTIR can thus be used to identify surface groups and adsorbed molecules in the aqueous interfacial region. This technique has successfully been applied

qualitatively and quantitatively in aqueous solution (41,42) and
aqueous colloidal suspensions (37).

Our studies have focused on using CIR-FTIR to examine the
aqueous goethite interface and adsorption of organic substrates
with IR active vibrations (simple substituted benzoic acid and
phenolic compounds). Except for investigations of electrode
surfaces (43), very little research has been published to date
concerning in situ aqueous-solid interfaces studied by FTIR. This
paucity of data is especially significant in both geochemistry and
environmental chemistry, where soil matrix components, such as
goethite, and organic adsorbates are (and have been) receiving
much attention (3,44-48).

Parfitt and co-workers have published several IR studies which
examined oxalate and benzoate adsorption on goethite (49);
phosphate, oxalate and benzoate adsorption on gibbsite (50); and
humic-fulvic acid adsorption on goethite, gibbsite and imogolite
(51). However, they utilized a technique of drying the organic and
metal oxide on AgCl plates and then obtained IR spectra with a
dispersive IR instrument and an evacuated cell. Cornell and
Schindler (52) analyzed hydroxycarboxylic acid (e.g., lactate)
adsorption on goethite and amorphous Fe(III) hydroxides using a
dispersive IR system and solid KBr and AgCl discs, respectively
(except for lactate where a liquid film was used). Liquid films of
Na benzoates in water and deuterium oxide have been analyzed by
dispersive IR spectrometers with cells having path lengths less
than 100 µm (53,54). Grating spectrometers have also been used to
examine salicylic acid (55-57) and NO_2 substituted phenolic
compounds in water films (55), in carbon tetrachloride (56-58) and
in dioxane (57). A FTIR study of crystalline salicylic acid has
been performed using KBr discs (59). This brief literature review
illustrates that simple substituted benzoic acid and phenolic
compounds (like salicylic and benzoic acid) have not been
investigated in situ in aqueous solutions or in systems with soil
component solids using the advantageous FTIR method.

Methods. CIR-FTIR analyses utilized a Barnes Analytical (CIRCLE)
ZnSe rod, 80 mm long and 6 mm in diameter. The system is designed
so that the light beam strikes the crystal at a 35° average angle
of incidence, providing 5 reflections within the sample holder.
The holder is a stainless steel "open boat" with a 3 mL cavity
25 mm in length.

Spectra were recorded interferometrically with either a 60SX
or 170SX Nicolet Fourier Transform Infrared spectrometer and a
Hg-Cd-Te (MCT) detector. Single beam spectra were obtained by
adding either 1000 or 2000 individual scans, giving 4 wave number
resolution in the 600-4000 cm^{-1} range. Typical spectrometer
parameters included setting the iris aperture at 8.7mm and using
1.57 cm/sec total optical mirror velocity. Happ-Genzel apodization
was used for all spectra.

Spectra of KCl solutions (reference) were subtracted from
spectra of aqueous organic solutions (sample), both at the same pH
value and both previously ratioed against the empty cell spectrum,
to yield aqueous solution spectra of the organic compounds.
Goethite spectra were obtained similarly by subtracting supernatant

spectra (reference) from suspension spectra (sample), both
previously ratioed against spectra of the empty cell. Each
reference and corresponding sample were closely matched in pH and
ionic strength. The cell remained untouched through all spectral
scans to keep its transmittance and average angle of incidence
constant.

Concentrations of the benzoic acid or phenolic compounds were
0.05 to 0.1 M, with 1M KCl used as a swamping inert electrolyte.
This high ionic strength was used to better match reference and
sample solutions, thus minimizing ionic strength effects on the
resulting aqueous solution spectra of the organic compounds.
Goethite spectra, however, were obtained using suspensions between
30 and 100 g/L concentration with background electrolyte
concentrations of 0.001 to 0.01 M KCl. Here the ionic strength was
kept as low as possible because we observed that less coagulated
systems provided more intense spectra (37). Also, goethite
suspension concentrations needed to be higher than for the
calorimetric titration study (10 g/L) in order to provide enough
goethite particles in close proximity to the ZnSe reflecting
crystal to produce good quality spectra. All pH adjustments were
made far enough in advance, using HCl or KOH, that the samples
were at steady state conditions before CIR-FTIR analysis.

Goethite suspensions were also examined by CIR-FTIR using D_2O
as the solvent to facilitate group assignments in regions where
H_2O is a strong absorber. For D_2O suspensions, the solid was
separated by centrifugation from the first D_2O suspension
prepared, then treated again with another D_2O aliquot to eliminate
residual H_2O. All D_2O experimental manipulations were performed
under nitrogen atmosphere.

Results. Infrared absorbing goethite surface groups were easily
observable in H_2O and D_2O using CIR-FTIR (37). Figure 3 (a and b)
shows the predominant IR peaks of H_2O (3305 and 1633 cm^{-1}), while
the subtracted spectrum (Figure 3c) contains bands obtained from
bulk (or subsurface lattice) goethite groups as well as hydrated
surface groups. The band assignments are (in cm^{-1}): 3400-3500
surface OH stretching from Fe-OH groups; 3045 bulk OH stretching;
1600-1650 surface H_2O bending; 895 and 800 bulk Fe-O-Fe and/or
δ(Fe-OH) vibrations (37,60). Russell et al. (61) and Rochester
and Topham (62,63) have reported IR bands at 3660 and 3480 cm^{-1}
for surface OH groups on samples of dry (evacuated) goethite.
These frequencies are higher than the 3400-3500 cm^{-1} band found
for goethite in aqueous suspension (Figure 3c) because hydrogen
bonding of OH surface groups to interfacial water molecules in our
system reduces stretching vibrational energy (37). The presence
of the 1600-1650 cm^{-1} band is important for two reasons: it shows
structured water on the surface and it overlaps important
absorption peaks for our test organic adsorbates.

Figure 4 illustrates the positions and relative intensities of
absorption bands given by functional groups when D_2O is the
solvent. Although the bulk Fe-O-Fe and/or δ(Fe-OH) vibration
peaks at 895 and 800 cm^{-1} are the same, three new bands appear
(1188-1219, 2256, and 2604 cm^{-1}). The band at 3145 cm^{-1} caused by
bulk OH stretching is now in closer agreement with the position and

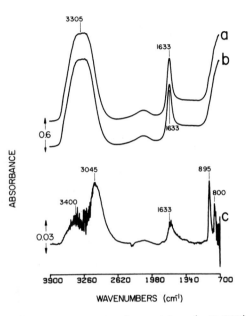

Figure 3. CIR–FTIR spectra of goethite (α-FeOOH) at 100 g/L solid concentration suspended in water at pH 6.5: (a) spectrum of suspension, (b) spectrum of supernatant, (c) difference spectrum a–b. Reproduced with permission from Ref. 37. Copyright 1986; American Chemical Society.

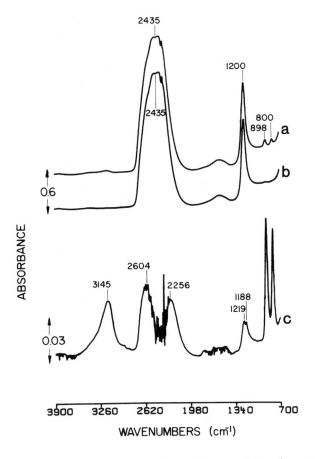

Figure 4.   CIR-FTIR spectra of goethite at 100 g/L solid concentration suspended in D_2O at pD 7:   (a) suspension, (b) supernatant, (c) difference a-b.   Reproduced with permission from Ref. 37, Copyright 1986; American Chemical Society.

relative intensity (compared to 800 cm^{-1} region peaks) of the
similar band seen in dry goethite transmission spectra (37). The
three new bands produced by D$_2$O treatment result from interactions
at the solid-liquid interface, since bands from the solution phase
have been eliminated by subtracting the supernatant (reference)
spectrum (37). These frequencies agree with reported hydrogen
bonded FeO–D stretching vibration and hydrogen bonded D$_2$O
stretching and bending vibrations (respectively the 2604, 2256, and
1188-1219 cm^{-1} bands shown in Figure 4c) (62,64).
     Figure 5 (upper) shows a CIR-FTIR spectrum of the aqueous
salicylate anion (SAL) at pH 5.5. Although Volovsek et al. (57)
attributed 1386 cm^{-1} to a phenolic O-H inner plane bend of
salicylic acid in CCl$_4$, the same strong peak is observed in
aqueous benzoate solution (Figure 5, middle) which has no phenolic
group, indicating that 1386 cm^{-1} is more likely a carboxylic
vibration. We assign the 1253 cm^{-1} peak to a C-O phenolic stretch
after analyzing aqueous phenol solutions at two pH values. Figure
5 (lower) shows the phenol spectrum at pH 5.5, where phenol is
uncharged. However, a spectrum taken at pH 11.4, where phenol
exists as the phenolate ion, shows no band near 1378 cm^{-1} while the
1240 cm^{-1} band shifts to 1270 cm^{-1}. These two observations
indicate that the 1378 cm^{-1} peak corresponds to a C-OH bending,
since the O-H proton is not present at pH 11.4, while the 1240 cm^{-1}
peak is a C-O phenolic stretch in which the C-O$^-$ stretching
vibrational energy is greater than for the C-OH group. Therefore,
if the 1240 cm^{-1} band is the phenolic OH in phenol, the 1253 cm^{-1}
band should be the corresponding vibration in SAL (in this case
more energy is needed for vibration due to internal hydrogen
bonding). By studying other aqueous solutions of benzoic acid and
phenolic compounds, we have assigned the SAL doublet at 1487 and
1458 cm^{-1} to C=C skeletal ring stretchings. The former assignment
matches research performed on solid salicylate (65) as well as a
study of salicylic acid in CCl$_4$ and dioxane solutions (57),
both of which attributed the 1487 cm^{-1} peak to a ring vibration.
     When the aqueous SAL spectrum (repeated in Figure 6, upper) is
compared to the spectrum of SAL adsorbed on goethite (Figure 6,
middle), where D$_2$O is used as the solvent to reduce water
interference in the 1570-1650 cm^{-1} region, a clear difference in
the two spectra can be observed, indicating that some interaction
between SAL and the goethite surface has occurred. Further
information about the interaction can be obtained by comparing
Figure 6 (middle) with the spectrum of SAL in aqueous Fe(III)
solution at pH 1.8 (Figure 6, lower). Even though the supernatant
pH values differ by 3 units in the two systems, their associated
spectra are very similar. Essentially identical bands occur at
1369 cm^{-1} (middle) and 1359 cm^{-1} (lower) (probable carboxylic
group), 1239 cm^{-1} (phenolic group observed when SAL adsorbed on
goethite in H$_2$O), and the 1456, 1467 cm^{-1} doublet (C=C ring
vibrations). Even the 1540, 1573, and 1601 cm^{-1} bands are similar,
although the aqueous Fe(III)-SAL system 1601 cm^{-1} band does have a
lower relative intensity.
     There is a slight discrepancy between the 1369 and 1359 cm^{-1}
peak positions since our system resolution is 4 cm^{-1}. This
discrepancy may be due to the difference in residual iron charge

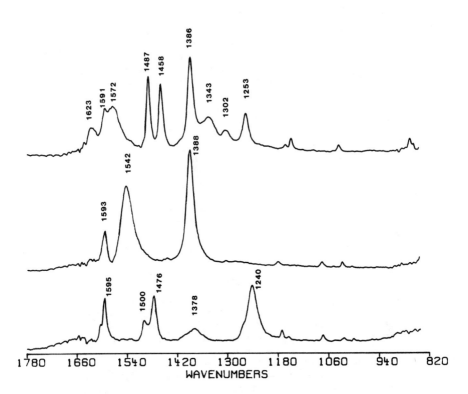

1780   1660   1540   1420   1300   1180   1060   940   820
                          WAVENUMBERS

Figure 5.   CIR-FTIR spectra of three 0.1 M organic compounds
in 1 M KCl at pH 5.5.   (upper) salicylate (SAL) = $C_6H_4OHCOO^-$;
(middle) benzoate = $C_6H_5COO^-$; (lower) phenol = C_6H_5OH.

between a solution complex and a SAL-goethite surface complex or
may be related to the 3 pH unit difference between supernatant pH
values. However, this small shift in peak positions is not a
solvent effect, because SAL-goethite spectra in H_2O have a
1370 cm^{-1} peak as compared to the 1369 cm^{-1} peak in D_2O.
     Previous studies propose that Fe(III) and SAL form a 1:1
chelate at low pH in aqueous systems (66,67). Such a chelate would
be highly conjugated and have the following structure:

In order to explain the similarity between the SAL-goethite
spectrum and the aqueous SAL-Fe(III) spectrum, a chelate structure
similar to that drawn above must be assumed to form during SAL
adsorption on goethite.

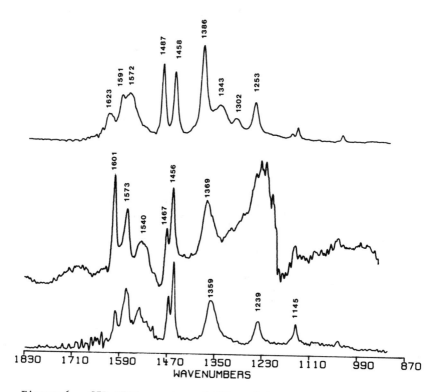

Figure 6. CIR-FTIR spectra of SAL and iron:SAL complex in
solution and SAL on goethite. (upper) 0.1 M SAL in 1 M KCl
at pH 5.5; (middle) SAL on goethite in D_2O with goethite
surface groups subtracted at pD 4.5, 0.01 M KCl and 100 g/L
solid concentration; (lower) aqueous iron:SAL complex at pH
1.6.

To facilitate such chelate formation, each SAL ion would have
to displace two surface hydroxyl groups from one iron atom in the
goethite crystal. This does assume, however, that the goethite
surface is fully hydrated before SAL adsorption occurs, so that all
iron atoms near the surface are coordinated to six oxygen atoms.
This assumption seems reasonable considering that all suspensions
used to obtain spectra in D_2O were equilibrated with D_2O for at
least four days before use, and some ultrasonic dispersion was
performed during this time. Steric constraints might be expected
to limit the number of iron atoms at the goethite surface which can
form such chelation structures. This would explain the relatively
low maximum adsorption densities for SAL on goethite reported
earlier. Also, if SAL-goethite surface chelates do form, possible
stoichiometries for this reaction are limited because they must
provide a proton abstracted to anion adsorbed ratio less than one,
as discussed previously in the titration calorimetry section under
anion adsorption results. Since this adsorption reaction is
occurring in a deuterated system already preequilibrated at pD 4.5,

one possible surface reaction which provides a proton abstracted to anion adsorbed ratio less than one is the following:

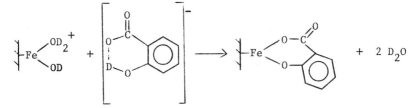

Similar bonding mechanisms have been proposed previously for adsorption of benzoic acid and phenolic derivatives on iron oxides (51) and aluminum oxides (14). Cornell and Schindler (52) have also suggested that COOH groups are involved in aliphatic hydroxycarboxylic acid adsorption on goethite. Parfitt et al. (49) proposed that oxalate adsorption (pH 3.4) on goethite involved replacing two singly coordinated oxygens by ligand exchange, while benzoate adsorption required replacing only one singly coordinated oxygen. However, most of these earlier studies did not investigate aqueous suspensions due to limitations in the techniques used.

As we have shown, the CIR-FTIR technique can be used to extend these mechanistic binding investigations into the aqueous solution or suspension phase. By examining CIR-FTIR peak assignments and shifts, in situ bonding of an organic ion (salicylate) to a metal oxide (goethite) can be investigated directly, complementing thermodynamic information obtained from an entirely separate technique, titration calorimetry.

<u>Summary</u>

Titration calorimetry and CIR-FTIR are valuable tools for studying processes which affect adsorption reactions at solid-liquid interfaces. Since CIR-FTIR can be used for direct, in situ investigations of metal oxide-water interfaces, it has tremendous potential for elucidating adsorption reaction mechanisms. We have used these techniques to obtain evidence suggesting that salicylate ions bind to goethite in a chelated structure by displacing two surface hydroxyls from single iron atoms. However, neither method should be considered a "stand-alone" technique. Just as Forni's statement (1) implies, information obtained using several methods (e.g., titration calorimetry, electrophoresis, adsorption isotherms, CIR-FTIR) must be combined in order to develop a reasonably complete picture of interfacial reactions in metal oxide and other mineral systems. Our challenge now is to apply these techniques in geochemical investigations of the factors controlling such processes as mineral dissolution and precipitation, adsorption of organic and inorganic molecules on mineral surfaces, and interfacial redox reactions. Many of these techniques can also be extended to studies of the interfacial chemistry of biological systems as well.

<u>Acknowledgments</u>

This research was funded by grants from the U.S. Department of Energy under Contract DE-AC02-80EV10467 and the U.S. Department of

Interior, U.S. Geological Survey, State of Wisconsin Water Resources Research Institute Program 14-08-0001-G890. We also wish to thank Jean Schneider and Helen Crogan for typing the manuscript.

Literature Cited

1. Forni, L. Catalysis Rev. 1973, 8, 67.
2. Westall, J.; Hohl, H. Adv. Colloid Interface Sci. 1980, 12, 265-294.
3. Sposito, G. "The Surface Chemistry of Soils"; Oxford: New York, 1984; Chap. 5.
4. Atkinson, R.J.; Posner, A.M.; Quirk, J.P. J. Inorg. Nucl. Chem. 1968, 30, 2371-2381.
5. Bolt, G.H. J. Phys. Chem. 1957, 61, 1166-1169.
6. Parks, G.A.; de Bruyn, P.L. J. Phys. Chem. 1962, 66, 967-973.
7. James, R.O.; Parks, G.A. Surf. Colloid Sci. 1982, 12, 119-216.
8. Atkinson, R.J.; Posner, A.M.; Quirk, J.P. J. Phys. Chem. 1967, 71, 550-558.
9. Yates, D.E. Ph.D. Thesis, University of Melbourne, Australia, 1975.
10. Balistrieri, L.S. M.S. Thesis, University of Washington, Seattle, Washington, 1977.
11. Kavanagh, B.V.; Posner, A.M.; Quirk, J.P. J. Colloid Interface Sci. 1977, 61, 545-553.
12. Martell, A.E.; Smith, R.M. "Critical Stability Constants Vol. 3: Other Organic Ligands"; Plenum: New York, 1977; p. 186.
13. Davis, J.A.; Leckie, J.O. Environ. Sci. Tech. 1978, 12, 1309-1315.
14. Kummert, R.; Stumm, W. J. Coll. Interface Sci. 1980, 75, 373-385.
15. Barthel, J. "Thermometric Titrations"; Wiley: New York, 1975; Chap. 2.
16. Hansen, L.D.; Lewis, E.A.; Eatough, D.J. In "Analytical Solution Calorimetry"; Grime, J.K., Ed.: Wiley: New York; 1985; Chap. 3.
17. Clearfield, A.; Kullberg, L.H. J. Phys. Chem. 1974, 78, 152-159.
18. Clearfield, A.; Kullberg, L.H. J. Phys. Chem. 1974, 78, 1812-1817.
19. Clearfield, A.; Tuhtar, D.A. J. Phys. Chem. 1976, 80, 1302-1305.
20. Kullberg, L.; Clearfield, A. J. Phys. Chem. 1980, 84, 165-169.
21. Clearfield, A.; Day, G.A. J. Inorg. Nucl. Chem. 1981, 43, 165-169.
22. Guarido, C.G.; Suarez, M.; Garcia, R.; Llavona, R.; Rodriguez, J. J. Colloid Interface Sci. 1985, 17, 63-68.
23. Goulding, K.W.T.; Talibudeen, O. J. Colloid Interface Sci. 1980, 78, 15-24.
24. Talibudeen, O.; Goulding, K.W.T. Clays and Clay Minerals 1983, 31, 37-42.
25. Khosla, N.K.; Bhagat, R.P.; Gandhi, K.S.; Biswas, A.K. Colloids and Surfaces 1984, 8, 321-336.

26. Pradip; Fuerstenau, D.W. Colloids and Surfaces 1985, 15, 137-146.
27. Eatough, D.J.; Christensen, J.J.; Izatt, R.M. "Experiments in Thermometric Titrimetry and Titration Calorimetry"; Brigham Young Univ. Press: Provo, UT, 1974.
28. Machesky, M. Ph.D. Thesis, University of Wisconsin, Madison, Wisconsin, 1985.
29. Onoda, G.Y., Jr.; de Bruyn, P.L. Surf. Sci. 1966, 4, 48-63.
30. Berube, Y.G.; Onoda, G.Y., Jr.; de Bruyn, P.L. Surf. Sci. 1967, 7, 448-461.
31. Mikani, N.; Sasaki, M.; Hachiya, K.; Astumian, R.D.; Ikeda, T.; Yasunaga, T. J. Phys. Chem. 1983, 87, 1454-1458.
32. Mikani, N.; Sasaki, M.; Kikuchi, T.; Yasunaga, T. J. Phys. Chem. 1983, 87, 5245-5248.
33. Atkinson, R.J.; Parfitt, R.L.; Smart, R.St.C. J. Chem. Soc. Faraday Trans. I 1974, 70, 1472-1479.
34. Parfitt, R.L. Adv. Agronomy 1978, 30, 1-50.
35. Taylor, R.W.; Ellis, B.G. Soil Sci. Soc. Am. J. 1978, 42, 432-436.
36. Yost, E.C., unpublished data.
37. Tejedor-Tejedor, M.I.; Anderson, M.A. Langmuir 1986, 2, 203-210.
38. Skoog, D.A. "Principles of Instrumental Analysis", 2nd Ed.; Saunders College: Philadelphia, PA, 1980; pp. 241-254.
39. Wilks, P., Jr. Industrial Research and Development 1982, September, 132.
40. Harrick, N.J. "Internal Reflection Spectroscopy"; Harrick Scientific Corp.: Ossining, NY, 1979.
41. Messerschmidt, R.G. Scan Time 1983, 2, 3.
42. Wong, J.S.; Rein, A.J.; Wilks, D.; Wilks, P., Jr. Appl. Spectros. 1984, 38, 32.
43. Foley, J.K.; Pons, S. Anal. Chem. 1985, 57, 945A-956A.
44. Mortland, M.M. Adv. Agr. 1970, 22:75-117.
45. Bailey, G.W.; White, J.L. Residue Rev. 1970, 32, 29-92.
46. Morrill, L.G.; Mahilum, B.C.; Mohiuddum, S.H. "Organic Compounds in Soils: Sorption, Degradation, Persistence"; Ann Arbor Science, Ann Arbor, MI., 1982; Chap. 4-5.
47. Davis, J.A. Geochim. Cosmochim. Acta 1982, 46, 2381-2393.
48. Ward, C.H.; Giger, W.; McCarty, P.L. "Ground Water Quality"; Wiley: New York, 1985; Chap. 4,8,19,23.
49. Parfitt, R.L.; Farmer, V.C.; Russell, J.D. J. Soil Sci. 1977, 28, 29-39. 50. Parfitt, R.L.; Fraser, A.R.; Russell, J.D.; Farmer, V.C. J. Soil Sci. 1977, 28,40-47.
51. Parfitt, R.L.; Fraser, A.R.; Farmer, V.C. J. Soil Sci. 1977, 28, 289-296.
52. Cornell, R.M.; Schindler, P.W. Coll. Polym. Sci. 1980, 258, 1171-1175.
53. Lindberg, B.T. Acta Chem. Scand. 1968, 22, 571-580.
54. Dunn, G.E.; McDonald, R.S. Can. J. Chem. 1969, 47, 4577-4588.
55. Mori, N.; Asaro, Y.; Irie, T.; Tsuzuki, Y. Bull. Chem. Soc. Japan 1969, 42, 482-487.
56. Czarnecka, E.; Tramer, A. Acta Phys. Polonica 1969, 36, 133-143.
57. Volovsek, V.; Colombo, L.; Furic, K. J. Raman Spec. 1983, 14, 347-352.

58. Thijs, C.; Zeegers-Huyskens, Th. Spectroscopy Letters 1977,
     10, 593-602.
59. Wojcik, M.J. Chem. Phys. Letters 1981, 83, 503-507.
60. Marshall, P.R.; Rutherford, D. J. Colloid Interface Sci.
     1971, 37, 390.
61. Russell, J.D.; Parfitt, R.L.; Fraser, A.R.; Farmer, V.C.
     Nature 1974, 248, 220-221.
62. Rochester, C.H.; Topham, S.A. J. Chem. Soc. Faraday Trans. I
     1979, 75, 591-602.
63. Rochester, C.H.; Topham, S.A. J. Chem. Soc. Faraday Trans. I
     1979, 75, 872-882.
64. Nakamoto, K. "Infrared and Raman Spectra of Inorganic and
     Coordination Compounds"; Wiley: New York, 1978; pp. 119-123.
65. Ansari, A.K.; Verna, P.K. Indian J. Pure Appl. Phys. 1979, 17,
     632-634.
66. Bertin-Batsch, P.C. Ann. de Chim. 1952, 12, 481-494.
67. Ernst, Z.L.; Menashi, J. Trans. Faraday Soc. 1963, 59,
     1794-1802.

RECEIVED May 20, 1986

9

Macroscopic Partitioning Coefficients for Metal Ion Adsorption
Proton Stoichiometry at Variable pH and Adsorption Density

Bruce D. Honeyman[1] and James O. Leckie

Environmental Engineering and Science Group, Department of Civil Engineering, Stanford University, Stanford, CA 94305-4020

This paper addresses one aspect of the problem of extending laboratory data to field situations. The emphasis is on macroscopic, semi-empirical descriptions of metal-ion adsorption and their use in the "predictive" modeling of adsorption over a range of pH values and adsorption densities. Two methods for determining the net release or consumption of protons during adsorption, Kurbatov plots and isotherm analysis, are critically evaluated. Isotherm analysis shows that the net proton stoichiometry is dependent on pH and metal adsorption density. Kurbatov's method is insensitive to these dependencies. Macroscopic proton coefficients, χ, are only rarely observed to have integer values although geochemists have often, arbitrarily, given χ a value of 1.0 or 2.0 in adsorption models. Comparison of macroscopic adsorption stoichiometries with the specific microscopic subreactions of a surface coordination model (Triple Layer) indicates that protolytic and electrolyte ion reactions appear important in determining the macroscopic proton coefficient. In addition, macroscopic partitioning coefficients provide ambiguous information on the existence and nature of surface-site heterogeneity.

Several different approaches can be used to model the interaction of solutes with reactive mineral surfaces. The conceptual approaches differ in the degree to which they account for observed or postulated solution and surface reactions. Whatever the approach, the description of interactions at the particle/solution interface must inevitably take into account the effect of pH on solute adsorption.

The advent of surface-complexation models brought the concepts and formalisms of coordination chemistry to the study of

[1]Current address: Swiss Federal Institute for Water Resources and Water Pollution (EAWAG), CH-8600 Dübendorf, Switzerland.

interactions at the solid/solution interface. The objective is an understanding of observations of macroscopic adsorbate behavior through the use of multiple, competing microscopic subreactions, both for the surface and bulk solution. The emphasis is on elementary equilibrium reactions with an accounting for the effects of the electrified interface either by explicitly considering electrostatic effects (1-3) or by the incorporation of field effects into the rational activity coefficients (4). For example, a subset of the suite of microscopic reactions needed to describe the macroscopic removal of a metal ion from solution by a hydrous metal oxide may include the surface and solution reactions shown in Table I (SO⁻ represents a broken oxygen bond on the oxide surface and M a metal of charge $^{z+}$).

Table I.   Examples of Sub-Reactions Needed to Describe Overall
Metal Adsorption Using Surface Coordination Models.

---

adsorbate surface reactions

$$SO^- + M^{z+} = SOM^{z-1}$$

$$SO^- + MOH^{z-1} = SOMOH^{z-2}$$

other surface reactions

protolysis

$$SOH = H^+ + SO^-$$

$$SOH + H^+ = SOH_2^+$$

exchange

$$SOH + Na^+ = SO^- --Na^+ + H^+$$

$$SOH + NO_3^- + H_2O = SOH_2^+--NO_3^- + OH^-$$

$$SO^---Na^+ + MOH^{z-1} = SO^---MOH^{z-1} + Na^+$$

solution-phase reactions

hydrolysis

$$M^{z+} + H_2O = MOH^{z-1} + H^+$$

complex formation

$$M^{z+} + Cl^- = MCl^{z-1}$$

Surface-complexation models require a high degree of detail about the heterogeneous systems. Unfortunately, the chemical detail required to use surface-complexation models will often exceed our knowledge of interactions taking place in natural systems. Consequently, geochemists have often resorted to semi-empirical, macroscopic descriptions, which are more easily utilized.

The semi-empirical descriptions of adsorbate/solid interactions are based on net changes in system composition and, unlike surface complexation models, do not explicitly identify the details of such interactions. Included in this group are distribution coefficients (K_D) and apparent adsorbate/proton exchange stoichiometries. Distribution coefficients are derived from the simple association reaction

$$M_{aq} \rightleftarrows M_{ads} \tag{1}$$

and

$$K_D = \frac{\{M_{ads}\}}{[M_{aq}]} = \frac{[M_{ads}]}{[M_{aq}]} \cdot \frac{1}{C_P} \cdot (m\ell \cdot g^{-1}) \tag{2}$$

where [ ] denotes concentration, { } the mass of metal removed from
solution per mass of particles, C_p the particle concentration and M
represents all species of the metal of interest either in solution
or associated with the particle surface. Because a K_D does not
include a functional relationship between $[H^+]$ and $[M_{ads}]/[M_{aq}]$, a
K_D cannot be utilized at pH values other than for which it was
determined. A modified form of the distribution coefficient can be
derived from a macroscopic (i.e., net) metal/proton exchange
reaction (e.g., (5))

$$SOH_\chi + M \; \overset{\rightarrow}{\underset{\leftarrow}{}} \; SOM + \chi H \tag{3}$$

$$K = \frac{[SOM][H]^\chi}{[SOH_\chi][M]} \tag{4}$$

SOH represents any surface site unassociated with any species of M,
SOM a metal/surface-site complex and χ the apparent ratio of moles
of protons released or consumed per mole of adsorbate removed from
solution.

In surface-complexation models, the relationship between the
proton and metal/surface-site complexes is explicitly defined in the
formulation of the proposed (but hypothetical) microscopic subreactions. In contrast, in macroscopic models, the relationship between
solute adsorption and the overall proton activity is chemically less
direct: there is no information given about the source of the
proton other than a generic relationship between adsorption and
changes in proton activity. The macroscopic solute adsorption/pH
relationships correspond to the net proton release or consumption
from all chemical interactions involved in proton tranfer. Since it
is not possible to account for all of these contributions directly
for many heterogeneous systems of interest, the objective of the
macroscopic models is to establish and 'calibrate' overall partitioning coefficients with respect to observed system variables.

In this paper we compare two graphical methods for determining
macroscopic proton release (i.e., χ in Equation 4): the adsorption
edge linearization method of Kurbatov et al. (6) and the isotherm
analysis technique of Perona and Leckie (7). We will examine their
usefulness in describing metal adsorption over a range of pH and
adsorption densities. We will also demonstrate that, while Kurbatov
plots may be useful for describing the characteristics of individual
pH/metal fractional adsorption edges, they are inappropriate for use
over a range of solution conditions. This is particularly important
in light of a recent renewal of interest in Kurbatov plots (8-10).
Finally, the macroscopic proton coefficient and mass action expression are compared to the subreactions used in the Triple-Layer Model
of Davis et al., (1) and it is demonstrated that i) macroscopic
proton coefficients do not represent the actual microscopic
adsorbate/H^+ stoichiometry, and ii) overall partitioning expressions
are poor diagnostic tools to demonstrate the existence of surface
site heterogeneity.

Macroscopic Descriptions of Solute Adsorption and the Net Proton
Coefficient. The macroscopic proton coefficient plays two important
roles in our macroscopic descriptions of surface processes. First,

the magnitude of χ determines the extent to which net partitioning
of an adsorbate between solid and solution phases changes in
response to system pH and site concentration. Second, what is known
about χ is intimately related to phenomenological models for surface
interactions.

χ and the Partitioning of a Metal Between Solid and Solution. As an
illustration of the first point, consider Figure 1. This figure
shows the hypothetical adsorption of a metal, M, onto either hydrous
oxide A or B. That is, each set of M-distribution lines represents
the net solid/solution partitioning of M in a system containing only
solid A or solid B. In either case, the solids are present in
equivalent total site concentrations (SOH_T). The macroscopic proton
coefficient, χ, for the interaction of M with S_A is 1.0 and for M
with S_B it is 2.0. Adsorption is described according to an overall
exchange reaction (Equation 3). The adsorption equivalence point
($[SOM] = 0.5\ M_T$) is equal to pH 8 for both systems at the initial
conditions (Figure 1a). The slopes of the distribution lines are
equal to the respective net proton coefficients.

An interesting characteristic of the modeling is that the
response of the net solid/solution partitioning of M to changes in
total site concentration is also a function of χ. This is indicated
by shifts in the system equivalence point. For example, as the site
concentration of each solid is increased by an order of magnitude,
from 10^{-4} to 10^{-3}M, the equivalence point of system M/S_A shifts by
one pH unit ($1/\chi=1$) and that of M/S_B by ($1/\chi=2$) pH units. This
point is illustrated more directly in Figure 2 where the equivalence
point is shown as a function of site concentration. Again, the
systems have an initial equivalence point of pH 8. In practical
application this means that adsorptive reactions which produce a
greater change in the net proton coefficient will exhibit less
dependence on particle concentration in determining overall partitioning between the water and solid phases. For example, a step
change in the concentration of amorphous iron coated particles in a
lake water column will produce a much smaller change in the net
solid/solution partitioning of Pb(II), e.g., adsorbed Pb/dissolved
Pb, ($\chi \approx 2.0$, 7) than it will for Ag(I) ($\chi \approx 0.9$, 7).

General Observations About χ, its Relationship to the Overall
Partitioning Coefficient and to the Concept of Surface-Site
Heterogeneity. One approach to metal/particle surface interactions
which has been developed, historically, in a variety of forms, is a
conceptual model that assumes only two conditions for surface
sites: occupied by an adsorbate or unoccupied. In applying this
approach to the solid/aqueous solution interface, the adsorption
reaction is modeled as analogous to a simple solution-phase reaction

$$M + SO = SOM \tag{5}$$

and

$$K = \frac{[SOM]}{[M][SO]} \tag{6}$$

The surface-site mass balance is given by

$$S_T = [SO] + [SOM] \tag{7}$$

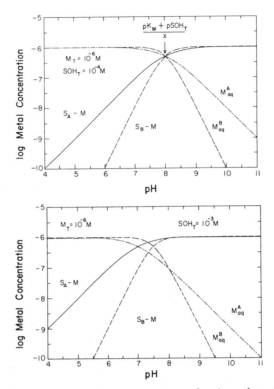

Figure 1. Log concentration-pH diagram showing the effect of the macroscopic proton coefficient on metal adsorption onto two hypothetical solids in separate, single adsorbent systems. Both adsorbents are present at the same site concentration: top, 10^{-4}M, and bottom, 10^{-3}M.

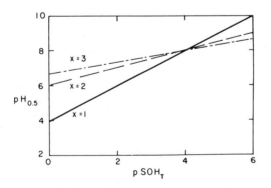

Figure 2. Shift in system equivalence points (pH of 50% fractional metal adsorption) as a function of site concentration and macroscopic proton coefficient. Initial equivalence point of pH 8 and $SOH_T = 10^{-4}$M are arbitrary reference conditions.

M and SOM represent all species of an adsorbate, M, which are dissolved or associated with the surface, respectively. SO represents
all surface sites not associated with any adsorbed species of M and
[ ] denotes concentration. A combination of Equations 6 and 7
yields the Langmuir isotherm

$$[SOM] = \frac{S_T \ K \ [M]}{1 + K \ [M]} \tag{8}$$

where K is the metal ion/surface association constant.

Two assumptions used to derive the Langmuir isotherm are that
all surface sites are equivalent (surface-site homogeneity) and that
the energy of interaction between an adsorbate and the surface does
not change with the adsorption density (Γ = moles M adsorbed/mole
sites). The distribution coefficient, K_D (Equation 2), is often
mistakenly assumed to behave Langmuirian at all adsorption densities, i.e., that K_D is constant for a given adsorbate and surface
type and general solution conditions. However, K_D is only constant
when the concentration of adsorbed metal is well below S_T. Likewise, in its simplest form, the overall exchange reaction and mass-
action expression (Equation 4) is also derived from assumptions of
Langmuirian behavior: the particle surface is homogeneous and K is
independent of adsorption density.

Benjamin and Leckie (5) tested the applicability of extending
the Langmuirian model to metal ion adsorption onto hydrous metal-
oxide surfaces by examining metal ion adsorption over a wide range
of adsorption densities (Γ). At very low adsorption density, when
unoccupied surface sites were in great excess ([SOH] $\cong$ SOH$_T$ or [SOM]
<< SOH$_T$), the intensity of metal ion/surface site interactions (as
described by K, Equation 4) appeared independent of surface coverage, and metal ion adsorption was considered to exhibit Langmuirian
behavior. However, at higher adsorption densities (above some critical adsorption density, Γ^*) K decreased monotonically with increasing (Γ). Benjamin and Leckie attributed this phenomenon to heterogeneity of surface sites on oxide surfaces, which produced a range
of metal ion/surface reaction energies.

In their description of metal ion adsorption, Benjamin and
Leckie used an apparent adsorption reaction which included a generic
relationship between the removal of a metal ion from solution and
the release of protons. The macroscopic proton coefficient was
given a constant value, suggesting that χ was uniform for all site
types and all intensities of metal ion/oxide surface site interaction. Because the numerical value of χ is a fundamental part of the
determination of K, discussions of surface site heterogeneity, which
are formulated in terms similar to Equation 4, cannot be decoupled
from observations of the response of χ to pH and adsorption density.
As will be discussed later, it is not the general concept of
surface-site heterogeneity which is affected by what is known of χ;
instead, it is the specific details of the relationship between K,
pH and Γ which is altered.

To what extent are assumptions of a constant χ valid? Table II
shows the observed macroscopic proton coefficients for cation and
anion adsorption in a variety of heterogeneous systems. The coefficients were determined by Kurbatov plots (6) or by isotherm analysis
(7), unless otherwise indicated. In all cases, χ is not an integer.

Table II.   Overall Conditional Proton Coefficients (after Schindler
(17))

ADSORBATE	ADSORBENT	pH	$p\Gamma(-\log \frac{mol}{mol})$	χ	Ref.
Pb(II)	(am)$Fe_2O_3 \cdot H_2O$	4.5-5.0	2.05	1.98	7
	"FeOOH"	5.0	--	1.18	11
		6.0	--	1.59	11
	δ-MnO	6.0	--	1.40	12
	γ-Al_2O_3	4.0-7.0	--	1.50	2
Cd(II)	(am)$Fe_2O_3 \cdot H_2O$	6.0-7.5	1.8-5.4	1.4-2.5	13
	α-TiO_2	5.0-7.0	1.8-4.3	0.9-1.1	13
	α-Al_2O_3	6.0-8.0	2.5-4.5	1.2-2.0	13
	α-FeOOH	5.8-9.1	2.5-4.5	0.4	10
	Na-Mont.	6.5	2.6-4.4	0.2-.44	13
Mn(II)	δ-MnO_2	4.5-8.2	--	1.0-1.7	14
Cu(II)	(am)$Fe_2O_3 \cdot H_2O$	5.1-5.8	2.17	1.9	7
	α-FeOOH	3.8-5.9	2.0-3.5	1.1	10
		5.0-6.0	1.2-3.0	1.3-1.7[†]	15
Ag(I)	(am)$Fe_2O_3 \cdot H_2O$	6.5-10.0	2.0	0.9	7
Sb(III)	(am)$Fe_2O_3 \cdot H_2O$	7.0-9.0	3.0	0.4	7
Cr(VI)	(am)$Fe_2O_3 \cdot H_2O$	6.0-8.0	2.5-5.0	-0.7 - -2.7	10
	α-TiO_2	7.0-8.5	2.0-4.5	-1.7 - -2.6	10
	α-Al_2O_3	6.0-8.0	1.2-5.0	-0.3 - -0.7	10
Se(VI)	(am)$Fe_2O_3 \cdot H_2O$	4.5-6.4	1.5	-0.9	7
As(VI)	(am)$Fe_2O_3 \cdot H_2O$	9.5-10.5	1.5	11.5	7
V(V)	(am)$Fe_2O_3 \cdot H_2O$	9.5-11.5	3.2-4.0	-0.6 - -1.0	16
	α-TiO_2	8.5-11.5	3.0-3.8	-.59 - -.61	16
	α-Al_2O_3	8.0-11.0	4.3-4.7	-0.7 - -1.0	16

[†]titrations

In addition, χ is generally not constant and appears to exhibit some dependency on pH and adsorption density (Γ). Cation coefficients are generally greater than one, although there are some marked exceptions (e.g., Cd/α-FeOOH or Na-montmorillonite). In contrast, the absolute value of net proton coefficients for anions are generally between zero and one. The negative value of χ for anions indicates that anion adsorption results in an overall removal of protons from solution.

Non-integer, net proton coefficients are reasonable considering the complexity of heterogeneous systems (q.v., Table I). Although integer stoichiometric coefficients are appropriate for microscopic subreactions, arbitrarily extending stoichiometric relationships used in microscopic reactions to macroscopic partitioning expressions is unwarranted.

It also follows from analogy to coordination chemistry of solutions that the apparent macroscopic stoichiometric coefficient for $[H^+]$ should be affected by pH. For example, the mole-fraction averaged proton release from the two competing surface reactions

$$SOH + M^{z+} \overset{{}^*K^{int}_M}{=} SOM^{z-1} + H^+ \qquad (9)$$

$$SOH + M^{z+} + H_2O \overset{{}^*K^{int}_{MOH}}{=} SOMOH^{z-2} + 2H^+ \qquad (10)$$

is a function of the pH-dependent hydrolysis of M^{z+} in solution in addition to the relative magnitude of ${}^*K_M$ and ${}^*K_{MOH}$.

The relationship between χ and Γ is less obvious; however, the most straightforward explanation is that the relative importance of Equations 9 and 10 varies as a function of Γ. This would indicate that the dependence of ${}^*K^{int}_M$ on Γ at a fixed pH is different from that of ${}^*K^{int}_{MOH}$ since the solution speciation and, therefore, the ratio $[M^{z+}]/[MOH^{z-1}]$, are fixed once the pH is known. Another possibility is that the average number of protons associated with surface sites varies on a surface with heterogeneous sites.

Whatever the specific cause of the observed range in χ, the fact that the numerical value of the net partitioning coefficient is very sensitive to differences in χ makes knowing χ accurately an important issue. Consequently, χ must be known, i.e., "calibrated," in terms of major, observable variables.

Determination of $\chi(pH,\Gamma)$ as a Macroscopic Model Parameter

The macroscopic proton coefficient may be determined by graphical analysis of observed system variables according to two different procedures: fractional adsorption edge linearization (6) and isotherm analysis (7). The procedures for calculating the macroscopic proton coefficients according to these two methods are discussed in detail below, as are their relative advantages and disadvantages for use in semi-empirical descriptions of adsorption.

Kurbatov Plots.

Kurbatov plots (6) are made by applying the linearization of a general overall mass action expression (e.g., Equation 4)

$$\log \left\{ \frac{[M_{ads}]}{[M_{aq}][SOH]} \right\} = \chi_K \cdot pH + \log K_K \qquad (11)$$

to experimental data for metal-ion adsorption as a function of pH.
The variables in the argument of the log in the left hand side (LHS)
of Equation 11 are all readily determined from experimental observation. The LHS of Equation 11 plotted as a function of pH yields a
line of slope χ_K and intercept of log K_K. The subscript 'K' denotes
Kurbatov constants.

The application of this method to typical experimental results
is illustrated for Cd(II) adsorption onto α-Al$_2$O$_3$. Cadmium (II)
fractional adsorption as a function of pH and alumina concentration
is shown in Figure 3a. At constant Cd(II)$_T$, the position of the
pH/Cd(II) fractional "adsorption edge" shifts to higher pH as the
surface site concentration decreases from 7.3×10^{-3} M (50 g/ℓ
alumina to 2.9×10^{-4} M (2 g/ℓ alumina). Such a shift is expected by
analogy to metal-ligand complexation reactions in solution. In
addition to the shift, the slope of the fractional adsorption edge
decreases with increasing $[SOH_T]/[M]_T$. Figure 3b contains Kurbatov
plots corresponding to the fractional adsorption data shown in
Figure 3a. For example, linearized edge data for the 6 g/ℓ α-Al$_2$O$_3$
systems (circles, Figure 3b) have a slope of 1.09 ± 0.09 moles
protons released per mole Cd(II)$_T$ adsorbed and an intercept of -4.89
± 0.66. The observed decrease in the fractional adsorption edge
with decreasing $[SOH]_T/[Cd(II)]_T$ ratio corresponds to a decrease in
χ_K from 1.31 ± 0.12 to 0.83 ± 0.083.

Surface site densities used in the computation of the oxide
site concentrations presented in this paper were determined by
either rapid tritium exchange or acquired from published values
(18). Reported total site densities for hydrous metal oxides show
relatively little variation; generally they range by less than a
factor of 3. Since [M], [SOM], [H] and χ are known or can be
determined from experimental data, uncertainties in estimates of the
total site concentration are directly translated into uncertainties
in the calculated partitioning coefficient.

In spite of their usefulness as curve-fitting parameters,
Kurbatov coefficients (i.e., proton and partitioning) have deficien-
cies sufficient to restrict their use in macroscopic adsorption
models. The compilation of proton coefficients presented in Table
II suggests that χ has a dependency on either pH or Γ or both.
However, each set of fractional adsorption data included in a
Kurbatov plot includes a range of pH values and adsorption densities. Consequently, Kurbatov coefficients cannot be used to
resolve the question of whether or not χ should have a clear functional relationship to pH or Γ. For example, Figure 4a shows χ_K
plotted as a function of the adsorption density which corresponds to
50% fractional adsorption from an extended Cd(II)/α-Al$_2$O$_3$ data set
(17). χ_K exhibits a general decrease with increasing surface coverage, although the scatter in the data obscures any clear functional
dependence of χ_K on Γ. Since the uncertainty in χ_K (from multiple
linear regression at the 95% confidence interval) is generally less
than 10% (approximately the diameter of a datum point symbol), the
scatter in Figure 4a suggests that χ_K is a function of another

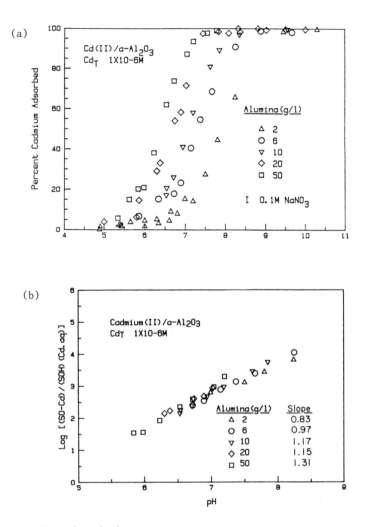

Figure 3.   a) Cd(II) fractional adsorption onto α-alumina as a
function of pH and site concentration; b) Kurbatov plots of the
Cd(II)/α-alumina system.

variable, in addition to Γ. The scatter in χ_K produces a wild variation in K_K with adsorption density (Figure 4b).

Isotherm Subtraction. A second method (7) of determining the net proton coefficient from adsorption data is an adaptation of the thermodynamics of linked functions as applied to the binding of gases to hemoglobin (19). The net proton coefficient determined by this method is designated, χ_p. The computational procedure makes a clear distinction between the influence of adsorption density and pH on the magnitude of the net proton coefficient. The fundamental equation used in the calculation of χ_p is

$$\chi_p = (\frac{\partial \bar{n}_H}{\partial \bar{n}_M})_{\mu_H} = (\frac{\Delta \log[M]}{\Delta pH})_\Gamma \tag{12}$$

It carries with it the restriction that the chemical potential of all other species must remain constant and the value of the coefficient must be estimated at constant surface coverage. The appendix to this paper contains a straightforward and intuitive derivation of Equation 12.

The protocol for determining χ from adsorption data is as follows. Using the metal ion adsorption data as a function of pH (e.g., that of Figure 3a), two adsorption isotherms are generated for a narrow pH interval around the pH of interest (Figure 5). The smaller the pH interval the more representative χ_p is of the target pH. Kurbatov coefficients (χ_K and K_K) are useful for interpolating to pH values for which there are no adsorption data. The proton coefficient determined from isotherm analysis may be calculated in accordance to Equation 12 in the following manner:

$$\chi_p = \{(\Gamma - y_2)/m_2 - (\Gamma - y_1)/m_1\}/(pH_2 - pH_1) \tag{13}$$

where m is the slope of the isotherm and y is the surface coverage at an arbitrarily chosen dissolved equilibrium metal ion concentration. The isotherm proton coefficients given in Figure 6 were determined by evaluating Equation 13 at different surface coverages for isotherms at the pH values pH_1 and pH_2. The pH at which the $\chi_p(\Gamma)$ relationship is shown is the pH at the center of the evaluation interval.

At constant pH, χ_p increases with increasing surface coverage and, at constant $\Gamma(p\Gamma)$ decreases with increasing pH (Figure 7). For example, for Cd(II) adsorption onto α-Al_2O_3 at pH 7

$$\chi_p = 0.105 \log \Gamma + 1.840 \tag{14}$$

This combined dependency of χ_p on pH and surface coverage has the net effect of decreasing the slope of the pH-fractional adsorption edge as the ratio of adsorbate to adsorbent sites increases, as in Figure 3a.

Consequences of $\chi_p = f(pH, \Gamma)$. In previous sections it was demonstrated that the net proton coefficient plays an important role in macroscopic models of metal adsorption. However, its relationship to major system variables, such as pH and Γ, is poorly understood.

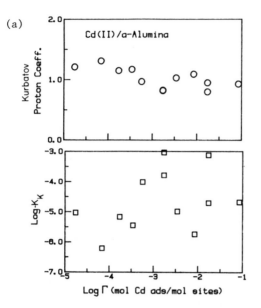

Figure 4.   Kurbatov: a) proton and b) partitioning coefficients as a function of adsorption density.

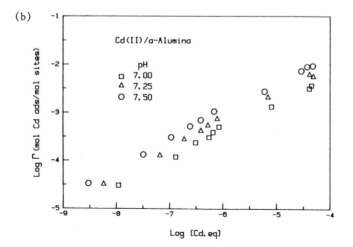

Figure 5.   Isotherms for Cd(II) adsorption onto α-Al$_2$O$_3$.   Isotherm slopes are approximately 0.7.

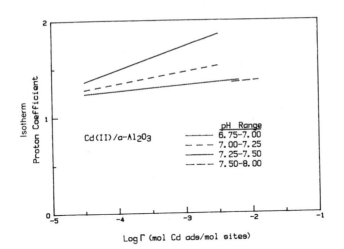

Figure 6. χ_p as a function of log Γ for Cd(II)/α-Al$_2$O$_3$ for different pH values.

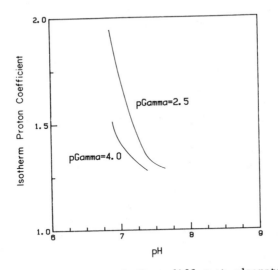

Figure 7. χ_p as a function of pH at different adsorption densities for the Cd(II)/alumina system.

Two methods for determining net proton coefficients from adsorption data were also described. In the following sections the pH and adsorption-density dependency of χ are discussed and implications for the macroscopic modeling of metal adsorption are presented. The effect of including the pH- and pΓ-dependence of χ_p in the calculation of a macroscopic partitioning coefficient is shown in Figure 8. The line designated 'surface reaction' represents the calculated partitioning coefficient when the contribution of the proton is excluded from the mass-action expression, i.e.,

$$K_{sr} = \frac{[SOCd]}{[Cd][SOH]} \tag{15}$$

The decrease in log K_{sr} with increasing surface coverage, i.e., non-Langmuirian behavior, comes directly from experimental observation. If Cd^{2+}/H^+ exchange is included, but with a constant proton coefficient ($\chi \neq f(pH,\Gamma)$, the resultant log K/log Γ line in Figure 8 is parallel to that of K_{sr} but lower in magnitude by $[H^+]^{\chi}$. For example, data at pH 7 and $\chi = 1.2$, the difference in the partitioning coefficients would be $\sim 2 \times 10^9$. However, χ is not constant; instead, it increases with surface coverage. When the specific relationship between χ_p and Γ is included in the calculation of K, the slope of the log K/log Γ line becomes more negative (Figure 8). The partitioning coefficient which is sensitive to the variation of χ with Γ (i.e., it includes χ_p) is designated P and

$$P = K_{sr} [H]^{\chi_p} \tag{16}$$

For example, for Cd(II) adsorption onto α-Al$_2$O$_3$ at pH 7

$$\log P = 1.50 \log \Gamma - 12.85 \tag{17}$$

The relationship between P and Γ also changes with pH for two reasons (compare lines 3 and 4 of Figure 8): i) dissolved/solid partitioning of the metal (i.e., K_{sr}) changes with pH, and ii), χ_p is a function of pH (Figures 6 and 7).

The observation that the macroscopic proton coefficient is a function of adsorption density and pH has several implications for macroscopic modeling of cation and anion adsorption. The dependency of χ_p on pH and Γ affects: 1) the relationship of the macroscopic partitioning coefficient to pH and adsorption density, 2) the notion of metal ion preferences for a particular surface in systems with multiple solid phases, 3) the accuracy of predictive models when used over a range of adsorption density and pH values, and 4) conclusions about site heterogeneity based upon partitioning expressions which use constant proton coefficients.

$P=f(\chi_p, pH)$. As described above, the magnitude of P is inexorably linked to the variations of χ with pH and adsorption density. However, the response of χ (and P) to Γ and pH varies among hydrous oxides. For example, Figure 9a shows the instantaneous (isotherm) proton coefficient (χ_p) "zones" determined for Cd ion adsorption onto (am)Fe$_2$O$_3$·H$_2$O, α-Al$_2$O$_3$ and α-TiO$_2$. The zones are defined by the calculated proton coefficients determined for a range of pH values and adsorption density. The "thickness" of each zone gives a qualitative comparison of the pH dependency of χ_p at each adsorption

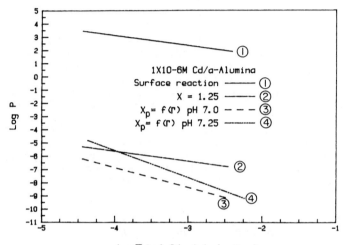

Figure 8. Log P as a function of log adsorption density for different χ_p adsorption density dependencies.

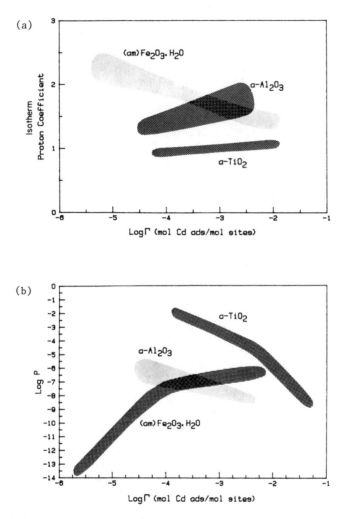

Figure 9.   Comparison of a) net proton coefficient and b) log P pH and surface coverage "zones" for Cd(II) adsorption onto α-Al$_2$O$_3$, (am)Fe$_2$O$_3$.H$_2$O and α-TiO$_2$.

density. Titania shows little change in χ_p with either pH or surface coverage while amorphous iron oxyhydroxide and alumina exhibit large ranges in the observed proton coefficients. The corresponding partitioning coefficient zones are shown in Figure 9b. Because χ_p is an exponent in the numerator of the overall mass action expression describing metal adsorption (e.g., Equation 4) the slopes of the χ_p and log P vs log Γ functions (and the predominance zones) have opposite signs (this is not the case for anion adsorption; see Honeyman, 1984). The dependency of χ_p on pH produces a number of possible partitioning coefficients at each adsorption density. It is also clear from Figure 9b that the relative ability of these three metal oxides to remove Cd(II) from solution varies considerably with adsorption density which, in part, is due to changes in χ_p (pH,Γ). For example, consider each oxide to be present at equal site concentrations. At low surface coverages, α-Al$_2$O$_3$ should be more effective at removing Cd(II) from solution than is (am)Fe$_2$O$_3$ H$_2$O. At moderate adsorption densities (pΓ ~ 3 to 4), the similarity of partitioning coefficients will make them indistinguishable in their ability to remove Cd(II) from solution. At still higher adsorption densities, amorphous iron oxyhydroxide may successfully compete with α-TiO$_2$. This example suggests that metal ion preference for a particular surface which is determined at one point in pH/Γ "space" should not be extended to other conditions unless P = f(pH,Γ) is explicitly known. This illustrates one of the problems in extending laboratory data for metal ion preferences to field situations.

Metal Ion Adsorption in Mixtures of Multiple Solid Phases. One of the arguments put forth for extending the concepts of solution coordination chemistry to heterogeneous systems is the hypothesis that the mineral components of soils or sediments can be considered as ligands which compete for complexation of adsorbates. To this end, it is important to know the relative ability of different mineral surfaces to complex solutes. For example, consider the adsorption of Cr(VI), an oxyanion, onto α-Al$_2$O$_3$ and (am)Fe$_2$O$_3$.H$_2$O (Table III). With each oxide in suspension present at the same total site concentration, approximately the same fractional adsorption of Cr(VI) was observed at pH 7 from a one micromolar solution of Cr(VI). Consequently, the affinity of alumina and amorphous iron oxyhydroxide surfaces at pH 7 for chromate ion are the same when the affinity is defined in terms of a distribution coefficient (K_D or K_{sr}). However, the pH dependence of Cr(VI) fractional adsorption is different for the two adsorbents, as indicated by the Kurbatov proton coefficients, i.e., the slope of the pH/fractional adsorption edge. Therefore, at pH values less than 7, K_{sr} for Cr(VI)/α-Al$_2$O$_3$ will be less than the K_{sr} for Cr(VI)/(am)Fe$_2$O$_3$ H$_2$O. At pH values greater than 7 the converse is true. When χ_K is used to account for the partitioning of Cr(VI) with pH, calculated affinities differ by approximately two and one half orders of magnitude. Use of χ_p markedly widens the apparent affinity preference and, as discussed next, improves our ability to describe solute adsoprtion over a range of adsorption densities.

Table III.   Cr(VI)Adsorption onto α-Al$_2$O$_3$ and (am)Fe$_2$O$_3$·H$_2$O

Cr(VI) 1.0 x 10^{-6}M; SOH$_T$ 8.8 x 10^{-4}M; pH 7

	f	log K$_{sr}$	χ_K	p^{K}K	χ_p	p^P
αAl$_2$O$_3$	0.56	3.16	-0.70	8.09	0.72	7.68
Fe$_2$O$_3$ H$_2$O	0.47	3.01	-1.09	10.65	-1.46	13.26

<u>Predicting Adsorption Over a Range of Adsorption Densities.</u>
Kurbatov coefficients (χ_K,K$_K$) are useful curve-fitting parameters
for fractional adsorption edges, but their use in macroscopic models
of solute adsorption should be carefully considered. This is illustrated in two examples. Figure 10 compares the observed adsorption
densities for Cd(II) on α-Al$_2$O$_3$ with the corresponding adsorption
densities calculated from a single pair of Kurbatov coefficients
(χ_K,K$_K$). The coefficients used were for a Cd(II) fractional adsorption edge representing a relatively low adsorption density (pΓ ≈ 4;
log(mol ads/mol sites); Cd(II)$_T$ = 1 x 10^{-6}M; 7.3 x 10^{-3}M total
sites). Except at lowest adsorption density, there is a significant
departure between the observed and calculated adsorption: the calculations overestimated the actual Cd(II) removal.

An invariant partitioning coefficient will not account for
decreasing metal/surface affinities with increasing adsorption
density. The isotherms in Figure 5 have slopes of approximately
0.7. This means that the macroscopic partitioning coefficient
decreases with increasing surface coverage (Langmuirian behavior
should exhibit a slope of 1.0). In fact, because of the way in
which Kurbatov coefficients are determined (refer to the discussion
of Figure 4), it is very unlikely that it would be possible to
determine an unambiguous relationship between K$_K$ and Γ. Thus, the
use of Kurbatov coefficients requires that adsorption be Langmuirian
in behavior (that is, that the metal ion/surface association parameter must be independent of Γ.

As a second example, consider the partitioning of Cd(II)
between two adsorbents--α-TiO$_2$ and (am)Fe$_2$O$_3$·H$_2$O. Figure 11 shows
Cd(II) fractional adsorption as a function of pH for binary mixtures
of these adsorbents under experimental conditions such that Cd(II)$_T$
and SOH$_T$ are constant: only the surface site mole fraction varies
from one end-member to the next. As the site mole fraction shifts
between the end-members, the fractional adsorption edges for the
binary adsorbent mixtures varies between the limits defined by end-
members. In the absence of particle-particle interactions, the
adsorbents should act as independent ligands competing for complexation of Cd(II). If this is the case, then the distribution of
Cd(II) in such binary mixtures can be described by a composite mass-
action expression (<u>13</u>) which includes a separate term for the interaction of Cd(II) with each adsorbent.

The results of two sets of computations are shown in Figure
12. Open circles represent calculations for the binary adsorbent
systems which used the Kurbatov coefficients for the end-member
systems. This is the approach used by Davies-Colley et al., (<u>9</u>) in
their examination of metal ion adsorption in mixtures of model

GEOCHEMICAL PROCESSES AT MINERAL SURFACES

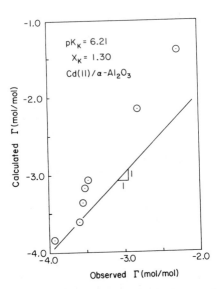

Figure 10. Comparison of calculated and observed adsorption densities for Cd(II) adsorption on α-Al₂O₃. A single set of Kurbatov coefficients is used for the entire range of adsorption densities.

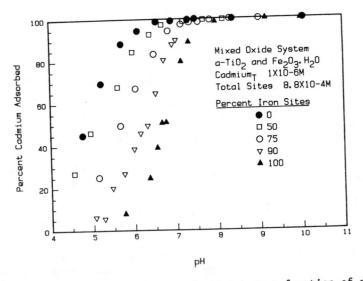

Figure 11. Cd(II) fractional adsorption as a function of pH in binary mixtures of α-TiO₂ and (am)Fe₂O₃·H₂O at constant SOH_T.

adsorbents. The Γ-invariant Kurbatov coefficients are inadequate to describe the general adsorption behavior at the relatively high adsorption densities of the experimental systems. However, the use of proton coefficients which depend on pH and Γ results in good agreement between the experimental data and calculations.

Macroscopic Coefficients and Surface-Site Heterogeneity. Benjamin and Leckies' model (5) of heterogeneous metal oxide surface sites includes two observations of metal ion/surface site interactions. At very low adsorption densities, metal ion adsorption exhibits Langmuirian behavior. However, above some critical adsorption density, Γ^* (i.e., when approximately 0.01% of surface sites are occupied), the intensity of metal ion/interactions decreases with increasing surface-site occupancy. A corollary of this model is that K has a unique value at adsorption densities below Γ^* for a specific adsorbate and adsorbent. By extension, ΔG_{ads} is unique and, therefore, independent of Γ. This is one of the basic tenents of Langmuirian adsorption.

The incorporation of $\chi_p(pH, \Gamma)$ into macroscopic mass-action expressions for adsorption has shown not only that K is not a unique function of Γ at adsorption densities greater than Γ^* but that K is also not unique below Γ^*. In both cases it is due to the dependence of the macroscopic partitioning coefficient on pH (Figure 13).

If K is taken as a surrogate measure of the intensity of metal ion/surface reactions, then the observation that K is not unique at adsorption densities below Γ^* suggests that surface site/adsorbate interactions are still of variable intensity (ΔG_{ads} is not constant). Hence, one of the basic tenents of Langmuirian behavior is not met.

Microscopic Subreactions and Macroscopic Proton Coefficients. The macroscopic proton coefficient may be used as a semi-empirical modeling variable when calibrated against major system parameters. However, χ has also been used to evaluate the fundamental nature of metal/adsorbent interactions (e.g., 5). In this section, macroscopic proton coefficients (χ_K and χ_p) calculated from adsorption data are compared with the microscopic subreactions of the Triple-Layer Model (1) and their inter-relationships are discussed.

The macroscopic proton stoichiometry of an adsorption reaction represents the net consumption or release of protons by all subreactions which result from the removal of an adsorbate from solution. Macroscopic partitioning expressions do not identify the details of such interactions; the stoichiometric expressions merely show a generic relationship between adsorbate removed from solution and a transfer of protons. Schindler (17,23), for example, has discussed the relationship between the expected maximum proton release and the observed overall stoichiometry of surface complex formation due to competing surface protolytic and electrolyte ion surface reactions. He observed that the proton release will be equal to the actual net metal species/proton exchange only when the contribution of competing reactions, e.g., those of Table IV, are negligible or are specifically taken into account. Surface coordination models, in contrast, explicitly define a proposed suite of reactions which can be used to describe net system behavior. In the following comparison of microscopic subreactions and macroscopic coefficients it is shown

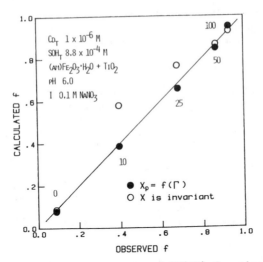

Figure 12. Comparison of the observed Cd(II) fractional removal
f = [Cd(II)$_{ads}$/Cd(II)$_T$] with model results incorporating different
characteristics of χ. Open circles: Kurbatov coefficients for
end-members (Titania: χ_K = 0.99, pK_K = 1.77; Amorphous Iron:
χ_K = 1.3, pK_K = 5.37). Closed circles: P(pH,Γ) for end-members
(Titania: log P = 2.54 log Γ - 8.841; Amorphous Iron: log P =
2.54 log Γ + 1.98).

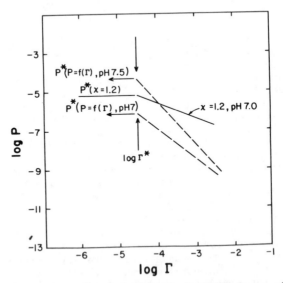

Figure 13. Schematic diagram showing the dependence of log P* on
the χ$_p$ functionality. For surface coverages below Γ*, no unique
set of P* and Γ can describe adsorbate behavior.

that electrolyte ion reactions have an important effect on the net proton transfer.

It is not currently possible to examine the configuration of the adsorbed species unambiguously. However, since thermodynamic arguments do not require a specific model at the molecular level, it is still possible to analyze equilibrium data within a thermodynamic context. Most surface reactions are inferred from experimental observations of reaction stoichiometries and perhaps only in a limited range of Γ. Consequently, the choice of specific surface species is dependent on two considerations: (1) the need to explain the observed measurements in terms of reaction stoichiometries, and (2) the selection of a model to allow the representation of metal/ surface site interaction intensities.

Table IV.   Subreactions and Constants Used in Triple-Layer Model Dalculations for $Cd(II)/\alpha\text{-}Al_2O_3$

reaction	K^{int}	source
$SOH + Na^+ = SO^-\text{-}Na^+ + H^+$	-9.2	20
$SOH + NO_3^- + H^+ = SOH_2^+\text{-}\text{-}NO_3^-$	8.3	21
$SO^- + H^+ = SOH^o$	$6.3^\dagger$	22
$SOH + H^+ = SOH_2^+$	$-12.0^\dagger$	22

$C_1 = 120$, $C_2 = 20$ $\mu F/cm^2$:   The capacitances for the inner and outer regions of the compact layer.

†for γ-alumina

A specific example of the relationship between the microscopic subreactions required to model experimental observations of metal removal and the macroscopic proton coefficient is shown for the case of $Cd(II)$ adsorption onto $\alpha\text{-}Al_2O_3$ (Figure 3). One variation of the surface coordination concept is used to describe the system subreactions: the Triple Layer Model of Davis et al., (1,20). The specific subreactions which are considered, the formation constants and compact layer capacitances, are shown in Table IV. Protons are assigned to the o-plane (the oxide surface) and $Cd(II)$ surface species and electrolyte ions to the β-plane located a distance, β, from the o-plane.

The information in Figure 14 was produced in the following way. The slope (or the Kurbatov coefficients shown in Table V) and position of the fractional adsorption edges in Figure 3 were used as the criteria of model fit. $^*K_{Cd}^{int}$ and $^*K_{CdOH}^{int}$ were used as the fitting parameters and all other parameters were held constant. Consequently, the intrinsic constants shown in Figure 14a represent best fit parameters and, given that all other surface and solution association constants are invariant, constitute a unique solution set for each adsorption density.

At low surface coverages (high SOH_T) only the $SO^-\text{-}CdOH^+$ surface species is required to fit the data. For example, decreasing SOH_T from 7.4×10^{-3} to 2.9×10^{-3}M increases the $Cd(II)$ adsorption

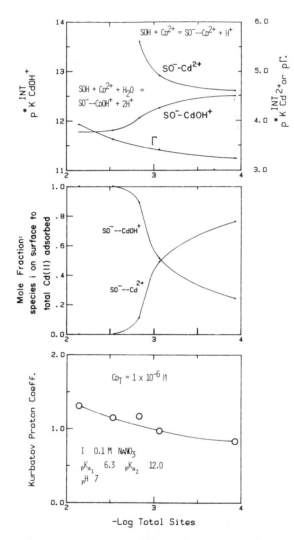

Figure 14. Triple-layer model (1) results for Cd(II) adsorption onto α-alumina at different site/adsorbate ratios. Top, Cd(II) surface reaction best fit constants; middle, Cd(II) surface species mole fractions; and bottom, slopes of fractional adsorption edges used as the criteria of fit.

density but only the SO^--$CdOH^+$ surface species is needed to fit the experimental data. Furthermore, $p^*K_{CdOH}^{int}$ is the same for both adsorption densities at 11.8. At lower alumina concentrations and greater values of Γ the SO^--Cd^{2+} species must be included for two reasons: 1) to reduce the overall proton release (the formation of the SO^--Cd^{2+} species produces one proton compared to 2 for the SO^--$CdOH$ surface species), and 2) to produce the observed shift in the pH/metal-fractional adsorption edge with changing $SOH_T/Cd(II)_T$ ratio.

The distribution of Cd(II) surface species is shown as a function of SOH_T in Figure 14b. The mole fraction of Cd(II) surface species is defined as the ratio of the concentration of a Cd(II) surface species relative to the total Cd(II) on the surface, e.g., $[SO^-$-$CdOH^+]/[\sum Cd(II)_{adsorbed}]$.

To what extent is the macroscopic proton release the direct expression of the metal/surface site reactions? Table V compares the macroscopic proton coefficients (χ_p, χ_K) with the coefficient expected if only the Cd(II) surface reactions are considered: $\chi_{m.f.}$ is the proton coefficient determined by considering the mole fraction of Cd(II) surface species and their formation reactions (Figure 14b). For example, when pSOH is 2.84, $\chi_{m.f.} = 0.11 \times 1 + 0.89 \times 2 = 1.89$. At high alumina concentrations (pSOH 2.14-2.53) the single surface reaction required to fit the data sets a limiting proton release of 2.0.

Table V.   Comparison of Proton Coefficients

mole fraction		pSOH	$\chi_{m.f.}$	χ_K	χ_p
SO--Cd	SO--CdOH				
0.77	0.23	3.53	1.23	0.83	1.48
0.49	0.51	3.06	1.51	0.97	1.48
0.11	0.89	2.84	1.89	1.17	1.47
0.00	1.00	2.53	2.00	1.15	1.46
0.00	1.00	2.14	2.00	1.31	1.43

It is clear, for the Cd(II)/α-Al_2O_3 system at pH 7, that other reactions are not negligible in their contribution to the macroscopic proton coefficient; at low surface coverage, when SO^--$CdOH^+$ is the only postulated metal-containing surface species, the macroscopic proton coefficient is less than 2. χ_p, however, does approach 2.0 at low pH and high Γ (Figure 7).

Furthermore, although other electrostatic models for the oxide/water interface may yield different relationships among postulated system components, it appears unlikely that either χ_p nor χ_K alone will adequately represent the postulated 'true' adsorbate/proton exchange ratio.

Evaluations of Surface-Site Heterogeneity. An interesting aspect of these modeling results relates to the concept of site heterogeneity. When alumina site concentrations were greater than 2.9×10^{-4}M (or log $\Gamma < -3.2$) a single surface complexation reaction was needed to describe Cd(II) adsorption. $^*K_{CdOH}^{int}$ was constant and independent of surface coverage. Under these conditions, Cd(II) adsorption (in terms of the specific subreaction) exhibits Langmuirian behavior. In contrast, at the macroscopic level, χ_K and χ_p (and K_K and P) are not constant with Γ, suggesting non-Langmuirian behavior. Such a contradiction between the macroscopic description of adsorption and microscopic sub-reactions again suggests that conclusions about the existence or extent of surface-site heterogeneity, when based solely on macroscopic partitioning coefficients, may be over estimated. Although it is reasonable that multiple site types could exist on oxide surfaces, the specific relationship between a macroscopic partitioning coefficient and adsorption density may not be an accurate reflection of changing intensities for metal/surface reactions. Instead, variations in macroscopic partitioning coefficients with surface coverage may represent changes in surface speciation as well as changes in interaction intensity.

Summary

The chemical complexity of most natural systems often requires that adsorption reactions be described using semi-empirical, macroscopic models. A common approach is to describe the net transfer of an adsorbate from the solution phase to the solid/water interface with a single stoichiometric expression. Such stoichiometries include a generic relationship between the adsorption of a solute and the release or consumption of protons.

Kurbatov plots (6) have often been employed to determine the net metal ion/proton exchange, χ, from adsorption data. Although Kurbatov constants are convenient curve-fitting parameters, they are insensitive to the variation of χ with pH and adsorption density and should be dispensed with for use in adsorbate partitioning calculations, particularly when high adsorption densities are expected (9).

It is possible to factor out the macroscopic pH- and Γ-dependence of χ through isotherm analysis (7), however, the value of χ derived from such a technique will still reflect the net change in H^+ from all subreactions (including surface protolytic and electrolyte ion/surface interactions), not just the formation of metal ion/surface site complexes (i.e., $\chi_{m.f.}$ compared to χ_p). In spite of this, partitioning coefficients determined from isotherm analysis may be useful for describing metal behavior in natural environments since much of the variation in metal adsorption with pH and adsorbate/adsorbent ratio may be accounted for by variations in the macroscopic proton coefficient. This includes adsorption behavior which cannot be predicted through the use of deterministic models. For example, the apparent existence of heterogeneous surface sites, or where the use of equilibrium speciation models are unwieldly, such as describing metal adsorption over a wide range of M_T/SOH_T.

Although macroscopic partitioning coefficients will provide accurate empirical descriptions of the variation in adsorption intensity with adsorption density, they are unlikely to yield sufficient information about the origin and fundamental physical/chemical nature of surface heterogeneity. For example, in certain instances, what is interpreted from an analysis of P or K as a range of surface-complexation energies (log K is not constant, therefore ΔG_{ads} varies) may instead be the consequence of changing surface speciation.

APPENDIX I

THE MATHEMATICAL DEVELOPMENT OF ISOTHERM ANALYSIS

FOR THE MACROSCOPIC PROTON COEFFICIENT, χ_P

This method is an adaptation of the thermodynamic analysis of linked functions, the binding of multiple ligands to macromolecules (e.g., Wyman, 19) and the binding of gases to hemoglobin.

The Gibbs Free Energy change accompanying the transfer of dn_B moles of B from a reservoir in which it is present in its standard state to the equilibrium mixtures is

$$dG = (\mu_b - \mu_B^o)dn_B \tag{A-1}$$

where μ_B^o is the chemical potential of B in its standard state. The Gibbs-Duhem Equation for species in solution is written as

$$n_B d\mu_B + n_H d\mu_H + \Sigma n_i d\mu_i = 0 \tag{A-2}$$

n_H and $d\mu$ refer to protons in solution and the summation extends to all other species in solution.

The total differential of the Gibbs Free Energy of B in solution $(n_B \mu_B)$ is

$$dG_B = n_B d\mu_B + \mu_B dn_B \tag{A-3}$$

Substitution of Equation A-2 into Equation A-3 gives

$$dG_B = \mu_B dn_B - n_H d\mu_H - n_i d\mu_i \tag{A-4}$$

Since dG_B is an exact differential

$$\left(\frac{\partial \mu_B}{\partial \mu_H}\right)_{n_B} = -\left(\frac{\partial \mu_H}{\partial n_B}\right)_{\mu_H} \tag{A-5}$$

The variables in Equation A-5 refer to the solution phase. If $\bar{n}_B$ represents the moles of B on the oxide surface, then

$$dn_B = -d\bar{n}_B = -d\Gamma \tag{A-6}$$

where Γ is the moles of adsorbed B per mole of adsorbent. Equation A-5, evaluated at constant n_B, can be rewritten, according to Equation A-6, as

$$\left(\frac{\partial \mu_B}{\partial \mu_H}\right)_\Gamma = -\left(\frac{\partial n_H}{\partial n_B}\right)_{\mu_H} \tag{A-7}$$

The right-hand side of Equation A-7 is the number of moles of H^+ lost or gained by the solution per mole of B adsorbed. Thus, in terms of the general stoichiometric relationship for cation adsorption

$$SOH_\chi + M = SO-M + {}_\chi H$$

or anion adsorption

$$SOH + An + \chi H = SOH_{(1+\chi)} - An$$

the definition of the isotherm (net) proton coefficient becomes

$$\chi_p = (\frac{\partial n_H}{\partial n_B})_{\mu_H} \tag{A-8}$$

Since $dG = (RT \ln(a_B))dn_B$, the effect of the H^+ activity on dG can be found by substituting Γ for n_B and taking the partial derivative of dG with respect to μ_H at constant Γ. Thus,

$$(\frac{\partial(dG)}{\partial\mu_H})_{\Gamma,T,P} = -RT(\frac{\partial \ln(a_B)}{\partial\mu_H})_{\Gamma,T,P} \cdot d\Gamma \tag{A-9}$$

Also, from Equation A-1

$$(\frac{\partial(dG)}{\partial\mu_H})_{\Gamma,T,P} = - [(\frac{\partial\mu_B}{\partial\mu_H})_{\Gamma,T,P} - (\frac{\partial\mu_B^o}{\partial\mu_H})_{\Gamma,T,P}]d\Gamma \tag{A-10}$$

Since μ_B^o is dependent of μ_H

$$(\frac{\partial(dG)}{\partial\mu_H})_{\Gamma,T,P} = - (\frac{\partial\mu_B}{\partial\mu_H})_{\Gamma,T,P} \cdot d\Gamma \tag{A-11}$$

Consequently, a combination of Equations A-7, A-9, and A-11 yields

$$(\frac{\partial \ln(a_B)}{\partial\mu_H})_{\Gamma,T,P} = - \frac{1}{RT} (\frac{\partial\mu_H}{\partial n_B})_{\mu_H} \tag{A-12}$$

Since $\mu_H = \mu_H^o - RT(2.303)(pH)$, Equation A-12 can be rewritten in terms of pH as

$$(\frac{\partial \ln(a_B)}{\partial(pH)})_{T,P,\Gamma} = (\frac{\partial \ln(a_B)}{\partial\mu_H})(\frac{\partial\mu_H}{\partial(pH)})_{T,P,\Gamma} = -2.303RT(\frac{\partial \ln(a_B)}{\partial\mu_H})_{T,P,\Gamma} \tag{A-13}$$

Substitution of Equation A-13 into Equation A-12 yields

$$\frac{1}{2.303} (\frac{\partial \ln(a_B)}{\partial(pH)})_{T,P,\Gamma} = (\frac{\partial n_H}{\partial n_B})_{\mu_H,T,P} \tag{A-14}$$

and

$$\chi_p = (\frac{\Delta n_H}{\Delta n_B})_{\mu_H, T,P} = (\frac{\Delta \log_{10}(B)}{\Delta pH})_{T,P,\Gamma} \tag{A-15}$$

where the activity of B (a_B) is replaced by the equilibrium molar concentration. Since Equation A-15 is based upon the overall system behavior, χ_p includes contributions from all reactions which result in a net change in the bulk-solution H^+ activity.

Symbols

a	activity
An	anionic adsorbate
B	generic adsorbate: anion or cation
C	capacitance
C_p	particle concentration (g ℓ^{-1})
ΔG_{ads}	Gibbs free energy of adsorption
K	generic metal ion/surface association coefficient
K_D	distribution coefficient (ml g^{-1})
$^*K^{int}$	intrinsic surface complexation constant
K_K	Kurbatov partitioning coefficient
K_{sr}	solution/solid distribution coefficient (mol^{-1})
M	a generic metal element; or, all species of element M
m.f.	mole fraction
P	macroscopic partitioning coefficient: a specific function of pH and adsorption density (Γ)
$p\Gamma$	$-\log\Gamma$
P^*	P when $\Gamma < \Gamma^*$
SO	unoccupied surface site
SOH	surface site occupied by a proton
SOH_T	total surface sites
SOM	surface site occupied by any species of M
S_T	total surface site concentration
μ	chemical potential
Γ	adsorption density [(moles M adsorbed/moles of surface sites)]
χ	generic macroscopic proton coefficient
χ_p	isotherm proton coefficient = f(pH, Γ)
χ_K	Kurbatov proton coefficient
[ ]	concentration (mol ℓ^{-1})
{ }	mass solute adsorbed/mass adsorbent (mg g^{-1})

Acknowledgments

The authors gratefully acknowledge the financial support given by the National Science Foundation (grant #CME 80-09028) for the work done at Stanford University and to the Joint Institute for the Study of the Atmosphere and Oceans (JISAO) at the University of Washington for additional support given to B.D.H. The critical reviews by Jim Davis, George Redden, Peter Santschi and three anonymous reviewers were very valuable for clarification of the manuscript. The authors also acknowledge Mike Perona for the derivation given in the Appendix.

Literature Cited

1. Davis, J. A.; James, R. O.; Leckie, J. O. J. Colloid Interface Sci. 1978, 63, 480-498.
2. Hohl, H.; Stumm, W. J. Colloid Interface Sci. 1976, 55, 281-288.

3. Schindler, P. W.; Gamsjager, H. Kolloid Z.Z. Polym. 1972, 250, 759–763.
4. Sposito, G. J. Colloid Interface Sci. 1983, 91, 329–340.
5. Benjamin, M. M.; Leckie, J. O. J. Colloid Interface Sci. 1981, 79, 209–221.
6. Kurbatov, M. H.; Wood, G. B.; Kurbatov, J. D. J. Phys. Chem. 1951, 55, 1170–1182.
7. Perona, M.; Leckie, J. O. J. Colloid Interface Sci. 1985, 106, 64–69.
8. Tessier, A.; Rapin, F.; Carignan, R. Geochim. Cosmochim. Acta 1985, 49, 183–194.
9. Davies-Colley, R. J.; Nelson, P. O.; Williamson, K. J. Environ. Sci. Technol. 1984, 18, 491–499.
10. Balistrieri, L. S.; Murray, J. W. Geochim. Cosmochim. Acta 1983, 47, 1091–1098.
11. Gadde, R. P.; Laitinen, H. A. Environ. Lett. 1973, 5, 223–235.
12. Gadde, R. P.; Laitinen, H. A. Anal. Chem. 1974, 46, 2022–2026.
13. Honeyman, B. D.; Leckie, J. O. Geochim. Cosmochim. Acta 1985a, submitted for publication.
14. Morgan, J. J.; Stumm, W. J. Colloid Interface Sci. 1964, 19, 334–359.
15. Altmann, R. S. Ph.D. Thesis, Stanford University, Stanford, Ca., 1984.
16. Honeyman, B. D.; Leckie, J. O. Geochim. Cosmochim. Acta 1985b, submitted for publication.
17. Schindler, P. W. in "Adsorption of Inorganics at Solid/Solution Interfaces"; Anderson, M.; Rubin, A., Ed.; Ann Arbor Science: Ann Arbor, 1981; Chapter I.
18. Honeyman, B. D. Ph.D. Thesis, Stanford University, Stanford, Ca., 1984.
19. Wyman, J. Advances in Protein Chemistry, 1964, 19, 223–286.
20. Davis, J. A.; Leckie, J. O. J. Colloid Interface Sci. 1978, 67, 90–106.
21. Davis, J. A. Ph.D. Thesis, Stanford University, Stanford, Ca., 1977.
22. Huang, C. P. Ph.D. Thesis, Harvard University, Cambridge, Mass., 1971.
23. Schindler, P. W.; Furst, B.; Dick, R.; Wolf, P.U. J. Colloid Interface Sci. 1976, 55, 469–475.

RECEIVED June 18, 1986

Sorption of Hydrophobic Organic Compounds by Sediments

Gary P. Curtis, Martin Reinhard, and Paul V. Roberts

Environmental Engineering and Science Group, Department of Civil Engineering, Stanford University, Stanford, CA 94305-4020

Thermodynamic and kinetic principles which govern the uptake of nonionic, hydrophobic organic chemicals by sediments in aqueous systems are summarized. Sorption onto
organic-rich sediments can be modeled as a process where
the hydrophobic compound partitions into the organic
matter associated with the sediments analogous to the
partitioning in the octanol water system resulting in a
linear free energy relationship between the two partition
coefficients. The influence of dissolved organic matter
can be accounted for by considering a binding isotherm
between the hydrophobic solute and dissolved macromolecules. In the case of mixed cosolvents, partition coefficients can be correlated with predictions based on the
solvophobic model. Adsorption at the mineral-water
interface becomes important when the adsorbate contains
polar functional groups and/or when the adsorbate contains quite small concentrations of organic matter.
Sorption can require more than a month to reach equilibrium for highly hydrophobic compounds, but can be adequately described by a radial diffusion model accompanied
by the retarding influence of sorption.

Hydrophobic contaminants, such as halogenated hydrocarbons and polynuclear aromatic hydrocarbons are one of the most important classes
of environmental pollutants. The hydrophobicity of these compounds
is generally characterized by their physical properties, such as low
aqueous solubility, high octanol/water partition coefficient, and
high air/water partition coefficient (i.e., high Henry's constant).
These properties vary over several orders of magnitude for the contaminants encountered in the ecosphere. Research on the sorptive
interactions between sediments and hydrophobic organic solutes in
aqueous solution has focused primarily on determining the solute and
sediment properties that govern the sorption process. Information
on such processes is necessary for modeling the environmental fate
of these pollutants.

0097-6156/86/0323-0191$07.50/0
© 1986 American Chemical Society

Previous research has demonstrated that sorption of hydrophobic
compounds by sediments can be estimated from (1) molecular parameters
indicating the sorbate's hydrophobicity, such as octanol/water
partition coefficient (K_{ow}), and (2) from the mass fraction of the
organic carbon (f_{oc}) in the sorbent. The chemical principles of
hydrophobic sorption by sediments have been reviewed most recently
by Karickhoff (1). In general, the empirical relationship between
sorption, K_{ow}, and f_{oc} has been developed using organic-rich sediments or soils, whereas relatively little attention has been given
to low-organic, sandy aquifer materials. The purpose of this paper
is to summarize the concepts of hydrophobic sorption and to address
the limitations imposed by the influence of dissolved macromolecular
material, sorption kinetics, the nonsingularity in the sorption and
desorption isotherms and the conditions under which mineral surfaces
influence sorption.

Adsorption Mechanisms

The mechanisms of adsorption of organic solutes--including hydrophobic, polar, and ionic species--onto surfaces have been summarized
previously (2-5). Assuming that the various adsorptive mechanisms
act independently, the free energy of adsorption (ΔG_{ads}) can be
expressed as the sum of the individual contributions as follows (4):

$$\Delta G_{ads} = \sum \Delta G_i \tag{1}$$

where ΔG_i corresponds to the free energy contributions of the possible adsorption mechanisms. The various mechanisms by which an
organic solute may adsorb include (1) interactions between an organic ion and the electrical double layer, (2) ion exchange,
including protonation followed by ion exchange, (3) coordination by
surface metal cations, (4) ion-dipole interactions, (5) hydrogen
bonding, and (6) hydrophobic interactions (2, 3). The first and
second of these mechanisms are important only for ionizable compounds, which are not considered here. Coordination by exchanged
and structural metal cations is important when the organic is a good
electron donor (Lewis base) relative to water, such as amines (2).
Similarly, the ion-dipole interactions between the charged surface
and the uncharged adsorbate are also expected to be negligible in
aqueous solution. Finally, because the carbon-hydrogen and the
carbon-halogen bonds in halogenated hydrocarbons are only weakly
acidic and basic, respectively (6), hydrogen bonding should account
for only a small contribution to the total adsorption free energy.
In addition, the relative insignificance of the hydrogen bond
between a clay surface and an OH group of an adsorbed molecule has
been demonstrated by IR spectroscopy (2).
The arguments presented above lead to the conclusion that the
adsorption of nonionic compounds such as halogenated hydrocarbons
results primarily from "hydrophobic bonding" or, perhaps more appropriately, the hydrophobic interaction (7). The thermodynamic
driving force for hydrophobic interactions is the increase in
entropy resulting from the removal, or decrease, in the amount of
hydration water surrounding an organic solute in water. Studies
have shown that the adsorption of aliphatic amines onto clays (8)

and surfactants onto oxides (9) increased linearly with increasing hydrocarbon chain length, with each additional methylene group contributing approximately -1.8 to -2.5 kJ/mol to the total free energy of adsorption. These values are slightly less negative than the -2.9 to -3.7 kJ/mol observed for the transfer of a methylene group in a hydrocarbon chain from an aqueous phase to an organic solvent (10, 11). The dependence of the adsorption intensity on the chain length indicates that hydrophobic interactions are responsible for the increased adsorption for the longer chain lengths.

In considering the sorption of nonionic solutes onto sediments, it is generally found that the sorption isotherm for hydrophobic organic solutes are approximately linear over a substantial concentration range and can be adequately described by a constant distribution coefficient, K_d (1). K_d tends to be variable over several orders of magnitude for a given solute with different solids, but it generally correlates with the organic carbon content of the sediment. By attributing all the sorption to the organic matter, an organic carbon partition coefficient, K_{oc}, can be defined by

$$K_{oc} = \frac{K_d}{f_{oc}}$$ (2)

where f_{oc} is the mass fraction of organic carbon of the sorbent. K_{oc} generally varies only by factors of three to five for a given solute (12). K_{oc} typically correlates well with physico-chemical properties of the sorbate, such as aqueous solubility (S) or the octanol-water partition coefficient (K_{ow}), again suggesting that hydrophobic interaction predominates. The correlation of K_{oc} with K_{ow} has led to the definition of linear free-energy relationships (LFER) of the form

$$\log K_{oc} = a \log K_{ow} + b$$ (3)

or,

$$\log K_{oc} = c \log S + d$$ (4)

where a, b, c, and d are constants resulting from a regression analysis. K_{ow} is most widely used for correlations with K_{oc} because of the availability of an extensive data base and estimation methods (9, 11). Other correlations have been proposed, most notably with the aqueous solubility, S, (14-16).

In addition to the dependence of sorption on the organic fraction of the sorbent, and the K_{ow} of the sorbate, Chiou et al. (13) cite the following observations as support for the hypothesis that the sorptive mechanism is hydrophobic partitioning into the organic (humic) fraction of the sediments: (1) the linearity of the isotherms as the concentration approaches solubility, (2) the small effect of temperature on sorption, and (3) the lack of competition between sorbates for the sorbent. These arguments also illustrate the applicability of the K_d approach for modeling sorption on hydrophobic compounds; an approach which has been criticized when used in the context of adsorption of trace metals onto oxides (17).

Partitioning Thermodynamics

The LFER that results when correlating partitioning in the octanol-
water system and the humic substances-water system implies that the
thermodynamics of these two systems are related. Hence, much can be
learned about humic substances-water partitioning by first considering partitioning in the simpler octanol-water system. The thermodynamic derivation that follows is based largely on the approach
developed by Chiou and coworkers (18-20), Miller et al. (21), and of
Karickhoff (1, 22). In the subsequent discussion, we will adopt the
pure liquid as the standard state and, therefore, use the Lewis-
Randall convention for activity coefficients, i.e., $\gamma = 1$ if the
mole fraction $x = 1$.

Partitioning in Octanol-Water Systems. At equilibrium, the chemical
potential of a solute (defined as $\mu_i = \mu_i^\ominus + RT \ln \gamma_i x_i$) is equal in
the octanol and the water phase. Hence, we may write

$$\gamma_w^* x_w^* = \gamma_o^* x_o^*$$

(5)

where γ_w^* and γ_o^* are the activity coefficients of the solute in the
aqueous phase and the octanol phase, respectively, and x_w^* and x_o^* are
the mole fractions of the solute in the water and the octanol
phases, respectively. The asterisks denote that the two phases are
not pure but are saturated with the other "immiscible" solvent. In
dilute solutions, the mole fractions x can be approximated by $x =$
$C \cdot V_s$, where C is the molar concentration and V_s is the molar volume
of the solvent. Hence we can write Equation 5 in terms of the molar
concentrations and obtain an expression for the octanol/water partition coefficient, $K_{ow} = C_o/C_w$ in terms of the activity coefficient
and the molar volumes of octanol and water (19):

$$K_{ow} = \frac{\gamma_w^* \; V_w^*}{\gamma_o^* \; V_o^*}$$

(6)

where V_w^* and V_o^* are the molar volumes of the water saturated with
octanol and the octanol saturated with water, respectively. The
asterisks in Equation 6 denote that the phases are not pure but in
fact are saturated with the other "immiscible" solvent.

We may consider the aqueous solubility as partitioning of a
compound between its pure state in liquid form and the saturated
aqueous phase. Then the partitioning coefficient (Equation 6) may
be written as

$$S = \frac{1}{V_w \gamma_w}$$

(7)

where the asterisks have been omitted since the cosolvent (octanol)
is absent and the organic compound is assumed to be a pure liquid
($x_o = 1$), which in turn implies that the organic solute is at the
standard state or $\gamma_o = 1$ (23).

A correction to Equation 7 is required for organic solutes that
are solids at the temperature and pressure of the partitioning

system. This correction is required, because the intermolecular forces in the crystal must be overcome before the chemical enters solution. However, it cancels out in octanol/water systems, since the correction applies to both phases equally (23). One method that accounts for the crystalline energy effect is to use an estimate of the solubility of the supercooled liquid given by (24):

$$\log S_\ell = \log S_s + \frac{\Delta S_f}{2.303R} \left(\frac{T_m - T}{T}\right) \tag{8}$$

where S_ℓ and S_s are the molar solubilities of supercooled liquid and the solid, respectively, ΔS_f is the entropy of fusion, T_m and T are the melting temperature and the temperature of interest, and R is the gas constant. Since the entropy of fusion is positive for non-ionic organic compounds (24), the solubility of the supercooled liquid is larger than the solubility of the solid as shown by Equation 8. The subsequent discussion assumes that the solubility of a solid organic solute is expressed as the solubility of the super-cooled liquid.

Since the solubility of octanol in water is only 0.0045M (9), the molar volume of the aqueous phase saturated with pure octanol can be approximated by the molar volume of pure water ($V_w^* = V_w$). With this approximation, Equations 6 and 7 give

$$K_{ow} = \frac{1}{SV_o^*} \frac{\gamma_w^*}{\gamma_w^* \gamma_o^*} \tag{9}$$

which can be rewritten as

$$\log K_{ow} = - \log SV_o^* + \log \frac{\gamma_w^*}{\gamma_w} - \log \gamma_o^* \tag{10}$$

If an ideal solution is formed in the octanol phase, and the solute in the aqueous phase is not affected by the dissolved octanol, then the last two factors in Equation 10 equal zero. Under these assumptions, an ideal octanol/water partition coefficient ($K_{ow}^\ominus$) can be defined by (19):

$$\log K_{ow}^\ominus = - \log S - \log V_o^* \tag{11}$$

Equation 11 implies that under these idealized conditions a log-log plot of S versus K_{ow} will have a slope of -1 and an intercept of $-\log V_o^*$ (19). The factors γ_w^*/γ_w and $1/\gamma_o^*$ can be viewed as corrections for the activity change in the aqueous phase caused by the octanol dissolved in the water, and for nonideality in the octanol caused by the incompatibility of the solute with water-saturated octanol, respectively.

Nonidealities in Water-Saturated Octanol. Chiou et al. (19), Tewari et al. (25) and Banerjee et al. (26) presented solubility data and octanol/water partition coefficients for 125 organic solutes. Figure 1 shows data from these three sources, the least-squares

regression line for all of the data, and the ideal line as defined
by Chiou et al. (19). The regression line has a slope of -0.860 and
differs from the ideal slope of -1 at the 99.5% significance level.
These deviations are caused by either (1) a linear increase in
log γ_o^* with decreasing log S, (2) a linear decrease in log γ_w^*/γ_w
with increasing log S, or (3) a combination of both. Chiou et al.
(19) reported that both factors were significant: the solubility of
the hydrophobic compounds hexachlorobenzene (log K_{ow} = 5.50) and DDT
(log K_{ow} = 6.36) were enhanced by dissolved octanol to the extent of
80% and 160%, respectively, such that log(γ_w^*/γ_w) accounted for
approximately one-third of the deviation from the ideal line. In
contrast, Miller et al. (21) observed that γ_w^* and γ_w were indistinguishable for compounds with log K_{ow} up to 5.8. Calculations
based on the solvophobic model (27), discussed below, suggest a
negligible effect of dissolved octanol on the solubility of pyrene
in water (log K_{ow} = 5.18). Therefore, it appears that the variation
in log γ_o^* accounts for the majority of the deviations from the ideal
line.

If all of the deviations between the ideal line (Equation 11)
and the observed data are attributed to log γ_o^*, the empirical correlation between log S and log K_{ow} (Figure 1) can be combined with
Equation 9 to give:

$$\log \gamma_o^* = 0.16 \log K_{ow} + 0.08 \tag{12}$$

Equation 12 shows that γ_o^* increases by a factor of only 7 as K_{ow}
increases by 10^5.

The most important factor in determining K_{ow} is the aqueous-
phase activity coefficient (aqueous solubility) of the organic solute. The observed partition coefficients are less than the ideal
partition coefficients (K_{ow}^o) as result from 1) the incompatibility
of the solute in water-saturated octanol and, to a lesser degree,
2) the depression of the aqueous-phase activity coefficient by
octanol. The values of log γ_o^* and log γ_w^*/γ_w increase linearly with
decreasing log S.

Partitioning in Natural Organic Matter-Water Systems. For K_{oc} we
may write, by analogy to Equation 6,

$$K_{oc} = \frac{\gamma_{w/oc}^* V_w^*}{\gamma_{oc/w}^* V_{oc}^*} \cdot \frac{1}{\rho_{oc}} \tag{13}$$

where $\gamma_{w/oc}^*$ is the activity coefficient of the solute in water saturated with dissolved organic carbon derived from humic substances
and $\gamma_{oc/w}^*$ is the activity coefficient in humic substances saturated
with water. This analogy is justifiable since sorption to organic
matter proceeds by a process similar to hydrophobic liquid-liquid
partitioning (1, 13, 15, 20). The density of the organic carbon
phase is included in Equation 12 so that K_{oc} is expressed in the
conventional units of cm^3/g. By multiplying the numerator and
denominator of Equation 13 by $V_o^* \gamma_w^* \gamma_o^*$, substituting K_{ow} as defined in
Equation 6, and assuming $\gamma_w^* = \gamma_w$ as before, it can be shown that

$$K_{oc} = C_1 K_{ow} \left(\frac{\gamma_o^*}{\gamma_{oc/w}^*}\right)\left(\frac{\gamma_{w/oc}^*}{\gamma_w^*}\right) \tag{14}$$

The quotient $V_o^*/V_{oc}^* \rho_{oc}$ has been substituted by C_1, since this quantity should be constant for a given sorbent. Equation 14 illustrates that K_{oc} is proportional to K_{ow} but is modified by two terms, $(\gamma_o^*/\gamma_{oc/w}^*)$ and $(\gamma_{w/oc}^*/\gamma_w^*)$, which account for the differences in activity coefficients in the aqueous and organic phases, respectively. In the following discussion, we examine the factors that influence the activity coefficients in the organic phases $(\gamma_o^*/\gamma_{oc/w}^*)$ and the activity coefficients in the aqueous phase $(\gamma_{w/oc}^*/\gamma_w^*)$.

Sorption data from several different studies are correlated with K_{ow} in Figure 2. The data presented include halogenated aliphatics and aromatics, alkyl-substituted aromatics, and polynuclear aromatic hydrocarbons. By selecting data for these classes of compounds, we have minimized possible influences of complexation, hydrogen bonding, strong dipole interactions, and charge effects. Therefore the correlation indicated in Figure 2 represents hydrophobic partitioning into the organic matter and water. In view of Equation 14, the observed slope of less than unity must arise from a negative linear dependence of $\log\left(\gamma_o^*/\gamma_{oc/w}^*\right)\left(\gamma_{w/oc}^*/\gamma_w^*\right)$ on $\log K_{ow}$. In the subsequent discussion, we examine this dependence, which defines the coefficient a of Equation 3.

Nonidealities in Humic Polymers

Organic-Phase Nondealities. To assess the impact of nonideality in the organic matter, we assume for illustrative purposes that $\gamma_{w/oc}^*/\gamma_w^*$ equals unity. Consequently, Equation 14 simplifies to

$$K_{oc} = C_1 K_{ow} \frac{\gamma_o^*}{\gamma_{oc/w}^*} \tag{15}$$

Substituting the previously derived expression for γ_o^* (Equation 12) gives

$$K_{oc} = C_2 K_{ow}^{1.16} \frac{1}{\gamma_{oc/w}^*} \tag{16}$$

where C_2 equals $C_1 \times 10^{0.08}$. This substitution yields an expression of K_{oc} as a function of the molecular property K_{ow} and the activity coefficient of the solute in the water-saturated organic matter. Equivalently, K_{oc} could be expressed as a function of the solubility of the solute in water by substituting K_{ow} with $1/S^{0.84}$ (see Figure 1).

In logarithmic form, Equation 16 rearranges to

$$\log K_{oc} = 1.16 \log K_{ow} + \log C_2 - \log \gamma_{oc/w}^* \tag{17}$$

To account for the experimentally observed slope of 0.92 (Figure 2), $\gamma_{oc/w}^*$ must be positively correlated with K_{ow} if C_2 is a constant; otherwise the slope of the correlation would equal 1.16.

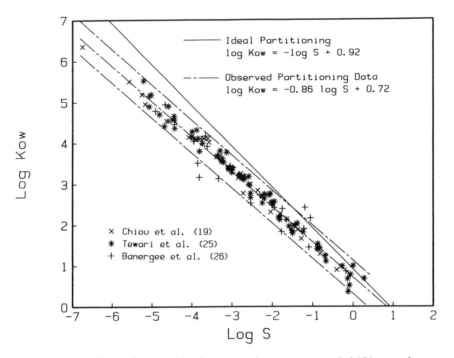

Figure 1. The relationship between the aqueous solubility and octanol-water partition coefficient for nonpolar organic solutes.

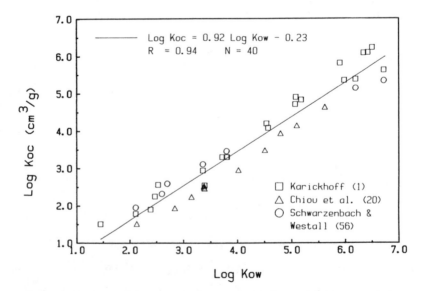

Figure 2. The correlation between the organic carbon partition coefficient and the octanol water partition coefficient.

It is well known in polymer science that a small amount of organic solvent dissolved in a polymer results in large negative deviations from Raoult's law because of the large entropic effects without the associated enthalpic effects (28). This is a consequence of the mixing of molecules of greatly different sizes (29). Therefore, it is unreasonable to assume an activity coefficient equal to one in the organic phase. Chiou et al. (20) have treated the humic material as a non-swelling amorphous polymer and used the Flory-Huggins theory to estimate the activity $(x_o^* \gamma_o^*)$ of a solute partitioned into the humic matter from (28, 30)

$$\ln x_o^* \gamma_o^* = \ln \phi + \phi_p \left(1 - \frac{\overline{V}}{\overline{V}_p}\right) + \chi \phi_p^2 \tag{18}$$

where ϕ and ϕ_p are the volume fractions of the solute and humic phase, respectively, $\overline{V}$ and $\overline{V}_p$ are the molar volume of the solute and an estimated average molar volume of the humic polymers, respectively. The Flory-Huggins interaction parameter (χ) characterizes the interaction energy between a solute molecule and the polymer.

With the assumption of a dilute solution in the humic phase (i.e., $\phi_p \approx 1$, $\overline{V} \ll \overline{V}_p$, and $\phi \approx n_1 \overline{V}/n_2 \overline{V}_p$, Equation 18 simplifies to

$$\ln \gamma_{oc/w}^* = \ln(\overline{V}/\overline{V}_p)) + \chi + 1 \tag{19}$$

Equation 19 illustrates that $\ln \gamma_{oc/w}^*$ (or $\log \gamma_{oc/w}^*$) will be less than zero for $\overline{V} \ll \overline{V}_p$ and small χ. Under these conditions, the large disparity in molecular size favors partitioning to the humic material relative to organic solvent phases with smaller $\overline{V}$.

The Flory-Huggins interaction parameter, χ, is the sum of enthalpic (χ_H) and entropic (χ_S) contributions to the polymer-solute interactions (28). χ_S is an empirical constant related to the coordination of the polymer subunits (29). Chiou et al. (20) have selected a value of 0.25 for χ_S of humic matter. From regular solution theory, χ_H is given by

$$\chi_H = \frac{\overline{V}}{RT} (\delta - \delta_p)^2 \tag{20}$$

where δ and δ_p are the solubility parameters of the solute and humic material, respectively (29).

For a constant $\overline{V}_p$ and χ_S, Equations 17, 19, and 20 can be combined to give

$$\log K_{oc} = 1.16 \log K_{ow} - \log \overline{V} - \frac{\overline{V}}{RT} (\delta - \delta_p)^2 + C_3 \tag{21}$$

This equation illustrates that K_{oc} depends on the solute parameters K_{ow}, $\overline{V}$, and δ. Since molar volume and $\log K_{ow}$ are positively correlated ($\log K_{ow} = 0.49 + 0.020\overline{V}$; 21), K_{oc} could be expressed as a function of the solute parameters K_{ow} and δ, and polymer parameter δ_p.

Table I presents an evaluation of Equation 21 based on previously reported values for $\log K_{ow}$, $\overline{V}$, and δ. The constant C_3 was calculated from the $\log K_{oc}$ of benzene estimated from Figure 2. Estimates of $\log K_{oc}$ for the remaining compounds in Table I are presented for δ_p equal to 10.3 and 11.5. The value of 10.3 corresponds to the δ_p implied from solvent extraction studies (31). In this case,

Table I. K_{oc} Estimated from K_{ow}, $\overline{V}$, δ, and δ_p

Compound	log K_{ow}^*	$^*\overline{V}(\frac{cm^3}{mol})$	$^\dagger\delta(\frac{cal}{cm^3})^{1/2}$	log K_{oc} $\delta_p = 10.3$	log K_{oc} $\delta_p = 11.5$
benzene	2.13	88.7	9.2	1.51	1.51
trichloroethylene	2.53	89.0	9.2	1.96	1.96
toluene	2.65	106.3	8.9	1.83	1.55
chlorobenzene	2.98	116.9	9.5	2.42	2.36
ethylbenzene	3.13	122.5	8.8	2.19	1.71
m-xylene	3.20	123.2	8.8	2.27	1.78
naphthalene	3.33	148.0	9.9	2.84	2.85
1,2-dichlorobenzene	3.38	137.8	10.0	2.92	3.04
propyl benzene	3.69	139.4	8.6	2.54	1.77
phenanthrene	4.54	197.0	9.9	4.05	3.84
anthracene	4.57	199.0	9.8	4.05	3.74
			slope	1.04	0.92
			intercept	−0.82	−0.60
			R^2	0.92	0.70

* Ref. 21. † Ref. 32.

the estimated log K_{oc} values do correlate well with log K_{ow}. The slope of the correlation (1.04) is less than 1.16 and closer to observed data. A close match of the experimental and estimated slope is obtained with $\delta_p = 11.5$. If $\delta_p = 11.5$, which is approximately equal to the average of previously suggested values (20, 31), then the slope from the estimated log K_{oc} values is nearly identical to that presented in Figure 2. However, the correlation coefficient decreased markedly to 0.70. This decrease in r^2 may be due to the limited applicability of the Flory-Huggins theory to hydrophobic partitioning into sediment organic matter, which may not exhibit ideal polymer properties, as assumed by Flory-Huggins. Although more complex thermodynamic models have been proposed (30), their use is unwarranted at this time given that solubility parameters for the humic material are not well known and that numerous assumptions were required in deriving Equation 21. Nevertheless, the calculations based on the Flory-Huggins theory are encouraging in that the trends predicted in terms of the LFER slope do agree with observed data.

Aqueous-Phase Nonidealities

The Effect of Dissolved Humic Material. Several investigators have reported that the solubilities of strongly hydrophobic solutes increase in the presence of dissolved organic matter (DOC) (33-38). This solubilization phenomenon is most likely a consequence of the solute binding to dissolved macromolecules (39). Such binding may alter rates of photolysis, hydrolysis, volatilization, or biodegradation, all of which can affect the environmental fate of trace organic solutes (40). In the context of this work, the most important effect of binding is a lowering of the thermodynamic activity of the solute when compared with the activity in the absence of the macromolecule ($\underline{39}$). By analogy to Equation 7, we may express S_T, the total solubility of a solute in the presence of dissolved humic material, as

$$S_T = \frac{1}{V_w \gamma_{w/oc}^*} \qquad (22)$$

where $\gamma_{w/oc}^*$ is the activity coefficient of the solute in water in the presence of humics. It is therefore clear that the reported solubilization phenomenon is directly related to changes in $\gamma_{w/oc}^*$.

Chiou et al. ($\underline{41}$) proposed that a linear equilibrium model may be used to describe the effect of DOC on the solubility of an organic solute. The model regards a hydrophobic solute in solution as being either "free" or truly dissolved, on the one hand, or "bound" or complexed with the humic material on the other hand. The model relates the free and DOC-bound concentrations, C_w and C_b respectively, by a simple partitioning equilibrium constant

$$C_b = C_w C_{doc} K_{doc} \qquad (23)$$

where C_{doc} is the concentration of the DOC and K_{doc} is the DOC-water distribution coefficient ($\underline{41}$, $\underline{42}$). The total measured concentration C_T of organic solute in solution is merely the sum of C_w and C_b which is given by

$$C_T = C_w(1 + C_{doc} K_{doc}) \qquad (24)$$

The value of K_{doc} is presently uncertain, but recent studies report that K_{doc} varies with the source of the humic material, the solution pH, and the ionic strength ($\underline{42}$, $\underline{43}$). As an approximation, K_{doc} can be estimated from K_{oc}, using one of the correlations developed between K_{ow} and K_{oc}. Alternatively, as an upper bound, K_{doc} may be assumed equal to the K_{ow} of the compound ($\underline{1}$).

Equation 24 may be used to predict the effect of DOC on the total solubility of a hydrophobic compound in the presence of DOC (S_T) as a function of S and the above binding model. At solubility ($C_T = S_T$), the ratio S/S_T is equal to $\gamma_{w/oc}^*/\gamma_w$ (Equations 7 and 22), and therefore, by considering Equation 24 we may write:

$$\frac{\gamma_{w/oc}^*}{\gamma_w} = (1 + C_{doc} K_{doc})^{-1} \qquad (25)$$

where the assumption of constant molar volume and dilute solution have been made.

By analogy to Equations 10 and 11 we may express the observed partition coefficient K_{oc} as

$$K_{oc} = K_{oc}^o \frac{\gamma_{w/oc}^*}{\gamma_w} \qquad (26)$$

where K_{oc}^o is an ideal partition coefficient, which applies only when DOC has no effect on the dissolved solute ($\gamma_{w/oc}^* = \gamma_w$). As discussed previously, $\gamma_{w/oc}^*/\gamma_w$ is the correction factor which accounts for the

solubilizing effect of DOC. Combining Equations 25 and 26 results
in

$$K_{oc} = K_{oc}^{o}(1 + C_{doc}K_{doc})^{-1} \qquad (27)$$

This equation, which is identical to equation 9 of Gschwend and Wu
(44), allows one to estimate the effect of DOC on the observed partition coefficients. If we assume that K_{doc} can be related to K_{ow}
by a LFER (log K_{doc} = a log K_{ow} + b), we obtain

$$\frac{\gamma_{w/oc}^{*}}{\gamma_{w}} = \left(1 + C_{doc} \, a \, K_{ow}^{b}\right)^{-1} \qquad (28)$$

where a and b are empirical constants. Figure 3 shows $\gamma_{w/oc}^{*}/\gamma_{w}$ as a
function of K_{ow} at various DOC concentrations, using the correlations between K_{ow} and K_{oc} reported by Karickhoff et al. (15). These
model calculations show that at DOC concentrations representative of
natural systems, (e.g., C_{doc} < 50 mg/l), binding of organic solutes
to DOC should be insignificant for compounds with log K_{ow} < 4. For
a compound with log K_{ow} equal to 6.0 and a C_{doc} equal to 10 mg/l,
however, the apparent partitioning coefficient is approximately an
order of magnitude smaller than the ideal value. These predictions
have been supported by experimental solubility data showing that
moderately hydrophobic compounds such as toluene (44) and lindane
(45) are not influenced by DOC, whereas DOC markedly affects highly
hydrophobic compounds such as DDT (42), PCB (46) and polynuclear
aromatic hydrocarbons (47).

Calculations which consider binding of sorbing solutes to dissolved humic matter show that the previously reported inverse relationship between effect of sediment concentration and sorption
equilibrium can be accounted for by desorption of humic material
followed by binding to the DOC (46, 48).

One of the important assumptions made in deriving Equations 24
through 28 is that K_{doc} is independent of C_{doc}. However, it has
been observed that K_{doc} is inversely related to C_{doc} (42, 43, 45,
47) in a manner that is analogous to the sediment concentration
effect on K_{oc}, although the dependence is substantially weaker.
Thus, questions still remain regarding the cause of the sorbent
concentration effect, since the binding of solutes to macromolecules, which was used to interpret the effect of sorbent concentration, also varies with loading of sorbent sites.

Influence of Organic Cosolvents. Rao et al. (49) have recently
presented a solvophobic model for estimating the sorption of a
hydrophobic solute from a mixed solvent. This model is based on the
work of Yalkowsky et al. (27), who developed an empirical relationship between the solubility in a mixed solvent system, S_m, and that
in pure water given by

$$\ln S_m = \ln S + \sigma_c f_c \qquad (29)$$

where f_c is the volume fraction of cosolvent. For hydrophobic
solutes, the parameter σ_c is given from empirical relationships by

$$\sigma_c = \frac{\Delta\gamma^c(HSA)}{kT} \qquad (30)$$

where $\Delta\gamma^c$ is the difference in surface tension of pure water and pure cosolvent, HSA is the hydrophobic surface area of the solute, k is the Boltzmann constant, and T is the absolute temperature. Based on the inverse relationship between K_{oc} and S (Equation 4), the following relationship can be derived (49, 50):

$$\ln(K_{oc}^m/K_{oc}^w) = -\alpha \, \sigma_c f_c \qquad (31)$$

where the superscripts m and w denote the mixed solvent and water, respectively, and α is an empirical constant. Equation 31 predicts that (1) K_{oc}^m should decrease exponentially with increasing f_c for a given solute and (2) the decrease in K_{oc}^m caused by a given solvent should be greatest for the least soluble (largest HSA) compounds. These predictions have been verified experimentally for water-methanol, and water-acetone cosolvent systems (50, 51).

Typically, K_{oc} is expressed in units of volume of solution per mass of organic carbon. Fu and Luthy (50) observed that a plot of $\ln(K_{oc}^m/K_{oc}^w)$ versus f_c was nonlinear when K_{oc} was expressed in the common units of volume per mass of organic carbon. A linear plot was obtained when K_{oc} was expressed in units of moles of solvent per gram of organic carbon.

Using values of σ_c determined from solubility measurements, Fu and Luthy (50) found an average value of α equal to 0.51, which shows that the decrease in K_{oc}^m was half of that expected from the increase in solubility. This was interpreted as evidence that the cosolvent was swelling the organic fraction of the soil and consequently increasing the accessibility to the organic matter (50) in accordance with the gel-partition model (31).

Frequently it is convenient, when conducting sorption experiments, to use a cosolvent such as methanol to facilitate transfer of a volatile or insoluble solute. Equation 31 may be used to estimate the effect of the cosolvent on measured sorption coefficient. If f_c equals 10^{-3} and α is conservatively chosen as unity, the methanol is predicted to decrease the sorption of anthracene ($\sigma_c = 9.76$; log K_{ow} = 4.54; 50) by 1% relative to the case of no cosolvent. The effect of the cosolvent should be smaller for solutes with smaller HSA (or K_{ow}), which is consistent with the observed negligible impact of methanol at $f_c \leq 10^{-3}$ on the sorption of hexachloroethane (log K_{ow} = 3.6; 52).

Limitations of Partitioning Theory

Inefficiency of Partitioning. Equations 2-4 imply that the sorptive capacity (K_{oc}) of organic matter for hydrophobic organic solutes is independent of origin and of the composition of the mineral matrix. However, investigators studying individual particle size fractions found that the organic matter associated with a coarse or sand fraction is apparently less efficient in sorbing organics than that associated with the finer materials (15, 53). To account for the relative inefficiency of the f_{oc} of the sand fraction, Equation 2 may be rewritten as

$$K_{oc} = \frac{K_d}{\Omega f_{oc}} \qquad (32)$$

where Ω is an efficiency factor similar to that originally defined by Lambert (54), which always equals unity with the exception of sandy soils, for which Ω would be less than unity. Either kinetic limitations or fundamentally different properties of the organic matter associated with the coarse grain fraction could account for the inefficiency factor.

Adsorption on Inorganic Surfaces. McCarty et al. (55) and Karickhoff (1) proposed a two-phase model for sorption onto mineral surfaces and organic matter partitioning. This model hypothesizes that sorption to inorganic surfaces may occur simultaneously with the accepted partitioning into the organic matter. Assuming additivity of these two contributions, K_d can be estimated from

$$K_d = f_{oc}K_{oc} + f_{io}K_{io} \tag{33}$$

where f_{io} and K_{io} are the fraction and the distribution coefficient for the inorganic material, while f_{oc} and K_{oc} are as previously defined. If only the organic and one inorganic phase is considered, the sum of f_{io} and f_{oc} equals one. For a given mineral, K_{io} is hypothesized to be the product of the specific surface area, S_A, and a specific surface adsorption constant, K_s. Therefore, the overall distribution coefficient can be given by

$$K_d = f_{oc}K_{oc} + f_{io}S_AK_s \tag{34}$$

The inorganic contribution, K_{io}, for chromatography-grade silica was found to depend weakly on K_{ow} (log $K_s = 0.16\ K_{ow}$) (55), which is consistent with the results for silica and alumina obtained by Schwarzenbach and Westall (56).

Figure 4 presents distribution coefficients predicted from the two-phase model for two values of f_{oc} and S_A. Figure 4 shows that the dominance of one or the other contribution to sorption depends on two factors: K_{ow} and f_{oc}. For a given f_{oc}, compounds with low K_{ow} are influenced primarily by K_{io}, while compounds with higher K_{ow} are influenced primarily by K_{oc}. Alternatively, at a given K_{ow} and S_A, the model predicts the dominance of inorganic adsorption below a certain f_{oc} and dominance of organic-phase partitioning above this f_{oc}. The range of organic and inorganic sorption dominance depends on both the K_{ow} and the S_A.

The two-phase model proposed by McCarty et al. (55) is based in part on the assumption of independently acting organic and inorganic sorption sites. Shin et al. (57) and Pierce et al. (58) reported that the distribution coefficient for sorption of DDT increased after the oxidation or removal of some of the organic matter from the sorbent. Karickhoff (1) has interpreted this increase as evidence that the humic material had blocked clay sorption sites. Conversely, Chiou et al. (41) proposed that the oxidation process may have preferentially removed soluble organics which, if present in the aqueous phase, would decrease the sorption intensity due to binding of the organic solute to the DOC. This difference emphasizes the need for characterizing the sorption properties of the dissolved as well as sorbed organic matter and of the mineral matrix.

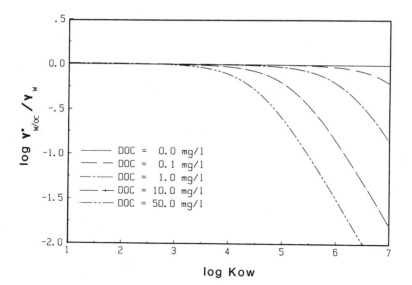

Figure 3. The effect of dissolved organic carbon on the aqueous activity coefficient estimated from the octanol-water partition coefficient.

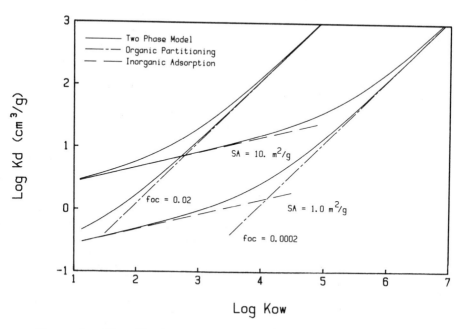

Figure 4. Distribution coefficients predicted from the two-phase sorption model.

Karickhoff (1) presented data that show more sorption than can be explained by simple partitioning theory when the expanding clay matter to organic matter ratio (cm:om) is "high". The term high is a relative term that cannot be easily quantified, but appears to be inversely related to K_{ow}, as predicted by the two-phase model. For the nitrogen heterocycles simazine (log K_{ow} = 2.16) and biquinoline (log K_{ow} = 4.31), the limiting cm:om ratio is 30:1. Pyrene, which is more hydrophobic (log K_{ow} = 5.18) and contains no polar moieties unlike the previous two compounds, was not strongly influenced by the clays. Similar data have been observed for sorption of amino- and carboxy-substituted polynuclear aromatic hydrocarbons (59), acetophenone (60), and α-naphthol (61).

The data from these various sources are presented in Figure 5 along with partitioning data previously presented in Figure 2. The partitioning data for the polar solutes are in agreement with general predictions of the two-phase model; specifically, the deviations are greatest for the least hydrophobic compounds. All of these compounds exhibit larger K_{oc} values than predicted from the organic matter partitioning theory, probably because of the presence of polar moieties which can participate in nonhydrophobic bonding, such as complexation of exchanged cations by the amino or carboxyl groups (2), hydrogen bonding (1), and even covalent bonding (62).

Organic compounds containing polar functional groups are not the only organic solutes that can participate in sorption interactions in excess of organic matter partitioning. Schwarzenbach and Westall (56) found that silicon and aluminum oxides, non-expanding clay minerals, and a low f_{oc} aquifer material, all exhibited more sorption capacity for halogenated and aromatic solutes than could be explained by partitioning into the organic matter. Furthermore, all of these samples with this "anomalously" high sorption capacity had f_{oc} values less than 0.001, which led them to caution against using any of the numerous correlations when the sorbent has an f_{oc} below 0.001.

In support of this finding, Curtis et al. (63) found enhanced sorption for five halogenated organic solutes onto an aquifer material with both a low surface area (0.8 m^2/g) and a low f_{oc} (0.0002). These data, along with those of Schwarzenbach and Westall (56) are presented in Figure 6 along with the partitioning data presented in Figure 2. Unlike the data presented in Figure 5, all of the data presented in Figure 6 are for halogenated aliphatic and aromatic compounds, and hence specific sorption mechanisms, such as complexation interactions or hydrogen bonding, should be negligible. The higher than predicted K_{oc} values for the sorbents with low f_{oc} suggest that sorption onto the mineral surfaces is a significant factor. The partition constant correlates positively with K_{ow}, which is expected, since both the free energy of sorption and partitioning in octanol/water increase linearly with the number of methylene groups (8-11).

Since adsorption at a mineral surface is a replacement process, we would expect mineral surfaces with weak affinity for water to have the strongest affinity for hydrophobic solutes. Infrared spectroscopy shows that siloxane surfaces on clays with little iso-morphic substitution form weaker hydrogen bonds than water forms with itself (64), which corresponds to one of the definitions of a hydrophobic surface offered by Texter et al. (65). Therefore,

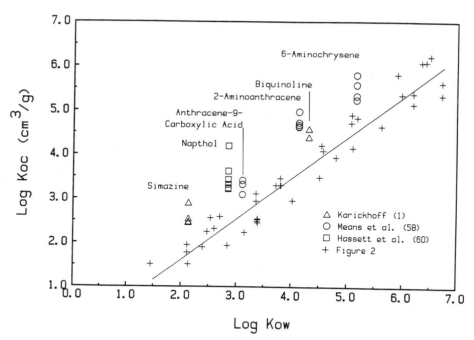

Figure 5. The influence of mineral surfaces on the sorption of organic solutes containing polar functional groups.

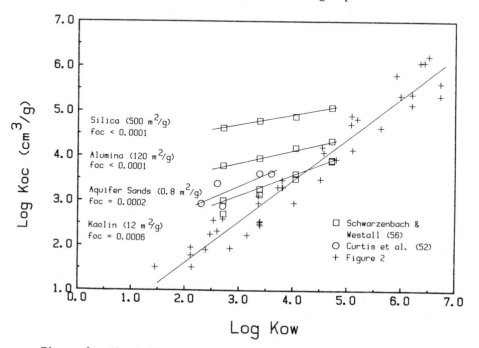

Figure 6. The influence of mineral surfaces on the sorption of organic solutes containing no functional groups.

mineral surfaces of sediments need to be characterized with respect
to their affinity for water in order to predict adsorption of hydrophobic solutes.

Sorption Kinetics

The rate at which pollutants sorb to sediments has frequently been
assumed to occur rapidly and consequently equilibration studies have
often been conducted by mixing sample for 24 hours. Karickhoff
($\underline{1}$, $\underline{66}$) has reported that sorption may require up to two months to
reach an apparent equilibrium. Similarly, desorption has also been
observed to require on the order of months to reach completion
($\underline{67}$, $\underline{68}$). McCall and Agin ($\underline{67}$) observed that the desorption rate of
picloram was inversely related to the contact time.
    Two models, a first-order kinetic model and a hindered diffu-
sion model, have been proposed to describe the sorption and desorption rates of hydrophobic molecules. The possible origin of slow
sorption rates include (1) the hindering of diffusion by the
polymer-like organic macromolecules ($\underline{31}$), and (2) diffusion in
microporous minerals or into the interlamellar regions of expanding
clays. These two processes may interact synergistically in the case
of carbonate minerals, which can be both porous and contain significant amounts of organic carbon ($\underline{69}$).

The First-Order Kinetic Model. Karickhoff ($\underline{1}$, $\underline{68}$) has proposed a
two-compartment equilibrium-kinetic model for describing the solute
uptake or release by a sediment. This model is based on the assumption that two types of sorption sites exist: labile sites, S_L,
which are in equilibrium with bulk aqueous solution, and hindered
sites, S_H, which are controlled by a slow first-order rate process.
Conceptually, sorption according to this model can be considered
either as a two-stage process:

$$C \underset{K_{eq}}{\overset{}{\rightleftharpoons}} q_\ell \underset{k_{21}}{\overset{k_{12}}{\rightleftharpoons}} q_h \tag{35}$$

or, alternatively, as a parallel process:

$$q_\ell \underset{K_{eq}}{\overset{}{\rightleftharpoons}} C \underset{k_{21}}{\overset{k_{12}}{\rightleftharpoons}} q_h \tag{36}$$

where C is the solution concentration and q represents the sorbed
concentration. Karickhoff and Morris ($\underline{68}$) observed that the fraction of labile sites decreased with increasing sorption and, more
importantly, the rate of desorption was inversely proportional to
K_{ow}.
    The two-phase kinetic model developed by Karickhoff ($\underline{65}$) is
capable of fitting either the sorption or desorption of a sorbing
solute. For linear isotherms, the mathematical description given by
Karickhoff ($\underline{1}$) and others ($\underline{67}$, $\underline{70}$, $\underline{71}$) is virtually identical to
that of a mass transfer process ($\underline{72}$).
    We may also postulate that the rate of sorption or desorption
can be described by a diffusive mechanism. The application of diffusion equations to describe the uptake of a sorbing solute can be

quite restrictive because of the need to satisfy rigid boundary
conditions (68, 73). However, solutions for the diffusive uptake
into such different geometries as a plane sheet, a cylinder, or a
sphere deviate only slightly (73). Therefore it is useful to examine the solution to the diffusion equation to evaluate important
parameters controlling the diffusive uptake of a solute.

The Diffusion Model. The uptake of a solute by a sorbent can be
analyzed by a diffusion model, which has been used successfully to
model adsorption rates onto activated carbon (74, 75), ion exchangers (72), heterogeneous catalysts (76), and soil columns (77). For
the purpose of illustration, we can consider the diffusion of a
compound into a spherical sorbent grain under conditions of linear
sorption and no exterior mass transfer limitations (73), which is
described by

$$\frac{\partial C}{\partial t} = \frac{D_e}{1 + R_p} \left(\frac{\partial^2 C}{\partial r^2} + \frac{2}{r} \frac{\partial C}{\partial r} \right) \tag{37}$$

where D_e = effective diffusion coefficient, (L^2/T); R_p = a dimensionless equilibrium partition coefficient, r = radial distance
from the center of the sorbing grain, (L), and t = time. R_p is a
dimensionless partition coefficient equal to $\rho/\varepsilon \ K_d$, where ρ =
sorbent grain density (M/L^3), ε = sorbent internal porosity (L^3/L^3),
K_d = sorbent partition coefficient (L^3/M). The effective diffusion
coefficient accounts for the molecular diffusivity and tortuosity in
the sorbent grain.

When the sorbent is initially free from solute, Equation 37 can
be solved analytically (73) to give the ratio of the mass sorbed at
time t to the mass sorbed at equilibrium (i.e., the fractional
approach to equilibrium). The mathematical solution depends on the
mass fraction ultimately sorbed from the aqueous phase (F), and is
most conveniently presented in terms of τ, a dimensionless time
parameter given by

$$\tau = \frac{1}{1 + R_p} \frac{D_e \ t}{a^2} \tag{38}$$

where a is the radius of the sorbent grain. The quantity $[a^2(1 +
R_p)]/D_e$ can be considered a relaxation time for diffusion accompanied
by linear sorption. Three solutions to Equation 38 corresponding to
values of F equal to 0.2, 0.5, and 0.8 are illustrated in Figure 7.

Several important assumptions were necessary to achieve the
solution given in Equation 38. First, it is assumed that the rate
of sorption is much faster than the rate of diffusion. This implies
that all solute in the internal pores of the sorbent is at equilibrium with its immediate surroundings. In addition, it is assumed
that the sorption sites are homogeneous and that once a compound is
sorbed to the solid, it does not migrate along the surface.
Finally, to solve the diffusion equation, the inherent assumption
required is that the geometry or physical dimensions of the sorbent
do not change during the course of the experiment due to abrasion,
dissolution, swelling, or other process.

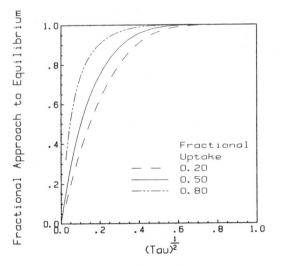

Figure 7. The fractional approach to equilibrium (after 73).
Fractional uptake = mass of solute sorbed at equilibrium/mass of
solute added to system; fractional approach to equilibrium = mass
of solute sorbed at time τ/mass of solute sorbed at time ∞.

Figure 7 shows that for F = 0.5, a sorbing (or desorbing) solute will have reached 95% of the ultimate sorption capacity at
$(\tau_{95})^{1/2}$ = 0.38. If we assume R_p ≫ 1, then Equation 38 can be
rearranged to give

$$t_{95\%} = \frac{(0.38)^2 R_p a^2}{D_e} \qquad (39)$$

Thus we see that $t_{95\%}$ is proportional to R_p and square of particle
radius.

For a hypothetical sorbent with properties listed in Table II,
we assume that experiments have shown that sorption of dichloroben-
zene (log K_{ow} = 3.3) is 95% complete within one day. R_p equals 1000
on the basis of the correlation of Karickhoff et al. (15). If we
estimate R_p for other solutes from the same correlation and assume
negligible mineral sorption, we may estimate the time required to
reach 95% of equilibrium for other compounds. These calculations
imply that approximately 0.1 days will be required to reach 95%
equilibrium for benzene (log K_{ow} = 2.1), approximately 10 days for
1,2,4,5-tetrachlorobenzene (log K_{ow} = 4.7) and approximately 100
days will be required for hexachlorobenzene (log K_{ow} = 5.5).
Clearly the time required to approach equilibrium can be substantially longer than provided in usual laboratory protocols, depending
on the strength of sorption.

Table II.   Assumed Sorbent Properties Used to
Estimate the Time to Equilibrium

Property	Assumed Value
Bulk density (ρ)	1700 (kg/m^3)
Internal porosity (ϵ)	0.05 (-)
f_{oc}	0.02 (-)
Molecular diffusivity (D_{mol})	10^{-9} (m^2/s)
Tortuosity (T)	10
Particle radius (a)	2.4×10^{-4}m

Since the same general solution holds for desorption under conditions of linear equilibrium, similar arguments can be employed to describe the diffusion out of the sorbent (73).

Previously, it was noted that organic matter associated with the coarse fraction of a sorbent was a substantially weaker sorbent than the organic matter associated with smaller materials (15). This discrepancy is consistent with a radial diffusion model: since $\tau_{95\%}$ is proportional to a^2 (Equation 39), the time required to reach equilibrium should increase with the square of the particle diameter. Clearly if the same mixing time were used for both fine-grained and coarse-grained sorbents, the coarser material could have been far from equilibrium.

Wu and Gschwend (78) successfully employed a radial diffusion model to describe laboratory observed sorption and desorption kinetics. Their data show that sorption and desorption rates were slower for more hydrophobic compounds and sorbents with a larger grain size in a manner consistent with the radial diffusion model.

In many early experiments, hysteresis was observed for highly hydrophobic compounds such as PCBs (79, 80). Since the time to reach equilibrium can be quite long for strongly hydrophobic compounds, a solute may have never reached equilibrium during the sorption isotherm experiment. Consequently, K_d would be underestimated, which leads to the discrepancy between the sorption and desorption coefficients that was attributed to hysteresis. The case for hysteresis being an artifact is supported by recent data for tetrachlorobenzene (log K_{ow} = 4.7), illustrating that sorption and desorption require approximately two days to reach equilibrium with approximately equal time constants (78). Finally, the diffusion model is consistent with the observation that the extent of hysteresis was inversely related to particle size (81).

Summary and Conclusions

In general, the driving force for the sorption of nonpolar organic compounds by sediments is their incompatibility with water as a solvent, i.e., their hydrophobicity, and their affinity for the lipophilic environment of the sediment organic matter. If the fraction of organic matter is low and the specific surface of the mineral matrix is high relative to the hydrophobicity of the solute, sorption by mineral surfaces may become significant, and organic sorption may become insignificant. Practical approaches for evaluating dominance

ranges for organic and inorganic sorptions have been evaluated. If
the aqueous phase contains organic cosolvents or organic macromolecules, the aqueous-phase activity of the hydrophobic solute is
lowered. The effect of organic macromolecules (usually characterized as DOC) has been modeled as a binding between the hydrophobic
solute and the macromolecule. The effect of organic cosolutes has
been modeled by semi-empirical correlations. The significance of
both effects increases with increasing hydrophobicity of the organic
solute and the concentrations of DOC and cosolvents. DOC may be
responsible for hysteretic adsorption isotherms in batch adsorption
experiments.
        The kinetics of sorption can be considered as the sum of two
processes: 1) rapid sorption by labile sites which are in equilibrium with solutes dissolved in bulk solution, and 2) hindered
sorption by sites which are accessible only by slow diffusion.
Alternatively, sorption kinetics can be modeled by a radial diffusional process into spherical sorbents. The slow sorption process
prevents complete equilibration within one day, the time used in
typical batch experiments. Because the apparent rate of diffusion
decreases with increasing hydrophobicity, time to equilibrium is
longer for highly hydrophobic compounds.
        The description of the sorption process is largely based on
empirical correlations, without knowledge of the detailed structure
of the sediments. No doubt, in the future a greater effort will
have to be made to understand sorption behavior in terms of sediment
constituents. It will not suffice to consider sorption onto sediments simply in terms of partitioning into a uniform, thermodynamically ideal, stationary organic phase.

List of Symbols and Abbreviations

Symbols:
a radius of a sorbent grain (L)
C aqueous concentration (ML^{-3})
C_b aqueous concentration for solute bound to dissolved macro-
           molecules (ML^{-3})
C_{doc} aqueous concentration of dissolved organic carbon (ML^{-3})
C_o organic-phase concentration of a solute (ML^{-3})
C_w aqueous concentration of a freely dissolved solute
C_T $C_b + C_w$ (ML^{-3})
D_e effective diffusivity (L^2T^{-1})
f fractional approach to equilibrium $(-)$
f_{io} fraction of inorganic material $(-)$
f_{oc} fraction of organic carbon $(-)$
F fractional uptake $(-)$
G Gibbs free energy
HSA hydrophobic surface area (L^2)
k Boltzmann constant (JK^{-1})
k_{12} empirical rate constant (T^{-1})
K equilibrium constant (varies)
K_d distribution coefficient (L^3M^{-1})
K_{doc} distribution coefficient between free and DOC-bound solute
           (L^3M^{-1})

K_{io} distribution coefficient for solute bound to inorganic surfaces (L^3M^{-1})

K_{oc} organic-carbon-normalized distribution coefficient (L^3M^{-1})

K_{ow} octanol-water partition coefficient $(-)$

$K_{oc}^{\ominus}$ K_{oc} excluding the effect of C_{doc} (L^3M^{-1})

$K_{ow}^{\ominus}$ ideal K_{ow} (L^3M^{-1})

K_s surface specific partition coefficient (L^2M^{-1})

n porosity $(-)$

q sorbed concentration (M/M)

r radial distance from the center of a sorbing grain (L)

R gas constant $(JK^{-1}mol^{-1})$

R_p equilibrium partition coefficient $(-)$

S aqueous solubility (ML^{-3})

S_A specific surface area (L^2M^{-1})

S_ℓ aqueous solubility of a supercooled liquid (ML^{-3})

S_m aqueous solubility in a mixed solvent (ML^{-3})

S_s aqueous solubility of an organic solid (ML^{-3})

t time (t)

T absolute temperature (T)

T_m melting temperature (T)

V_i molar volume of solute in phase i (L^3M^{-1})

x mole fraction concentration $(mol\ mol^{-1})$

Greek Symbols:

α empirical constant $(-)$

γ_i activity coefficient of a solute in phase i $(-)$

γ_i^* activity coefficient of a solute in phase i saturated with a cosolvent $(-)$

δ solubility parameter of a solute $(cal\ M^{-1}L^{-3})^{1/2}$

δ_p solubility parameter of a polymer $(cal\ M^{-1}L^{-3})^{1/2}$

$\Delta\gamma^c$ difference in surface tension (Mt^{-2})

ϵ sorbent internal porosity (L^3L^3)

ρ sorbent grain bulk density (ML^{-3})

ρ_{oc} bulk density of organic matter (ML^{-3})

τ time constant for diffusion accompanied by linear sorption $(-)$

ϕ volume fraction of solute in polymer $(-)$

ϕ_p volume fraction of polymer in solution $(-)$

χ Flory-Huggins interaction parameter $(-)$

χ_H enthalpic contribution to χ $(-)$

χ_S entropic contribution to χ $(-)$

Ω efficiency factor for sorption $(-)$

Acknowledgments

We thank William Ball, Samuel Karickhoff, and Phil Gschwend for helpful suggestions and comments. This study was supported by the Robert S. Kerr Environmental Research Laboratory of the U.S. Environmental Protection Agency under assistance agreements R-808851 and R-812462-01-0. Although the information described in this article has been funded in part by the U.S. EPA, it has not been subjected to the Agency's required peer and administrative review and therefore does not necessarily reflect the views of the Agency and no official endorsement should be inferred.

Literature Cited

1.   Karickhoff, S.W.   J. of Hydraulic Engineering 1984, 110,
      707-35.
2.   Mortland, M.M.   Adv. Agron. 1970, 22, 75-114.
3.   Mortland, M.M.   In "Ground Water Quality"; Ward, C.H.; Giger,
      W.; McCarty, PL., Eds.; John Wiley & Sons: New York, 1985;
      pp. 370-86.
4.   Healy, T.W.   In "Organic Compounds in Aquatic Environments";
      Faust, S.J.; Hunter, J.V., Eds.; Marcel Dekker, Inc.: New
      York, 1971; Chap. 7.
5.   Fuerstenau, D.W.   In "The Chemistry of Biosurfaces"; Hau,
      M.L., Ed.; Marcel Dekker, Inc.: New York, 1971; Vol. 1,
      Chap. 4.
6.   March, J.W.   "Advanced Organic Chemistry"; John Wiley & Sons:
      New York, 1985; p. 73.
7.   Franks, F.   In "Water: A Comprehensive Treatise"; Franks,
      F., Ed.; Plenum Press: New York, 1975; Vol. 4, p. 4.
8.   Cowan, C.T.; White, D.   Trans. Fard. Soc. 1958, 54, 691-7.
9.   Leo, A.; Hansch, C.; Elkins, D.   Chem. Rev. 1971, 71,
      525-616.
10.   Tanford, C.   "The Hydrophobic Effect"; 2nd ed.; Wiley-
      Interscience: New York, 1980; p. 14.
11.   Hansch, C.; Leo, A.   "Substituent Constants for Correlation
      Analysis in Chemistry and Biology"; Wiley Interscience: New
      York, 1979.
12.   Hamaker, J.W.; Thompson, J.M.   In "Organic Chemicals in the
      Soil Environment"; Goring, C.A.I.; Hamaker, J.W., Eds.;
      Marcel Dekker, Inc.: New York, 1972.
13.   Chiou, C.T.; Shoup, T.D.; Porter, P.E.   Org. Geochem. 1985,
      8, 9-14.
14.   Chiou, C.T.; Peters, L.J.; Freed, V.H.   Science 1979, 206,
      831-2.
15.   Karickhoff, S.W.; Brown, D.S.; Scott, T.A.   Water Res. 1979,
      13, 241-8.
16.   Briggs, G.G.   J. Agric. Food Chem. 1981, 29, 1050-9.
17.   Reardon, E.J.   Groundwater 1981, 19, 279-86.
18.   Chiou, C.T.   In "Hazard Assessment of Chemicals: Current
      Developments"; Saxena J.E.; Fisher, F., Eds.; Academic Press:
      New York, 1981; Vol. 1, Chap. 2.
19.   Chiou, C.T.; Schmedding, D.W.; Manes, M.   Env. Sci. Technol.
      1982, 16, 4-10.
20.   Chiou, C.T.; Porter, P.E.; Schmedding, D.W.   Env. Sci.
      Technol. 1983, 17, 227-31.
21.   Miller, M.M.; Wasik, S.P.; Huang, G.; Shiu, W.; Mackay, D.
      Env. Sci. Technol. 1985, 19, 522-9.
22.   Karickhoff, S.W.   Chemosphere 1981, 10, 833-46.
23.   MacKay, D.   Env. Sci. Technol. 1977, 11, 1219.
24.   Yalkowsky, S.H.; Valvani, S.C.   J. Chem. Eng. Data 1979, 24,
      127-9.
25.   Tewari, Y.B.; Miller, M.M.; Wasile, S.P.; Martine, D.E.
      J. Chem. Eng. Data 1982, 27, 451-4.
26.   Banerjee, S.; Yalkowsky, S.H.; Valvani, S.C.   Env. Sci.
      Technol. 1980, 14, 1227-9.
27.   Yalkowsky, S.H.; Valvani, S.C.; Amidon, G.L.   J. Pharm. Sci.
      1976, 65, 1488-94.

28. Flory, P.J. "Principles of Polymer Chemistry"; Cornell University Press: Ithaca, NY, 1956; Chap. 8.

29. Tanford, C. "Physical Chemistry of Macromolecules"; J. Wiley & Sons: New York, 1961; Chap. 4.

30. Prausnitz, J.M.; Lichtenthaler, R.N.; Azevedo, E.G. "Molecular Thermodynamics of Fluid Phase Equilibria"; Prentice-Hall: Englewood Cliffs, NJ, 1985; pp. 306-16.

31. Freeman, D.H.; Cheung, L.S. Science 1981, 214, 790-2.

32. Weast, R.C. "CRC Handbook of Chemistry and Physics"; CRC Press, Inc.: Boca Raton, FL, 1982; pp. C-732-3.

33. Wershaw, R.L.; Burcan, P.J.; Goldberg, M.C. Env. Sci. Technol. 1969, 3, 271-3.

34. Matsuda, K.; Schnitzer, M. Bull. Environ. Contam. Toxicol. 1971, 6, 200-3.

35. Poirrier, M.A.; Bordelon, B.R.; Laseter, J.L. Env. Sci. Technol. 1972, 6, 1033-5.

36. Boehm, P.D.; Quinn, J.G. Geochim. Cosmochim. Acta 1973, 37, 2459-77.

37. Hassett, J.J.; Anderson, M.A. Env. Sci. Technol. 1979, 13, 1526-9.

38. Carlberg, G.E.; K. Martinsen, K. Sci. of Total Environ. 1982, 25, 245-54.

39. Molyneux, P. In "Water: A Comprehensive Treatise"; F. Franks, Ed.; Plenum Press: New York, 1975; Vol. 4, p. 678.

40. Kahn, S.U. In "Dynamics, Exposure and Hazard Assessment of Toxic Chemicals"; Hague, R., Ed.; Ann Arbor Science: Ann Arbor, MI, 1980; pp. 224-9.

41. Chiou, C.T.; Porter, P.E.; Shoup, T.D. Env. Sci. Technol. 1984, 18, 295-7.

42. Carter, C.W.; Suffet, I.H. Env. Sci. Technol. 1982, 16, 735-40.

43. Landrum, P.F.; Nihart, S.R.; Eadie, B.J.; Gardner, W.S. Env. Sci. Technol. 1984, 18, 187-92.

44. Gschwend, P.M.; Wu, S. Env. Sci. Technol. 1985, 19, 90-6.

45. Voice, T.C.; Weber, Jr., W.J. Env. Sci. Technol. 1985, 19, 789-96.

46. Haas, C.N.; Kaplan, B.M. Env. Sci. Technol. 1985, 19, 643-5.

47. Caron, G.; Suffet, I.H.; Belton, T. Chemosphere 1985, 14, 993-1000.

48. MacCarthy, J.F.; Jimenez, B.D. Env. Sci. Technol. 1985, 19, 1072-7.

49. Rao, P.S.C.; Hornsby, A.G.; Kilcrease, D.P.; Nkedi-Kizza, P. J. Environ. Qual. 1985, 14, 376-83.

50. Fu, J.K.; Luthy, R.G. "Pollutant Sorption to Soils and Sediments in Organic/Aqueous Solvent Systems"; EPA/600/3-85/050, 1985.

51. Nkedi-Kizza, P.; Rao, P.S.C.; Hornsby, A.G. Env. Sci. Technol. 1985, 19, 975-9.

52. Curtis, G.P. Engineers Thesis, Stanford University, Stanford, CA, 1984.

53. Nkedi-Kizza, P.; Rao, P.S.C.; Johnson, J.W. J. Environ. Qual. 1983, 12, 195-7.

54. Lambert, S.H. J. Agr. Food Chem. 1968, 16(2), 340-343.

55. McCarty, P.L.; Reinhard, M.; Rittman, B.E. Env. Sci. Technol. 1981, 15, 40-51.

56.   Schwarzenbach, R.P.; Westall, J. Env. Sci. Technol. 1981, 15, 1360-7.
57.   Shin, V.O.; Chodan, J.J.; Walcott, A.R. J. Agric. Food Chem. 1970, 18, 1129-33.
58.   Pierce, R.H., Jr.; Olney, C.E.; Felbeck, Jr., G.T. Geochim. Cosmochim. Acta 1974, 38, 1061-73.
59.   Means, J.C.; Wood, S.G.; Hassett, J.J.; Banwart, W.C. Env. Sci. Technol. 1982, 16, 93-8.
60.   Kahn, A.; Hassett, J.J.; Banwart, W.L. Soil Sci. 1979, 128, 297-302.
61.   Hassett, J.J.; Banwart, W.L.; Wood, S.G.; Means, L.C. Soil Sci. Soc. Am. J. 1981, 45, 38-42.
62.   Parris, G.E. Env. Sci. Technol. 1980, 14, 1099-106.
63.   Curtis, G.P.; Roberts, P.V.; Reinhard, M. Water Resour. Res. 1986 (in press).
64.   Farmer, V.C.; Russell, J.D. Clays Clay Miner. 1967, 15, 121-42.
65.   Texter, J.; Klier, K.; Zettlemoyer, A.C. In "Progress in Surface and Membrane Science, Vol. 12"; Academic Press: New York, 1978; p. 339.
66.   Karickhoff, S.W. In "Contaminants and Sediments, Vol. 2"; R. Baker, Ed.; Ann Arbor Science: Ann Arbor, MI, 1980.
67.   McCall, P.J.; Agin, G.L. Environ. Toxicol. Chem. 1985, 4, 37-44.
68.   Karickhoff, S.W.; Morris, K.R. Environ. Toxicol. Chem. 1985, 4, 469-79.
69.   Blatt, H.; Middleton, G.; Murray, R. "Origin of Sedimentary Rocks"; Prentice Hall, Inc.: Englewood Cliffs, NJ, 1981; Chap. 13.
70.   Hassett, J.P.; Milicic, E. Env. Sci. Technol. 1985, 19, 638-43.
71.   Hawker, D.W.; Connell, D.W. Env. Sci. Technol. 1985, 19, 643-6.
72.   Helfferich, F. "Ion Exchange"; McGraw-Hill Book Co.: New York, 1962; pp. 259-62.
73.   Crank, J. "The Mathematics of Diffusion"; Oxford University Press: Ely House, London, 1975; Chaps. 6, 14.
74.   Weber, W.J. Jr.; Rumer, R.R. Water Resour. Res. 1965, 1, 361-9.
75.   Weber, T.W.; Chakravorti, R.K. AIChE J. 1974, 20, 228-38.
76.   Satterfield, C.N. "Mass Transfer in Heterogeneous Catalysis"; MIT Press: Cambridge, MA, 1970; Chap. 1.
77.   Rao, P.S.C.; Davidson, J.M.; Jessup, R.E.; Selim, M.M. Soil Sci. Soc. Am. J. 1979, 43, 22-8.
78.   Wu, S.; Gschwend, P. Submitted to Env. Sci. Technol. 1986.
79.   DiToro, D.M.; Horzempa, L.M. Env. Sci. Technol. 1982, 16(9), 594-602.
80.   Horzempa, L.M., DiToro, D.M. Water Res. 1983, 17(8), 851-9.
81.   Rao, P.S.C.; Berkheiser, V.E.; Ou, L.T. "Estimation parameters for modeling behavior of selected pesticides and ortho-phosphate"; EPA-6001, U.S. Envionmental Protection Agency, Athens, GA, 1984.

RECEIVED June 24, 1986

Distinguishing Adsorption from Surface Precipitation

Garrison Sposito

Department of Soil and Environmental Sciences, University of California, Riverside, CA 92521

Measurements of the chemical composition of an aqueous
solution phase are interpreted commonly to provide
experimental evidence for either adsorption or surface
precipitation mechanisms in sorption processes. The
conceptual aspects of these measurements vis-à-vis
their usefulness in distinguishing adsorption from
precipitation phenomena are reviewed critically. It
is concluded that the inherently macroscopic, indirect
nature of the data produced by such measurements limit
their applicability to determine sorption mechanisms in
a fundamental way. Surface spectroscopy (optical or
magnetic resonance), although not a fully developed
experimental technique for aqueous colloidal systems,
appears to offer the best hope for a truly molecular-
level probe of the interfacial region that can
discriminate among the structures that arise there
from diverse chemical conditions.

The loss of a chemical species from an aqueous solution phase to a
contiguous solid phase may be termed a sorption process. Among the
mechanisms by which sorption processes occur, the three principal
ones are: precipitation, the growth of a solid phase exhibiting a
primitive molecular unit (a complex) that repeats itself in three
dimensions; adsorption, an accumulation of matter at the interface
between an aqueous solution phase and a solid adsorbent without the
development of a three-dimensional molecular arrangement; and absorption, the diffusion of an aqueous chemical species into a solid
phase (1,2). A precipitation mechanism may be initiated by either
homogeneous or heterogeneous nucleation, may involve the formation
of a solid mixture either by inclusion or by coprecipitation, or may
take place on the surface of a pre-existent solid phase (surface precipitation). Regardless of these variations, the essential characteristic of precipitation is the development of a solid phase whose
molecular ordering is intrinsically three-dimensional (2). An adsorption [strictly speaking, positive adsorption (1)] mechanism, on
the other hand, involves only two-dimensional molecular arrangements

0097-6156/86/0323-0217$06.00/0
© 1986 American Chemical Society

on a surface. This latter restriction does not preclude mixed
adsorbates ["two-dimensional solid solutions" (2)], but it does
eliminate solid phases whose structure is inherently three-dimensional, even if they form on surfaces and are hindered in their
growth for stereochemical reasons [e.g., interlayer metal hydroxides
on 2:1 layer type aluminosilicates (1)]. From this point of view,
"multilayer adsorption" must refer to a succession of adsorbate
layers, each of whose molecular ordering can be influenced only by
the layer on which it forms and not by any other previously adsorbed
layers. In the same vein, "absorption" must refer to the penetration
of a chemical species into a solid phase beyond the nanometer depth
from its periphery that operationally defines the interfacial region.
    A central problem in the chemistry of natural water systems is
the establishment of experimental methods with which to distinguish
adsorption from surface precipitation (1-3). Corey (2) has written
a comprehensive review of this problem which should be read as an
introduction to the present essay, particularly for his set of six
conclusions that set out general conditions likely to result in
adsorption or precipitation. The discussion to follow is not a comprehensive review, but instead focuses on three popular approaches
to the adsorption/surface precipitation dichotomy. The emphasis here
is on the conceptual relationship of each approach to the defining
statements made above: To what extent is an approach capable of
distinguishing adsorption from surface precipitation?

Solubility Methods

Adsorption isotherms. The quantity of a chemical species i adsorbed
per unit mass of a solid material contacting an aqueous solution
phase is calculated with the equation (1):

$$q_i^{(w)} = n_i - M_w m_i \tag{1}$$

where n_i is the total number of moles of species i in the suspension
per kilogram of solid, M_w is the mass of water per unit mass of
solid, and m_i is the molality of the adsorptive i in the aqueous
solution phase. In batch experiments M_w is the inverse of the
suspension density, whereas in column experiments M_w is the gravi-
metric water content. Equation 1 represents the surface excess of
species i assigned to an interface where there is no net accumulation
of water (1); hence, the superscript w on the left side of the equation.
    Adsorption phenomena frequently are studied by measuring solely
the change in concentration of a species i in the aqueous solution
phase. Simple mass-balance considerations (1) show that Equation 1
can be rewritten in a form compatible with this methodology:

$$q_i^{(w)} = \Delta m_i M_{Tw} \tag{2}$$

where $\Delta m_i = m_i^o - m_i$ and m_i^o is the molality of species i in M_{Tw}
kilograms of water in the aqueous solution phase prior to its being

brought into contact with 1 kg of solid material. Equation 2 provides a basis for calculating $q_i^{(w)}$ as a loss of mass from aqueous solution. The critical assumption underlying its equivalence to Equation 1 is that the species $\underline{i}$ (e.g., $H_2PO_4^-$ or Cd^{2+}) preserves its chemical identity after it accumulates on the adsorbent. This assumption seldom is checked experimentally by chemical species analysis of the solid material after reaction. Often even the aqueous solution phase is not speciated and $\underline{i}$ refers, for example, to Cd(II) instead of $Cd^{2+}(aq)$, with a concomitant lack of specificity in what is actually being adsorbed. When $q_i^{(w)}$ and m_i refer to total concentrations instead of species concentrations, the task of distinguishing adsorption from surface precipitation becomes correspondingly more difficult.

A graph of $q_i^{(w)}$ against m_i, or an equivalent concentration variable, at fixed temperature and pressure is an adsorption isotherm. Data of this kind typically have been fitted numerically to special cases of the equation (3):

$$q_i^{(w)} = \sum_{m=1}^{n} \frac{b_m \; K_m^{\beta_m} \; c_i^{\beta_m}}{[1 + B_m \; c_i^{\delta_m}]^{\gamma_m}} \tag{3}$$

where c_i is the concentration of an adsorptive $\underline{i}$ in the aqueous solution phase (e.g., the molality) and b_m, K_m, B_m, β_m, δ_m, and γ_m ($m = 1, \ldots, n$) are adjustable parameters. Equation 3 represents an adsorption isotherm equation. Popular special cases of this expression include (3,4) the Langmuir equation ($\beta_m = \delta_m = \gamma_m = 1$, $B_m = K_m$; $n = 1$ or $\overline{2}$), the van Bemmelen–Freundlich equation ($B_m = 0$, $0 < \beta_m < 1$; $n = 1$), and the Tóth equation ($\beta_m = 1$, $B_m = K_m^{\delta_m}$, $\gamma_m = 1/\delta_m$; $n = 1$). In general, the larger is the number of adjustable parameters in an adsorption isotherm equation, the better its fit to experimental data is likely to be.

The provenance of expressions like that in Equation 3 has never been shown to be uniquely an adsorption mechanism. On the contrary, it is possible to derive special cases of Equation 3, such as the classical Langmuir equation

$$q_i^{(w)} = \frac{b \; K \; c_i}{1 + K \; c_i} \tag{4}$$

on the basis of sorption mechanisms for cations and anions involving only precipitation reactions (1,5). The situation for the "two-surface" Langmuir equation, a four-parameter version of Equation 3,

$$q_i^{(w)} = \frac{b_1 \; K_1 \; c_i}{1 + K_1 \; c_i} + \frac{b_2 \; K_2 \; c_i}{1 + K_2 \; c_i} \tag{5}$$

is yet more ambiguous. On strictly mathematical grounds (6), it has been shown that Equation 5 can be fit to any set of sorption data for which a plot of $q_i^{(w)}/c_i$ (the distribution coefficient) against $q_i^{(w)}$ is convex toward the x-axis and has finite (extrapolated) y- and x-intercepts. This result is independent of the mechanism of the sorption process. Unfortunately, many sorption data — indeed, most — meet the mathematical criteria required in order to apply Equation 5. These kinds of difficulties make evident the point that adsorption isotherm equations should be regarded simply as curve-fitting devices without a priori chemical significance, but with predictive capability under limited conditions. The mechanistic implication of this conclusion can be formalized in the following rule (1):

> The adherence of experimental sorption data to an adsorption isotherm equation provides no evidence as to the actual mechanism of a sorption process.

The scope of this general rule extends to adsorption isotherms which turn convex to the concentration axis at higher adsorptive concentrations (7) and to adsorption "edges" or "envelopes" observed in plots of $q_i^{(w)}$ against pH at fixed total adsorptive i concentration (1). Adsorption isotherms that show a "monolayer knee" at lower concentrations followed by an upward turn at higher concentrations of adsorptive can be modeled by retaining two or more terms in Equation 3 and choosing the adjustable parameters judiciously (e.g., a Langmuir first term and a van Bemmelen-Freundlich second term (7)). This possibility does not imply that adsorption alone governs the process described by isotherms that grow continually with the adsorptive concentration, any more than fitting the data by an adsorption surface precipitation model (7) would imply that surface precipitation was indeed the controlling process at higher concentrations. The same conclusion applies to the modeling of adsorption "edges" or "envelopes". That an adsorption "edge" for a metal reacting with a hydrous oxide exhibits its sharp rise at lower pH values than it does when the adsorbent is absent can mean either that an adsorption process has occurred or that surface precipitation of the metal as a hydrous oxide has been induced by the presence of the adsorbent. No direct evidence favoring one interpretation or the other can be provided by these data alone. Similarly, the enhancement of metal sorption sometimes observed when a strongly sorbing anion has been reacted previously with a hydrous oxide adsorbent may be modeled either as a metal-anion surface precipitate effect (8) or a metal-anion surface complex effect (9). Sorption data themselves do not provide for a choice of model, unless the ion-activity product for a proposed surface precipitate exceeds the corresponding solubility product constant.

Ion-Activity Products. As in the determination of the amount sorbed through Equation 2, the characterization of surface precipitates often utilizes measurements made solely on the aqueous solution phase. Solubility studies limited in this way run a risk of being ambiguous as to mechanism because of the lack of direct information about the solid phase (10). In respect to the aqueous solution phase, ambiguity can be minimized if equilibrium is approached both from supersaturation and from undersaturation; if the equilibration time is varied systematically; if the aqueous solution phase is monitored through two or more concentration (or activity) variables susceptible to quantitation with high precision; and if the stoichiometry of the assumed precipitation-dissolution reaction is verified experimentally (10).

Once the composition of the aqueous solution phase has been determined, the activity of an electrolyte having the same chemical formula as the assumed precipitate can be calculated (11,12). This calculation may utilize either mean ionic activity coefficients and total concentrations of the ions in the electrolyte, or single-ion activity coefficients and free-species concentrations of the ions in the electrolyte (11). If the latter approach is used, the computed electrolyte activity is termed an ion-activity product (12). Regardless of which approach is adopted, the calculated electrolyte activity is compared to the solubility product constant of the assumed precipitate as a test for the existence of the solid phase. If the calculated ion-activity product is smaller than the candidate solubility product constant, the corresponding solid phase is concluded not to have formed in the time period of the solubility measurements. This judgment must be tempered, of course, in light of the precision with which both electrolyte activities and solubility product constants can be determined (12).

The difficulty here is that the ion-activity product includes not only the Gibbs energy change in a solid dissolution process but also the activity of the solid itself. Consider, as a simple example, the dissolution of $CdCO_3(s)$, for which the ion-activity product (IAP) is (12):

$$IAP = [Cd^{2+}][CO_3^{2-}] = K_{so}[CdCO_3(s)] \qquad (6)$$

where $K_{so} = 10^{-11}$ at 298 K (10) and [ ] represents a thermodynamic activity. If Cd(II) has coprecipitated with another metal (e.g., Ca(II)) to form a solid solution, then $[CdCO_3] < 1$ and $IAP < K_{so}$. Thus a homogeneous, mixed surface precipitate typically can be expected to produce low IAP values and, if the chemical element of interest is in the mixture only in trace amounts, the discrepancy between IAP and K_{so} easily can be an order of magnitude (1,10). A low IAP value of this kind then might be interpreted to mean that surface precipitation had not occurred, and that adsorption had occurred, because undersaturation existed in the aqueous solution phase. The error of such a conclusion, in the absence of a direct examination of the solid phase, is apparent from Equation 6: Only precipitates whose activity equals or exceeds 1.0 have been eliminated. The inference to be drawn from this discussion can be formalized conservatively in the rule (1):

The experimental observation that an ion-activity product is smaller than a corresponding solubility product constant by an order of magnitude or less provides no evidence as to the general mechanism of a sorption process.

In systems where surface precipitation has been verified on the
basis of a direct examination of the solid phase, it sometimes is
true that, because of epitaxial or other stereochemical constraints,
the activity of even an aged precipitate is larger than 1.0, relative
to a standard state in which the solid phase is macrocrystalline and
free of inclusions (12). The larger solid-phase activity leads to
an IAP larger than $\overline{K_{so}}$ (cf. Equation 6). A typical example of
this effect is in the precipitation of aluminum hydrous oxides onto
the interlayer siloxane surfaces of 2:1 layer type aluminosilicates
during weathering. When an aluminum hydroxy-solid precipitates
onto the interlayer surface of smectite or vermiculite, IAP =
$[Al^{3+}][OH^-]^3 \sim 10^{-32}$, whereas for macrocrystalline gibbsite,
$K_{so} = 10^{-34}$ (13). The fact that IAP $>$ K_{so} implies a lower degree
of crystallinity or crystalline size exists in the surface precipitate.

Kinetics Methods

Electrokinetic behavior. A shearing stress applied to or induced in
an aqueous solution phase contacting a charged adsorbent produces a
response at the solid-liquid interface known as an electrokinetic
phenomenon (1,14). The principal electrokinetic phenomena of relevance to sorption experiments are: electrophoresis, the response of
a charged adsorbent to an applied, constant electric field; electro-
osmosis, the response of an electrolyte solution near a stationary,
charged adsorbent to an applied electric field, and the streaming
potential, the response of an electrolyte solution near a stationary,
charged adsorbent to an applied, uniform pressure gradient. For all
three phenomena, experimental data can be summarized in calculations
of the zeta potential ζ, the inner electrostatic potential near the
adsorbent surface at the plane of shear induced by an applied electric field or produced by an applied pressure gradient (1,14).

The significance of ζ for distinguishing adsorption from surface
precipitation has been brought into clear focus by James and Healy
(15). They pointed out that ζ often decreases to a minimum value,
followed by a rise to a maximum value then decline toward negative
values, as the pH is increased in an aqueous suspension containing a
hydrous oxide adsorbent and a hydrolyzable metal cation adsorptive.
This behavior can be interpreted as the result of a gradual accumulation of hydrolytic species of the metal on the surface of the
adsorbent (producing a net increase in surface charge and an increase in ζ with increasing pH) which culminates in the formation of
a hydroxy-polymer coating of the metal on the adsorbent (producing
ultimately a net decrease in both ζ and the surface charge, which
gradually reflects that of the coating, not the adsorbent). This
interpretation applies to any bivalent, trivalent, or tetravalent
metal cation that hydrolyzes to some extent above pH 6 in aqueous
solution (16). The magnitude of the concentration of hydrolytic
metal species in solution is not relevant, even if the concentration
is very small, since aqueous solutions are effectively open systems
with respect to these species. If an adsorbent exhibits a high
enough affinity for a hydrolytic species, it can be adsorbed at once
and be replaced in the aqueous solution phase through hydrolysis of a

solvated species until the availability of the latter has been ex-
hausted. The concentration of the hydrolytic species may remain
quite small, but its adsorption is determined by the affinity of the
adsorbent for it and the total metal concentration in aqueous solution.

If a surface precipitate of metal hydroxy-polymer has formed on
an adsorbent, the ζ-pH relationship for the coated adsorbent should
resemble closely that observed for particles consisting purely of the
hydroxy-polymer or the hydrous oxide of the metal (15). This kind of
evidence for Co(II), La(III), and Th(IV) precipitation on silica
colloids was cited by James and Healy (15). It should be noted,
however, that the increase in ζ toward a maximum value often occurs
at pH values well below that required thermodynamically to induce
bulk-solution homogeneous precipitation of a metal hydrous oxide (15,
16). If surface precipitation is in the incipient stage under these
conditions, it must be a nucleation phenomenon. James and Healy (15)
argue that the microscopic electric field at the surface of a charged
adsorbent is sufficiently strong to lower the vicinal water activity
and induce precipitation at pH values below that required for bulk-
solution precipitation of a metal hydrous oxide.

Both Schindler (17) and Fuerstenau et al. (18) have called
attention to the point that the surface precipitation concept need
not be invoked to explain the ζ-pH relationship described above. If
only solvated metal cations adsorb in inner-sphere surface complexes,
their adsorption will be enhanced by decreasing the adsorbent charge
through increasing the pH and they will concomitantly increase ζ by
bringing positive charge to the solid-liquid interface. At high pH
values, the metal cations will begin to hydrolyze significantly in
aqueous solution and these hydrolytic species can form at the expense
of adsorbed species, with the result that ζ decreases as the metal
cations desorb to hydrolyze. The qualitative form of the ζ-pH relationship produced by this mechanism resembles experimental observations for the bivalent metal cation-silica system closely (17). The
implication of this fact is that an observed ζ-pH relationship does
not provide an unambiguous method of distinguishing adsorption from
surface precipitation.

Reaction kinetics. The time-development of sorption processes often
has been studied in connection with models of adsorption despite the
well-known injunction that kinetics data, like thermodynamic data,
cannot be used to infer molecular mechanisms (19). Experience with
both cationic and anionic adsorptives has shown that sorption reactions typically are rapid initially, operating on time scales of
minutes or hours, then diminish in rate gradually, on time scales of
days or weeks (16,20-25). This decline in rate usually is not
interpreted to be homogeneous: The rapid stage of sorption kinetics
is described by one rate law (e.g., the Elovich equation), whereas
the slow stage is described by another (e.g., an expression of first
order in the adsorptive concentration). There is, however, no pro-
found significance to be attached to this observation, since a consensus does not exist as to which rate laws should be used to model
either fast or slow sorption processes (16,21,22,24). If a sorption
process is initiated from a state of supersaturation with respect to
one or more possible solid phases involving an adsorptive, or if the

adsorbents present are either poorly crystallized or well hydrated, it is likely that multiple sorption mechanisms will operate right from the beginning (1,2). The time-development of the sorption process then should reflect this multiplicity and defy any simple interpretation in terms of an adsorption mechanism.

When the kinetics of a sorption process do appear to separate according to very small and very large time scales, the almost universal inference made is that pure adsorption is reflected by the rapid kinetics (16,21,22,26). The slow kinetics are interpreted either in terms of surface precipitation (20) or diffusion of the adsorbate into the adsorbent (16,24). With respect to metal cation sorption, "rapid kinetics" refers to time scales of minutes (16,26), whereas for anion sorption it refers to time scales up to hours (1, 21). The interpretation of these time scales as characteristic of adsorption rests almost entirely on the premise that surface phenomena involve little in the way of molecular rearrangement and steric hindrance effects (16,21).

An illustration of the reaction kinetics approach to distinguishing adsorption from surface precipitation is provided by the sorption of o-phosphate by calcite (27-30). The loss of o-phosphate from aqueous solution in the presence of calcite is pronounced on a time scale of tens of minutes and is enhanced by increasing temperature or pH (27-29). Thereafter, on a time scale of hours or days, the o-phosphate solubility decreases gradually, then drops sharply again (27,28). This behavior is interpreted mechanistically as adsorption of o-phosphate at selective sites on calcite followed by the nucleation of a calcium phosphate solid on the surface (27,29, 30). The gradual decline in o-phosphate solubility, which persists longer the smaller is the initial o-phosphate concentration (28), represents the period of rearrangement of adsorbed o-phosphate clusters into calcium phosphate nuclei (27,29). Epitaxial, three-dimensional growth of calcium phosphate crystals then follows. Scanning electron micrographs of calcite taken during the rearrangement period (30) show hemispherical growths of o-phosphate (identified by microprobe analysis) at edge sites and dislocations on the crystal surface. Griffin and Jurinak (29) found the adsorption kinetics to be second-order, whereas the rearrangement kinetics were first-order. Similar results for the slower kinetics of o-phosphate sorption by metal hydrous oxides and soils have been reported (20, 21), but no consensus exists (23,25,31). The principal criterion, however, is not homogeneity of the rate law, but a clear separation of the kinetics according to time scale (1).

Surface Spectroscopy

Solubility and kinetics methods for distinguishing adsorption from surface precipitation have the common features of being essentially macroscopic in nature and of not utilizing a direct examination of sorbed material. The essential difference between an adsorbate and a surface precipitate lies with molecular structure, however, and it is inevitable that methodologies not equipped to explore that structure directly will produce ambiguous results requiring ad hoc assumptions in order to interpret them. The principal technique for investigating molecular structure is spectroscopy (32,33). Surface spectroscopy, both optical and magnetic, offers the best opportunity at present to elucidate the structures of chemical species at the solid-liquid interface.

Surface spectroscopic techniques must be separated carefully into those which require dehydration for sample presentation and those which do not. Among the former are electron microscopy and microprobe analysis, X-ray photoelectron spectroscopy, and infrared spectroscopy. These methods have been applied fruitfully to show the existence of either inner-sphere surface complexes or surface precipitates on minerals found in soils and sediments (13b,30,31-37), but the applicability of the results to natural systems is not without some ambiguity because of the dessication pretreatment involved. If independent experimental evidence for inner-sphere complexation or surface precipitation exists, these methods provide a powerful means of corroboration.

X-ray diffraction, Raman spectroscopy, and magnetic resonance spectroscopy (nuclear and electron), on the other hand, do not require dehydration of the sample. X-ray diffraction is a method of long standing for the detection of surface precipitates that is usually — but need not be — applied to dried materials (30,38-41). Raman spectroscopy (42)and nuclear magnetic resonance spectroscopy (32,33) have not often been used to distinguish surface species in aqueous systems comprising natural colloids, but their potential for this kind of investigation remains significant (33,37). Perhaps the most useful technique has been electron spin resonance spectroscopy (32,43), particularly the applications of it made by McBride and his coworkers (44-48). Although limited intrinsically (32) to only three metals of major interest in natural colloidal systems — Fe(III), Mn(II), and Cu(II) — this method has been uniquely successful in providing a general understanding of the conditions under which adsorption or surface precipitation is likely to occur.

A prototypical example of electron spin resonance (ESR) spectroscopy applied to surface speciation on a layer silicate is Cu(II) sorption by hectorite (44b). The ESR spectrum of a Cu-hectorite suspension shows a gradual decrease in the intensity of the line characteristic of the $Cu(H_2O)_6^{2+}$ species as pH increases, consistent with the gradual appearance of hydrolytic species of Cu(II) which do not produce an ESR signal. At pH < 5.2, the fraction of Cu(II) in hydrolytic species on the hectorite surface is much larger than in aqueous solution, indicating a high affinity of the surface for these species. Air-dried Cu-hectorite presents ESR spectra assignable to Cu(II) bound in inner-sphere surface complexes in the hectorite interlayers, by contrast with the diffuse-ion swarm or outer-sphere surface complex Cu(II) inferred from the suspension spectra. As pH increases, this adsorbed Cu(II) converts gradually to hydrolytic species. Kinetics data taken with the ESR spectra in suspension were consistent with the formation of surface precipitates at pH > 6. A similar study of Cu(II) sorption by microcrystalline gibbsite (47) produced spectral evidence for inner-sphere surface complexes of Cu^{2+} predominating below pH 5 and Cu-hydroxy species — eventually $Cu(OH)_2(s)$ — predominating above pH 5. For both hectorite and gibbsite, the IAP = $[Cu^{2+}][OH^-]^2$ for the surface hydroxy species was

smaller than the solubility product constant of macrocrystalline
$Cu(OH)_2(s)$.

Concluding Remarks

Solubility and kinetics methods for distinguishing adsorption from
surface precipitation suffer from the fundamental weakness of being
macroscopic approaches that do not involve a direct examination of
the solid phase. Information about the composition of an aqueous
solution phase is not sufficient to permit a clear inference of a
sorption mechanism because the aqueous solution phase does not determine uniquely the nature of its contiguous solid phases, even at
equilibrium (49). Perhaps more important is the fact that adsorption
and surface precipitation are essentially molecular concepts on which
strictly macroscopic approaches can provide no unambiguous data (12,
21). Molecular concepts can be studied only by molecular methods.
It is for this reason that spectroscopy offers the only experimental
method for characterizing the interfacial region that is not automatically destined to run into basic conceptual difficulties. This
is not to say that difficulties of a technical nature will not arise
(40-48), nor that the conceptual difficulty of differing time scales
among spectroscopic techniques will cause no problems (50). Nonetheless, it is to be hoped that future investigations of sorption
reactions will focus more on probing the molecular structure of the
mineral/water interface than on attempting simply to divine what the
structure may be.

Acknowledgments

Gratitude is expressed to Drs. M. E. Essington, C. T. Johnston, and
R. L. Mikkelson for helpful written discussions of sorption mechanisms. The undertaking of this review was supported in part by Grant
No. I461-82 from BARD — The United States-Israel Binational Agricultural Research and Development Fund.

Literature Cited

1. Sposito, G. "The Surface Chemistry of Soils"; Oxford Univ.
    Press: New York, 1984; Chap. 4.
2. Corey, R. B. In "Adsorption of Inorganics at Solid-Liquid
    Interfaces"; Anderson, M. A.; Rubin, A. J., Eds.; Ann Arbor
    Science: Ann Arbor, 1981; Chap. 4.
3. Sposito, G. Crit. Rev. Environ. Control 1985, 15, 1-24.
4. Kinniburgh, D. G.; Barker, J. A.; Whitfield, J. J. Colloid
    Interface Sci. 1983, 95, 370-384.
5. Veith, J. A.; Sposito, G. Soil Sci. Soc. Am. J. 1977, 41, 697-
    702.
6. Sposito, G. Soil Sci. Soc. Am. J. 1982, 46, 1147-1152.
7. Farley, K. J.; Dzombak, D. A.; Morel, F.M.M. J. Colloid Inter-
    face Sci. 1985, 106, 226-242.
8. Benjamin, M. M.; Bloom, N. S. In "Adsorption from Aqueous
    Solutions"; Tewari, P. H., Ed.; Plenum Publ. Corp.: New York,
    1981; pp. 41-60.

9.   Bolland, M.D.A.; Posner, A. M.; Quirk, J. P. Aust. J. Soil Res. 1977, 15, 279-86.

10. Sposito, G. In "Applied Environmental Geochemistry"; Thornton, I., Ed.; Academic Press: London, 1983; Chap. 5.

11. Sposito, G. Soil Sci. Soc. Am. J. 1984, 48, 531-536.

12. Sposito, G. "The Thermodynamics of Soil Solutions"; Clarendon Press: Oxford, 1981; Chap. 1-3.

13. (a) Kittrick, J. A. Clays & Clay Minerals 1983, 31, 317-318.
     (b) Harsh, J. B.; Doner, H. E. Geoderma 1985, 36, 45-56.

14. Hunter, R. J. "Zeta Potential in Colloid Science"; Academic Press: London, 1981; Chap. 1-3.

15. James, R. O.; Healy, T. W. J. Colloid Interface Sci. 1972, 40, 53-64.

16. Kinniburgh, D. G.; Jackson, M. L. In "Adsorption of Inorganics at Solid-Liquid Interfaces"; Anderson, M. A.; Rubin, A. J., Eds.; Ann Arbor Science: Ann Arbor, 1981; Chap. 3.

17. Schindler, P. W. In "Adsorption of Inorganics at Solid-Liquid Interfaces"; Anderson, M. A.; Rubin, A. J., Eds.; Ann Arbor Science: Ann Arbor, 1981; Chap. 1.

18. Fuerstenau, D. W.; Manmohan, D.; Raghavan, S. In "Adsorption from Aqueous Solutions"; Tewari, P. H., Ed.; Plenum Publ. Corp.: New York, 1981; pp. 93-117.

19. Denbigh, K. "The Principles of Chemical Equilibrium"; Cambridge Univ. Press: Cambridge, 1981; Chap. 15.

20. Chen, Y.-S.R.; Butler, J. N.; Stumm, W. Environ. Sci. Technol. 1973, 7, 327-332.

21. Berkheiser, V. E.; Street, J. J.; Rao, P.S.C.; Yuan, T. L. Crit. Rev. Environ. Control 1980, 10, 179-224.

22. Hingston, F. J. In "Adsorption of Inorganics at Solid-Liquid Interfaces"; Anderson, M. A.; Rubin, A. J., Eds.; Ann Arbor Science: Ann Arbor, 1981; Chap. 2.

23. van Riemsdijk, W. H.; de Haan, F.A.M. Soil Sci. Soc. Am. J. 1981, 45, 261-266.

24. Bolan, N. S.; Barrow, N. J.; Posner, A. M. J. Soil Sci. 1985, 36, 187-197.

25. van Riemsdijk, W. H.; Lyklema, J. Colloids & Surfaces 1980, 1, 33-44.

26. Sparks, D. L. Advan. Agronomy 1985, 38, 231-266.

27. Leckie, J.; Stumm, W. In "Water Quality Improvement by Physical and Chemical Processes"; Gloyna, E. F.; Eckenfelder, W. W., Eds.; Univ. of Texas Press: Austin, 1970; pp. 237-249.

28. Griffin, R. A.; Jurinak, J. J. Soil Sci. Soc. Am. J. 1973, 37, 847-850.

29. Griffin, R. A.; Jurinak, J. J. Soil Sci. Soc. Am. J. 1974, 38, 75-79.

30. Freeman, J. S.; Rowell, D. L. J. Soil Sci. 1981, 32, 75-84.

31. van Riemsdijk, W. H.; Lyklema, J. J. Colloid Interface Sci. 1980, 76, 55-66.

32. Stucki, J. W.; Banwart, W. L. "Advanced Chemical Methods for Soil and Clay Minerals Research"; D. Reidel: Boston, 1980.

33. Fripiat, J. J. "Advanced Techniques for Clay Mineral Analysis"; Elsevier: Amsterdam, 1982.

34. Loeppert, R. H.; Hossner, L. R. Clays & Clay Minerals 1984, 32, 213-222.
35. Clarke, E. T.; Loeppert, R. H.; Ehrman, J. M. Clays & Clay Minerals 1985, 33, 152-158.
36. Dillard, J. G.; Koppelman, M. H.; Crowther, D. L.; Murray, J. W.; Balistrieri, L. In "Adsorption from Aqueous Solutions"; Tewari, P. H., Ed.; Plenum Publ. Corp.: New York, 1981; pp. 227-240.
37. Goldberg, S.; Sposito, G. Commun. Soil Plant Anal. 1985, 16, 801-821.
38. Rengasamy, P.; Oades, J. M. Aust. J. Soil Res. 1978, 16, 53-66.
39. Harsh, J. B.; Doner, H. E. Soil Sci. Soc. Am. J. 1984, 48, 1034-1039.
40. Carr, R. M. Clays & Clay Minerals 1985, 33, 357-361.
41. Brindley, G. W.; Brown, G. "Crystal Structures of Clay Minerals and Their X-ray Identification"; Mineralogical Society: London, 1980; Chap. 3.
42. Johnston, C. T. Ph.D. Thesis, University of California, Riverside, 1983.
43. Motschi, H. Colloids and Surfaces 1984, 9, 333-347.
44. (a) McBride, M. B. Clays & Clay Minerals 1982, 30, 21-28.
    (b) McBride, M. B. Clays & Clay Minerals 1982, 30, 200-206.
45. McBride, M. B.; Bouldin, D. R. Soil Sci. Soc. Am. J. 1984, 48, 56-59.
46. Harsh, J. B.; Doner, H. E.; McBride, M. B. Clays & Clay Minerals 1984, 32, 407-413.
47. McBride, M. B.; Fraser, A. R.; McHardy, W. J. Clays & Clay Minerals 1984, 32, 12-18.
48. Bleam, W. F.; McBride, M. B. J. Colloid Interface Sci. 1985, 103, 124-132.
49. Stumm, W.; Morgan, J. J. "Aquatic Chemistry"; John Wiley: New York, 1981; Chap. 9.
50. Sposito, G.; Prost, R. Chem. Rev. 1982, 82, 553-573.

RECEIVED June 18, 1986

ION EXCHANGE

12

Adsorption–Desorption Kinetics at the Metal–Oxide–Solution Interface Studied by Relaxation Methods

Tatsuya Yasunaga[1] and Tetsuya Ikeda[2]

[1]Institute of Science and Technology, Kinki University, Higashi-Osaka 577, Japan
[2]Department of Chemistry, Faculty of Science, Hiroshima University, Hiroshima 730, Japan

Chemical relaxation methods can be used to determine mechanisms of reactions of ions at the mineral/water interface. In this paper, a review of chemical relaxation studies of adsorption/desorption kinetics of inorganic ions at the metal oxide/aqueous interface is presented. Plausible mechanisms based on the triple layer surface complexation model are discussed. Relaxation kinetic studies of the intercalation/deintercalation of organic and inorganic ions in layered, cage-structured, and channel-structured minerals are also reviewed. In the intercalation studies, plausible mechanisms based on ion-exchange and adsorption/desorption reactions are presented; steric and chemical properties of the solute and interlayered compounds are shown to influence the reaction rates. We also discuss the elementary reaction steps which are important in the stereoselective and reactive properties of interlayered compounds.

The fast reactions of ions between aqueous and mineral phases have been studied extensively in a variety of fields including colloidal chemistry, geochemistry, environmental engineering, soil science, and catalysis (1-6). Various experimental approaches and techniques have been utilized to address the questions of interest in any given field as this volume exemplifies. Recently, chemical relaxation techniques have been applied to study the kinetics of interaction of ions with minerals in aqueous suspension (7). These methods allow mechanistic information to be obtained for elementary processes which occur rapidly, e.g., for processes which occur within seconds to as fast as nanoseconds (8). Many important phenomena can be studied including adsorption/desorption reactions of ions at electrified interfaces and intercalation/deintercalation of ions with minerals having unique interlayer structure.

In this paper, a review of the mechanistic information that has been obtained in chemical relaxation studies of reactions of ions with metal oxide minerals in aqueous suspensions is discussed. The

0097-6156/86/0323-0230$07.00/0

review is devided into five sections: (1) Principles of Chemical
Relaxation Method, (2) Experimental Methods and Materials, (3)
Surface Reaction Kinetics, (4) Intercalation Kinetics, and (5)
Summary and Conclusions.

Principles of the Chemical Relaxation Methods

Chemical relaxation methods involve small perturbations of equilibrium (8). The dependences of the equilibrium constant K on temperature, pressure, and the electric field are given by

$$\left(\frac{\partial \ln K}{\partial T}\right)_P = \frac{\Delta H}{RT^2} \tag{1}$$

$$\left(\frac{\partial \ln K}{\partial P}\right)_T = -\frac{\Delta H}{RT} \tag{2}$$

$$\left(\frac{\partial \ln K}{\partial E}\right)_{P,T} = \frac{\Delta M}{RT} \tag{3}$$

where ΔH, ΔV, and ΔM are the standard enthalpy, standard volume, and
electric moment of reaction. In jump-relaxation methods, the equi-
librium constant is changed by rapidly altering an external parameter (temperature, pressure, or electric field strength). The re-
equilibration or relaxation can be observed by some suitable means
of detection such as conductivity (Figure 1).

Experimental Methods and Materials

(a) Apparatus. Pressure-jump relaxation with electric conductivity
detection can be applied to reactions occuring on the order of
seconds to milliseconds. The details of the measurement of the
relaxation signals using the pressure-jump apparatus with conductivity detection have been described elsewhere (7). It has a time
constant of 80 μs at a bursting pressure of 200 atm.
    The electric-field-jump method with electric conductivity
detecting system can be applied to reactions occuring on the order
of milliseconds to microseconds. The rise time of the applied
electric field is much faster than 0.1 μs. The strength of the
electric field is 20 kV/cm. The details of the electric-field-jump
apparatus can be found elsewhere (9).
    Reactions which cannot be perturbed by changing an external
parameter may be detected by the stopped-flow method. The detection
system of this apparatus is the same as that of the pressure-jump
apparatus described previously (10). For this system, aqueous
electrolyte solution and an aqueous metal-oxide suspension are mixed
rapidly by operating an electric solenoid valve under nitrogen gas
of 7 atm. The dead time of this apparatus is 15 ms.
    The ζ-potential of the metal-oxide particles was measured by
means of the micro-electrophoresis method (11).

(b) Materials and Sample Preparation. The metal oxides used were
Fe_2O_3, Fe_3O_4, α-FeOOH, TiO_2, silica-alumina, and γ-Al_2O_3. Layered
compounds used were α- and γ-zirconium phosphates (α-, γ-ZrP), TiS_2,
hydrotalcite-like compound (HT), montmorillonite (Mont), zeolite 4A

(Z-4A), and zeolite H-ZSM-5. The interlayer distance varied by the intercalation was determined from X-ray diffraction patterns. The interlayer space of the crystalline zeolite is separated by the three-dimensional cage structures. The mean diameters of particles were approximately 1 μm. Such small particles formed very stable suspensions with no sign of sedimentation over the time course of the kinetic measurements. The analytical techniques used to obtain the equilibrium concentration are described elsewhere (10-22). All samples were equilibrated for 24-72 h after preparation. The temperature was controlled at 25 °C.

Surface Reaction Kinetics

In aqueous suspensions, metal oxides have amphoteric properties (4). One can describe the adsorption/desorption of H^+ and counter-ion on surface hydroxyl groups (SOH) using the following mass action equations:

in the pH range below the pH_{zpc}

$$SOH_2^+ \underset{k_{-1}}{\overset{k_1}{\rightleftharpoons}} SOH + H^+ \tag{I}$$

$$SOH_2^+ + A^- \underset{k_{-2}}{\overset{k_2}{\rightleftharpoons}} SOH_2^+\text{---}A^- \tag{II}$$

in the pH range above the pH_{zpc}

$$SOH \underset{k_{-3}}{\overset{k_3}{\rightleftharpoons}} SO^- + H^+ \tag{III}$$

$$SO^- + B^+ \underset{k_{-4}}{\overset{k_4}{\rightleftharpoons}} SO^-\text{---}B^+ \tag{IV}$$

where pH_{zpc}, A^-, and B^+ stand for the pH of the zero point charge of the metal oxide, an anion, and a cation, respectively, and k_i (i = 1, 2, 3, 4) are the rate constants. For the above reactions, the equilibrium constants are given by

$$K_1 = \frac{[SOH][H^+]_s}{[SOH_2^+]} \exp(\frac{e\psi_0}{k_B T}) = K_1^{int} \exp(\frac{e\psi_0}{k_B T}) \tag{4}$$

$$K_{anion} = \frac{[SOH_2^+\text{---}A^-]}{[SOH_2^+][A^-]_\beta} \exp(\frac{e\psi_\beta}{k_B T}) = K_{anion}^{int} \exp(\frac{e\psi_\beta}{k_B T}) \tag{5}$$

$$K_2 = \frac{[SO^-][H^+]_s}{[SOH]} \exp(\frac{e\psi_0}{k_B T}) = K_2^{int} \exp(\frac{e\psi_0}{k_B T}) \tag{6}$$

$$K_{cation} = \frac{[SO^-\text{---}B^+]}{[SO^-][B^+]_\beta} \exp(-\frac{e\psi_\beta}{k_B T}) = K_{cation}^{int} \exp(-\frac{e\psi_\beta}{k_B T}) \tag{7}$$

where the subscripts 0 and β refer to the planes of adsorbed H^+ and counterion (A^- and B^+), respectively. ψ_0 and ψ_β are the electrostatic potential at the surface and β-plane, respectively. K_i^{int} are the intrinsic equilibrium constants and k_B is the Boltzmann constant. According to the surface complexation model described by Davis et al. (4), the potential and charge relationships in the electric double layer are given by

$$\psi_0 - \psi_\beta = \sigma_0/C_1 \tag{8}$$

$$\psi_\beta - \psi_d = -\sigma_d/C_2 \tag{9}$$

$$\sigma_0 + \sigma_\beta + \sigma_d = 0 \tag{10}$$

where ψ_d is the mean potential at the slipping plane, which is assumed equal to the ζ-potential in these studies, and σ_0, σ_β, and σ_d are the charge density at the surface plane, the β-plane, and the slipping plane, respectively; C_1 and C_2 are the integral capacitances in the inner and outer regions of the immobile layer, respectively. From Gouy–Chapman diffuse layer theory, σ_d ($\mu C/cm^2$) is

$$\sigma_d = -11.74 \ C^{1/2} \ \sinh(\frac{e\psi_d}{2k_BT}) \tag{11}$$

where C is the equilibrium concentration for a symmetrical electrolyte in the bulk phase. The expressions of σ_0 and σ_β are

$$\sigma_0 = A([SOH_2^+] + [SOH_2^+ \text{---} A^-] - [SO^-] - [SO^- \text{---} B^+]) \tag{12}$$

$$\sigma_\beta = A([SO^- \text{---} B^+] - [SOH_2^+ \text{---} A^-]) \tag{13}$$

with

$$A = 10^6 F/([P]S) \tag{14}$$

where F is the Faraday constant, [P] is the particle concentration, and S is the specific surface area.

(a) Adsorption/Desorption Kinetics of Potential Determining Ions.
Application of the pressure-jump method to the acidic suspensions of metal oxides shifts equilibrium for reactions I and II towards dissociation. Under the condition pH $<$ pH$_{zpc}$, reactions III and IV above can be neglected because the counterion of SO^- is vanishingly small. Since no relaxation is observed in acid-free suspensions, homogeneous $HClO_4$ solution, or the supernatant solution of centrifuged samples, the single relaxation found in the pressure-jump studies has been attributed to proton adsorption/desorption on the suspended particles (12).

When counterion binding reaction II is extremely rapid, the concentration dependence of the relaxation time is given by (11)

$$\tau^{-1} = k_1^{int}\exp(-\frac{e\psi_0}{2k_BT})([SOH] + [H^+] + K_1\frac{K_{anion} + [SOH_2^+]}{K_{anion} + [SOH_2^+] + [A^-]}) \tag{15}$$

In order to verify Equation 15, one must determine the equilibrium concentrations of [SOH], [SOH$_2^+$], [H$^+$], and [A$^-$]. The surface concentrations are determined by potentiometric titrations while solute concentrations are measured using standard analytical techniques (12). The surface potential ψ_0 can be evaluated by using the following relation (4):

$$\frac{e\psi_0}{k_B T} = 2.303(pK_1^{int} - pK_1) \tag{16}$$

The value of K_{anion}^{int} for the perchlorate ion used in this study was not known, but numerical values of the constant for various other anions have been reported to be 50-100. Since τ^{-1} in Equation 15 is quite insensitive to the value of K_{anion}^{int} in this range, $K_{anion}^{int} = 50$ was used in the plot of τ^{-1} vs. the concentration term on the left hand side of Equation 15. The values of the rate constants obtained in the TiO$_2$, Fe$_2$O$_3$, Fe$_3$O$_4$, silica-alumina, and α- and γ-zirconium phosphate systems are listed in Table I.

Table I. Rate Constants of the Adsorption/Desorption
and Equilibrium Acidity Constants

system	k_1^{int} $\overline{\text{mol}^{-1} \text{ dm}^3 \text{ s}^{-1}}$	k_{-1}^{int} $\overline{\text{s}^{-1}}$	$-pK_1^{int}$
TiO$_2$/H$^+$	6.2×10^5	1.3×10^1	4.7
Fe$_2$O$_3$/H$^+$	2.4×10^5	1.6×10^{-1}	6.2
Fe$_3$O$_4$/H$^+$	1.4×10^5	3.4×10^{-1}	5.6
silica-alumina/H$^+$	2.9×10^4	4.6×10^1	2.8
α-ZrP/H$^+$	1.8×10^5	1.9×10^2	3.0
γ-ZrP/H$^+$	2.4×10^4	1.3×10^2	2.3

To compare the rate constants of the adsorption/desorption of H$^+$ in aqueous suspensions of the metal oxides, we plotted log k_1^{int} and log k_{-1}^{int} against pK$_1$ in Figure 2. In the systems of TiO$_2$, Fe$_2$O$_3$, Fe$_3$O$_4$, and α-zirconium phosphate, the values for log k_{-1}^{int} are related linearly to pK$_1$. This suggests that the intrinsic acidity of the metal oxide compounds is related to the desorption step. As has been pointed out by Astuminan and Schelly (13), due to reduction of dimensionality during adsorption, rate constants with values as low as 10^5 mol^{-1} dm^3 s^{-1}, may be indicative of diffusion control. Hence, the nearly constant value of the adsorption rate constant k$_1$ ($\sim 10^5$ mol^{-1} dm^3 s^{-1}) may be due to diffusion limitation. On the other hand the desorption step having a much higher activation (k$_{-1}$ << 10^5 mol^{-1} dm^3 s^{-1}) reflects differences in the chemical nature of the acid sites of minerals. However, in the silica-alumina and γ-zirconium phosphate systems, the adsorption rate constants are one order of magnitude smaller, indicating the possibility of a chemical reaction controlled adsorption step in these systems. This may be due to the higher acidity of the surface sites (see K^{int}) values in Table I) and hence higher activation energy of proton adsorption for the silica-alumina and γ-zirconium phosphate.

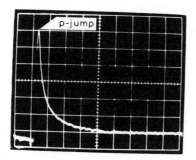

Figure 1.   Typical relaxation curve in the aqueous γ-Al_2O_3 – $Cu(NO_3)_2$ suspension observed by using the pressure-jump method. [P] = 30 g/dm^3, and I = 7.5×10^{-3} at 25 °C; sweep 20 ms/div.

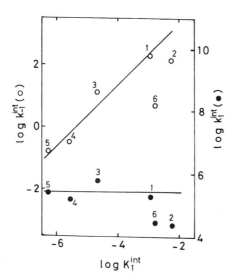

Figure 2.   Relationships between k_1^{int} ($\bullet$), k_{-1}^{int} (o), and acidity constant K_a^{int}:   (1) α-ZrP, (2) γ-ZrP, (3) TiO_2, (4) Fe_3O_4, (5) Fe_2O_3, (6) silica-alumina.

(b) Adsorption/Desorption Kinetics of Counterions. For the fast
process in the acidic suspensions of α-FeOOH - HCl and HClO$_4$ systems,
electric-field-jump studies were carried out (14). A single relaxation caused by the second Wien effect was observed. It has been
shown that the relaxation time for surface protonation reaction I
is on the order of milliseconds, and thus it may be assumed that the
reaction I cannot possibly contribute to relaxations of microsecond
order observed in these electric-field experiments. In reaction II,
it is noted that the interaction between the protonated surface
hydroxyl group and the counterion results in a decrease of electrostatic potential in the vicinity of surface and an increase in the
amount of proton adsorbed.

For the counterion binding reaction II, the concentration
dependence of the relaxation time is derived as (12)

$$\tau^{-1} = k_2^{int} \exp\left(\frac{e\psi_\beta}{2k_B T}\right)\{[SOH_2^+] + [A^-] + (K_{anion})^{-1}\exp\left(-\frac{e\psi_\beta}{k_B T}\right)\} \tag{17}$$

In Equation 17, ψ_β was calculated from the ζ-potential by using
Equations 5-11, $C_1 = 140$, and $C_2 = 20$ as reported by Davis et al.
(4). The value of ψ_0 was estimated from the acidity constant pK_1^{int}
and titration data using Equation 16. However, when the experimental values of τ^{-1} are plotted against the concentration term, the
result does not even remotely resemble a straight line through the
origin, and hence reaction II which has been suggested by Davis et
al. (4) may be eliminated from consideration. Taking into account
a possible, though somewhat more complicated explanation of the
surface reaction, i.e., assuming reaction II involves the diffusion
of an anion with three degrees of translational freedom to the β-plane
followed by the surface reaction IV as shown below, leads to

$$A^- + \text{surface} \rightleftharpoons A_\beta^- \quad \text{(diffusion process)} \tag{V}$$

$$SOH_2^+ + A_\beta^- \underset{k_{-2}'}{\overset{k_2'}{\rightleftharpoons}} SOH_2^+\text{---}A^- \tag{VI}$$

$$K_{anion}' = \frac{[SOH_2^+\text{---}A^-]\,[A^-]_\beta}{[SOH_2^+][A^-]_\beta\,[A^-]} = K_{anion}^{int}\, K_d \tag{18}$$

where K_d is the equilibrium constant of the diffusion reaction.
According to the rate theory for diffusion of an ion from bulk
solution to an oppositely charged particle surface (8), the
relaxation time is expected to be shorter than 0.1 μs. Thus the
relaxation time for reaction VI coupled with reaction V is given
by (14)

$$\tau^{-1} = k_2'^{int}\left(\frac{K_d}{K_d + 1}[SOH_2^+] + [A^-]_\beta\right) + k_{-2}'^{int} \tag{19}$$

where K_{anion}^{int} $(= k_2'^{int}/k_{-2}'^{int})$ was treated as a variable parameter. The values of the rate constants obtained in the α-FeOOH $-$ HCl and HClO$_4$ systems are listed in Table II.

Table II. Rate Constants of the Adsorption/Desorption of the Counterion

system	k_2^{int}	k_{-2}^{int}	K_{anion}^{int}
	mol^{-1} dm^3 s^{-1}	s^{-1}	mol^{-1} dm^3
α-FeOOH/Cl$^-$	6.0×10^5	1.0×10^4	60 (50)[a]
α-FeOOH/ClO$_4^-$	1.4×10^5	2.0×10^4	6

[a] The values reported by Davis et al. (4). K_{anion}^{int} taken equal to 60 mol^{-1} dm^3 for Cl$^-$ and 6 mol^{-1} dm^3 for ClO$_4^-$.

It should be noted that the value of K_{anion}^{int} in the α-FeOOH $-$ HCl system is in good agreement with that reported by Davis et al. (4). The K_{anion}^{int} shown in Table II indicate that the ion-pair surface complex of Cl$^-$ is one order of magnitude stronger than ClO$_4^-$.

Figure 3 shows the pH dependence for phosphate adsorption on γAl$_2$O$_3$ for a phosphate concentration of 8×10^{-3} mol dm^{-3}, at I= 1.5×10^{-2}M, [P] = 30 g dm^{-3}, and 25°C. It is well known that below the pH$_{zpc}$ in the absence of phosphate, both the surface- and ζ-potential in the γ-Al$_2$O$_3$ suspension are positive; however in the presence of phosphate, the surface is negatively charged at pH values below the zpc (pH = 8.5 indicating specific adsorption of phosphate (see Figure 4). A mechanism to explain phosphate adsorption based on pressure-jump studies has been postulated by Mikami et al. (15)

$$AlOH_2^+ + H_2PO_4^- \underset{k_{-1}}{\overset{k_1}{\rightleftharpoons}} AlOH_2^+ ---H_2PO_4^- \tag{VII}$$

$$AlOH_2^+ + HPO_4^{2-} \underset{k_{-2}}{\overset{k_2}{\rightleftharpoons}} AlOH_2^+ ---HPO_4^{2-} \tag{VIII}$$

where $AlOH_2^+ ---H_2PO_4^-$ and $AlOH_2^+ ---HPO_4^{2-}$ are the adsorbed states of $H_2PO_4^-$ and HPO_4^{2-}, respectively. Assuming that the adsorbed anions locate on the same plane as that of the adsorbed electrolyte counterion, K$_1$ and K$_2$ are defined as

$$K_1 = \frac{[AlOH_2^+ ---H_2PO_4^-]}{[AlOH_2^+][H_2PO_4^-]_\beta} \exp(\frac{e\psi_\beta}{k_B T}) = K_1^{int} \exp(\frac{e\psi_\beta}{k_B T}) \tag{20}$$

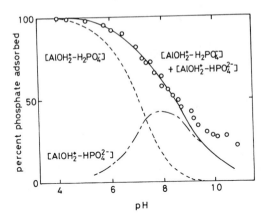

Figure 3. pH dependence of the amount of the phosphate adsorbed (o) at [P] = 30 g/dm^3, phosphate concentration = 8×10^{-3} mol dm^{-3}, I = 1.5×10^{-3}, and 25 °C. The solid curve indicates the sum of the calculated amounts of adsorbed $H_2PO_4^-$ (---) and HPO_4^{2-} (—·—).

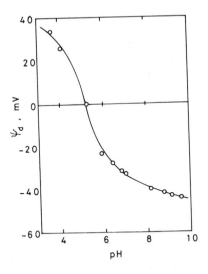

Figure 4. ζ-potential as a function of pH at I = 1.5×10^{-2} and 25 °C. Concentration of phosphate is 8×10^{-3} mol dm^{-3}.

$$K_2 = \frac{[AlOH_2^+ \text{---} HPO_4^{2-}]}{[AlOH_2^+][HPO_4^{2-}]_\beta} \exp(\frac{e\psi_\beta}{k_B T}) = K_2^{int} \exp(\frac{2e\psi_\beta}{k_B T}) \tag{21}$$

For reactions VII and VIII, the concentration dependences of the relaxation times are given by

$$\tau_1^{-1} = k_1^{int}\{\exp(\frac{e\psi_\beta}{2k_B T})\{[AlOH_2^+] + \frac{(K_2^{-1} + [AlOH_2^+])[H_2PO_4^-]}{K_2^{-1} + [AlOH_2^+] + [HPO_4^{2-}]}$$

$$+ (K_1^{int})\exp(-\frac{e\psi_\beta}{k_B T})\} \equiv k_1^{int} F_1 \tag{22}$$

$$\tau_2^{-1} = k_2^{int}\{\exp(\frac{e\psi_\beta}{k_B T})([AlOH_2^+] + [HPO_4^{2-}])$$

$$+ (K_2^{int})^{-1}\exp(-\frac{e\psi_\beta}{k_B T})\} \equiv k_2^{int} F_2 \tag{23}$$

To calculate the concentration terms in Equations 22 and 23, the intrinsic equilibrium constants K_1^{int} and K_2^{int}, ψ_β, and the equilibrium concentrations of surface species have to be obtained. These values were determined by solving a set of simultaneous equations, i.e., Equations 4-14, 20 and 21. In a series of calculations, the values, $K_{anion}^{int} = 80$, $K_{cation}^{int} = 50$, and $C_2 = 20$ $\mu F/cm^2$, which have been reported by Davis et al. (5), were used.

The plots of τ_1^{-1} vs. F_1 and τ_2^{-1} vs. F_2 in Equations 22 and 23 are shown in Figures 5 and 6, respectively, where the value of the inner layer capacitance C_1 was varied. Both plots yield straight lines passing through the origin at $C_1 = 180$. The rate constants of adsorption/desorption of phosphate and chromate on γ-Al$_2$O$_3$ are listed in Table III. The adsorption rate constants for $HCrO_4^-$ and CrO_4^{2-} are

Table III.   Rate Constants of Adsorption/Desorption
of the Phosphate and Chromate

system	k_1^{int} mol^{-1} dm^3 s^{-1}	k_{-1}^{int} s^{-1}	k_2^{int} mol^{-1} dm^3 s^{-1}	k_{-2}^{int} s^{-1}
γ-Al$_2$O$_3$ phosphate	4.1×10^5	2.3	1.1×10^7	2.7
γ-Al$_2$O$_3$ chromate	5.3×10^4	1.9×10^1	9.9×10^4	5.2×10^1

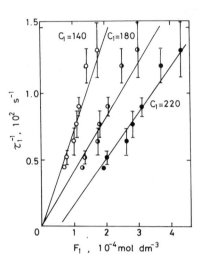

Figure 5. Plots of τ_1^{-1} vs. F_1 in Equation 22 using the various values of C_1 = 140 (o), 180 (ʘ), and 220 (●).

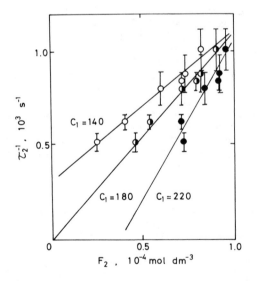

Figure 6. Plots of τ_2^{-1} vs. F_2 in Equation 23 using the various values of C_1 = 140 (o), 180 (ʘ), 220 (●).

one and two orders of magnitude smaller that those for $H_2PO_4^-$ and HPO_4^{2-}, respectively, while the values of desorption rate constant for the former are one order of magnitude larger than those for the latter. This indicates that the interaction of phosphate with $AlOH_2^+$ is stronger than that of chromate with $AlOH_2^+$. On the other hand, the value of $C_1 = 180$ determined is slightly larger than that reported by Davis et al. (5). This difference may result from an increase of the electrical double layer dielectric constant caused by the large amount of anion adsorbed.

(c) Adsorption/Desorption Kinetics of Divalent Metal Ions. Based on pressure-jump kinetic studies the mechanism of adsorption/desorption of Cu^{2+}, Mn^{2+}, Zn^{2+}, Co^{2+}, and Pb^{2+} on the γ-Al_2O_3 has been postulated (16,17)

$$
\text{AlOH} \underset{k_{-1}}{\overset{k_1}{\rightleftharpoons}} \text{AlOM(H}_2\text{O)}_{n-1}^+ \underset{k_{-2}}{\overset{k_2}{\rightleftharpoons}} \text{AlOMOH(H}_2\text{O)}_{n-2} \qquad (XI)
$$
$$
\text{M(H}_2\text{O)}_n^{2+} \qquad \text{H}^+ + \text{H}_2\text{O} \qquad \text{H}^+
$$

where $\text{AlOM(H}_2\text{O)}_{n-1}^+$ and $\text{AlOMOH(H}_2\text{O)}_{n-2}$ denote two kinds of surface complexes formed by adsorption of hydrated metal ions, $M(H_2O)^{2+}$. The dependences of pH and ζ-potential on the adsorbed amount of $M(H_2O)^{2+}$ at the total metal ion concentrations of 3×10^{-3} mol dm^{-3} are shown in Figures 7 and 8, respectively. The amount adsorbed for each M^{2+} increases with the pH, and the inflection points are shifted toward the lower pH region in the order of Co^{2+}, Zn^{2+}, Pb^{2+}, Cu^{2+}, which corresponds to the order of the hydrolysis constant of metal ions. To explain the M^{2+}-adsorption/desorption, Hachiya et al. (16,17) modified the treatment of the computer simulation developed by Davis et al. (4). In this model, M^{2+} binds coordinatively to amphoteric surface hydroxyl groups. The equilibrium constants are expressed as

$$
K_1 = \frac{[\text{AlOM(H}_2\text{O)}_{n-1}^+][\text{H}^+]}{[\text{AlOH}][\text{M(H}_2\text{O)}_n^{2+}]} = K_1^{int} \exp\left(-\frac{e\psi_0}{k_BT}\right) \qquad (24)
$$

$$
K_2 = \frac{[\text{AlOMOH(H}_2\text{O)}_{n-2}][\text{H}^+]}{[\text{AlOM(H}_2\text{O)}_{n-1}^+]} = K_2^{int} \exp\left(\frac{e\psi_0}{k_BT}\right) \qquad (25)
$$

M^{2+} is not adsorbed at the β-plane of adsorbed counterions but at the surface plane of adsorbed protons. The surface potential ψ_0 and surface charge density are defined by

$$
\psi_0 = \psi_M + \psi_H \qquad (26)
$$

$$
\sigma_0 = \sigma_M + \sigma_H
$$
$$
= \frac{F}{[P]S}([\text{AlOH}_2^+] + [\text{AlOM(H}_2\text{O)}_n^+]) \qquad (27)
$$

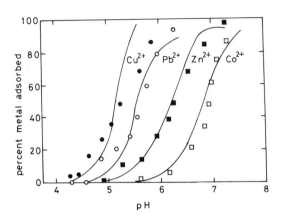

Figure 7. pH dependences of adsorption of divalent metal ions to γ-Al_2O_3 of $[P] = 30$ g/dm^3, the total metal ion concentration of 3×10^{-3} mol dm^{-3}, and $I = 7.5 \times 10^{-3}$ at 25 °C. The solid curves are the theoretical ones on the basis of the coordinate adsorption model.

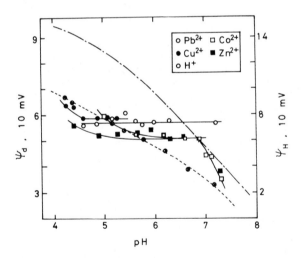

Figure 8. pH dependences of the ζ-potentials and ψ_H in the aqueous γ-Al_2O_3-$M(NO_3)_2$ and γ-Al_2O_3 - HNO_3 systems of $[P] = 30$ g/dm^3 and $I = 7.5 \times 10^{-3}$ at 25 °C. The solid and dashed curves show the ζ-potentials in γ-Al_2O_3 - $M(NO_3)_2$ and γ-Al_2O_3 - HNO_3 systems, respectively. The dash-dotted curve represents ψ_H.

where ψ_M and ψ_H are the surface potential created by M^{2+} and H^+-adsorptions, respectively; σ_M and σ_H are the corresponding charge densities. Using the above relationships, pH dependences of the amount of adsorbed M^{2+} were numerically evaluated with K_1^{int}, K_2^{int}, C_1, and C_2 treated as variable parameters. As can be seen from Figure 7, good model fit to data was obtained using the values of K_1^{int}, K_2^{int}, C_1, and C_2 shown in Table IV. From Figure 9a, one can see that the values of K_1^{int} and $K_1^{int}K_2^{int}$ correlate with the hydrolysis constants K_h of metal ions.

Table IV.   Intrinsic Constants and Capacitance Values
for Divalent Metal Ion Binding

system	pK_1^{int}	pK_2^{int}	C_1 $\mu F\ cm^{-2}$	C_2 $\mu F\ cm^{-2}$
$\gamma-Al_2O_3/Cu^{2+}$	1.6	6.6	250	40
$\gamma-Al_2O_3/Pb^{2+}$	1.8	7.6	200	40
$\gamma-Al_2O_3/Zn^{2+}$	2.4	8.6	200	30
$\gamma-Al_2O_3/Co^{2+}$	3.7	8.1	200	40
$\gamma-Al_2O_3/Mn^{2+}$	4.7	7.6	200	40

Considering the possible steady–state intermediate and mechanism shown below

$$AlOH \underset{k_{-1a}}{\overset{k_{1a}}{\rightleftharpoons}} AlO\ \underset{H_2O}{\overset{H}{M(H_2O)_{n-1}^+}} \underset{k_{-1b}}{\overset{k_{1b}}{\rightleftharpoons}} AlOM(H_2O)_{n-1}^+ \qquad (X)$$

$$M(H_2O)_n^{2+} \qquad H_2O \qquad H^+$$

and assuming $k_{-1a} \gg k_{1b}$, the relaxation time is given by

$$\tau^{-1} = k_{-1}'^{int}\{[AlOM(H_2O)_{n-1}^+] + [H^+]$$

$$+ K_1'^{int} \exp(-\frac{e\psi_0}{k_B T})([AlOH] + [M(H_2O)_n^{2+}])\} \qquad (28)$$

with

$$k_1'^{int} = k_{1a}^{int} \qquad (29)$$

$$k_{-1}'^{int} = k_{-1a}^{int} k_{-1b}^{int} / k_{1b}^{int} = k_{-1a}^{int} / K_{1b}^{int} \qquad (30)$$

$$K_1'^{int} = k_1'^{int} / k_{-1}'^{int} = K_{1a}^{int} K_{1b}^{int} \qquad (31)$$

where the k_i^{int} are rate constants accounting for interfacial potential affects as described by Mikami et al. (15) and Hayes and Leckie (this volume). Figure 9b shows that the values of k_{-1}^{int}

(= k_{1a}^{int}) obtained correlate with those for a release of a water
molecule from a hydrated metal ion, k_{H_2O}, in the metal complex
formation in homogeneous systems. This correlation is consistent
with the coordinative adsorption/desorption of M^{2+}. In addition,
Hachiya et al. (16,17) reported that γ-Al$_2$O$_3$ has at least two kinds
of the surface sites with different adsorption energies, i.e., one
is the strong surface sites of the largest fraction and the others
are the remaining weak sites of the small fraction among multisites.
Similar results for Pb^{2+} ion adsorption/desorption on goethite are
reported by Hayes and Leckie (this volume).

Intercalation Kinetics

Layered compounds such as montmorillonite and zirconium phosphate
intercalate adsorbing molecules into their two dimensional interlamellar spacing. Intercalation accompanies ion exchange in which
exchangeable ions are bound on the host lattice in the interlayer
and are exchanged with intercalating ions without large change of
the surface potential. Stereoselective and catalytic intercalation
processes are governed mainly by steric properties of the intercalating molecule which may depend on both the interlayer distance
and the size of the guest molecule. Recently, the mechanisms of
intercalation have been clarified by the authors using the chemical
relaxation method (18-22). Here we present the results on the
intercalation kinetics and discuss how differences in the stereoselectivity and acid strength affect intercalation phenomena.

Two chemical relaxation processes (involving decreasing
conductivity) having a characteristic duration of the order of
milliseconds were observed in α-ZrP - alkali metal ion, TiS$_2$ - alkali
metal ion, Z-4A - alkylammonium ion, HT - L- and D-histidines, and
Mont - L-arginine systems by using the pressure-jump method (18-22).
Single relaxations were observed in Z-4A - alkali metal ion system
and several H-ZSM-5 - NH$_4^+$ systems with silica-alumina ratios of 40,
80, and 160 using the stopped-flow and the pressure-jump methods,
respectively. The kinetic experimental data in each system show
that the relaxation times depend on the concentration of added
intercalating species. The static experimental data show that
intercalation of each ion into the lamellar layer of the layered
compound takes place and that in some systems ion-exchange reaction
may accompany intercalation.

For intercalation of molecules into the interlamellar layer
region, the reaction scheme is written as

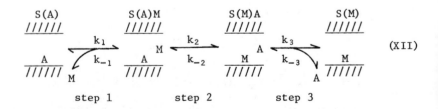

where S(A), S(A)M, S(M)A, S(M) denote the exchangeable ion, A, bound
in the interlayer; intercalating ion, M, adsorbed at the entrance of

the interlayer; M intercalated deeply by the intracrystalline exchange; and M intercalated with A released from the interlayer, respectively. Exchangeable ions in the interlayer of the layered compounds are H^+ for α-ZrP and H-ZSM-5, Na^+ for Z-4A and Mont, and Cl^- for HT. In the case of TiS_2, the interlayer is initially vacant. In typical intercalation, step 3 is the rate determining step, requiring 24–72 h for equilibration. Under the assumption that step 1 is much faster than step 2, the relaxation times are given by

$$\tau_1^{-1} = k_1([S(A)] + [M]) + k_{-1} \tag{32}$$

$$\tau_2^{-1} = k_2([M] + [S(A)]\ \frac{[S(A)] + [M]}{K_1 + [S(A)] + [M]}) + k_{-2} \tag{33}$$

The rate constants and interlayer distances determined from X-ray diffraction patterns for the intercalation studies described above are given in Table V. In those systems where intercalation causes large changes in the interfacial potential (ZrP and TiS_2), Equations 32 and 33 were modified using intrinsic rate constants. In cases where steady state reactive intermediates were postulated, the rate constants in Equations 32 and 33 were modified as shown in Table V.

(a) Stereoselective Intercalation. The values of rate constants and interlayer spacing shown in Table V yield important information with respect to the stereoselectivity of layered, channeled, and cage-structured minerals. In pressure-jump kinetic studies of the intercalation of alkali metal ions into α-ZrP, a strong correlation between the size of the metal ion and reaction rate was found (20). This is shown in Figure 10 where there is a linear relationship between $\log k_1$ and $\log k_2$ and the interlayer distance. The interpretation is that the alkali metal ion size determines the relative interlayer spacing in α-ZrP which in turn influences the rates of the intercalation steps 1 and 2 in the reaction scheme of Equation XII shown above. In studies of the intercalation of alkali metal ions into TiS_2 and Z-4A (10), the interlayer distance is less influenced by the intercalating ion and as a result the intercalation rates for each alkali metal ion are less dependent on the size of the intercalating ion (Table V). Apparently, the layered TiS_2 compound and the cage-structured Z-4A zeolite compounds are more rigid and their structure are less affected by the intercalating ions.

The effect of ion size on the rates of intercalation of NH_4^+, $CH_3NH_3^+$, $C_2H_5NH_3^+$, n-$C_3H_7NH_3^+$, i-$C_3H_7NH_3^+$, $(CH_3)_2NH_2^+$, $(CH_3)_3NH^+$, and $(CH_3)_4N^+$ into Z-4A zeolite has also been investigated using the pressure-jump technique (18). The results shown in Table V illustrate that Z-4A acts like an ion-sieve. For ions with a volume (estimated from van der Waals radii) greater than 120 $\mathring{A}^3$ no exchange was observed, i.e., neither relaxations or any ion-exchange was found in the intercalation studies of i-$C_3H_7NH_3^+$, $(CH_3)_3NH^+$, and $(CH_3)_4N^+$ (all whose estimated volumes are greater than 120 $\mathring{A}^3$). Figures 11 and 12 show the dependence of the ion-exchange and kinetic rate constants on the volume of the alkylammonium ions. As

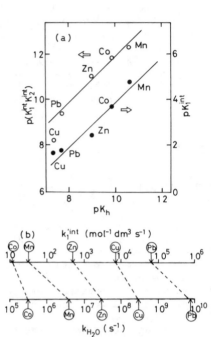

Figure 9. (a) Logarithmic plot of $K_1^{int}K_2^{int}$ (o) and K_1^{int} (•) vs. the hydrolysis constants of the metal ions, K_h. (b) Relationship between the adsorption rate constants, $k_1'^{int}$, and the rate constants for a release of water molecule from hydrated metal ion in the metal complex formation k_{H_2O}, for the divalent metal ions.

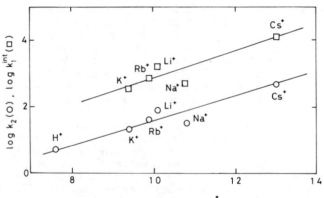

Figure 10. Plots of $\log k_1^{int}$ and $\log k_2$ vs. interlayer distance.

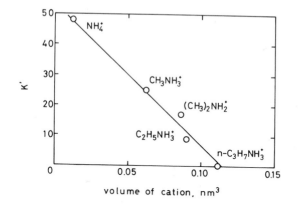

Figure 11.   Plot of K' vs. volume of cation. K' is the ion-exchange constant ($[M_{ads}][Na^+]/[S(Na)][M]$).

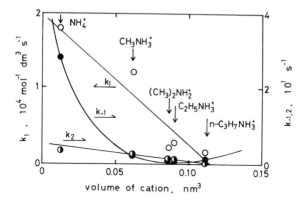

Figure 12.   Plots of k_1 (o), k_{-1} (•), and k_2 (o) vs. volume of cation.

Table V. Rate Constants and Interlayer Distance in Intercalation Studies [a]

system	k_1 $\overline{M^{-1} s^{-1}}$	k_{-1} $\overline{s^{-1}}$	k_2 $\overline{s^{-1}}$	k_{-2} $\overline{s^{-1}}$	k_3 $\overline{s^{-1}}$	k_{-3} $\overline{M^{-1} s^{-1}}$	interlayer distance Å	transition [b] state
α-ZrP [c]								
Li⁺/H⁺	1600	150	83	3.7			10.1	—
Na⁺/H⁺	3800	100	26	1.0		very slow	10.8	—
K⁺/H⁺	3300	55	18	0.7			9.4	—
Rb⁺/H⁺	400	220	36	2.0			9.9	—
Cs⁺/H⁺	1200	2400	550	30			13.0	—
TiS₂ [c]								
Li⁺	31000	120	13	11			8.6	—
Na⁺	13000	32	13	2.8			8.7	—
K⁺	8200	17	9.8	3.2			8.8	—
Rb⁺	5900	24	8.7	4.8			8.9	—
Cs⁺	6700	7.9	19	5.5			9.1	—
Z-4A								
Li⁺/Na⁺	—	—	—	—	3.7	310	2.2 [e]	S(A)M
K⁺/Na⁺	110	0.45					4.5 [e]	or
Rb⁺/Na⁺	140	1.0					4.5 [e]	S(M)A
Cs⁺/Na⁺	140	1.9					4.5 [e]	
HT								
L-his/Cl⁻	140	54			500	27000	7.8	S(A)M or
D-his/Cl⁻	98	25			500	27000	7.8	S(M)A

Table V (continued)

Z-4A						
NH_4^+/Na^+	18000	28	3.7	3.7	——	4.5 e
$CH_3NH_3^+/Na^+$	12000	1.8	2.3	——	——	4.5 e
$C_2H_5NH_3^+/Na^+$	2800	0.42	1.3	——	——	4.5 e
$n-C_3H_7NH_3^+/Na^+$	1600	1.3	0.28	——	——	4.5 e
$(CH_3)_2NH_2^+/Na^+$	2100	0.73	1.8	0.48	——	4.5 e
Mont						
L-arg	85000	580	200	220	——	15
H-ZSM-5						
NH_4^+/H^+(40) d	440	——	——	——	380	6.5 e S(A)M
NH_4^+/H^+(80) d	650	——	——	——	800	6.5 e and
NH_4^+/H^+(160) d	700	——	——	——	1800	6.5 e S(M)A

a The dash —— represents not applicable.

b When the steady-state approximation has been made, an intermediate appears in this column and the expression for the rate constants should be modified as follows: S(A)M or S(M)A; $k_{-1} \to k_{-1}k_{-2}/(k_2+k_{-1})$, or $k_3 \to k_2k_3/(k_{-2}+k_3)$; S(A)M and S(M)A; $k_1 \to k_1k_2/(k_2+k_{-1})$, $k_{-3} \to k_{-1}k_{-2}k_{-3}/(k_2k_3+k_{-1}k_{-2}+k_{-1}k_3)$, $k_1 \to k_1k_2k_3/(k_2k_3+k_{-1}k_{-2}+k_{-1}+k_3)$,

c Rate constants for these systems are intrinsic rate constants.

d Silica-alumina ratio.

e Aperture distance.

is apparent from these data, the smaller ions are able to intercalate more favorably into the solid phase. From Figure 12 it is
seen that the trend in the overall reaction is a function of both
steric factors and chemical properties of the exchanging ion. The
intercalation rate constants k_1 and k_2 (the intercalation step
through the aperture to an adsorption site and the subsequent ion-
exchange step) correlate with the cation volume, reflecting steric
effects. On the other hand, the rate constant k_{-1} (deintercalation
step) seems to correlate better with the pK_a value of the alkylammonium ion as shown in Figure 13, suggesting that the chemical
properties of the exchanging ion may also be important.

A possible explanation for the preference of living systems
for the L (levorotatory) over the D (dextrorotatory) optical isomer
may be associated with the stereoselective properties of layered
minerals. To test this hypothesis, the rates of L- and D-histidine
intercalation into HT layered compound was investigated using
the pressure-jump relaxation technique (21). The rate constants
and interlayer spacing based on this investigation are summarized
in Table V. As shown the slightly enhanced rate for L-histidine
suggests that relative chemical reactivity may be associated
with natural selection of the L-form of amino acids in nature.

(b) Catalytic Intercalation Phenomena. The ability of mineral
surfaces to catalyze hydrolysis reactions of organic contaminants
is a topic of current interest and discussed elsewhere in this
volume (see, e.g., Voudrias and Reinhard). The hydrolysis of L-
arginine to L-ornithine and urea, catalyzed by arginase, is well known.
However, this reaction takes place only very slowly in the absence
of a catalyst. The mechanism for the hydrolysis reaction in the
presence of montmorillonite using the pressure-jump technique was
determined to be as follows (19):

$$S(Na) \underset{k_{-1}}{\overset{k_1}{\rightleftarrows}} S(Na)Arg^{+-} \underset{k_{-2}}{\overset{k_2}{\rightleftarrows}} S(Na \cdot Orn \cdot Urea) \underset{k_{-3}}{\overset{k_3}{\rightleftarrows}} S(Na) \quad (XIII)$$

$$Arg^{+-} \qquad\qquad H_2O \qquad\qquad\qquad Orn \ or \ Urea$$

The rate constants from this study are given in Table V. The
kinetic data show that intercalation steps (steps 1 and 2) are
responsible for the observed catalyzed hydrolysis. The much slower
product release step (step 3), occuring over hours and not associated with the observed relaxations, is the rate limiting step.
This study suggests that interlayer spacing and molecular size of
contaminant may be important in determining the catalytic potential
of minerals in the environment.

(c) Effect of Acidic Properties on Intercalation Phenomena. In the
channels of the H-ZSM-5 zeolite, protons bound on the aluminosilicate
framework play an important role in their catalytic and acidic
activities (23). The catalytic and acidic properties of H-ZSM-5 in
turn depend on the silica-alumina ratio (23). The rate constants
obtained from the relaxation kinetic studies of exchange of NH_4^+ for
H^+ in H-ZSM-5 at three different silica-alumina ratios (40, 80, 160)
(22) are summarized in Table V. The mechanism that was proposed is
as follows:

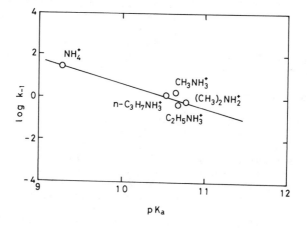

Figure 13.   Plot of $\log k_{-1}$ vs. pK_a.

$$S(H) \quad + \quad NH_4^+ \quad \underset{k_{-3}}{\overset{k_1}{\rightleftarrows}} \quad S(NH_4) \quad + \quad H^+ \qquad\qquad (XIV)$$

where the intermediates $S(H)NH_4^+$ and $S(NH_4)H^+$ in the reaction scheme of Equation XII are not required to interpret the kinetic results in this study, i.e., the relaxation observed was found to be result of the ion-exchange reaction. As shown in Table V, the interlayer spacings were the same for all silica-alumina ratios studied; hence the trend of increasing values of k_1 and k_{-3} corresponds to the increasing acidity (more silica) in going from a silica-alumina ratio of 40 to 160, and is directly related to the acidity of the ion-exchange site. This implies that in systems where the inter-calating ion is not appreciably sterically hindered (e.g., NH_4^+) the chemical properties of the ion-exchange site play an important role in the relative reactivity of the minerals.

Summary and Conclusions

In this review we have shown that chemical relaxation methods can be successfully applied to study the dynamics of reactions occuring at the mineral/water interface. For reactions of inorganic ions with charged, essentially nonporous metal oxides, electrical double layer theory and equilibrium and rate data can be combined to obtain plausible reaction mechanisms. Mechanisms of adsorption/desorption of proton, electrolyte counterion, and specifically adsorbing anions and cations have been discussed. Relaxation kinetic studies of intercalation/deintercalation and ion-exchange of inorganic and organic ions with minerals having predominately interlayer surface have also been reviewed. The mechanistic information obtained has illustrated that steric and chemical properties of ions and minerals have varying degrees of influence on the rates of reaction. Minerals with stereoselective, catalytic and ion-sieve properties have been discussed.

Acknowledgments

The authors are grateful to Kim F. Hayes at Stanford University for critical reading of the manuscript and for helpful discussion.

Literature Cited

1. Allen, L. H.; Matijevic, E.; Meites, L. J. Inorg. Nucl. Chem.
    1971, 33, 1293-1299.
2. Huang, C. P.; Stumm, W. J. J. Colloid Interface Sci. 1973, 43,
    409-420.
3. James, R. O.; Healy, T. W. J. Colloid Interface Sci. 1972, 40,
    42-52, 53-64, 65-81.
4. Davis, J. A.; Leckie, J. O. J. Colloid Interface Sci. 1978,
    67, 90-107.
5. Davis, J. A.; Leckie, J. O. J. Colloid Interface Sci. 1980,
    74, 32-43.

6. Whittinham, M. S.; Jacobson, A. J., Eds., "Intercalation Chemistry"; Academic Press: New York, 1982.
7. Hachiya, K.; Ashida, M.; Sasaki, M.; Kan, H.; Inoue, T.; Yasunaga, T. J. Phys. Chem. 1979, 83, 1866–1871.
8. Bernasconi, C. F. "Relaxation Kinetics"; Academic Press: New York, 1976.
9. Tsuji, T.; Yasunaga, T.; Sano, T.; Ushio, H. J. Am. Chem. Soc. 1976, 98, 813–818.
10. Ikeda, T.; Nakahara, J.; Sasaki, M.; Yasunaga, T. J. Colloid Interface Sci. 1984, 97, 278–283.
11. Dolzhenkova, A. N.; Gevorkyan, B. A.; Vishnyakova, G. V. Obogasch. Rud. (Leningrad), 1973, 18, 31.
12. Astumian, R. D.; Sasaki, M.; Yasunaga, T.; Schelly, Z. A. J. Phys. Chem. 1981, 85, 3832–3835.
13. Astumian, R. D.; Schelly, Z. A. J. Am. Chem. Soc. 1984, 106, 304–308.
14. Sasaki, M.; Moriya M.; Yasunaga, T.; Astumian, R. D. J. Phys. Chem. 1983, 87, 1449–1453.
15. Mikami, N.; Sasaki, M.; Hachiya, K.; Astumian R. D.; Ikeda, T.; Yasunaga, T. J. Phys. Chem. 1983, 87, 1454–1458.
16. Hachiya, K.; Sasaki, M.; Saruta, Y.; Mikami, N.; Yasunaga, T. J. Phys. Chem. 1984, 88, 23–27.
17. Hachiya, K.; Sasaki, M.; Ikeda, T.; Mikami, N.; Yasunaga, T. J. Phys. Chem. 1984, 88, 27–31.
18. Ikeda, T.; Sasaki, M.; Yasunaga, T. J. Phys. Chem. 1983, 87, 745–749.
19. Ikeda, T.; Yasunaga, T. J. Phys. Chem. 1984, 88, 1253–1257.
20. Mikami, N.; Sasaki, M.; Yasunaga, T.; Hayes, K. F. J. Phys. Chem. 1984, 88, 3229–3233.
21. Ikeda, T.; Amoh, H.; Yasunaga, T. J. Am. Chem. Sco. 1984, 106, 5772–5775.
22. Ikeda, T.; Yasunaga, T. J. Colloid Interface Sci. 1984, 99, 183–186.
23. Kokotailo, G. T.; Lowton, S. L.; Olson, D. H.; Meier, W. M. Nature (London), 1978, 272, 437–438.

RECEIVED June 18, 1986

13

Highly Selective Ion Exchange in Clay Minerals and Zeolites

A. Maes and A. Cremers

K. U. Leuven, Laboratorium voor Colloïdchemie, Kard. Mercierlaan 92, B-3030 Leuven, Belgium

The charge density of the clay mineral and the polarizability difference between the exchanging pair of ions are the two important factors governing the ion exchange behaviour of alkali, alkaline earth and organic ammonium cations. Highly selective ion exchange behaviour is discussed in terms of these parameters. It can essentially be ascribed to high enthalpic contributions resulting from enhanced electrostatic interactions. Complexation of transition metal ions with uncharged ligands leads to significant enhancements – up to three orders of magnitude – of ion selectivities in smectites and these effects prevail over the entire range of surface compositions. Similar enhancements in selectivity occur upon complexation in the rigid zeolite pores but are of a lesser extent. The extremely high Cs-selectivity observed in a small fraction of the charge of illites, illite-smectite interlayers and reduced charge montmorillonites, is a thermodynamically reversible ion exchange process and is discussed in terms of a multi-site ion exchange model. In zeolites, the combination of a typical crystallographic configuration and cation properties may in certain cases (Ag, Na) lead to high selectivities in a limited number of exchange sites.

The distribution of elements between the solid and the liquid phase is of primary importance for the transport processes in the environment. In addition, the uptake of elements in plants and other living organisms is determined by the speciation of the element in that phase.

The distribution of the major elements (Ca, Mg, Na, K, ...) in soils is well known to be governed by ion-exchange processes (1). The behaviour of transition elements such as Co, Ni, Cd, Cu, etc. in natural systems (soils, sediments) often results from a combination of different effects such as precipitation, sorption in oxides, exchange in clay minerals and complexation with organic

0097-6156/86/0323-0254$11.50/0
© 1986 American Chemical Society

matter. These effects are largely influenced by the pH and redox conditions (2-3). The binding capacity and binding strength by ion exchange in clay minerals is considered to be weak relative to other metal associations under oxidizing conditions, and negligible under reducing conditions (2-3). In pure clay minerals and at sufficiently low pH ion exchange mechanisms are involved which are ruled by selectivity patterns comparable to the exchange of the major elements. With increasing pH a gradual shift occurs form predominant ion exchange towards a specific adsorption behaviour on 'oxide-like' surfaces involving either broken bonds at clay mineral edges or oxide coatings (4).

Extremely high ion exchange affinities are however sometimes observed for alkali metals (e.g. Cs) and transition metal ion complexes in clay minerals and zeolites. The objective of this paper is to give an account of the factors which are involved in these high selectivity phenomena. The discussion will be focussed mostly on montmorillonites and faujasites as representatives of the phyllosilicate and tectosilicate groups.

Thermodynamic background

The free energy change for the general cation exchange reaction between A^{z_A} and B^{z_B} cations on an exchanger X^-, represented by

$$z_B A^{z_A} X + z_A B^{z_B} \rightleftarrows z_A B^{z_B} X + z_B A^{z_A}$$

is written as

$$\Delta G^o_{ex} = (z_A \bar{\mu}^o_B - z_B \bar{\mu}^o_A) - (z_A \mu^o_B - z_B \mu^o_A) \qquad (1)$$

and is governed by the difference in standard chemical potentials of both exchanging species in the surface (indicated by a bar) and solution phase.

From a thermodynamic viewpoint the theoretical framework (5-6) and the experimental measurements of such equilibria are well established. Two approaches are currently in use. They only differ in the definition of the ion activity at the surface, which is either expressed as an equivalent fraction (5) or as a molar fraction (6-8).

Both approaches lead to identical standard thermodynamic values of exchange (9-10). Such a difference in the choice of the surface concentration scale is of course only important for heterovalent exchange equilibria. For the heterovalent case the numerical value for both selectivity coefficients, K^c_G (Gaines & Thomas) and K^c_V (Vanselov) differ and, consequently, their variation with surface composition also differs.

Although both methods are thermodynamically equivalent it appears that K^c_V remains independent of surface composition for exchanges of Na^{+1} versus Ca^{+2} (11-12), versus Co^{+2}, Cu^{+2}, Ni^{+2}, Cd^{+2} (13) and versus Cu^{+2} (14) in montmorillonites. In contrast K^c_G values increase with occupancy in all these equilibria.

Straightforward thermodynamic data relating to the exchange

of these ions are only obtained from measurements in ClO_4^- media. In the case of Cu and Ca exchange in Wyoming Bentonite (14-15), it was shown that ion pair formation with Cl^- or NO_3^- may lead to higher selectivities compared to values obtained in ClO_4^-, caused by the selective adsorption of the $CuCl^+$ or $CaCl^+$ ion-pairs. Since most ion exchange data were obtained in Cl^- or NO_3^- media they may be somewhat inaccurate due to the presence of a third (ion-pair) component.

Exchange in clay minerals

Influence of interlayer charge density and interlayer hydration

Rather small selectivity differences are observed for homovalent- and heterovalent exchanges involving alkali, alkaline earth, bivalent transition metal ions, aluminium and rare earth cations, as is amply evidenced from the extensive compilation by Bruggenwert and Kamphorst (16). This compilation includes various clay minerals : illite, montmorillonite, vermiculite and kaolinite.

Homovalent exchange of inorganic cations. The difference in ΔG^o_{ex} among the monovalent cations ranges from about 0.6 (Na-Li) to about 10 (Na-Cs) kJ Eq^{-1} in a typical montmorillonite from Camp Berteau. Differences observed among the various clay minerals (illites, kaolinites, vermiculites, montmorillonites) are ascribed to differences in mineral charge density and/or to the mineralogical heterogeneity or purity. The charge density of the mineral is undoubtedly a very important parameter. This is unequivocally verified in the Na-Cs exchange (17) measured on an isostructural series of montmorillonites obtained by charge reduction of Camp Berteau montmorillonite using the Hofmann-Klemen effect (18). Fig. 1a shows that exchange data on several montmorillonites follow the reduced charge montmorillonite series (R.C.M.). Small deviations in Wyoming Bentonite and Chambers montmorillonite may be ascribed to the influence of tetrahedral charge substitution (19) and/or to the presence of micaceous components (20).

In general the thermodynamic functions of exchange for different pairs increases with increasing difference in hydration energy of the ions (See table 1).

Focusing, on the Na-Cs pair, the ΔG^o_{ex} is less pronounced with decreasing charge density and tends to vanish at zero charge density, corresponding to a tendency of equal differences in surface and solution terms in eq. (1). This situation is possible if the hydration status of the adsorbed cations tends to equal that of solution cations. It follows therefore that the action of forces that tend to dehydrate the interlamellar cations such as the increase in charge density of the mineral or the increase in electrolyte concentration (32), enhance the selectivity of the least hydrated cation.

The observation that the cation with the smallest hydration energy is increasingly preferred with increasing charge density is

TABLE I. Thermodynamic data for alkali metal cation exchange at 25 °C (in kJ equiv^{-1}) in Chambers, Camp Berteau and Wyoming Bentonite

	Li/Na	Na/K	Na/Rb	Na/Cs
Chambers[25]				
1.2 mequiv g^{-1}				
$-\Delta G^o$	0.33	3.03	5.63	7.88
$-\Delta H^o$	0.46	4.84	8.03	11.08
$-T\Delta S^o$	0.13	1.81	2.40	3.20
Camp Berteau[30]				
1 mequiv g^{-1}				
$-\Delta G^o$	0.58	3.47	8.11	10.40
$-\Delta H^o$	-0.58	4.51	9.57	16.05
$-T\Delta S^o$	-1.16	1.04	1.04	5.65
Camp Berteau[21]				
1 mequiv g^{-1}				
$-\Delta G^o$				8.65
$-\Delta H^o$				18.20
$-T\Delta S^o$				9.55
Wyoming Bentonite[31]				
0.8 mequiv g^{-1}				
$-\Delta G^o$	0.20	1.28	2.65	4.50
$-\Delta H^o$	0.63	2.53	6.30	10.70
$-T\Delta S^o$	0.43	1.25	3.65	6.20

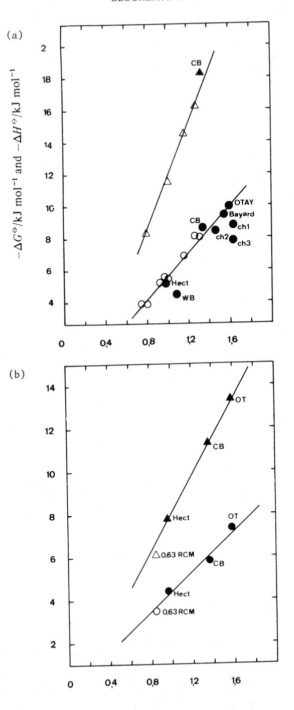

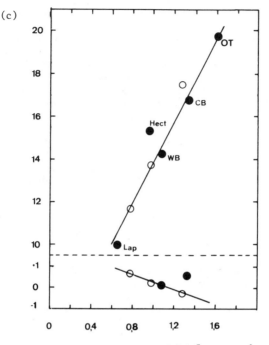

(c)

$\Gamma/10^{-7}$ meq cm^{-2}

Figure 1.   Variation of the ΔG^o_{ex} (circles) and ΔH^o_{ex} (triangles) (in kJ mol^{-1}) with the charge density of the clay mineral (in meq. cm^{-2}) for

a) the Na$^+$ → Cs$^+$ exchange in different RCM's (empty symbols) (17) and clays as indicated, viz. Otay, Hectorite (17), Camp Berteau (21), Bayard (22), Wyoming Bentonite (23), Chambers montmorillonite (ch1, (24); ch2, (22); ch3, (25)).

b) the Ca^{+2} → enH$_2^{+2}$ exchange in Otay, Camp Berteau, Hectorite and 0.63 RCM (Reduced charge montmorillonite retaining 63 % of it's original charge) (26).

c)the Ca^{+2} → Cu(en)$_2^{+2}$ (upper part) and Ca^{+2} → Cu^{+2} (lower part) exchange in different RCM's (empty circles) (27) and clays as indicated, viz. Otay (27), Hectorite (27), Laponite (27), Camp Berteau (28) and Wyoming Bentonite (29). Reproduced with permission from Ref. 17, 26 and 27. Copyright 1978, 1981 and 1983, The Chemical Society.

semi-quantitatively predicted (19) from double layer theory corrected for hydration effects (33-34) by incorporating the potential energy difference between hydrated and unhydrated cations.

Eberl (35) predicted the opposite variation of ΔG^o_{ex} (Na → Cs) with charge density in both the dry (Eisenmann model (36)) and the wet interlayer (Cruickshank model (37)) as shown in fig. 2. By plotting the experimental data of Maes and Cremers (17) in fig. 2 it is again demonstrated that the decreasing hydration status of the interlamellar cation (increasing interlayer molality) with increasing charge density corresponds with increasing ΔG^o_{ex} (Na → Cs).

The high hydration energies of the bivalent and trivalent cations makes them more reluctant to dehydrate in the interlamellar region at these charge densities in montmorillonite and consequently only small ΔG^o_{ex} (A^{2+} → B^{2+}) values are expected and observed for alkaline earth (32,38) and bivalent transition metal ions (39).

The variation of bivalent-bivalent exchanges with charge density are expected to be similar to, but smaller (27) than monovalent-monovalent exchanges.

Exchange of organic ammonium cations. Exchange selectivity of monovalent alkyl ammonium cations in montmorillonites (40-41) and octahedrally substituted synthetic clay minerals (laponite) increases with their chain length (42) and along the series primary < secondary < tertiary < quaternary ammonium cations. This selectivity rise is ascribed to the lowering of the alkylammonium ion hydration upon adsorption in the interlamellar space. This relationship was shown by the linear correlation between the ΔG^o_{ex} and the gas phase basicity of the amines (42-43). The gas phase basicity refers to the free energy of interaction of a proton and the amine in the gas phase (44-45). The gas phase basicity itself is logically correlated with the electron density distribution on the amine head group under varying conditions of chain length and replacement of hydrogen in the headgroup (45) without any interfering effect of hydration forces.

The true gas phase basicity order is observed in the montmorillonite (43), whereas in solution the well known amine anomaly exists, i.e. the expected inductive effects of the organic carbon chain are screened off by the solvent. For example, identical ΔG^o values of protonation are found in solution for methyl- and butylammonium.

The inductive effect of the carbon chain in the clay phase amounts to (only) 5 to 7 % of the effect in the gas phase. Ammonium cations in the interlamellar region of clay minerals are therefore less hydrated than in equilibrium solution. The free energy of alkylammonium exchange increases with charge density from Laponite (42) < Red Hill montmorillonite (40) < Camp Berteau montmorillonite (41) in line with the smaller interlamellar hydration status of the adsorbed cation at higher charge density.

The exchange of Ca-ethylenediammonium in a series of montmorillonites (26) shows ΔG^o_{ex} values which increase with charge

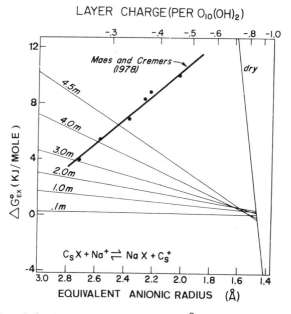

Figure 2. Calculated variation of ΔG^o_{ex} (Na → Cs) (kJ mol^{-1}) with the equivalent anionic radius (Å) at different interlayer molalities (35). Data of Maes and Cremers (17) are also shown. (Figure supplied by D.D. Eberl). Reproduced with permission from Ref. 35. Copyright 1980, The Clay Minerals Society.

density increase (see fig. 1b) similar to the Na-Cs exchange. Relatively high ΔG^o_{ex} values are observed as expected from the small hydration energy of one of the exchanging cations (ethylenediammonium). The decreasing and small d_{001} spacing with charge density increase (table II) are in line with the preceding picture of smaller interlamellar alkylammonium hydration at higher charge density.

Correlation of exchange data with electron density distribution in organic ammonium cations indicate that spreading of the charge (delocalization) over increased volumes leads to higher selectivities. This was shown for alkylammonium cations (see table III) and bisquaternary ammonium cations (47) of different chain length. The thermodynamic data for the exchange (ΔG^o, ΔH^o, ΔS^o) among alkali- and alkaline earth metal cations are linearly related to the polarizability difference between both exchanging cations (38). This observation fits in nicely with the concept of charge delocalization.

Heterovalent exchange. Although double layer theory predicts increasing ΔG^o_{ex} of mono-bivalent exchange with charge density increase, such a relationship was not always obvious (20,48).

By using an isomorphic series of reduced charge montmorillonites, the variable parameter charge density can be isolated. It was shown (46) that ΔG^o_{ex} (Na → Ca) in such a series of samples increases linearly with the logarithm of the charge density. The experimental free energy loss is less dependent on charge density than double layer theory predicts. This might well be due to the involvement of Ca-NO$_3^+$ ion pairs (15), which are indeed expected to contribute more to the total free energy change with decreasing exchange capacity.

The small discrepancy between double layer predictions of ΔG^o_{ex} and ΔH^o_{ex} values, assuming interactions of point charges, and the experimental observations were ascribed to the partial dehydration of the Ca^{2+}- exchanged form, which is more pronounced at higher charge density as verified by the decreasing d_{001} spacing (table II) and by the increasing endothermicity and entropy of exchange.

Goulding and Talibudeen (20) observed the reverse relationship with charge density for the K → Ca exchange in that the selectivity increased from New Mexico montmorillonite (1.35 meq/g) < Camp Berteau (1.15 meq/g) < Red Hill (1.12 meq/g) ≅ Wyoming bentonite (1.0 meq/g). The selectivity order is governed by the decreasing mica and hydrous mica character of the different montmorillonites. The contribution of mica and hydrous mica interlayers is obtained from measurements of the differential enthalpy of K-Ca exchange (see later). Non-specific coulombic interactions (of point charges) are expected to hold irrespective of the charge origin and density. Other factors are therefore involved. Detailed analysis of the thermodynamic data (ΔH and ΔS terms) shows that specific interactions between the cations and the surface as well as entropic factors determine the overall selectivity and its variation with surface composition (20).

TABLE II. d_{001} spacings (nm) of concentrated suspensions of different clay minerals and reduced charge Camp Berteau montmorillonites (RCM) fully exchanged with Ca^{+2}, Cu^{+2}, $Cu(en)_2^{+2}$ and ethylenediammonium cations

	0.59 RCM	0.63 RCM	Hectorite	0.74 RCM	Wyoming bentonite	0.95 RCM	Camp Berteau	Otay	Reference
NaCEC (meq g^{-1})	0.59	0.63	0.72	0.74	0.82	0.95	1.00	1.21	(26,46)
CaCEC (meq g^{-1})	0.68	0.74	0.85	0.86	0.84	1.08	1.10	1.30	(26,46)
Ca^{+2}	2.01-2.05	-	2.01	-	-	-	1.86-1.88	1.86	(46)
Cu^{+2}	-	-	2.27	2.34	-	-	2.00	1.97	(27)
$Cu(en)_2^{+2}$	1.52	-	1.48	1.47	1.50	-	1.33	1.31	(27)
enH_2^{+2}	-	1.63	1.62	-	-	-	1.60	1.55	(26)

TABLE III. Comparison of the Gibbs energies of Na$^+$-alkylammonium exchange in different clay minerals with their charge distribution from CNDO (complete neglect of differential overlap) calculations

	Total charge on NH$^+$ group Ref. (45)	ΔG^o_{ex} versus Na$^+$ (kJ mol^{-1})		
		Laponite(42) (0.7-0.8 meq g^{-1})	Red Hill(40) (0.9 meq g^{-1})	Camp Berteau(41) (1.0 meq g^{-1})
Methylamine	0.72	+3.13	-1.76	-3.01
Ethylamine	0.69	+1.88	-1.96	-3.51
Propylamine	0.68	+0.83	-2.22	-3.68
Butylamine	0.67	+0.87	-2.84	-3.97
Methylamine	0.72	+3.13	-1.76	-3.01
Dimethylamine	0.49	+1.01	-2.84	
Trimethylamine	0.30	-0.58	-4.81	
Tetramethylamine	-	-	-5.02	

<u>Cation selectivity dependency on loading</u>. The mass action selectivity coefficient is generally found to depend on the ionic composition of the exchanger (<u>49</u> and references therein) and on the ionic strength of the solution (<u>32</u>). The variation of the $K_c(M_1^{+n} \rightarrow M_2^{+n})$ with surface composition is related to the difference in the hydration status of the exchanging cations in that the preference for the least hydrated cation decreases with its increase in occupancy. This dependency is more pronounced with increasing difference in hydration energy of the ions. In heterovalent exchange the selectivity for the ion with the highest charge increases with its occupancy. The functional dependency on surface composition is more pronounced for heterovalent (<u>30</u>, <u>50-54</u>) than for homovalent exchanges (<u>12</u>, <u>24-25</u>, <u>30</u>).

    Several factors have been invoked to explain the variation with surface composition. Heterogeneity of exchange sites is an important factor in view of the charge density dependency of the thermodynamic functions of exchange (<u>17</u>, <u>26-27</u>) and the measured heterogeneity of surface charge density distribution (<u>55-58</u>). The differential heat of exchange also decreases with loading in homovalent (<u>17</u>) and heterovalent (<u>20</u>) exchanges. In the latter case the variation was ascribed to the presence of a mixture of different clay minerals. Demixing (<u>54</u>, <u>59</u>), tactoid formation (<u>11</u>, <u>60-61</u>), and layer stacking, which is influenced by charge heterogeneity and density (<u>55</u>, <u>62-63</u>), are phenomena related to the hydration status of the exchanging cations and are a source of variation of K_c with surface composition.

    McBride (<u>49</u>, <u>64</u>) proposes entropy changes due to site localization of the exchanging cations to explain the selectivity variation of inorganic cations with composition. Exchange among complex cations (Ag(ethylenediamine)$_2$ $\rightarrow$ Cu(ethylenediamine)$_2$) in Wyoming bentonite (<u>29</u>), in which case the interlamellar space is collapsed, shows a pronounced but opposite variation with surface composition and is related to changes in electrostatic interactions, due to the heterogeneous surface charge distribution.

Exchange of cations of high polarizability

The former experimental observations allow one to make predictions on the conditions necessary to obtain highly selective exchange behaviour in two cases :
1. By increasing the charge delocalization in one of the exchanging cations;
2. By increasing the charge density of the mineral.

<u>Exchange of complex cations</u>. Complexation of transition metal cations with uncharged ligands such as with amines and with amino acids results in a selectivity enhancement compared to the selectivity of the aqueous metal cation (<u>27</u>, <u>65-72</u>). Fig. 3 shows an example for the Cu(ethylenediamine)$_2$ adsorption in montmorillonites of different charge density. Standard thermodynamic data for other cases are given in table IV. In all cases the free ligand concentration in equilibrium solution was

TABLE IV.   A. Thermodynamic data for the exchange of transition metal ions and transition metal ion complexes
     B. The stabilization factor ($\log \bar{\beta}_n/\beta_n$) for the formation of the indicated complexes in the clay interphase. All data refer to 25 °C and 0.01 total normality.

Reaction	Clay mineral	LnK	A ΔG^0_{ex} (kJ Eq^{-1})	Complex	B $\log \bar{\beta}_n/\beta_n$	Ref.
$Na^{+1} \rightarrow Ag^{+1}$	C.B.[a]	-0.24	0.60			(65)
$Na^+ \rightarrow Ag(thiourea)^{+1}_n$	C.B.	8.10	-20.10	$Ag(TU)^{+1}_n$	3.63	(65)
$Na^{+1} \rightarrow Ag^{+1}$	W.B.[b]	0.4	-0.99			(74)
$Na^{+1} \rightarrow Ag(Pyridine)^{+1}_2$	W.B.	7.46	-18.40	$Ag(Py)^{+1}_2$	3.07	(74)
$Cs^{+1} \rightarrow Ag^{+1}$	W.B.	-2.05	5.06			(29)
$Cs^{+1} \rightarrow Ag(en)^{+1}_2$	W.B.	4.60	-11.35	$Ag(en)^{+1}_2$	2.90	(29)
$Ca^{+2} \rightarrow Cu^{+2}$	W.B.	-0.04	0.05			(73)
$Ca^{+2} \rightarrow Cu(en)^{+2}_2$	W.B.	5.80	-7.15	$Cu(en)^{+2}_2$	2.54	(29)
$Ag^{+1} \rightarrow Cu^{+2}$	W.B.	0.78[c]	-0.96[c]			(29)
$Ag(en)^{+1}_2 \rightarrow Cu(en)^{+2}_2$	W.B.	-7.84	9.67			
$Ca^{+2} \rightarrow Cu^{+2}$	C.B.	0.25	-0.32			(28)
$Ca^{+2} \rightarrow Cu(en)^{+2}_2$	C.B.	7.00	-8.58	$Cu(en)^{+2}_2$	2.95	(28)
$Ca^{+2} \rightarrow Ni(en)^{+2}_3$	C.B.	6.65	-8.25	$Ni(en)^{+2}_3$	2.78	(75)

$Ni(en)_3^{+2} \rightarrow Zn(en)_3^{+2}$	C.B.	-1.05	$+1.30$	$Zn(en)_3^{+2}$	2.32	(75)
$Ni(en)_3^{+2} \rightarrow Cd(en)_3^{+2}$	C.B.	1.59	-2.00	$Cd(en)_3^{+2}$	3.47	(75)
$Ni(en)_3^{+2} \rightarrow Hg(en)_2^{+2}$	C.B.	2.35	-2.93	$Hg(en)_{2-3}^{+2}$	3.80	(75)
$Zn(en)_3^{+2} \rightarrow Cd(en)_3^{+2}$	C.B.	2.55	-3.18			(75)
$Ca^{+2} \rightarrow Cu(dien)^{+2}$	C.B.	≈ 7.00	-8.58	$Cu(dien)^{+2}$	≈ 3.04	(75)
$Ca^{+2} \rightarrow Cu(trien)^{+2}$	C.B.	≈ 7.00	-8.58	$Cu(trien)^{+2}$	≈ 3.04	(75)
$Ca^{+2} \rightarrow Cu(tetren)^{+2}$	C.B.	≈ 7.00	-8.58	$Cu(tetren)^{+2}$	≈ 3.04	(75)
$Ca^{+2} \rightarrow Cu^{+2}$	illite	$\approx 0^d$	$\approx 0^d$			(76)
$Ca^{+2} \rightarrow Cu(tetren)^{+2}$	illite	≈ 5	-6.16	$Cu(tetren)^{+2}$	≈ 2.68	(77)
$Ca^{+2} \rightarrow Zn(tetren)^{+2}$	illite	≈ 5	-6.16	$Zn(tetren)^{+2}$	≈ 2.68	(77)
$Ca^{+2} \rightarrow Ni(tetren)^{+2}$	illite	≈ 4	-4.93	$Ni(tetren)^{+2}$	≈ 2.14	(77)
$Ca^{+2} \rightarrow Cd(tetren)^{+2}$	illite	≈ 4	-4.93	$Cd(tetren)^{+2}$	≈ 2.14	(77)

a) C.B. = Camp Berteau montmorillonite; b) W.B. = Wyoming bentonite; c) calculated values; d) after correction for specific adsorption effects.

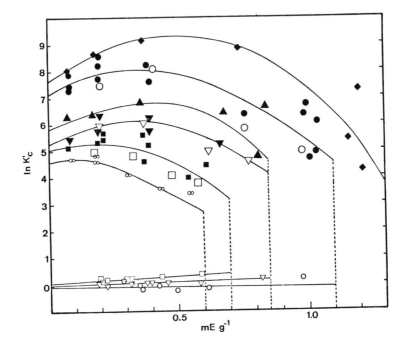

Figure 3. Logarithm of the selectivity coefficient for Ca-Cu(en)$_2$ against Cu content of the exchanger in Otay, ◆ ; 0.95 RCM, ●,o; 0.74 RCM, ▼ , ▽ ; hectorite, ▲; 0.59 RCM, ■ ,◻ and Laponite, ∞; (upper curves) at 10^{-2} total normality (closed symbols) and at high ionic strength (open symbols). The lower curves represent the log of the selectivity for Ca - Cu in 0.95 RCM, o; 0.74 RCM, ▽ and 0.59 RCM, ◻ . Reproduced with permission from Ref. 27. Copyright 1979, The Chemical Society.

sufficient to obtain the fully coordinated complex. Selectivity enhancements of up to three orders of magnitude are observed. The high affinity process was put to use in the determination of the cation exchange capacity of soils and clay minerals (70, 78-79).

The enhanced selectivity of the complexed transition metal cation compared to the uncomplexed aqueous form can be expressed as a gain in the stability constant of the adsorbed complex with respect to its stability constant in the solution phase (80). The complex formation reaction and corresponding stability constants of a transition metal cation M^{z+} with an uncharged ligand L in both the surface (indicated by bars) and solution phase are defined as

$$\overline{M}^{z+} + n\overline{L} \rightleftharpoons \overline{ML}_n^{z+} \quad \text{and} \quad M^{z+} + ML \rightleftharpoons ML_n^{z+}$$

$$\overline{\beta}_n = \frac{\overline{ML_n}^{z+}}{\overline{M}^{z+}(\overline{L})^n} \qquad \beta_n = \frac{ML_n^{z+}}{M^{z+}(L)^n}$$

The surface species $\overline{ML}_n^{z+}$ and $\overline{M}^{z+}$ are expressed as equivalent fractions of the exchanger capacity; $(\overline{L})$ is in mol/l. Three methods can be used to measure the complex formation constant of adsorbed complexes.

1) The ion exchange method, (28, 80) which consists in determining the ion exchange constants of a complexed and uncomplexed transition metal cation versus a non-complex forming reference cation, leading to the determination of

$$\frac{\overline{\beta}_n}{\beta_n} = \frac{ML_n}{M} K_{ex} \frac{(L)^n}{(\overline{L})^n}$$

Assuming unit partition of the uncharged ligand between the equilibrium and interlamellar solution allows one to determine the overall surface complex formation constant $\overline{\beta}_n$ from the hypothetical exchange constant $\frac{ML_n}{M}K_{ex}$ of the complex versus the uncomplexed transition metal ion, and the complex formation constant in solution (β_n). Unit partition was verified for ethylenediamine in the Ca-Cu-ethylenediamine system (28). The excess free energy loss for the formation of the coordinatively saturated complex in the interlayer phase is then defined as

$$\Delta G_{excess} = - 2.3 \, RT \, \log \frac{\overline{\beta}_n}{\beta_n}$$

2) The Bjerrum surface complex formation function (28, 67) consists in measuring the gradual formation of the transition metal complex at the surface, or the mean ligand number $\overline{n}^s$, with increasing ligand concentration. This method allows to determine the successive surface complex formation constants from

$$\bar{n}^s = \frac{\Sigma_1^n \; \bar{\beta}_n \; L^n}{1 + \Sigma_1^n \; \bar{\beta}_n \; L^n}$$

3) The selectivity-complex formation function (81) describes the variation of the selectivity coefficient of the transition metal ion versus a non complex forming reference cation with increasing ligand concentration as

$$\frac{K_c \; (L)}{K_c \; (aq)} = \frac{1 + \bar{K}_1 L + \bar{K}_1 \bar{K}_2 L^2}{1 + K_1 L + K_1 K_2 L^2}$$

The second and third method allow the measurement of surface complexation constants at various transition metal loadings and consequently yield apparent composition dependent constants. In the first method on the contrary a truly thermodynamic constant is obtained under standard state conditions.

All three methods lead to consistent surface complexation constants in clay minerals (27, 28) and zeolites (82). The surface protonation constants of the organic compounds such as amines can be defined and determined similarly (41, 83) and are also found to be enhanced in the clay interface.

The influence of the charge density on the selectivity of complex cations is similar to the case of Ca-ethylenediammonium (enH$_2^{+2}$) and Na-Cs exchanges as shown in fig. 1c for the case of the Ca-Cu(en)$_2$ exchange in the isostructural series of montmorillonites.

The stabilization factor which is identical to

$$\log \frac{Cu(en)_2}{Cu} K_{ex} \quad \text{is given by} \quad \log \frac{\bar{\beta}_n}{\beta_n} = 2.01 \; 10^7 \; \Gamma + 0.36$$

and vanishes at zero charge density (Γ = charge density in meq cm^{-2}). The similarity of complex ion stabilization and simple inorganic and organic cation exchange is evident.

The similarity in the exchange of complex cations, organic cations and simple inorganic cations is also apparent in that both Ca^{+2}-enH$_2^{++}$ (29) and Na$^+$-Cs$^+$ exchanges (17) in montmorillonites of varying charge density and Ca^{+2}-Cu(en)$_2^{+2}$ exchange in Camp Berteau montmorillonite (27) are all governed by exothermic enthalpy changes.

The thermodynamic functions of exchange in all cases also decrease with decreasing charge density which may be visualized as a "dilution" of the exchangeable cations in the interlamellar phase. This process is verified by the increasing d$_{001}$ spacings for Cu(en)$_2^{+2}$, enH$_2^{++}$, Ca^{+2} and Cu^{+2} (table II) and results in nearly vanishing ΔG^o, ΔH^o and ΔS^o functions at zero charge density. Judging from the small variation of the Cu-Ca exchange with charge density, the magnitude of the thermodynamic functions are clearly related to differences in the hydration status of the adsorbed cations. It is therefore more realistic to consider both hydration and electrostatic changes in the surface phase with charge density, as was already shown for the Na-Cs pair in fig. 2 and calculated for alkali metal cation exchanges (19, 33, 34).

Standard free energy changes of heterovalent exchange equilibria among complexes increase with charge density in agreement with double layer expectations as shown for the $Ag(en)_2^{+1}$ - $Cu(en)_2^{+2}$ exchange in fig. 4. For the case of Wyoming bentonite the $Ag(en)_2^{+1}$ - $Cu(en)_2^{+2}$ reaction is endothermic (12.5 kJ Eq^{-1}), while $\Delta G^{o} = 9.665$ kJ (see table IV), showing that enthalpic factors also rule the exchange among complex ions.

In general, the (scarce) thermodynamic data for exchanges involving complexes leads us to conclude that the selectivity enhancement upon complexing is enthalpy driven and may be ascribed to enhanced charge dependent (primarily coulombic) interactions with the surface as compared with the aqueous ions.

The energetic contribution to the extra stability which is gained upon adsorption of the complexes in the interface is confirmed by spectroscopic data obtained on air dry samples (84-86) showing crystal field stabilization energies (CFSE) exceeding those of the complexes in aqueous solution. In the case of $Cu(en)_2$, the CFSE values increase with the charge density of the clay mineral (87) and decrease with $Cu(en)_2$ loading, which is consistent with the heterogeneous distribution of the isomorphic charges over the surface (55, 58) and with the fact that the clay interlayers with the highest negative charge density are filled up first. In this process of stabilization the clay polyanion acts physically as a (heterogeneous) anionic solvent with solvation power smaller than water (85, 88). The clay favours the adsorption of planar complexes by stripping off two axially coordinated water molecules, thereby increasing the tetragonal distortion of the complex.

Considering the contribution of the solvent and surface terms to the total free energy change, it is apparent that the enhanced selectivity in the presence of e.g. ethylenediamine corresponds in sign to variations in the solution term and is (in part) due to smaller ΔG values of hydration of the complex cations. This is exemplified for the Ca-Cu and Ag-Cu cases in the presence and absence of ethylenediamine by the equations :

$$\Delta G_{(aq)} = (\bar{G}_{Cu} - \bar{G}_{Ca}) - (G_{Cu} - G_{Ca})$$

$$\Delta G_{(en)} = (\bar{G}_{Cu(en)_2} - \bar{G}_{ca}) - (G_{Cu(en)_2} - G_{Ca})$$

and

$$\Delta G_{(aq)} = (\bar{G}_{Cu} - 2\bar{G}_{Ag}) - (G_{Cu} - 2 G_{Ag})$$

$$\Delta G_{(en)} = (\bar{G}_{Cu(en)_2} - 2\bar{G}_{Ag(en)_2}) - (G_{Cu(en)_2} - 2G_{Ag(en)_2})$$

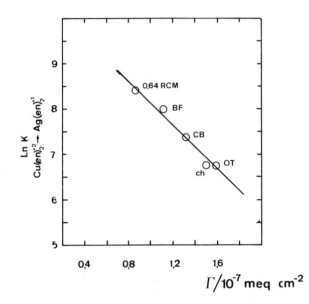

Figure 4. Variation of lnK (Cu(en)$_2$ → Ag(en)$_2$) with charge density (meq cm^{-2}) in Otay, Chambers (Na CEC = 1.13 meq g^{-1}), Camp Berteau, Belle fourche (Na CEC = 0.84 meq g^{-1}) and 0.64 RCM.

in which G and $\bar{G}$ stand for the free energy content in solution and the surface phase, respectively.

The contribution of the surface term to the total ΔG change is difficult to assess since no quantitative data on hydration of complexes are available. The importance of the surface term, however, can be judged from the surface charge density dependence of the Ca-Cu(en)$_2$ and Ca-Cu exchange equilibria. ΔG^{o}_{ex} (Ca^{+2} $\to$ Cu^{+2}) is almost independent of charge density, whereas ΔG^{o}_{ex} (Ca^{+2} $\to$ Cu(en)$_2^{+2}$) decrease linearly from about $-$ 10 kJ Eq^{-1} at 1.5 10^{-7} meq cm^{-2} to zero at vanishing charge density.

<u>Exchange of organic cations.</u> α, ω bistrimethylammoniumalkanes (BTM-n) of different chain length (indicated by the number of carbon atoms in the alkyl chain) are obtained by quaternizing the primary amino group in the corresponding alkanediamines by substituting H for CH$_3$ groups.

Important selectivity enhancements between the pairs of ions, Ca^{+2} $\to$ enH$_2^{+2}$ (ΔG^{ex}_o = $-$ 9.95 kJ mol$^{-1}_{+2}$) and Ca^{+2} $\to$ BTM-2^{+2} (ΔG^{ex}_o = $-$ 17.26 kJ mol^{-1}) or Ca^{+2} $\to$ BTM-10^{+2} (ΔG^{ex}_o = $-$ 22.2 kJ mol^{-1}) in Camp Berteau montmorillonite have been ascribed to the delocalization of the charge over increased volumes (<u>47</u>). CNDO/2 calculations of electron distribution in BTM cations explains their solution behaviour and this in turn is reflected in the changes in their ion exchange behaviour with chain length (<u>47</u>).

The exchange of BTM-n cations is mainly governed by enthalpic effects in contrast to alkylammonium exchange, the behaviour of which is primarily determined by entropy changes (ΔH^o being essentially zero). This suggests that electrostatic interactions, which are enhanced due to better delocalization, are also important. d$_{001}$ spacings of 1.45 nm confirm a flat orientation of these ions.

The introduction of unsaturated aromatic compounds also leads to better charge delocalization and enhanced ion exchange affinity. This is verified by 1) the higher free energy of exchange versus Ca^{+2} in biprotonated histammonium (<u>83</u>) (ΔG^o_{oex} = $-$ 12.33 kJ mol^{-1}) compared to ethylenediammonium (<u>38</u>) (ΔG^{ex}_o = $-$ 5.87 kJ mol^{-1}) and by 2) the highly exothermic enthalpies of exchange of mono- and bipyridinium compounds in Na$^-$ - montmorillonite (Clay Spur) (<u>89</u>).

The influence of the organocation structure on the exchange adsorption becomes evident from the data in table V. 4,4'Bipyridinium cations adsorb two times more energetically (ΔH^o_{ex} $\cong$ 22 kJ Eq^{-1}) than do 2,2'bipyridinium cations (ΔH^o_{ex} $\cong$ 11 kJ Eq^{-1}). The former adapt a planar orientation (d$_{001}$ = 1.26 nm) in contrast to the inclined position of the latter (d$_{001}$ = 1.4 nm), despite the fact that sufficient surface is available for adsorption in a flat configuration. Smaller enthalpy terms are consistent with smaller electrostatic interaction energies. The reason for the tilting is unknown however.

Flat-oriented bipyridinium compounds adsorb with additional stabilization compared to flat-oriented mono-pyridinium compounds, due to charge transfer interactions and resonance stabilization in linked aromatic rings versus single pyridinium ions. Although

4-phenylpyridinium compounds were not flattened in the interlayer space, they exhibit highly exothermic interactions which are ascribed to the involvement of Van der Waals forces between the ring systems, which are oriented perpendicular to the clay surface. In this case excess adsorption was also observed.

The interaction of the compounds, shown in table V, with vermiculite (90) are entirely different from their behaviour with montmorillonite in that endothermic effects are observed in the initial adsorption process. The endothermicity is attributed to the removal of water which is tightly bound by vermiculite. This is also corroborated by the fact that none of the organocations takes a flat orientation upon adsorption in the interlayer (d_{001} exceeds 1.45 nm in all cases).

The exchange of complex cations and organic ammonium cations follows the general rules for exchanges among alkali and alkaline earth metal cations. Homovalent exchange equilibria involving one complex cation, eg. Cs-Ag(en)$_2$ and Ca-Cu(en)$_2$ or involving the bisquaternary ammonium cations are exergonic and exothermic when the most polarizable cation exchanges for a less polarizable.

The charge delocalization or the polarizability difference explains the selectivity behaviour in cases of high polarizability differences (complex versus aqueous metal ion) or in a homologous series of ions (either inorganic cations or ammonium cations). The smaller hydration status of all types of interlamellarly adsorbed cations is ascribed to the mutual stabilization by charge delocalization over the planar oxygens and exchangeable cations and is caused by the electrostatic interaction forces.

Highly selective exchange in illite and modified montmorillonites

Alkali and alkaline earth metal ion exchange in illite. Fig. 5 shows typical Ca $\rightarrow$ Cs and Ca $\rightarrow$ Rb selectivity profiles in Morris illite (91). Adsorbed Cs and Rb ranges from about 10 meq/100 g (CEC = 20 meq/100 g) down to 10^{-4} meq/100 g. The selectivity coefficients appear to be highly dependent on loading and change by 4 orders of magnitude in a narrow loading range, indicating that crystallographically distinct sites with very different interaction energy are involved.

The selectivities for Cs ($\ln^{Cs}_{Ca}K_c \cong 20$) and Rb ($\ln^{Rb}_{Ca}K_c \cong 14$) versus alkaline earth metal ions are extremely high at trace loading and decrease to the usual values at high Cs and Rb content. The sites involved at trace loading highly differentiate between Cs and Rb, and the selectivity behaviour changes into the usual pattern in the higher loading range.

Notwithstanding the exceedingly high selectivities, the process occurring at trace loading is not an irreversible fixation but a reversible ion exchange reaction, as can be deduced from the internal consistency of the trace Cs and trace Rb adsorption equilibria in Na$^+$ - and K$^+$-illite shown in fig. 6.

The data in fig. 5 were simulated using the multi-site ion exchange model of Barrer and Klinowski (92). The model essentially consists in assigning intrinsic selectivity coefficients to the

TABLE V. Calorimetrically measured enthalpies of exchange of pyridinium and bipyridinium salts in Na-montmorillonite (89) (Clay Spur) and Na-vermiculite (90) (Palabora) and the C-spacings (nm) of the wet fully exchanged forms

Cation	Anion	Montmorillonite		Vermiculite	
		ΔH (kJ Eq^{-1})	C-spacing	ΔH (kJ Eq^{-1})	C-spacing
4,4'bipyridinium	2Cl⁻	-21.5	1.26	≡ 0	1.46
4,4'bipyridinium	2Br⁻	-21	1.26	≡ 0	1.46
4,4'bipyridinium	2I⁻	-23	1.26	≡ 0	1.46
1,1'Dimethyl-4,4'bipyridinium	2Cl⁻	-24	1.26	42	1.44
2,2'bipyridinium	2Br⁻	-11.5	1.40		1.45
1,1'Dimethyl-2,2'bipyridinium	2I⁻	-9	1.55	35	1.48
1,1'-Ethylene-2,2'bipyridinium	2Br⁻	-16	1.27	26	1.54
Pyridinium	Cl⁻	-11	1.26	18	1.42
1-Methylpyridinium	I⁻	-11	1.28	15	1.48
4-Methylpyridinium	I⁻	-10	1.29	23	1.45
1-Methyl-4-methylpyridinium	I⁻	-12	1.27	31	1.47
4-Phenylpyridinium	I⁻	-29	1.53	8	1.82
1-Methyl-4-phenylpyridinium	I⁻	-25	1.52	5	1.48

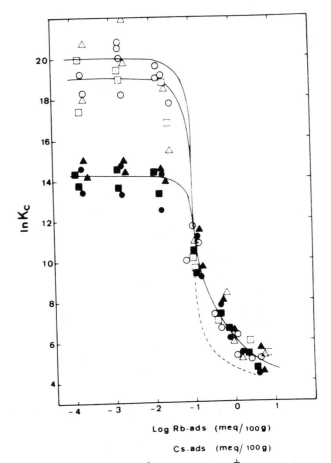

Figure 5.   LnK$_c$ vs.   Cs$^+$ – (and Rb$^+$ –) exchange levels (meq./100 g) in illite clay (25 °C) for the equilibria Cs$^+$ – Ca^{2+} (○), Cs$^+$ – Sr^{2+} (△), Cs$^+$ – Ba^{2+} (□), Rb$^+$ – Ca^{2+} (●), Rb$^+$ – Sr^{2+} (▲), and Rb$^+$ – Ba^{2+} (■). Dashed and full curves relate to a two-site and a three-site model respectively (see text). Reproduced with permission from Ref. 91. Copyright 1983, American Chemical Society.

separate site groups. Taking an exchanger consisting of only highly (H) and poorly (L) selective sites and choosing the equivalent fraction scale one writes :

$$K_c^H = \frac{(\overline{M_H^+})^2}{\overline{M_H^{2+}}} \frac{a_{M^{2+}}}{a_{M^+}^2}, \quad \text{and} \quad K_c^L = \frac{(\overline{M_L^+})^2}{\overline{M_L^{2+}}} \frac{a_{M^{2+}}}{a_{M^+}^2}$$

Taking H and L as the equivalent fraction of the high and low selectivity sites one obtains the overall selectivity from

$$K_c^{overall} = \frac{(H\overline{M_H^+} + L\overline{M_L^+})^2}{H\overline{M_H^{2+}} + L\overline{M_L^{2+}}} \cdot \frac{a_{M^{2+}}}{a_{M^+}^2}$$

The selectivity values at both ends of the surface composition scale are related to the intrinsic selectivity coefficients by (91) :

$$\lim k_c^{overall} \rightarrow H^2 K_c^H \text{ at } m_{M^+} \rightarrow 0, \text{ and } K_c^L/L \text{ at } m_{M^{++}} \rightarrow 0.$$

The full curves in fig. 5 (top to bottom) are calculated following a three-site model by using the sets of characteristic Lnk_c values for, respectively, site I (0.1 meq./100 g), site II (0.6 meq./100), and planar sites (19.3 meq./100 g) : 30.6 (I) – 13.8 (II) – 4.6 (planar); 29.6 (I) – 13.8 (II) – 4.6 (planar); 24.9 (I) – 13.8 (II) – 4.6 (planar). The dashed curve relates to a two-site model (site I : 0.1 meq./100 g; planar 20 meq./100 g) with Lnk_c values of 30.6 and 4.6.

Although no direct crystallographic evidence of ion-siting can be given, as was the case for the Na-Ag exchange in zeolite Y (93), the multi-site ion exchange approach is justified on the following grounds :
- the process involved in the high affinity sites is a reversible ion exchange reaction;
- the widely varying selectivities can hardly be assigned to non ideality effects resulting from ion-ion interactions, especially in view of the fact that, as shown below;
- such high affinity sites can be progressively generated in montmorillonite type clay minerals and are associated with collapsed 10 Å clay layers (94-96).

Notwithstanding the extremely high selectivities the selectivity order among the alkali metal cations follows the sequence commonly observed in clay minerals i.e. Cs > Rb > K > Na and corresponds to the selectivity sequence generated for low field strength sites using the Eisenmann approach (36). Exchange is therefore governed by the hydration energy difference of the cations. The extremely high selectivities were ascribed to the exchange of largely unhydrated cations, which is consistent with expectations based on the contents of fig. 2, showing increasing selectivities with increasing interlayer molality and charge density.

The high affinity sites were assigned to collapsed (10Å) edge-interlayer positions in the crystal, in line with generally accepted views (94), whereas the lowest selectivity sites were ascribed to the more abundant planar sites.

Differential enthalpy data obtained from microcalorimetry (97) is a powerful tool for measuring and interpreting exchange selectivities in terms of the underlying crystal structure components. Clay minerals appear to be composed of homogeneous groups of exchange sites which show characteristic constant exchange enthalpies. Typical enthalpy values (in kJ Eq^{-1}) of Ca-K exchange were identified, and could be ascribed to 'pure' clay minerals : Illite (-20.0), Vermiculite (-15.2), mica (-10.7), hydrous mica (-9.4 and -8.7) and montmorillonite (-7.5, -6.6 and -5.7). No one clay mineral appears to be 'pure' (except for muscovite). Montmorillonites (20) and kaolinites (98) show properties of mica (non-expanding) and hydrous mica components. Cicel and Machajdik (99-100) calculate the total layer charge of smectites in terms of the contribution of 10, 14 and 16.8 Å layers which are obtained after glycol treatment of K^+ or NH_4^+ exchanged samples. The 10, 14 and 16.8 Å spacings are typical for mica, vermiculite and montmorillonite layer charges.

Cs-Ca selectivity modelling is well suited to detect surface heterogeneity in an extremely small fraction of the CEC, which is difficult to measure by differential calorimetry. However, differential calorimetry detects site heterogeneity covering the whole range of CEC.

Cesium and Rubidium exchange in illite/smectite interlayers and in reduced charge montmorillonites. Further insight into the nature and location of the high affinity sites was recently (101) obtained from $Ca^{+2} \rightarrow Cs^+$ and $Ca^{+2} \rightarrow Rb^+$ selectivity measurements in two series of montmorillonite samples. The interlayer properties were modified either 1) by progressively reducing the montmorillonite charge density following the Hofmann-Klemen method (18) of Li incorporation after 24 hours heating at 240 °C of mixed Na-Li clays, or 2) by subjecting K-montmorillonite to alternate wetting-drying (W.D.) cycles. This process leads to K^+-fixation and to the progressive transformation of montmorillonite to illite through a series of random illite/smectites as verified by electron diffraction (102-104). The degree of reorganization of swelling K-montmorillonite into a collapsed more ordered mica-like structure, depends on the number of W.D. cycles (102-104) and on the layer charge of the smectite (104).

Fig. 7 shows the progressive transformation of montmorillonite to illite/smectite interlayers by the gradual development of both the characteristic Cs and Rb high selectivity profiles observed for 'pure' illite and the high Cs-Rb selectivity at trace loadings. The data can be simulated (see table VI for the $Ca^{+2} \rightarrow Cs^{+1}$ case) using a consistent set of intrinsic selectivity coefficients and identical site group capacities for the Ca-Cs and

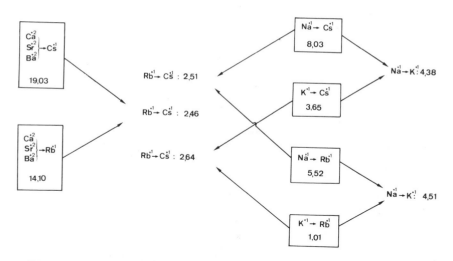

Figure 6. Reversibility test for the exchange of Cs^+, Rb^+, K^+, Na^+, Ca^{+2}, Sr^{+2} and Ba^{+2} on the high affinity sites in Morris illite (91). Values in frames are experimental; others are calculated from the indicated couples.

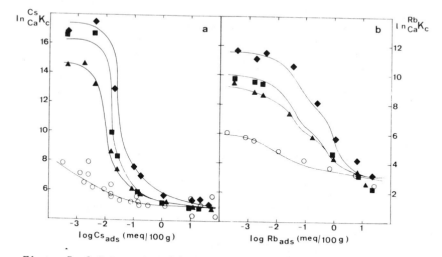

Figure 7. lnK_c vs. Cs- (a) and Rb-exchange (b) levels (meq/100 g) in original Camp Berteau montmorillonite (O) and Camp Berteau montmorillonite subjected to 2 (▲), 5 (■), and 10 (♦) wetting-drying cycles (101). Curves are a simulation. Table VI shows these data for the $(Ca^{2+} \rightarrow Cs^{+1})$ equilibrium. Reproduced with permission from Ref. 101. Copyright 1985, The Clay Minerals Society.

Ca-Rb equilibria. The increasing number of high energy sites (table VI) is associated with the formation of collapsed (10 Å) illite layers, which were observed (102-104) with increasing number of W.D. cycles.

High affinity sites of comparable magnitude to illite and illite/smectite series are also formed in reduced charge montmorillonites (R.C.M.). The Ca → Cs selectivity coefficients and site group capacities given in table VII simulate the experimental curves. Increasing amounts of high affinity sites were correlated with the residual Ca CEC of the RCM's, which again provides evidence for the association of high energy sites with collapsed layers. This suggests that any 10Å clay layer which for unspecified reasons is able to force the exchange of dehydrated cations, can produce extremely high selectivities. Stated alternatively, very high selectivity is not necessarily due to the interaction of cations with charged centers of sufficiently high charge density to provoke their dehydration.

The essential difference between the illite/smectite samples and the R.C.M. samples with respect to highly selective exchange is that the illite/smectite samples probably have K-collapsed interlayer regions similar to illites, whereas the in-situ charge neutralization in the R.C.M. samples leads to the collapse of very low charged interlayers or interlayer regions. This difference is verified by the fact that the selectivity in the lowest energy sites decreased with charge density in the R.C.M. samples (in contrast to the illite/smectite samples).

In addition to the very high affinity sites I (10Å phase) and the planar sites IV, at least two additional sites with intermediate selectivity are necessary to simulate the data in the two series of samples, as shown by sites II and III in table VI and VII. In both sample series, a range of sites with intermediate selectivity is probably formed (as visualized in figs. 9 and 10 from ref. 105) by the changing degree of hydration caused by steric hindrance at 'frayed edges'. The model representation of islands of high selectivity (10Å), surrounded by increasing areas of lower selectivity, is consistent with the contents of tables VI and VII in that the higher selectivities occur on the smaller site group capacities. Since high affinity sites are not a unique property of illites, but are merely present in or close to collapsed regions (94), the extremely high Cs selectivity cannot be used as a probe to identify small quantities of illite, without additional evidence. For example, the small number of high affinity sites observed in the original Camp Berteau montmorillonite may be assigned to micaceous impurities, as indicated from measurements of the partial enthalpy of Ca → K exchange (20). In the absence of such evidence, any sites located at the edge of low charged, collapsed layers can equally explain such behaviour.

Reversibility

Ion exchange equilibria involving alkali, alkaline earth (38) and transition metal ions (39) are reversible as a rule. Reversibility

TABLE VI. lnK values and site group capacities (meq 100/g) for simulating the $Cs^+ \rightarrow Ca^{2+}$ equilibria at 25°C of potassium-exchanged Camp Berteau montmorillonite subjected to alternate wetting-drying cycles

Wetting drying cycles	$lnK\,^{Cs+}_{Ca+2}$	Type I Site 33.8	Type II Site 25	Type III Site 13.8	Type IV Site 5
0	Site group capacity	0.0002	0.005	0.1	110
2		0.007	0.012	0.4	104
5		0.015	0.012	0.4	98
10		0.02	0.02	0.8	74

TABLE VII. lnK values and site group capacities (meq/100 g) used for simulating the Cs^+ – Ca^{2+} equilibria at 25°C of different reduced charge montmorillonites exposed or not to NH_3 vapour

Sample	$lnK_c{}^{Cs^+}_{Ca^{+2}}$	Type I Site 33.8	Type II Site 25.0	Type III Site 13.8	Type IV Site (given in parenthesis)
0.96 Ca-RCM		0.0025	0.012		
0.95 Ca-RCM		to	to	0.02–0.1	105 (5.15)
0.96 Ca-RCM-(NH_3)		0.001	0.008		
0.95 Ca-RCM-(NH_3)					
0.84 Ca-RCM		0.002			
0.84 Ca-RCM-(NH_3)		to	0.012	0.2	93 (3.85)
		0.001			
0.64 Ca-RCM-(NH_3)		0.003	0.04	0.4	70 (2.55)
0.54 Ca-RCM-(NH_3)					
0.50 Ca-RCM-(NH_3)		0.018	0.1	1.0	53.8 (2.35)
0.46 Ca-RCM-(NH_3)					
0.43 Ca-RCM-(NH_3)					
0.42 Ca-RCM		0.018	0.2	2.0	44 (2.50)
0.33 Ca-RCM-(NH_3)		0.005	0.13	1.3	34 (2.35)
0.29 Ca-RCM (NH_3)					
0.25 Ca-RCM		0.02	0.3	1.5	18 (2.35)
0.16 Ca-RCM					
0.13 Ca-RCM					

is evidenced from identical forward and reverse reactions or is verified when the Hess law is obeyed.

Extremely high selectivities are frequently interpreted as "ion fixation", which suggests an irreversible phenomenon. This is the case for exchanges of Cs, Rb and K in illite clay minerals (95-96) as well as for $Cu(NH_3)_4$ exchange in fluorhectorite (66). However, reversibility was verified from the Hess law for adsorption of Cs, Rb and K on the high affinity sites in illite (91) and modified montmorillonites (101) as well as for the exchange of transition metal complexes (29, 75).

Exchange in zeolites

Highly selective exchange as related to the presence of crystallographically different site groups

Zeolites present a porous structure of channels and cages of varying dimension (106-108). The exchangeable cations are located in different positions. Since more sites are available than charges to be neutralized, the neutralization pattern may vary with the kind of exchangeable cation (109-110).

Exchange in zeolites of alkali, alkaline earth, transition metal ions and small organic ammonium ions, has been reviewed (111), and in general, the exchange is characterized by small ΔG^o_{ex} values comparable to those found in clay minerals. Although identical selectivity orders for alkali and alkaline earth metal ions are obtained, as in montmorillonite, the opposite variation of ΔG^o_{ex} with charge density is found.

On account of the presence of crystallographically different site groups and their characteristic neutralization pattern for each type of cation, the entering ion does not necessarily take position of the leaving ions (111-112). This and the limited hydration in the narrow pore system leads to exchange selectivities which may vary strongly with occupancy in homovalent and heterovalent exchanges.

Two examples are discussed showing extreme selectivities : 1) the hexagonal prism for unhydrated Ag^+ in NaY zeolite, and 2) the cubooctahedron for hydrated Na ions in fully exchanged Co^{2+}-Y zeolite. Such very high selectivity differences cannot be assigned to non-ideality effects; instead they are ascribed to very different interaction energies with crystallographically distinct site groups.

Na-Ag exchange in NaY zeolite

Fig. 8 shows part of the Na → Ag exchange isotherm obtained in NaY (113). Using again the multi-site ion exchange model (92), the overall selectivity can be described in terms of the selectivities of different site groups. The set of parameters that simulate the experimental curve are given in the legend of fig. 8. A very high selectivity coefficient of the order of 2000 in site I (4 ions/u.c.) is necessary to describe the data. The remainder of the exchange sites show small Na → Ag selectivities. The behaviour of the different site groups is assumed ideal, i.e. site groups act independently and

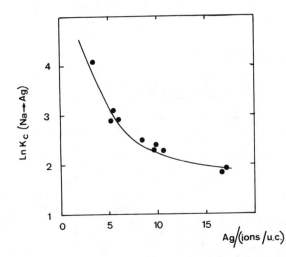

Figure 8. Comparison of the experimental and calculated Na^{+1} → Ag^{+1} selectivities in zeolite Y (113) as a function of the number of Ag$^+$ ions/unit cell. The set of selectivity coefficient values and maximum number of ions/u.c. (given in parenthesis) used in the simulation for respectively, site I (4 ions/u.c.), site I' (13.5 ions/u.c.), site II (10 ions/u.c.) and unlocalized ions (26.5 ions/u.c.) are : 2000 (I), 2(I'), 20(II) and 4 (unlocalized). Reproduced with permission from Ref. 113. Copyright 1978, The Chemical Society.

selectivities in each site group are assumed to be invariant with changing surface composition. This does not mean that these selectivity coefficients can be identified with thermodynamic equilibrium constants in each site group, as stated by Barrer (109, 110, 114), since it is impossible to assign different chemical potentials to the same component Ag on different site groups present in the same crystal (115).

As stated by Barrer (109), without additional information, a given isotherm cannot be analysed in terms of component isotherms because numerous empirical combinations of K_c and X_i may lead to the correct $K_c^{overall}$. Table VIII shows identical Ag ion distributions obtained from X-ray diffraction analysis and calculated from the selectivity data. Such correspondance allows to assign Na-Ag selectivity coefficients to various crystallographic sites.

Based on their radii, the theoretical $O^{2-} - K^+$ bond (0.273 nm distance) is better suited than the $O^{2-} - Ag^+$ (0.266 nm distance) to fit the observed O_3 – site I (hexagonal prism) distance of 0.275 nm. The very high selectivity for Ag is therefore ascribed to the formation of a coordination compound of Ag with 6 oxygen atoms in the hexagonal prism. The ability of Ag^+ to form complexes and its much higher hydration energy than K^+ support this hypothesis.

$Na^+ - K^+ - Co^{2+}$ exchange in Y zeolite. Heterovalent exchange reactions in zeolites generally show an even more pronounced dependency on loading (116-118). Rees (116) observed variations of the selectivity coefficient by a factor 1000 for the Na-Ca and Na-Mg exchange in zeolite A at 25 °C. An example of extreme variations is shown in fig. 9 for the K^+-Co^{+2} and Na^+-Co^{+2} selectivities in zeolite Y at 45 °C (117). The exchange temperature of 45 °C is necessary to obtain full exchange of Co in the small cavities. Heterovalent exchange at lower temperatures, suffer from incomplete exchange. Selectivities range from

$$\begin{matrix} Co \\ Na \end{matrix} K_c = 5 \quad \text{at} \quad Z_{Co} = 0 \text{ to } 2 \times 10^{-5} \text{ at } Z_{Co} = 1.$$

Judging from fig. 9, about 25-30% of the crystal charge exhibits an extremely high intrinsic selectivity of Na versus Co ($K_c^{intrin} = 2.5 \times 10^{-5}$). Such a high selectivity is in line with X-ray structural evidence (119) of the formation of a stable Na-hydrate in the sodalite cages in presence of bivalent (Ca) cations in the big cavities. Comparison with the K-Co exchange, which is less dependent on composition, shows (see fig. 9) that K is more selective in the supercages. The reverse effect is observed in the small cages, in agreement with the S-shaped K-Na exchange isotherm at 45 °C (120). The very high selectivities for Na in the small cages of zeolite Y are reminiscent of the 'apparent' difficulty of obtaining fully exchanged bivalent ion loaded zeolites, as well as the 'apparent' higher loadings obtained when starting from the K-Y zeolite (119). Kinetic effects related to the partial stripping of hydration water to allow migration into the small cages are superimposed on the aforementioned thermodynamic effect.

TABEL VIII. Comparison between the population of Ag^+ ions in each site group determined by X-ray diffraction techniques and calculated by use of the parameters given in fig. 8

Total Ag^+/ (ions/u.c.)	I		I'		II		U	
	fit	X-ray	fit	X-ray	fit	X-ray	fit	X-ray
2.0	2.00	1.9	0.01	-	0.1	-	0.05	-
7.25	3.84	3.9	0.3	0.3	1.90	1.8	1.19	1.25
14.00	3.92	4.4	1.16	1.8	4.8	6.0	4.08	1.8

Reproduced with permission from Ref. 113. Copyright 1978 Royal Society of Chemistry.

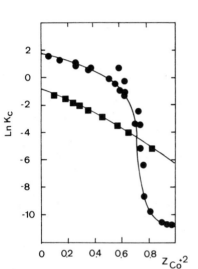

Figure 9. Surface composition dependency of the $Na^{+1} \rightarrow Co^{+2}$ (●) and $K^{+1} \rightarrow Co^{+2}$ (■) selectivities at 45 °C in zeolite Y. Reproduced with permission from Ref. 117. Copyright 1980, The Chemical Society.

Exchange of transition metal ion uncharged ligand complexes

The exchange and thermodynamic characterization of complex cations in montmorillonites present no problem with regard to the accessibility and the accomodation of the complexes, regardless of the extent of exchange.

Exchange of complexes in zeolites, however, is far more complicated, in that a) due to sieve effects, the exchange of complexes is restricted to the accessible part of the crystal; b) steric effects, due to the bulkiness of the complexes or due to the limited available pore space, may further limit the extent of exchange; c) competition between solvent, ligand and lattice oxygens for the coordinating metal ion may shift ion-positions in the crystal. Very few thermodynamic data are available for the adsorption of complexes in zeolites. The exchange of complex cations may or may not be enhanced with respect to their aqueous counterparts. Furthermore, exchange of complexes, is in general strongly dependent on composition due to steric effects. Increased affinity over the entire surface composition was observed for $Cu(NH_3)_4$ adsorption in mordenite ($\underline{121-122}$), X and Y zeolite ($\underline{123}$) (shown in fig. 10), and for Ag(TU)$_n$ adsorption in Y zeolite ($\underline{124}$). Cu (ethylenediamine)$_2$ on the contrary is only preferred versus $Cu(H_2O)$ at low loadings of the complex in X and Y zeolite ($\underline{125}$). Ag(pyridine)$_2$ complexes have smaller selectivity compared to aquo Ag in the big cavities of zeolite NaY ($\underline{82}$). The opposite behaviour is observed in Na-LaY zeolite in which 40 % of the charge is irreversibly neutralized in the small cages. This smaller selectivity for Ag(pyridine)$_2$ in NaY is obtained notwithstanding the higher polarizability of the complexed cation, and is very likely due to the fact that the colinear Py-Ag-Py complex (about 1 nm long) cannot be accomodated in the supercage without considerable strain. This arises from the electrostatic interactions with the framework charges. In Na-LaY this strain is lessened due to the smaller charge. In montmorillonite such strain is absent; the planar structure favours a 'double-sided' interaction and consequently a very high preference of Ag(pyridine)$_2$ over Ag is observed ($\underline{74}$).

The exchange of complex cations in zeolites is self-consistent with regard to charge density effects. Exchange involving ions of strongly differing size (polarizability) is favoured with charge density decrease (and consequently with internal dilution); e.g. Na-Cs in NaX and NaY zeolite ($\underline{126-128}$), Na-Ag(pyridine)$_2$ in NaY and Na-CaY ($\underline{82}$), and $NH_4-Cu(NH_3)_4$ in NH_4 Mordenite ($\underline{121-125}$), NH_4X and NH_4Y zeolite ($\underline{123}$). In montmorillonites the opposite occurs, both for Na-Cs ($\underline{17}$) and transition metal complexes with uncharged ligands ($\underline{27}$). (Charge density decrease in montmorillonites occurs with interlamellar dilution).

Selectivity enhancements observed when Co^{3+}, Cu^{+2} and Zn^{+2} are complexed with NH_3 in zeolites ($\underline{121-122}$), and the selectivity series of $Cu(NH_3)_4$ (Mordenite > Y > X) and Cu (X > Y > Mordenite) are rationalized in terms of dielectric theory ($\underline{122, 123, 126}$).

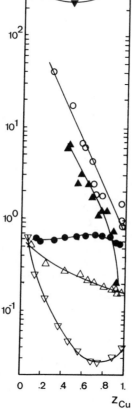

Figure 10.   Plots of the normalized selectivity coefficient (K_c^n) for Cu in zeolite X (●,o), Y (▲,△) and mordenite (▼,▽). Empty symbols : hydrated ion; filled symbols : amminated copper. Reproduced with permission from Ref. 123. Copyright 1980, Heyden & Son Ltd.

Reversibility

Exchange reactions are in general reversible. Apparent irreversibility is in fact an hysteresis effect ($\underline{118}$, $\underline{121}$, $\underline{130}$). This is obtained when the small cage sites of X and Y zeolites are forced to exchange with multivalent cations (La^{3+}, transition metal ions e.g.) by drying or freeze drying ($\underline{129}$), or by increasing the exchange temperature or the exchange time. Subsequent verification of the reverse exchange reaction at lower temperature or for shorter exchange times may lead to hysteresis effects.

Environmental Relevance

Highly selective ion exchange reactions described here in clay minerals and zeolites are reversible and occur on the constant charge fraction of these minerals. Interactions with a siloxane surface are therefore involved in contrast to the so-called 'specific adsorption' effects occuring on hydroxyl bearing surfaces.
   Important selectivity enhancements are observed upon complexing with neutral ligands. The stability constants of the adsorbed complexes exceed the values in aqueous solution by two to three orders of magnitude. Such observation may be relevant to the behaviour of transition elements in the environment in that stability constants of adsorbed organic matter complexes may differ from the values found for solution phase equilibria. Such effects are indeed observed for Cu ($\underline{139}$) and Ca ($\underline{132}$).
   Illites and illite/smectite interlayers show Cs ion-exchange selectivities ranging from extremely high values corresponding respectively to the exchange in collapsed (10Å) edge-interlayers and in hydrated smectite interlayers. In between these extremes a range of selectivities are observed which are associated with a range of binding sites formed by the changing degree of hydration caused by steric hindrance at frayed edges. The extreme affinity for Cs is relevant to the problem of the burial of radioactive waste where clay is used as a barrier to nuclide migration. The ion exchange and diffusion ($\underline{133}$) behaviour of Cs in Boom Clay (a potential repository site at -200 m depth located in Mol, Belgium) can be quantitatively explained in terms of the presence of illite.

Literature Cited

1. Sposito, G.; Mattigod, S. V. Soil Sci. Soc. Am. J., 1977, 41, 323-329.
2. Luoma, S.; Bryan, G. W. Sci. Total Environ., 1981, 17, 165-196.
3. Sposito, G. In "Applied Environmental Geochemistry"; Thornton, 1., Ed.; Academic Press, 1983; Chap. 5, p. 123.
4. Cremers, A.; Maes, A. In "Scientific Seminar on the application of distribution coefficients to radiological assessment models"; Louvain-La-Neuve 1985, (in press).

5. Gaines, G.; Thomas, H. C. J. Chem. Phys., 1953, 21, 714-718.
6. Sposito, G. "The thermodynamics of soil solutions"; Clarendon Press, 1981.
7. Ekedahl, E.; Högfeldt, E.; Sillen, L. G. Acta Chemica Scandinavica, 1950, 4, 556-558.
8. Argersinger, W. J.; Davidson, A. W.; Bonner, O. D. Trans. Kansan Acad. Sci., 1950, 53, 404-410.
9. Landelout, H. In "Plant Nutrient supply and movement" Tech. Report Ser. No. 48, I.A.E.A., Vienna, 1965, pp. 20-23.
10. Barrer, R. M.; Townsend, R. P. J. Chem. Soc., Faraday Trans. 2, 1984, 80, 629-640.
11. Shainberg, I.; Oster, J. D.; Wood, J. D. Soil Sci. Soc. Am. J., 1980, 44, 960-964.
12. Sposito, G.; Holtzclaw, K. M.; Jouany, C.; Charlet, L. Soil Sci. Soc. Am. J. 1983, 47, 917-921.
13. Sposito, G.; Mattigod, S. V. Clays and Clay Minerals 1979, 2, 125-128.
14. Sposito, G.; Holtzclaw, K. M.; Johnston, C. T.; Levesque-Madore, C. S. Soil Sci. Soc. Am. J. 1981, 45, 1079-10 .
15. Sposito, G.; Holtzclaw, K. M.; Charlet, L.; Jouany, C.; Page, A. L. Soil Sci. Soc. Am. J. 1983, 47, 51-56.
16. Bruggenwert, M. G. M.; Kamphorst, A. In "Soil Chemistry, B. Physico-Chemical Models" Bolt, G. H. Ed., Elsevier Sci. Publ. Amsterdam, 1979, Chap. 5.
17. Maes, A.; Cremers, A. J. Chem. Soc., Faraday Trans. I, 1978, 75, 1234-1241.
18. Hofmann, U.; Klemen, R. Z. Anorg. Chem. 1950, 262, 95-99.
19. Maes, A.; Cremers, A. In "Soil Chemistry, B. Physico-chemical Models". Bolt, G. H., Ed.; Elsevier Sci. Publ. Amsterdam, 1979, Chap. 6.
20. Talibudeen, O.; Goulding, K. W. T. Clays and Clay Minerals, 1983, 31, 37-42.
21. Cremers, A.; Thomas, H. C. Israel J. Chem. 1968, 6, 949-957.
22. Eliason, J. R. Am. Mineral. 1966, 51, 324-335.
23. Gast, R. G. Soil Ser. Soc. Am. Proc. 1969, 33, 37.41.
24. Lewis, R. J.; Thomas, H. C. J. Phys. Chem. 1963, 67, 1781-1983.
25. Gast, R. G. Soil Ser. Soc. Am. Proc. 1972, 36, 14-19.
26. Maes, A.; Cremers, A. J.C.S. Faraday Trans. I, 1981, 77, 1553-1559.
27. Maes, A.; Cremers, A. J. Chem. Soc., Faraday Trans. I, 1979, 513-524.
28. Maes, A.; Peigneur, P.; Cremers, A. J. Chem. Soc. Faraday Trans. I, 1978, 74, 182-189.
29. Maes, A.; Kasquin, R.; Cremers, A. J. Chem. Soc., Faraday Trans. I., 1982, 78, 2041-2049.
30. Martin, H.; Laudelout, H. J. Clim. Phys. 1963, 60, 1086-1099.
31. Gast, R. G.; Van Bladel, R.; Deshpande, K. B. Soil Sci. Soc. Am. Proc. 1969, 33, 661-664.
32. Laudelout, H.; Van Bladel, R.; Bolt, G.A.; Page, A.L. Trans. Faraday Soc., 1968, 64, 1477-1488.
33. Shainberg, I.; Kemper, W. D. Soil Sci. Soc. Am. Proc., 1966, 30, 707-713.

34. Shainberg, I.; Kemper, W. D. Soil Sci. 1967, 103, 4-9.
35. Eberl, D. Clays and Clay Minerals 1980, 28, 161-172.
36. Eisenman, G. Biophys. J. 2 pt. (supplement) 1962, 259-323.
37. Cruickshank, E. H.; Meares, P. Trans. Faraday Soc. 1957, 53, 1299-1308.
38. Landelout, H.; Van Bladel, R.; Gilbert, M.; Cremers, A. In "9th Int. Congr. Soil Sci. Transport" 1968, 1, 565-575.
39. Maes, A.; Peigneur, P.; Cremers, A. In "Proc. Int. Clay Conf. Mexico City" 1975, 319-329.
40. Theng, B. K. G.; Greenland, D. J.; Quirk, J. P. Clay Minerals 1967, 7,1-17.
41. Vansant, E. F.; Uytterhoeven, J. B. Clays and Clay Minerals 1972, 20, 47-54.
42. Vansant, E. F.; Peeters, O. Clays and Clay Minerals 1978, 26, 279-284.
43. Maes, A.; Marynen, P.; Cremers, A. Clays and Clay minerals 1977, 25, 309-310.
44. Aue, D. H.; Webb, H. M.; Bowers, M. T. J. Amer. Chem. Soc., 1972, 94, 4726-4728.
45. Aue, D. H.; Webb, H. M.; Bowers, M. T. J. Amer. Chem. Soc., 1976, 98, 311-317.
46. Maes, A.; Cremers, A. J. Chem. Soc., Faraday Trans. I, 1977, 73, 1807-1814.
47. Maes, A.; Van Leemput, L.; Cremers, A.; Uytterhoeven, J.B. J. Colloid & interface Sci. 1980, 77, 14-20.
48. Reddy, M. In "Ion exchange and solvent extraction". Marinsky, J.A.; Marcus, Y. Eds.; Marcel Dekker, Inc. New York, 1977, Vol. 6, Chap. 4.
49. McBride, M. B. Clays and Clay Minerals 1979, 27, 417-422.
50. Frysinger, G. R.; Thomas, H. C. J. Phys. Chem. 1960, 64, 224-228.
51. McBride, M. B.; Bloom, P. R. Soil Sci. Soc. Am. J. 1977, 41, 1073-1077.
52. Gaines, G. L.; Thomas, H. C. J. Chem. Phys. 1955, 23, 2322-2326.
53. McBride, M. B. Soil Sci. Soc. Am. J., 1978, 40, 452-456.
54. Inoue, A.; Minato, H. Clay and Clay Miner. 1979, 27, 393-401.
55. Lagaly, G. Clay Miner. 1981, 16, 1-21.
56. Lagaly, G.; Weiss, A. Proc. Int. Conf. Mexico City 1975, 157-172.
57. Stul, M. S; Mortier, W. J. Clays and Clay Miner. 1974, 22, 391-396.
58. Maes, A.; Cremers, A. Clays and Clay Miner. 1979, 27, 387-392.
59. Barrer, R. M.; Klinowski, J. Geochim. Cosmochim. Acta, 1979, 43, 755-766.
60. Banin, A. Israel J. Chem 1968, 6, 27-36.
61. Keren, R.; Shainberg, I. Clays and Clay Miner. 1979, 27, 145-151.
62. Frey, E.; Lagaly, G. In "Developments in sedimentology 1979, 27, 131-140.
63. Frey, E.; Lagaly, G. J. Coll. Int. Sci. 1979, 70, 46-55.

64. McBride, M. B. Clays and Clay Minerals 1980, 28, 255-261.
65. Pleysier, J.; Cremers, A. J. Chem. Soc., Faraday Trans. I, 1975, 71, 256-264.
66. Barrer, R. M.; Jones, D. L. J. Chem. Soc. (A), 1971, 3, 503-508.
67. El-Sayed, M. H.; Burau, R. G.; Babcock, K. L. Soil Science, 1970, 110, 202-207.
68. El-Sayed, M. H.; Burau, R. G.; Babcock, K. L. Soil Sci. Soc. Am. Proc. 1971, 35, 571-574.
69. Siegel, A. Geochimica et Cosmochimica Acta 1966, 30, 757-768.
70. Helsen, J. Bull. Soc. Chim. France 1966, 6, 1971-1974.
71. Das Kanungo, J. L.; Chakravarti, S. K. J. Coll. Interface Sci. 1971, 35, 295
72. Das Kanungo, J.L.; Chakravarti, S.K. Kolloid Z.u. Z. Polymere 1973, 251, 154-158.
73. El-Sayed, M. H.; Burau, R. G.; Babcock, K. L. Soil Sci. Soc. Am. Proc., 1970, 34, 397-400
74. Rasquin, E. Ph. D. Thesis Leuven, 1979.
75. Peigneur, P.; Maes, A.; Cremers, A. In "Developments in sedimentology" 1979, 27, 207-216.
76. Harmsen, K. Ph. D. Thesis Pudoc, Wageningen, 1979.
77. Smeulders, F.; Maes, A.; Sinnaeve, J.; Cremers, A. Plant and Soil, 1983, 70, 37-47.
78. Chabra, R.; Pleysier, J.; Cremers, A. Proc. Int. Clay Conf. 1976, 1, 439-449.
79. Barrett, R. B.; Wickham, C. S. Clays and Clay Min 1978, 26, 372.
80. Maes, A.; Marynen, P.; Cremers, A. J. Chem. Soc. Faraday Trans. I, 1977, 73, 1297-1301.
81. Maes, A.; Cremers, A. J. Chem. Soc. Faraday Trans. I, 1978, 74, 2470-2480.
82. Rasquin, E.; Maes, A.; Cremers, A. Proc. 6th Int. Zeolite Conf. Reno, Nevada, 1983, 641-650.
83. Maes, A.; Marynen, P.; Cremers, A. Progr. Colloid & Polymer Sci. 1978, 65, 245-250.
84. Velghe, F.; Schoonheydt, R. A.; Uytterhoeven, J. B.; Peigneur, P.; Lunsford, J.H. J. Phys. Chem 1977, 81, 1187-1194.
85. Schoonheydt, R. A.; Velghe, F. and Uytterhoeven, J. B. Inorg. Chem 1979, 18, 1842-1847.
86. Schoonheydt, R. A.; Velghe, F.; Baerts, R.; Uytterhoeven, J. B. Clays and Clay Min. 1979, 27, 269-278.
87. Maes, A.; Schoonheydt, R. A.; Cremers, A.; Uytterhoeven, J. B. J. Phys. Chem. 1980, 84, 2795-2799.
88. Velghe, F.; Schoonheydt, R. A. and Uytterhoeven, J. B. Clays and Clay Miner. 1977, 25, 375-380.
89. Hayes, M. H.; Pick, M. E.; Toms, B. A. J. Colloid Interface Sci., 1978, 65, 284-265.
90. Hayes, M. H.; Pick, M. E.; Toms, B. A. J. Colloid Interface Sci., 1978, 65, 266-275.
91. Brouwer, E.; Baeyens, B.; Maes, A.; Cremers, A. J. Phys. Chem. 1983, 87, 1213-1219.
92. Barrer, R. M.; Klinowski, J. J. Chem. Soc., Faraday Trans. I, 1972, 68, 73-87.

93. Costenoble, M.; Maes, A. J. Chem. Soc., Faraday Trans. 1
    1978, 74, 131-135.
94. Sawhney, B. L. Clays and Clay Miner. 1972, 20, 93-100.
95. Tamura, T.; Jacobs, D. G. Health Phys. 1960, 2, 391-398.
96. Tamura, T.; Jacobs, D. G. Health Phys. 1961, 5, 149-154.
97. Goulding, K. W. T.; Talibudeen, O. J. Colloid Interface Sci.
    1980, 78, 15-24.
98. Talibudeen, O.; Goulding, K. W. T. Clays and Clay Miner.
    1985, 31, 137-242.
99. Cicel, B.; Machajdik, D. Clays and Clay Miner. 1981, 29,
    40-46.
100. Machajdik, D.; Cicel, B. Clay and Clay Miner. 1981, 29,
    47-52.
101. Maes, A.; Verheyden, D.; Cremers, A. Clays and Clay Miner.
    1985, in press.
102. Gaultier, J. P.; Many, J. Clay Miner. 1978, 13, 139-146.
103. Plançon, A.; Besson, G.; Gaultier, J. P.; Many, J.; Tchoubar,
    C. In "Developments in sedimentology" 1979, 27, 45-54.
104. Srodon, J.; Eberl, D. In "Micas" Bailey, S. W. Ed., 1984,
    Chap. 12.
105. Kinniburgh, D. G.; Jackson, M. L. In "Adsorption of
    inorganics at solid-liquid interfaces" Anderson, M. A. and
    Rubin, A. L., Eds.; Ann Arbor Sci. 1981, Chap. 3.
106. Breck, D. W. "Zeolite Molecular Sieves - Structure, Chemistry
    and Use", John Wiley & Sons, 1974.
107. Meier, W. M.; Olson, D. S. "Atlas of zeolite structure
    types", Polycrystal Book Service, Pittsburg, 1978.
108. Mortier, W. J. 'Compilation of extra framework sites in
    zeolites" Butterworth and Co., 1982.
109. Barrer, R. M. In "Proc. 5th Int. Zeolite Conf." Rees, L.V.
    Ed., Heyden & Son Ltd., London, 1980, p. 273-290.
110. Barrer, R. M. Zeolites 1984, 4, 361-368.
111. Cremers, A. Molecular Sieves II, A.C.S. Symposium Series
    1977, 179-193.
112. Sherry, H. S. Adv. Chem. Ser. 1971, 101, 350-378.
113. Maes, A.; Cremers, A. J. Chem. Soc., Faraday Trans.1, 1978,
    74, 136-145.
114. Barrer, R. M.; Klinowski, J. J. Chem. Soc., Faraday Trans. 1,
    1979, 75, 247-251.
115. Mueller, R. F.; Ghose, S.; Saxena, S. K. Geochimica-
    Cosmochim. Acta 1970, 34, 1356-1360.
116. Rees, L. V. C. Proc. Sixth Int. Zeolite Conf., Butterworths
    1984, pp. 626-640.
117. Maes, A.; Verlinden, J.; Cremers, A. In "Properties and
    applications of zeolites" Townsend, R.P. Ed., The Chem. Soc.
    London, 1980, pp. 269-278.
118. Maes, A.; Cremers, A. J. Chem. Soc. Faraday Trans. 1 1975,
    71, 265
119. Costenoble, M.; Mortier, W.; Uytterhoeven, J. B. J. Chem.
    Soc. Faraday Trans. 1 1976, 72, 1877
120. Maes, A.; Verlinden, J.; Cremers, A. J. Chem. Soc. Faraday
    Trans. 1 1979, 75, 440-445.

121. Barrer, R. M.; Townsend, R. P. <u>J. Chem. Soc., Faraday Trans.</u> <u>1</u> 1976, 72, 661-673.

122. Barrer, R. M.; Townsend, R. P. <u>J. Chem. Soc., Faraday Trans.</u> <u>1</u> 1976, 72, 2650-2660.

123. Fletcher, P.; Townsend, R. P. In "Proc. 5th Int. Conf. Zeolites" Rees, L. V. C., Ed.; Heyden & Son Ltd London, 1980, pp. 311-320.

124. Pleysier, J.; Cremers, A. In "Proc. 3th Int. Conf. on Molecular Sieves" Uytterhoeven, J. B., Ed.; Leuven University Press, p. 206-210.

125. Peigneur, P. Ph. D. Thesis Leuven, 1976.

126. Barrer, R. M.; Rees, L. V. C.; Shamsuzoha, J. <u>J. Inorganic Nucl. Chem.</u>, 1966, 28, 629-

127. Barrer, R. M.; Davies, J. A.; Rees, L. V. C. <u>J. Inorganic Nucl. Chem.</u> 1968, 30, 3333-

128. Sherry, H. S. <u>J. Phys. Chem.</u> 1966, 70, 1158-

129. Barrer, R. M.; Townsend, R. P. <u>J. Chem. Soc. Faraday Trans. 1</u> 1978, 74, 745-755.

130. Maes, A.; Cremers, A. <u>Adv. Chem. Ser.</u> 1973, 121, 230-239.

131. Davis, J. A. <u>Geochim. Cosmochim. Acta</u>, 1984, 48, 679-691.

132. Tipping, E. <u>Geochim. Cosmochim. Acta</u>, 1981, 45, 191-199.

133. Baeyens, B.; Verheyden, D.; Maes, A.; Cremers, A. <u>J. Environ. Rad.</u> (in press).

RECEIVED August 5, 1986

14

Potassium Fixation in Smectite by Wetting and Drying

Dennis D. Eberl[1], Jan Środoń[2], and H. Roy Northrop[3]

[1]U.S. Geological Survey, Box 25046, Federal Center, Mail Stop 404, Denver, CO 80225
[2]Institute of Geological Sciences, Polish Academy of Sciences, 31–002 Krakow, Senacka 3, Poland
[3]U.S. Geological Survey, Box 25046, Federal Center, Mail Stop 963, Denver, CO 80225

Potassium-smectites with various layer-charge
densities and layer-charge locations were subjected
to as many as 100 wetting and drying cycles, thereby
producing randomly interstratified illite/smectite
that contained illite layers stable against exchange
by 0.1 $\underline{N}$ SrCl$_2$ or 1 $\underline{N}$ NaCl. The percentage of illite
layers formed by this process in montmorillonite was
proportional to layer-charge and, based upon very
limited data, their stability with respect to subsequent exchange was inversely proportional to alpha,
the angle of tetrahedral rotation. Most of the illite
layers were produced during the first 20 cycles.
Sodium-smectites treated in wetting and drying experiments that contained potassium-minerals (e.g.
feldspar) formed illite layers by fixing potassium
released from dissolution of the potassium-minerals.
The presence of NaCl, KCl, and HCl in the experimental
solutions had little effect on reaction rates, but
CaCl$_2$ decreased and KOH increased the rate of illitization. The reaction with KOH increased layer charge,
whereas, in all of the other experiments, layer charge
remained constant. Oxygen isotope data confirmed the
conclusion drawn from chemical data that the reaction
mechanism for illite formation at high pH (chemical
reaction of 2:1 layers) differed from that found at
more acid pH (mechanical rearrangement of 2:1 layers
around potassium). The wetting and drying process
may be responsible for producing mixed-layer illite/
smectite from smectite at surface temperatures, and
for accelerated dissolution of sparingly soluble
potassium-minerals in sediments and soils.

One method for studying reactions that occur at the mineral-water
interface is to concentrate the interface by drying a clay mineral
slurry. Certain reactions are promoted by this process, for
example, reactions that respond to Brønsted acidity (1, 2). Repeated

0097–6156/86/0323–0296$08.75/0
© 1986 American Chemical Society

wetting and drying (WD) cycles may be used to increase reaction yield.

Another type of reaction that responds to WD cycles is the fixation of K and NH_4 ions by smectite (3-7). The fixation of K in smectite has been studied extensively by soil scientists because of its effect on the availability of plant nutrients. The reaction also decreases smectite's ability to swell, decreases its cation exchange capacity (CEC), and modifies its Brønsted acidity. Therefore, an understanding of this phenomenon is applicable to many fields of study that are concerned with swelling clays, fields such as soil fertility, soil mechanics, waste disposal, clay catalysis, and the geochemistry of ground and surface waters.

Previous Work

A review of the literature concerning K-fixation by smectite reveals the following patterns of reaction:

(1) *Simple K-exchange of smectite without WD forms illite-like layers in the clay, but this reaction is completely reversible with respect to exchange by cations having substantial hydration energy.* For example, Weaver (8) reported that smectite derived from the weathering of muscovite (Womble shale) collapsed to 1.0-nm during K-saturation, but re-expanded completely during Ca-exchange as long as heat was not applied. K-Ca exchange isotherms for smectite (9-11) show complete reversibility, even though absorption sites with enthalpies similar to mica and vermiculite are present, and even after heating to 300°C. When K-exchanged smectite is exchanged by cations with minimal hydration energy, however, some of the K may remain fixed (i.e. non-exchangeable with respect to a standard cation exchange procedure) even though the clay has not been subjected to WD cycles. For example, NH_4, a cation with a hydration energy similar to that of K, also becomes fixed, and, therefore, replaces only part of the K, trapping the rest in collapsed interlayers. Both K and NH_4 are completely replaced, however, by exchange with Na, Mg, and Ca (12, 13). Simple K-exchange may produce an interlayering of three types of layers in smectite (14). For example, K- and NH_4-smectites investigated by direct Fourier analysis of their X-ray diffraction (XRD) patterns (15) were determined to be composed of randomly interstratified 1.0/1.26/1.56-nm layers (air-dry) or 1.0/1.4/1.68-nm layers (ethylene-glycol treated). More of the contracted (1.0-nm) layers were found in air-dry samples than in glycolated samples, thereby confirming earlier work on the sensitivity of the structure of K-smectite to solvent type (16). The structure also is sensitive to relative humidity (17).

(2) *The percentage of illite layers produced by K-exchange in smectite is proportional to the CEC of the clay (18-21).* Schultz (22) determined that among 83 smectites studied, only minerals of the Wyoming type [charge less than about -0.40 equivalents per $O_{10}(OH)_2$] do not change XRD characteristics after K-exchange and heat treatment, and that total layer charge, rather than charge location, is the important factor in determining the extent of layer expansion. Yaalon and Koyumdjisky (23) reported that high-charge smectites fix K by simple exchange with respect to subsequent NH_4-exchange, but that neither cations are fixed in low-charge materials.

(3) *Wetting and drying of K-exchanged smectite may fix K irreversibly, fixation being defined as a lack of K-exchange with respect to a standard-exchange procedure, such as shaking overnight in 1 N NaCl.* Wetting and drying experiments, performed with several different smectites and four cations (K, Na, Ca, Mg), indicated that only K is fixed, and that the original CEC of the smectite is decreased by as much as 52% by K-fixation (18). Small quantities of K are fixed without drying by illitic material, but WD is necessary for irreversible fixation by smectite (19). One complication is that interlayer Al- and Fe-complexes can block K-fixation (19).

(4) *The mechanism of K-fixation appears to be a trapping of dehydrated K-ions in "hexagonal" holes between 2:1 layers.* Wetting and drying may mechanically rearrange 2:1 layers around K into more stable configurations; i.e., line up "hexagonal" holes across interlayer space and change the coordination of interlayer K-ions from prismatic to octahedral by rotating adjacent 2:1 layers by ±60°. The reaction appears to result from an interaction between cation-hydration energy, 2:1 layer-charge density, and structural arrangements between adjacent 2:1 layers.

Page and Baver (24) suggested that the unique ability of K to be fixed by smectite is related to its ionic size, the dehydrated K-ion having the correct radius to fit into "hexagonal" holes in the basal oxygen planes of the 2:1 layers. However, potassium's minimal hydration energy also has been considered to be the major factor in K-fixation (25, 26). The latter hypothesis is consistent with experiments (27, 28) that determined that Cs and Rb also are are fixed by smectites and vermiculites, even though these ions are too large to fit comfortably into the "hexagonal" holes. In addition, others (29, 30) have determined that K may not fit well into these holes because of distortions related to tetrahedral rotations in the 2:1 layers. Eberl (31) presented a hypothesis for cation fixation and selectivity by smectite, based on a consideration of cation-hydration energies and smectite surface-charge density. The possibility that a mechanical rearrangement of 2:1 layers during WD is responsible for greater stability is supported by structural studies of K-smectites that showed that K-fixation by WD is accompanied by the development of tridimensional ordering of the initially turbostratic smectite structure (32, 33).

Aims of the Present Study

Previous studies have shown that elevated temperatures, such as those prevailing during deep burial, and long reaction times are required to produce illite/smectite (I/S) from smectite by chemical reaction (34-38). However, mixed-layer illite/smectites that have not been recycled from older materials occur in sediments that have never been exposed to elevated temperatures (39, 40), thereby indicating that another mechanism may be responsible for producing I/S at low temperatures. The studies discussed previously suggest that this mechanism is WD. Thus, WD was investigated systematically to discover how smectite-crystal chemistry and how chemical environment affect illite-layer formation by this mechanism. Recent methods for XRD analysis of mixed-layer clays (41-43) permitted a study in greater detail than was possible previously.

Materials and Methods

Smectites with a range of layer charges and a variation in layer-charge location were chosen for experimentation (Table I). Chemical analyses for some of these smectites, which had been purified and size-fractionated after the methods of Jackson (44), are given in Table II; structural formulae are presented in Table III. The formulae in Table III are average compositions for all of the smectite layers in a given smectite, although each smectite probably has a heterogeneous layer-charge distribution. The starting smectites were analyzed for 2:1 layer chemistry by X-ray spectroscopy (45); starting smectites and experimental products were analyzed for interlayer chemistry by atomic-absorption spectroscopy (46). Coarse-grain K-minerals (Table I) were ground, and then washed by shaking in distilled water overnight, then washed several more times to remove readily soluble K and to clean the samples of very fine (<5µm) material. K-, Na-, or Sr-saturation of the starting smectites was achieved by shaking overnight in 1 $\underline{N}$ chloride solutions, followed by two more 1 $\underline{N}$ chloride exchanges, by several washings in distilled water, and finally by dialysis, until chloride could not be detected by the $AgNO_3$ test. The first Sr-exchange of the experimental products was accomplished by shaking in 0.1 $\underline{N}$ $SrCl_2$ overnight; subsequent Sr-exchanges were accomplished by shaking for about 1 hour in 0.1 $\underline{N}$ solutions. Experimental products then were washed and dialyzed as discussed previously.

WD experiments were conducted at 30°, 60°, and 90°C in a drying oven, by putting 300 mg of clay together with 20 mL of solution in polyethylene-weighing boats. Most experiments were conducted without shaking; 2 WD cycles were completed each day at 60°C. In several experiments, clays were dispersed completely by shaking prior to each drying cycle, and 1 WD cycle was completed per day at 60°C. A Na-K exchange isotherm was determined for the Kinney montmorillonite by overnight shaking of 150 mg of clay dispersed in 100 mL of 0.05 $\underline{N}$ solutions having different NaCl/KCl ratios.

Solid-experimental products were investigated by powder XRD (automated Siemens D-500 system; the use of trade names in this paper is for identification only and does not constitute endorsement by the U.S. Geological Survey) of oriented, glass-slide preparations solvated with ethylene glycol (42). Clays subjected to WD cycles in distilled water were investigated with no additional treatment, and after exchange with Na or Sr. Clays processed in electrolyte solutions were washed prior to analysis, and then given additional exchanges by dispersion, centrifugation, and dialysis. Clays processed with K-minerals largely were separated from these K-minerals prior to XRD analysis.

Illite contents of the run products were measured by the techniques of Šrodoń (43, 44) using crystallite thickness determined from XRD characteristics of the starting smectites. Most illite contents were measured using the 003 and 005 XRD peaks (smectite reflections), but the 003 and 006 peaks (47) were used in the runs that contained muscovite.

Oxygen isotopes were analyzed using the technique of Clayton and Mayeda (69). Data are presented using the standard δ notation relative to Standard Mean Ocean Water (SMOW).

Table I. Description of Minerals[1] Used in Wetting and Drying Experiments

Mineral	Source	Size Fraction (μm)	Pretreatment
Smectites			
Black Jack	Smithsonian Institution	Bulk	None
Cheto	SCMR[2], SAz-1	<0.2	NaOAc buffer
Ferruginous	SCMR, SWa-1	<0.2	NaOAc buffer
Garfield "A"	Wards[3], #33B	<0.2	NaOAc buffer
Hectorite	SCMR, SHCa-1	<0.2	1 N NaCl
Kinney	D. D. Eberl (14)	<0.2	NaOAc buffer
Montmorillon	R. E. Grim	<0.2	NaOAc buffer
Otay	SCMR, S-Ca-2	<0.2	None
Texas	SCMR, STx-1	<0.2	1 N NaCl
Umiat	J. Hower	<2.0	None
Wyoming	SCMR, SWy-1	<0.2	1 N NaCl
K-Minerals[4]			
Microcline	C. V. Clemency	120 to 5.0	Washed in H_2O
Muscovite	C. V. Clemency	37 to 5.0	Washed in H_2O
Phlogopite	C. V. Clemency	37 to 5.0	Washed in H_2O
Glauconite	J. Hower, #39 (47)	<2.0	None
Illitic material	D. J. Morgan, #M8 (53)	<5.0	1 N $CaCl_2$

[1]For other data concerning the clays, see compilations in (57-59).
[2]Source Clay Mineral Repository, University of Missouri, Department of Geology, Columbia, Missouri 65211.
[3]Ward's Natural Science Establishment, Inc., Rochester, New York.
[4]For data on the K-minerals, see (60-62, 47, 53).

Table II. Chemical Analyses of Some Starting Smectites[1] Used in Wetting and Drying Experiments

Smectite	Weight Percent									
	SiO_2	Al_2O_3	Fe_2O_3	MgO	Li_2O	Na_2O	K_2O	SrO	TiO_2	LOI[2]
Cheto	56.0	16.2	1.49	5.25	NA[3]	0.12	0.39	6.17	0.23	16.9
Ferruginous	49.0	10.8	21.7	1.22	NA	0.04	0.89	4.31	0.58	15.0
Garfield "A"[4]	41.9	5.9	30.4	0.85	NA	0.25	0.09	4.71	0.05	20.0
Hectorite	52.3	0.52	0.17	23.2	1.31	0.13	0.93	3.49	0.03	20.7
Montmorillon	53.0	20.7	0.64	2.83	NA	0.13	0.09	5.25	0.02	19.4
Otay	54.2	15.6	1.55	6.86	NA	0.12	0.59	6.19	0.24	19.1
Texas	57.8	17.5	1.20	3.75	NA	0.09	0.09	5.01	0.26	17.3
Umiat	53.9	20.3	3.39	2.83	NA	0.07	0.17	4.92	0.17	18.8
Wyoming[4]	57.5	20.1	4.25	3.07	NA	0.31	1.11	3.28	0.09	11.7

[1]Sr-saturated. Analysis for Si, Al, Fe, Mg, and Ti by X-ray fluorescence, A. J. Bartel, K. Stewart, and J. Taggert, analysts. Analysis for Li by induction-coupled plasma spectroscopy, S. Wilson, analyst. Analysis for Na, K, and Sr by atomic-absorption spectroscopy, D. D. Eberl, analyst.

[2]Loss on ignition at 900°C.

[3]Not analyzed.

[4]Treated by WD for 100 cycles prior to Sr-saturation and to chemical analysis for interlayer cations.

Table III. Structural Formulae for Smectites Used in Wetting and Drying Experiments, Based on $O_{10}(OH)_2$ [Formulae Calculated from Data in Table II, Unless Noted Otherwise]

Smectite	Tetrahedral[1]			Octahedral				Interlayer			Octahedral Charge (Percent)	Total Layer Charge (Equivalents)
	Si (Equivalents)	Al	Fe^{3+}	Al (Equivalents)	Fe^{3+}	Mg	Li	Na (Equivalents)	K	Sr		
Black Jack[2]	3.48	0.52	0	1.99	0.02	0.01	NA^3	0.46	0	0	0	0.47
Cheto	4.00	0	0	1.36	0.08	0.56	NA	0.02	0.04	0.25	100	0.56
Ferruginous	3.68	0.32	0	0.64	1.22	0.14	NA	0.01	0.09	0.18	30	0.46
Garfield A	3.46	0.54	0	0.03	1.89	0.11	NA	0.04	0.08	0.23	4	0.56
Garfield B[2]	3.45	0.39	0.16	0.15	1.83	0.02	NA	0.57	0	0	4	0.57
Hectorite	3.95	0.05	0	0	0.01	2.61	0.40	0.02	0.09	0.15	85	0.41
Kinney[2]	3.97	0.03	0	1.47	0.07	0.46	0	0.47	0	0	96	0.49
Montmorillon	3.87	0.13	0	1.65	0.04	0.31	NA	0.02	0.01	0.22	67	0.44
Otay	3.92	0.08	0	1.25	0.08	0.74	NA	0.02	0.06	0.26	87	0.61
Texas[4]	4.00	0	0	1.52	0.07	0.41	NA	0.01	0.01	0.21	100	0.41
Uniat	3.83	0.17	0	1.53	0.18	0.30	NA	0.01	0.02	0.20	61	0.44
Wyoming	3.89	0.11	0	1.49	0.22	0.31	NA	0.04	0.09	0.13	69	0.36

[1]Ti is considered to be present as discrete anatase.
[2]Formulae for Black Jack from (63), for Garfield B from (51), and for Kinney from (14).
[3]Not analyzed.
[4]3.5% SiO_2, presumably present as amorphous material, subtracted from the analysis in Table II to give an ideal tetrahedral occupancy.

Experimental Results

Behavior of K-Exchanged Smectites Without Wetting and Drying. All of the smectites (Table I) gave XRD patterns characteristic for randomly interstratified (RO) mixed-layered minerals upon K-exchange. Those of low layer charge (e.g. Wyoming) gave patterns for I/S with a few percent illite layers. Those of higher layer charge were identified as 3-component 1.0/1.4/1.7-nm interstratifications. This heterogeneous swelling probably is related to heterogeneous layer-charge distributions in the smectites. XRD patterns for the 3-component interstratifications differed from those of the 2-component, in that the 001 was displaced toward higher angles; the 002 was much more diffuse; the 003 was displaced toward higher angles; and the 005 was displaced slightly towards lower angles of two-theta. The latter two peaks for the 3-component system plot on the extreme left side or outside of the graph used to identify mixed-layer I/S (see Figure 8A in 43). Saturation of K-smectite with an ion of higher hydration energy led to complete re-expansion, in agreement with previous studies. For example, XRD patterns for K- and Na-Black Jack smectite are given in Figure 1.

A Na-K exchange isotherm was determined using Na-Kinney smectite to study the evolution of 3-component mixed-layering with increasing K content. This evolution was followed by plotting the changing positions of the 003 and 005 XRD peaks as a function of K/Na ratio for the exchange solution on a graph used to determine illite contents and glycol spacings for I/S (Figure 2). From a K/Na ratio of 0 to about 25/75, the illite content of the sample increased from 0 to about 49%, and the glycol spacing of the expanded layers increased, as is expected for Na-K exchange (42). At K/Na ratios between 25/75 and 35/65, illite content remained nearly constant, and glycol spacing continued to increase, which is interpreted as a continued replacement of Na by K in expanded layers. At K/Na ratios greater than 40/60, the XRD spacings evolved out of the two-theta range possible for I/S; as discussed above, characteristics of 3-component interstratification could be observed. In particular, the 001 reflection began to migrate towards greater angles at Na/K of 35/65, as is expected for the development of 1.4-nm layers.

All of the K-exchanged smectites (Table I) re-expanded completely when exchanged four times with 1 N NaCl solution. Then it was determined that a single, overnight exchange with 0.1 N SrCl$_2$ is the mildest treatment sufficient to restore the K-Kinney to a 100% smectite structure, which means reexpanding to 1.7-nm on glycol saturation both the vermiculite-like (1.4 nm) and the unstable illite-like (1.0 nm) layers. This treatment subsequently was used as a standard test because, unlike Na, Sr is not replaced by hydronium ions during dialysis (49); therefore, Sr gives a better measure of CEC during chemical analysis. Illite layers formed by burial diagenesis are not affected significantly by either the Na- or the Sr-exchange procedures.

Response of K-Smectites to WD Cycles in Water at 60°C. Results of these WD experiments are given in Table IV; sample XRD patterns are given in Figures 3 and 4. The following patterns of reaction are evident from Table IV:

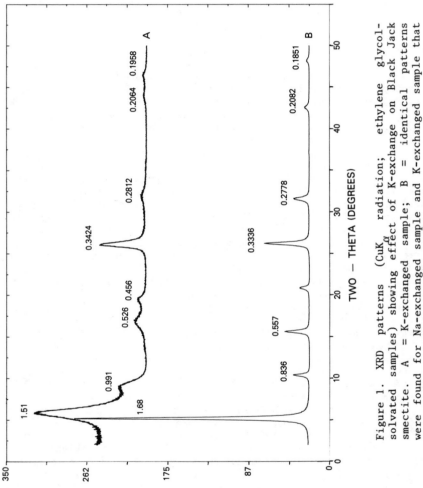

Figure 1. XRD patterns (CuK$_\alpha$ radiation; ethylene glycol-solvated samples) showing effect of K-exchange on Black Jack smectite. A = K-exchanged sample; B = identical patterns were found for Na-exchanged sample and K-exchanged sample that was then Na-exchanged. Peaks labeled in nm.

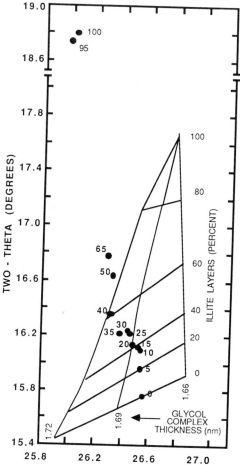

Figure 2. Plot of XRD peak positions (CuK$_\alpha$ radiation; ethylene glycol-solvated samples) for Kinney smectite treated with 0.05 $\underline{N}$ Na + K exchange solutions. Experimental points are labeled with percentages of K in solution. The graph, used to determine percentage illite layers and glycol-spacing for illite/smectites having crystallite thickness of 1-14 layers, is from (42).

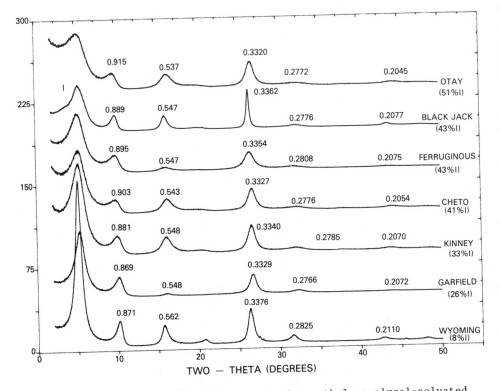

Figure 3. XRD patterns (CuK$_\alpha$ radiation; ethylene glycol-solvated samples) for Sr-exchanged K-smectites that have been subjected to 100 WD cycles in water at 60°C. A low-angle shoulder that indicates a trace of R1 ordering is marked by a tick on the Black Jack pattern. Peaks labeled in nm.

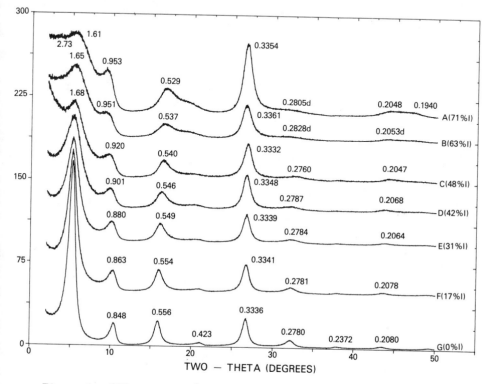

Figure 4. XRD patterns (CuK$_\alpha$ radiation; ethylene glycol-solvated
samples) showing a range of illite contents for K-Kinney smectite
that has been subjected to various treatments. A = 100 WD cycles
in 0.5 $\underline{N}$ NaCl; B = 40 WD cycles in 0.5 $\underline{N}$ NaCl; C = 100 WD cycles
in 0.5 $\underline{N}$ KOH, with 1 Sr exchange; D = 100 WD cycles in 0.5 $\underline{N}$
CaCl$_2$; E = 100 WD cycles in 0.5 $\underline{N}$ KCl, with 1 Sr-exchange; F =
100 WD cycles in 0.5 $\underline{N}$ CaCl$_2$, with 1 Sr-exchange; G = clay left
in suspension for a time equivalent to 100 WD cycles, with 1
Sr-exchange. Peaks labeled in nm.

Table IV. Percentage of Illite Layers and Interlayer Chemistry of K-Smectites Subjected to Wetting and Drying Cycles in Water at 60°C, and Then Exchanged with 0.1 N SrCl2

Reference Number	Sample	Number of WD Cycles	Number of Sr Exchanges	Meq per 100 g Oxide				Illite Layers (Percent)
				Na	K	Sr	Total	
1	K-Black Jack	100	1	0	52	80	132	43
2		100	3	6	9	122	137	0
3	K-Cheto	0	1	4	8	119	131	0
4		100	1	3	51	81	135	41
5		100	3	4	23	93	120	15
6	K-Ferruginous	0	1	1	19	83	103	0
7		5	1	0	31	71	102	33
8		10	1	1	32	72	105	38
9		15	1	5	28	64	97	36
10		20	1	0	30	66	96	36
11		25	1	0	33	68	101	37
12		30	1	1	34	66	100	41
13		40	1	3	35	68	106	41
14		50	1	0	34	66	100	41
15		75	1	0	36	68	104	41
16		100	1	0	35	68	103	43
17		100	3	6	21	78	105	29
18	K-Garfield	100	1	8	16	91	115	26
19	K-Hectorite	100	1	4	20	67	91	12
20	K-Kinney	5	1	4	29	92	125	22
21		10	1	3	33	89	125	26
22		15	1	3	36	87	126	26
23		20	1	4	34	80	118	28
24		25	1	4	35	83	122	28
25		30	1	4	35	86	125	29
26		40	1	4	37	83	124	30
27		50	1	4	37	82	123	30

	Name							
28		75	1	4	40	79	123	32
29		100	1	3	39	85	127	33
30	K-Montmorillon	100	3	NA¹	NA	NA	––	9
31		0	1	4	2	101	107	0
32		100	1	2	40	69	111	27
33		100	3	3	21	88	112	17
34	K-Otay	0	1	4	13	120	137	5
35		100	1	4	56	71	131	51
36	K-Texas	100	3	8	51	73	132	26
37		0	1	3	2	97	102	0
38		100	1	4	38	63	105	22
39	K-Umiat	100	3	3	22	81	105	11
40		0	1	2	4	95	101	0
41		100	1	6	29	74	109	10
42	K-Wyoming	100	3	3	20	79	102	3
43		100	1	10	24	63	97	8

¹Not analyzed.

(1) WD cycles stabilize 1.0-nm layers against exchange by a cation with substantial hydration energy.

(2) A regular decrease in CEC (meq Sr) occurs as the per- centage of fixed K and percentage of illite layers increases.

(3) The quantity of illite layers formed after 100-WD cycles and 1 Sr-exchange is proportional to layer charge (data from Tables III and IV plotted in Figure 5), although smectites with a large component of octahedral charge (montmorillonites) follow this relation better than do those that have a large component of tetrahedral charge (beidellites and one montmorillonite with 61% octahedral charge). A perfect correspondence between the quantity of illite layers formed by WD and layer charge is not expected, however, because this reaction may also depend on other factors, such as layer-charge distribution. Figure 5 may indicate that charge distribution is more heterogeneous for the beidellitic smectites.

(4) The total milliequivalents of interlayer cations is not changed significantly by the WD process, thereby indicating that 2:1 layer charge does not change. These data suggest that the reaction to illite is a transformation reaction caused by simple dehydration and rearrangement of smectite layers, rather than by an increase in 2:1 layer charge.

All of the I/S produced by WD are randomly interstratified, with the possible exception of the Black Jack sample (Figure 3) and the most illitic Kinney sample (Figure 4), both of which show signs of partial R1 ordering between illite and smectite layers. For these samples, the 001 XRD reflections are displaced towards larger angles, and a very weak superlattice reflection is visible at small angles.

The stability of the illite layers produced by WD was investigated further by applying two additional 0.1 $\underline{N}$ SrCl$_2$ exchanges to WD smectites for which sufficient sample was available (Table IV). The quantity of illite layers decreased significantly with this treatment, but measurable illite layers were preserved in all samples except the Black Jack smectite. Exchange experiments with a WD Kinney sample showed that the use of three 0.1 $\underline{N}$ SrCl$_2$ exchanges is a slightly stronger treatment than the use of four 1 $\underline{N}$ NaCl exchanges used to study diagenetic I/S ($\underline{44}$). Thus, the illite layers remaining in the WD clays after three Sr-exchanges may be of comparable stability to those formed by burial diagenesis.

The percentage change in illite layers between one and three exchanges with 0.1 $\underline{N}$ SrCl$_2$, which is a measure of the stability of the illite layers formed by WD, correlates neither to original smectite-layer charge, nor to layer-charge location, but, based on very limited data, does correlate roughly to the mean angle of tetrahedral rotation of the original smectite (Figure 6). Angle of tetrahedral rotation (α) was calculated from structural formulae (Table III) and measurements of b (Table V) by the following equations ($\underline{29}$, $\underline{50}$):

$$\text{Cos } \alpha = b_{observed} \div b_{ideal}$$

where

$$b_{ideal}(Si_{1-x}Al_x) = 9.15 + 0.74x$$

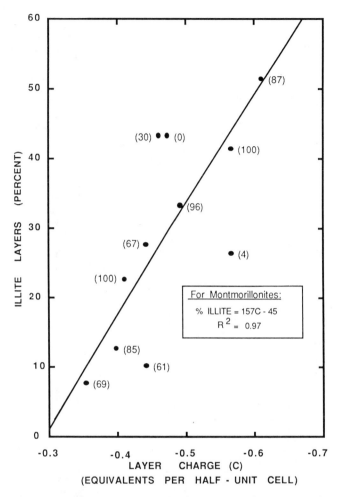

Figure 5. Percentage illite layers versus layer charge for K-smectites subjected to 100 WD cycles in water at 60°C and 1 Sr-exchange. Numbers in parentheses refer to percentage of octahedral charge. Best fit line is for montmorillonites having 69% or more octahedral charge. Data from Tables III and IV.

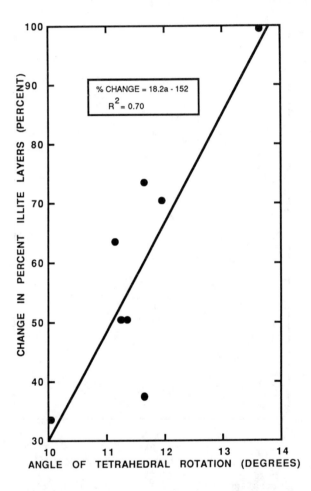

Figure 6. Stability of illite layers formed by WD mechanism: percentage of change in percentage illite layers between 1 and 3 Sr-exchanges is plotted against α, the angle of tetrahedral rotation. Data from Table V.

Table V.   Data Used to Construct Figure 6 for K-Smectites Submitted to
100 Wetting and Drying Cycles, and Then Exchanged Once and Three
Times With 0.1 $\underline{N}$ SrCl$_2$

| Clay | Original Smectite (nm) | | α | Illite Layers[1] |
	b$_{observed}$	b$_{ideal}$	(Degrees)	(Percent Change)
Black Jack	0.899	0.925	13.6	100
Cheto	0.898	0.915	11.1	63
Ferruginous	0.907	0.921	10.0	33
Garfield[2]	0.914	-----	6	39
Hectorite	0.909	0.916	7.1	NA[3]
Kinney	0.897	0.916	11.7	73
Montmorillon	0.898	0.917	11.7	37
Otay	0.899	0.917	11.4	50
Texas	0.897	0.915	11.4	50
Umiat	0.898	0.918	12.0	70
Wyoming	0.898	0.917	11.7	NA

[1] $\dfrac{(\% \text{ illite layers, 1 Sr-exchange}) - (\% \text{ illite layers, 3 Sr-exchanges})}{(\% \text{ illite layers, 1 Sr-exchange})}$.
[2] α measured directly ($\underline{51}$).
[3] Not analyzed.

This method for estimating α applies generally if α is greater than about 7°, and assumes an ideal O(apical)-Si-O(basal) angle of 109°28'. A relationship between α and illite layer stability is expected, because the larger the α is, the smaller the ditrigonal holes are in the basal oxygen planes of the 2:1 layers. The smaller the holes are, the less deeply dehydrated K ions can penetrate into 2:1 layers; thus, the more susceptible the K ions are to Sr-exchange. Alpha is only an approximate indicator of hole size; it would be better to measure directly the hole size, or the K-O bond lengths for the illite layers. But such a detailed structural analysis is beyond the scope of the present work.

Tetrahedral rotations in dioctahedral clays are required to adjust for a misfit in size between the normally larger tetrahedral sheet and the octahedral sheet. Rotation angles of 30° close the holes completely. The hole is so large for small angles of α that the hole approaches hexagonal symmetry; in this case, interlayer K enters more nearly into 12-fold coordination, rather than into the 6-fold coordination of ditrigonal holes (30). Thus, perhaps the reason that the Garfield sample, with a small, measured $\alpha = 6°$ (51), does not lie on the curve in Figure 6 is because it belongs to an energetically different system. But other factors, in addition to α, could influence, or even play a dominant role, in affecting the availability of K to Sr-exchange. Two such factors are the presence of heterogeneous layer charges, which would tend to bind K more strongly in highly-charged interlayers, and the presence or absence of three-dimensional ordering in the WD smectite-crystal structures, which could lead either to prismatic or octahedral coordination for K in the dehydrated interlayers, and to various distortions in the 2:1 layers that could affect K availability (V. A. Drits, personal communication).

The kinetics of the transformation towards illite for Ferruginous and Kinney smectites are shown in Figure 7. Most of the illite layers are formed in the first 20 WD cycles, although illite content appears still to be increasing slightly after 100 WD cycles. The Ferruginous smectite produces more illite layers for a given number of WD cycles than the Kinney does even though the Ferruginous smectite has a slightly smaller layer charge (Table III). This effect may be related to the much larger component of tetrahedral charge, to the slightly smaller degree of tetrahedral rotation for the Ferruginous sample, and to the other factors mentioned above. Similar experiments with other iron-rich smectites are needed to resolve this question.

Shaking in water prior to each drying cycle speeds reaction. For example, a K-Kinney smectite subjected to 64 WD cycles produced 42% illite layers, with shaking, compared with 30% for K-Kinney subjected to 50 WD cycles, and 32% for K-Kinney subjected to 75 WD cycles, without shaking.

The equivalents of fixed-interlayer cations per illite layer decrease with increasing illite content for the WD samples (Figure 8). Fixed cations are K, with a very small component of Na that was not removed by Sr-exchange (Tables IV and VI). The relation roughly fits a power curve, as depicted in the figure. Calculations for samples that plot on the extreme lower end of the curve (at 1.0 equivalents) actually gave equivalents of fixed-interlayer cations greater than one, which is impossible structurally, because there are not enough holes in the basal oxygen planes to accommodate more

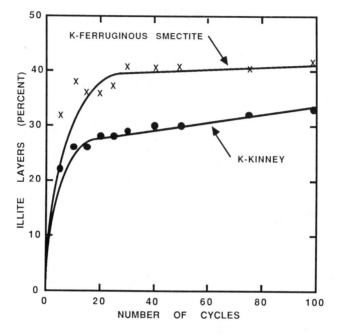

Figure 7.   Kinetics of WD illitization: percentage illite layers
versus number of WD cycles for Ferruginous and Kinney smectites
after 1 Sr-exchange.   Data from Table IV.

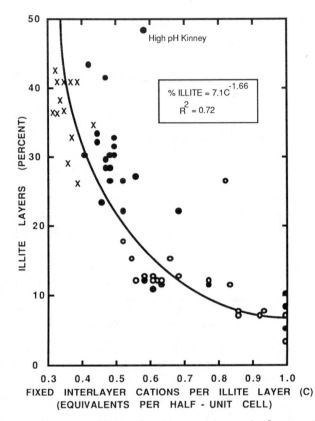

Figure 8. Percentage illite layers versus equivalents of fixed interlayer cations (Na + K) per illite layer [based on $O_{10}(OH)_2$]. Solid circles = aluminous smectites with 1 Sr-exchange. Open circles = aluminous smectites with 2 or 3 Sr-exchanges. X = iron-rich smectites with 1 Sr-exchange. Points calculated from data in Tables III, IV, VI, and VIII.

than 1.0 equivalents of K-ions. The excess is attributed to errors in determining illite contents in highly smectitic clays. The data in Figure 8 indicate, from a consideration of the increase in illitic content with increasing WD cycles (Table IV and Figure 7), that a greater number of WD cycles is required to stabilize low-charge illite layers.

The iron-rich samples plot mostly to the left of the curve in Figure 8, thereby indicating that iron-rich clays tested in these experiments can form illite layers by WD at a lower fixed cation content than aluminous clays can. A similar observation has been made for iron-rich illite layers (glauconite) produced by diagenesis (52). This effect may be related to the deeper penetration of K into the larger basal holes found in iron-rich 2:1 clays, as well as to the other factors mentioned above.

The pattern in Figure 8 is distinct from that for randomly interstratified I/S produced from bentonite by burial diagenesis (Figure 9). Fixed interlayer cation content for the latter clays is relatively constant at about 0.55 equivalents per illite layer for clays that contain less than 50% illite layers (53).

Effect of Solution Composition. The effect of changing solution composition on the reaction of K-Kinney is given in Table VI. The

Table VI. Effect of Solution Composition on the Reaction of K-Kinney Submitted to 40 Wetting and Drying Cycles at 60°C, and Then Exchanged Once With 0.1 N SrCl₂

Reference		Meq per 100 g Oxide				Illite Layers
	Solution					
Number		Na	K	Sr	Total	(Percent)
26	Water	4	37	83	124	30
44	0.5 N NaCl	5	35	82	122	32
45	0.5 N KCl	3	40	81	125	31
46	0.5 N CaCl₂	3	27	96	126	23
47	pH 3.6 (HCl)	3	32	85	121	30
48[1]	0.5 N KOH	6	71	77	154	48

[1]Average of two analyses.

presence of 0.5 N NaCl, 0.5 N KCl, and HCl (pH 3.6) had no effect on reaction rate as compared with experiments made in water. The 0.5 N CaCl₂ solution significantly slowed the rate of reaction, whereas the 0.5 N KOH solution increased the rate. With one exception, the data in Table VI indicate that total layer charge did not change significantly during reaction. The exception is the high pH run (KOH) in which layer charge increased, thereby indicating that the composition of the 2:1 layer was altered. The average charge for illite layers in this sample (0.57) is plotted in Figure 8, and is greater than that expected for clays of this expandability formed by WD. It fits better on the diagenetic curve (Figure 9). In addition, the average charge for expanded layers increased from -0.49 for untreated Kinney to about -0.52.

Oxygen isotope data (Table VII) support conclusions drawn from the chemical data (Tables IV and VI) that clay in the high pH, WD experiments underwent significant chemical reaction, whereas clay

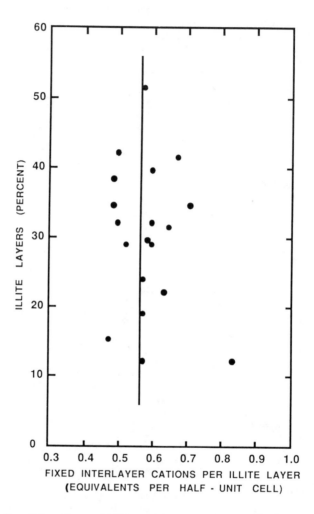

Figure 9.   Percentage illite layers versus equivalents of fixed interlayer cations per illite layer [based on $O_{10}(OH)_2$] for RO illite/smectites formed by diagenesis in bentonites.   Calculated from data in (53).

Table VII. Oxygen Isotope Data for Various Smectites

Smectite	Number of WD Cycles	Solution Composition	Measurements of $\delta^{18}O$	Average $\delta^{18}O$
Kinney	0	Water	13.8, 13.4, 14.6, 14.4	14.1
	64	Water	13.8	13.8
	100	Water	13.9, 14.1, 14.2	14.1
	40	0.5 N KOH	15.9, 16.5, 15.7	16.0
Black Jack	0	Water	1.6, 1.6, 1.4, 1.5, 2.1	1.6
	100	Water	2.0, 1.2, 1.3, 1.2, 2.0	1.5
Otay	0	Water	17.0, 17.8	17.4
	100	Water	16.9, 16.5	16.7
Montmorillon	0	Water	26.3	26.3
	100	Water	25.0	25.0
Texas	0	Water	26.2, 25.4, 25.5, 26.8	26.0
	100	Water	26.0, 25.5, 25.6, 25.8	25.7
Hector	0	Water	22.2, 22.3, 21.7, 21.7, 21.9, 22.0	22.0
	100	Water	21.5, 20.8, 21.7, 21.7, 21.7, 21.2	21.4
Umiat	0	Water	15.2, 14.5, 14.3	14.7
	100	Water	13.8, 14.2, 13.7	13.9
Cheto	0	Water	17.1, 17.2, 17.4	17.2
	100	Water	17.4, 16.9, 17.0	17.1
Ferruginous	0	Water	12.8, 12.7	12.8
	100	Water	12.7, 12.6	12.7

wetted and dried in other experimental solutions did not. Data in
Table VII indicate that clay subjected to WD cycles in distilled
water either remained unchanged isotopically, thereby indicating
that the 2:1 layer chemistry did not change, or became slightly
lighter, which is, perhaps, related to an exchange of oxygen atoms
on crystal edges (54), or to experimental difficulties that were
encountered in removing all of the interlayer water from the
Sr-exchanged samples. The latter possibility is favored by the
finding of larger oxygen yields (micromoles O_2 per mg clay reacted)
for the Sr-saturated clays, and of a decrease in scatter for $\delta^{18}O$
values for clays subjected to increased drying times. The Kinney
smectite, however, became 2 per mil heavier after 40 WD cycles in
the high pH experimental solution, thereby indicating substantial
alteration of 2:1 layer chemistry.

In other experiments, K-Wyoming and K-Otay smectite were
subjected to 20 WD cycles in 0.5 $\underline{N}$ KOH, and then Sr-exchanged. The
Wyoming reacted to form greater than 50% illite layers, giving a
weak XRD pattern after the treatment, and the Otay clay was
destroyed. Otay in suspension for the same length of time with 0.5
$\underline{N}$ KOH, but without WD, reacted to form about 42% illite layers,
thereby demonstrating that WD is not essential for the reaction to
illite at high pH. Previous studies (8, 55) have shown that a
significant number of illite layers can be formed in low-charge
smectites by boiling them in 1 $\underline{N}$ KOH. The increased solubility of
Al and Si at high pH may lead to a chemical reaction in which layer
charge is increased by Al for Si substitution. More work is needed
to confirm this suggestion.

The KCl, HCl, and KOH experimental products (Table VI), when
X-rayed prior to Sr-exchange, have characteristics of 3-component
1.0/1.4/1.7-nm interstratification. The NaCl and $CaCl_2$ experimental
products prior to Sr-saturation, however, were two component
1.0/1.7-nm interstratifications, containing 1.0-nm layers well
in excess of the number remaining after one Sr saturation. WD
experiments with more cycles than those given in Table VI showed
that illite content (no Sr-exchange) increased from 63 to 71%
between 40 and 100 WD cycles for the NaCl experiments, and from
30 to 42% for the $CaCl_2$ experiments. But the stability of these
additional illite layers did not increase with the number of WD
cycles: after one Sr-exchange, the 40- and the 100-cycle runs
differed by only a few percent illite.

Effect of Temperature. Temperature had little effect on the
percentage of illite layers formed from K-Kinney, which was
subjected to as many as 6 WD cycles at 30°, 60°, and 90°C, and then
saturated twice with 0.1 $\underline{N}$ $SrCl_2$ (Table VIII). Unfortunately, the
experimental products were not X-rayed after a single Sr-saturation;
therefore, results in Table VIII are not directly comparable to
those in Table IV. The erratic data resulting from the 90°C
experiments are unexplained.

Reaction with K-Bearing Minerals. WD experiments with mixtures of
Na-Kinney and sparingly soluble K-minerals were undertaken to
simulate natural conditions. When K-feldspar was shaken with
Na-Kinney at room temperature without WD for as long as 1 year, no
illite layers were found in the experimental product. Nor were
illite layers formed when muscovite was shaken with Na-Kinney for

Table VIII. Effect of Temperature on the Formation of Illite Layers in K-Kinney Submitted to Wetting and Drying Cycles, and Then Exchanged Twice With 0.1 N SrCl$_2$

Reference Number	Temperature (°C)	Number of WD Cycles	Meq per 100 g Oxide				Illite Layers (Percent)
			Na	K	Sr	Total	
49	30[1]	0	4	9	113	126	0
50		1	4	13	110	127	7
51		2	3	15	107	125	7
52		3	NA[2]	NA	NA	---	12
53		4	3	18	110	131	12
54		5	3	16	108	127	12
55		6	3	18	106	127	12
56	60	0	3	4	116	123	0
57		1	NA	NA	NA	---	4
58		2	3	16	110	129	11
59		3	3	17	106	126	11
60		4	3	17	106	126	12
61		5	NA	NA	NA	---	11
62		6	3	21	104	129	11
63	90	0	3	5	117	125	0
64		1	2	15	108	125	7
65		2	3	17	107	127	12
66		3	3	19	105	127	7
67		4	3	20	103	126	12
68		5	3	20	103	126	15
69		6	3	18	108	129	8

[1]Dried under vacuum.
[2]Not analyzed.

100 days. Phlogopite, however, formed about 4% 1.0-nm layers when
shaken with Na-Kinney for 100 days, but these layers disappeared
after Sr-exchange.
All WD experiments, however, produced a significant percentage
of illite layers after 100 WD cycles. The rate of illite-layer
formation was greatest for the potassium-feldspar systems, and
least for the systems that contained illitic material. The final
proportion of illite layers depended on the initial ratio of
K-mineral/Na-Kinney in the system, and on the number of subsequent
Sr-exchanges. With a ratio of 0.17, illite layers were unstable
with respect to Sr-exchange. With ratios of 1 or more, and with
the exception of the experiments containing illitic material and
phlogopite, these systems formed illite layers at a rate equivalent
to, or greater than, that found for K-smectite (Table IX).

Conclusions

The experiments indicate that WD can have a major effect on
reactions that occur at the mineral-water interface. This process
can rearrange clay 2:1 layers into more stable configurations, can
substantially alter smectite selectivity for competing exchange
cations, and can weather sparingly-soluble K-minerals in the
presence of smectite.
The experiments also indicate that WD may be an important
mechanism for producing I/S at low temperatures in nature by a
transformation mechanism (56). The percentage of illite layers
formed by this mechanism is proportional to the number of WD cycles,
and to the layer charge of the original smectite. Simple K-exchange
does not produce stable illite layers in smectite; therefore, these
layers probably form by WD prior to deposition in subaqueous
environments. The exception is found in high pH environments where
illite layers may form without WD by chemical reaction, as has been
reported previously for alkaline lakes (64, 65).
Illite layers form relatively quickly by WD (most in less than
20 WD cycles), and the reaction rate is not affected greatly by
changes in solution compositions or temperatures that are typical of
near-surface environments. Thus, that which has been studied in the
laboratory also may occur abundantly in nature.
The fixation of K is expected to occur wherever smectite and
K-minerals are subjected to the wetting and drying process. This
process occurs, for example, in soils, deltas, flood plains, and
playas. Thus, the patterns of reaction described here could be used
geologically in studies of sediment cores to discover when a lake
was dry, or to find the surface of a fluctuating, ancient water
table. From a geochemical perspective, WD concentrates K in the
unsaturated zone, whereas other cations are released to ground
water and surface runoff. Agriculturally, the WD process can free
K from sparingly soluble K-minerals, and then store it with varying
degrees of availability in expanding clay, with availability
depending on smectite-layer charge and on the angle of tetrahedral
rotation. Availability also may be influenced by factors other
than those investigated in these experiments, factors such as
heterogeneous layer-charge distribution (66), and three-dimensional
ordering of the 2:1 layers (67, 68). Finally, WD should be
considered in the design of swelling clay barriers used to contain
toxic chemical and radioactive wastes, because the process could

Table IX. Percentage of Illite Layers Found in Na-Kinney Subjected to 100 Wetting and Drying Cycles in the Presence of a Potassium Mineral at 60 $^{\circ}$C

Reference Number	Potassium Mineral	Ratio of weight K-Mineral to Weight Na-Kinney	Illite layers (%)	
			No Sr-exchange	1 Sr-exchange
70	Microcline	0.17	19	0
71		1.00	NA[1]	43
72		5.00	48	40
73	Muscovite	0.17	25	0
74		1.00	NA	37
75		5.00	47	NA
76	Glauconite	0.17	30	0
77		1.00	NA	37
78	Phlogopite	0.17	28	0
79		1.00	NA	31
80	Illitic	0.17	30	0
81	material	1.00	NA	24

[1] Not analyzed.

cause barriers to crack, or, at least, to lose some of their absorptive properties.

Acknowledgments

We thank C. V. Clemency, R. Grim, J. Hower, D. Morgan, and the Smithsonian Institution for supplying samples. The patient help of Eric Stenzel in conducting numerous wetting and drying cycles is appreciated. Rigel Lustwerk ran many of the oxygen isotope analyses. Gene Whitney and Len Schultz reviewed the original manuscript and offered helpful suggestions for its improvement. V. A. Drits reviewed a later version and offered many penetrating comments. Jan Šrodoń thanks the U.S. Geological Survey for supporting him as a visiting exchange scientist while this study was carried out.

Literature Cited

1. Mortland, M. M.; Raman, K. V. Clays Clay Miner. 1968, 16, 393-8.
2. Theng, B. K. G. "The Chemistry of Clay-Organic Reactions"; Adam Hilger: London, 1974; p. 157.
3. Kellner, O. Landwirtsch. Vers. St. 1887, 33, 359-69.
4. McBeth, I. G. J. Agr. Research. 1917, 4, 155.
5. Volk, N. J. At. J. Sci. 1933, 26, 114-129.
6. Volk, N. J. Soil Sci. 1934, 37, 267-87.
7. Volk, G. W. Soil Sci. 1938, 45, 263-77.
8. Weaver, C. E. Am. Mineral. 1958, 43, 839-61.
9. Hutcheon, A. T. J. Soil Sci. 1966, 17, 339-55.
10. Goulding, K. W. T.; Talibudeen, O. J. Colloid Interface Sci. 1980, 78, 15-24.
11. Inoue, A.; Minato, H. Clays Clay Miner. 1979, 27, 393-401.
12. Barshad, I. Am. Mineral. 1948, 69, 655-78.
13. Barshad, I. Soil Sci. 1951, 72, 361-71.
14. Khoury, H. N.; Eberl, D. N. Jb. Miner. Abh. 1981, 141, 134-141.
15. Cicel, B.; Machajdik, D. Clays Clay Miner. 1981, 29, 40-6.
16. Jonas, E. C.; Thomas, G. L. Clays Clay Miner. 1960, 8, 183.
17. Sayegh, A. H.; Harward, M. E.; Knox, E. G. Am. Mineral. 1965, 50, 490.
18. Truog, E.; Jones, R. J. Ind. and Eng. Chem. 1938, 30, 882-5.
19. Stanford, G. Soil Sci. Soc. Am. Proc. 1948, 12, pp. 167-71.
20. Bailey, T. A. Ph.D. Thesis, University of Wisconsin, Madison, 1942.
21. Weir, A. H. Clay Miner. 1965, 6, 17-22.
22. Schultz, L. G. Clays Clay Miner. 1969, 17, 115-49.
23. Yaalon, D. H.; Koyumdjisky, H. Soil Sci. 1968, 105, 403-8.
24. Page, J. B.; Baver, L. D. Soil Sci. Soc. Am. Proc. 1939, 4, pp. 150-5.
25. Norrish, K. Discuss. Faraday Soc. 1954, 18, 120-34.
26. Sawhney, B. L. Clays Clay Miner. 1972, 20, 93-100.
27. Barshad, I. Am. Mineral. 1950, 35, 225-38.
28. Wear, J. I.; White, J. L. Soil Sci. 1951, 1-14.
29. Radoslovich, E. W.; Norrish, K. Am. Mineral. 1962, 47, 599-616.

30. McCauley, J. W.; Newnham, R. E. Am. Mineral. 1971, 56, 1626-38.
31. Eberl, D. D. Clays Clay Miner. 1980, 28, 161-72.
32. Mamy, J.; Gaultier, J. P. Int. Clay Conf. Proc. 1975, pp. 149-55.
33. Plançon, A.; Besson, G.; Gaultier, J. P.; Mamy, J.; Tchoubar, C. Int. Clay Conf. Proc. 1979, pp. 45-54.
34. Weaver, C. E.; Beck, K. C. Geol. Soc. Am. Spec. Paper 134. 1971.
35. Perry, E.; Hower, J. Clays Clay Miner. 1970, 18, 165-77.
36. Hower, J.; Eslinger, E. V.; Hower, M. E.; Perry, E. A. Geol. Soc. Am. Bull. 1976, 87, 725-37.
37. Eberl, D.; Hower, J. Geol. Soc. Am. Bull. 1976, 87, 1326-30.
38. Środoń, J. Clay Miner. 1984, 19, 205-15.
39. Schultz, L. G. U.S. Geol. Surv. Prof. Paper 1064-A, 1978.
40. Nadeau, P. H.; Reynolds, R. C. Jr. Nature. 1981, 194, 72-74.
41. Reynolds, R. C. Jr. In "Crystal Structures of Clay Minerals and Their X-Ray Identification"; Brindley, G. W.; Brown, G., Eds.; Mineralogical Society: London, 1980; pp. 249-303.
42. Środoń, J. Clays Clay Miner. 1980, 28, 401-11.
43. Środoń, J. Clay Miner. 1981, 16, 297-304.
44. Jackson, M. L. "Soil Chemical Analysis-Advanced Course"; Published by the Author: Madison, Wisconsin, 1975.
45. Taggart, J. E.; Lichte, F. E.; Wahlberg, J. S. U.S. Geol. Sur. Prof. Pap. 1250, 1981, pp. 683-87.
46. Van Loon, J. C.; Parissis, C. M. Analyst. 1969, 94, 1057.
47. Środoń, J. Clays Clay Miner. 1984, 32, 337-49.
48. Clayton, R. N.; Mayeda, T. K. Geochim. Cosmochim. Acta. 1963, 27, 43-52.
49. Schramm, L. L.; Kwak, J. C. T. Clays Clay Miner. 1980, 28, 67-9.
50. Bailey, S. W. In "Micas"; Bailey, S. W., Ed.; Mineralogical Society of America: Washington, D. C., 1984, pp. 13-60.
51. Besson, G.; Brookin, A. S.; Dainyak, L. G.; Rautureau, M.; Tsipursky, S. I.; Tchoubar, C.; Drits, V. A. J. Appl. Crystallogr. 1983, 16, 374-83.
52. Thompson, G. R.; Hower, J. Clays Clay Miner. 1975, 23, 289-300.
53. Środoń, J.; Morgan, D. J.; Eslinger, E. V.; Eberl, D. D.; Karlinger, M. R. Clays Clay Miner. 1986 [in press].
54. Escande, M.; Decarreau, A.; Labeyrie, L. C. R. Acad. Sc. Paris. 1984, t. 299, Série II, n° 11, 707-10.
55. Caillère, S.; Hénin, S. Mineralogical Magazine. 1949, 28, 606-11.
56. Eberl, D. D. Philos. Trans. R. Soc. London, A., 1984, 311, 241-257.
57. Grim, R. E.; Güven, N. "Bentonites"; Elsevier: New York, 1978, p. 256.
58. Weaver, C. E.; Pollard, L. D. "The Chemistry of Clay Minerals"; Elsevier: Amsterdam, 1973, p. 213.
59. van Olphen, H.; Fripiat, J. J. "Data Handbook for Clay Materials and Other Non-Metallic Minerals"; Pergamon Press: New York, 1979, p. 346.
60. Busenberg, E.; Clemency, C. V. Geochim. Cosmochim. Acta. 1976, 40, 41-49.

61. Lin, F-C.; Clemency, C. V. Geochim. Cosmochim. Acta. 1981, 45, 571-6.
62. Lin, F-C.; Clemency, C. V. Am. Mineral. 1981, 66, 801-6.
63. Weir, A. H.; Greene-Kelley, R. Am. Mineral. 1962, 47, 137-46.
64. Singer, A.; Stoffers, P. Clay Miner. 1980, 15, 291-307.
65. Jones, B. F.; Weir, A. H. Clays Clay Miner. 1983, 31, 161-72.
66. Lagaly, G.; Weiss, A. Int. Clay Conf. Proc. 1975, pp. 157-72.
67. Tsipursky, S. I.; Drits, V. A. Clay Miner. 1984, 19, 177-93.
68. Drits, V. A.; Plancon, A.; Sakharov, B. A.; Besson, G.; Tsipursky, S. I.; Tchoubar, C. Clay Miner. 1984, 19, 541-61.
69. Clayton, R. N.; O'Neil, J. R.; Mayeda, T. K. Jour. Geophys. Research. 1972, 77, 3057-67.

RECEIVED June 18, 1986

Potassium–Calcium Exchange Equilibria in Aluminosilicate Minerals and Soils

Keith W. T. Goulding

Rothamsted Experimental Station, Harpenden, Herts AL5 2JQ, England

Potassium-calcium exchange equilibria in some selected layer silicate minerals and soils were studied using exchange isotherms and microcalorimetry. Groups of homogeneous exchange sites, identified by their differential enthalpies of Ca + 2K exchange, comprised eight distinct types. These, in turn, were tentatively associated with the surfaces of mica (one type of site), hydrous mica or illite (three types of site), vermiculite (one type of site), and montmorillonite (three types of site). If the identification is correct, the method can be used to estimate the amount of each mineral in a clay or soil and thus more precisely determine clay composition.

Plants take up the potassium (K) they require from the soil solution. Very little K is present in this form, however, perhaps 5-10 kg ha^{-1} in the surface soil (0-20 cm). Almost all of the K in soil is adsorbed on the surfaces of soil colloids, chiefly on aluminosilicate minerals; very little K is held on organic materials (see below). In temperate soils, about 500 kg K ha^{-1} is present in this exchangeable form in topsoil, and it is released into solution by a simple cation exchange reaction. By far the greatest proportion of K in soils of temperate regions, 80-99% or 2-50 Mg ha^{-1}, is held within the interlayers of partially-weathered and unweathered mica minerals, and within the crystal structure of feldspars. Although the term 'nonexchangeable' is widely used for this K, this is not strictly correct. The K is exchangeable, but only very slowly. It is released into more available forms - exchangeable or solution K - by slow weathering, which again involves cation exchange in the case of micas.

The release of K into solution from the solid phase is important in plant nutrition, and it has been extensively studied at Rothamsted as at other agricultural research stations throughout the world. We have been interested in the process both from the practical standpoint of determining the availability of K to crops, and from the more fundamental view of understanding its

0097-6156/86/0323-0327$06.00/0

adsorption and release properties and the nature of the surfaces
and sites on which K is adsorbed. We have thus examined both the
equilibria and kinetics of K exchange. Because calcium (Ca) is
the dominant cation in most British agricultural soils, we have
concentrated on K/Ca exchange. Also, soils are multicomponent
systems, containing aluminosilicate (phyllosilicates or layer
lattice clays, etc.) and organic exchangers. To determine
something of the contribution of each component to the exchange
characteristics of a soil we have therefore examined K/Ca exchange
reactions on soils and on some of the individual aluminosilicates
most commonly found in the clay fractions of temperate soils.

P.W. Arnold (1-3) was an early pioneer of the work at
Rothamsted. The present research programme was begun by O.
Talibudeen (4-6), with whom I began collaboration in 1974. This
paper summarises our results and my more recent work on exchange
equilibria (4-16).

Materials

The soils and clays used in the work were described in full in the
original reports (7,11-16). The soils used were taken
chiefly from long-term experiments on plant nutrition at
Rothamsted and at experimental stations and farms throughout
Britain. Those of contrasting mineralogy and constitution were
selected where possible and also those that had been subjected to
different fertilizer and cropping sequences, and which therefore
had very different K contents and exchange characteristics.
Generally, the whole, untreated soils were examined but also, in
one case, the various particle size separates.

The aluminosilicates examined were chosen as end members of
those groups of phyllosilicates that commonly occur in soils:
muscovite and biotite mica, Fithian and Morris illite, Montana
vermiculite, montmorillonites from Upton (Wyoming bentonite), Camp
Berteau, Redhill and New Mexico, and kaolinites from St. Austell,
England, and Georgia, U.S.A.

Methods

Equilibria. Potassium-calcium exchange equilibria were studied
using standard exchange isotherm techniques and by the
microcalorimetric measurement of enthalpies of exchange; the
methods were described in full by Goulding and Talibudeen (7).
Essentially the soils and clays were saturated with Ca and the
enthalpies of Ca $\rightarrow$ 2K exchange measured at 303K by adding
successive 5 µl aliquots of 0.5 M KCl to a suspension of about 50
µeq of the solid in 2-5 ml water in an LKB Batch microcalorimeter.
An injection system fitted to the calorimeter, described by
Talibudeen et al. (8) and Minter and Talibudeen (9), greatly
speeded up the experimental work. The enthalpy measured after
each step was summed to give the integral enthalpy of exchange
(ΔH_X), and the rate of change of this with K saturation was the
differential enthalpy of exchange, $d(\Delta H_X)/dx$ (Figure 1). The
ΔH_X:K saturation relationship appeared to be a series of
straight lines rather than a smooth curve. Thus, the derived

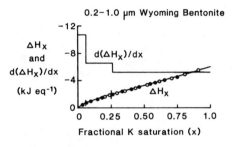

Figure 1. Integral (ΔH_x) and differential ($d(\Delta H_x)/dx$) enthalpies of Ca → 2K (●) and 2K → Ca (O) exchange on Wyoming bentonite as a function of the fractional K saturation of the exchange capacity.

$d(\Delta H_x)/dx$:K saturation relationship became a series of sharp
steps (Figure 1). To show if this was truly the case, and to
obtain the best fit for the data, the Rothamsted Maximum
Likelihood Program (17) was used to fit and compare a linear
spline (split line) with several smooth curves (quadratic,
exponential, exponential plus linear). In all the cases referred
to here, the linear spline gave the statistically better fit. The
extent of exchange following each injection was determined
separately in a scaled-up version of the calorimeter experiment.
The two procedures allowed all the thermodynamic parameters of
cation exchange to be calculated following the procedure of Gaines
and Thomas (18). The usual working standard states were used:
the homoionic exchanger in equilibrium with a solution of the
saturating cation of constant ionic strength (0.1M), and an ideal
solution at unit molarity. In this paper, only the differential
enthalpies of exchange, defined above, and the standard free
energy (ΔG_0), enthalpy (ΔH_0), and entropy (ΔS_0) of exchange
are considered. The latter functions represent the integration
over the whole exchange of the change in selectivity, bonding
strength and order, respectively, when changing from a
Ca/water/solid to a 2K/water/solid system.

Results and Discussion

Aluminosilicate minerals. Standard free energies of Ca → 2K
exchange, ΔG_0, were always negative, showing that all of the 2:1
phyllosilicate minerals (hereafter described simply as clays)
studied, and the 1:1 mineral, kaolinite, were selective for K over
Ca (Table I).

Table I. Standard free energies (ΔG_0), enthalpies (ΔH_0) and
entropies ($T\Delta S_0$) of Ca → 2K exchange in the soils and
phyllosilicate minerals (From references 7,15)

Material	Cation exchange capacity (Ca-form) $(\mu eq\ g^{-1})$	ΔG_0	ΔH_0 $kJ\ eq^{-1}$	$T\Delta S_0$
Muscovite mica, <1 μm	60	-19.35	-10.48	+8.87
Fithian illite, <2 μm	210	-5.10	-9.24	-4.14
Montana vermiculite, <2 μm	1200	-5.13	-7.67	-2.54
Wyoming bentonite, 0.2 - 1 μm	700	-1.32	-5.96	-4.64
St Austell kaolinite, < 2.5 μm	60	-4.42	-9.12	-4.70

Selectivity decreased through the weathering sequence mica >
illite = vermiculite > montmorillonite. Selectivity for K over Ca
has been ascribed to the low hydration number and polarizability
of K (19), to wedge sites at the weathered edge of clay

crystallites ([20], [21]), and to the characteristics of the adsorption site generally ([20], [22]). For the series of clays examined here, it must reflect differences in the adsorption sites, and therefore, in the surfaces at which exchange occurs. While ΔG_0 is a measure of selectivity, ΔH_0 is a measure of bond strength. The change in enthalpy reflects the breakage and formation of all bonds in the ion/water/clay system (ion-ion, ion-water, ion-clay, water-clay, clay-clay, water-water). A negative (exothermic) enthalpy change in going from Ca→2K implies stronger bonds in the 2K/water/clay system than in the Ca/water/clay system. As Sposito ([23]) states, thermodynamic quantities cannot be interpreted directly in terms of molecular structure, nor be assigned to any one component without further nonthermodynamic evidence. However, they can be seen to agree or disagree with a particular model. The data here agree with Sposito's ([23]) model of ion exchange in terms of surface complexes which is based on x-ray ([24]), infrared ([25]), and neutron scattering ([26]) data. According to the model, potassium adsorbed on 2:1 layer silicate minerals is in an inner-sphere complex with no solvent molecules between it and the surface functional group; the K is held by ionic or covalent bonds or some combination of the two. Good examples of this are the very stable K-mica and K-vermiculite surface complexes in soil ([24]). By comparison, adsorbed Ca is held by electrostatic bonds in an outer sphere complex with at least one water molecule between it and the surface functional group, e.g. the two-layer hydrate of Ca-montmorillonite.

The greater bond strength in the K-clay system decreases in the same order as the selectivity for K, suggesting at first that bond strength controls selectivity, as Brouwer et al. ([27]) stated. However, the data show the range of enthalpies to be very much less than that of free energies, and that entropies exert a strong influence, especially in the muscovite mica. The standard entropy change accompanying Ca → 2K exchange on the mica, expressed as $T\Delta S_0$, is positive, while that of all the other clays is negative (Table I). This implies greater order in the Ca than the K system in mica, but vice versa for the other clays. It is not easy in such a complex system to decide which component or components is influencing the entropy changes most. The overall order of the system includes that of the clay surfaces, the adsorbed cations and the water molecules - both those associated with the cations and those in the free solution. Following work of Plancon et al. ([28]) and Eberl and Srodon (this volume) which showed how the 2:1 layers of Na-smectites rotate and realign on adsorbing K, I think that this realignment is the principle component of entropy changes in illites, vermiculites and smectites, giving them greater order on adsorbing K. In mica, no rearrangement is possible, and the principal component of ΔS_0 is the increased randomness in distribution of adsorbed K over adsorbed Ca.

Differential enthalpies of exchange, plotted as a function of K saturation, exhibit a stepped character (Figure 1 and Table II). This can only reflect different types of adsorption sites on the clay surfaces, the group of sites with the most negative enthalpy

Table II. Groups of exchange sites in some layer silicate clays as distinguished by differential enthalpies of Ca → 2K exchange (From reference 7)

Material	Cation exchange capacity (Ca-form) ($\mu eq\ g^{-1}$)	Differential enthalpy values ($kJ\ eq^{-1}$) and, in brackets, μeq charge occupied by each type of site					
		Illite	Vermic.	Mica	Montmorillonite		
Muscovite mica <1 µm	60	—	—	-10.5 (60)	—	—	—
Fithian illite <2 µm	210	-20.0 (32)	—	-10.9 (51)	-7.6 (31)	-6.0 (96)	—
Montana vermiculite <2 µm	1200	—	-15.9 (156)	—	—	-6.3 (1044)	—
Wyoming bentonite (0.2-1 µm)	700	—	—	-10.7 (40)	—	-6.5 (164)	-5.2 (496)
St. Austell kaolin <2.5 µm	60	—	-14.4 (19)	—	-7.1 (41)	—	—
Mean (± SE)	—	-20.0 (0.5)	-15.2 (1.5)	-10.7 (0.6)	-7.4 (0.8)	-6.3 (0.4)	-5.2 (0.3)

(those which bind K most strongly) being completely exchanged before K begins to fill the next group of sites. Only the muscovite mica contained a single type of exchange site (cf 29), which was naturally ascribed to a mica-type surface. The 0.2-1 μm Wyoming bentonite contained a few sites just like those in the mica, but mostly contained sites which bind K weakly and which are attributed to montmorillonitic surfaces. The illite contained the mica-type sites and also sites with the same differential enthalpy as those found in Wyoming bentonite. These sites were ascribed to micaceous and montmorillonitic surfaces. The Fithian illite also contained a few sites that exhibited a very large bonding strength and selectivity for K over Ca and were not seen in any other mineral or soil. Such sites were also found by Brouwer et al. (27), and have recently been generated in montmorillonites by repeated wetting and drying and by charge reduction (30).

It is very difficult to link the observed groups of sites with known physical characteristics of the clays with certainty. As Talibudeen (10) states, it is unwise to assign them simply to the interlayer, edge, and planar surfaces of clays in the way that Bolt et al. (22) and Schouwenberg and Schuffelen (20) did, an interpretation that has become widely accepted. Several reasons can be adduced for the strong adsorption of K which involve physical (specific) and charge (coulombic effects). The occurrence of several types of sites in one sample could also reflect interstratification within relatively large crystallites (31) or, as has recently been suggested by Nadeau et al. (32), a random arrangement of thin fundamental particles. However, Maes et al. (30) thought that there were 'islands' of highly K selective sites corresponding to a collapsed 10Å core (i.e. wedge sites) in their altered (illitized) montmorillonite. They also thought that the highly K selective sites in illite were at the edges of crystallites. This agrees with the conclusions of Le Roux et al. (33) who, for Rb/Sr exchange on weathered micas, found the Rb (which behaves very much like K) to be concentrated at particle and step edges, at cracks, and, in the case of weathered biotite, at boundaries of vermiculite and mica zones (wedge zones). Altogether within the five clays, six groups of exchange sites can be identified on the basis of similar differential enthalpy values (Table II). The means of each group, excepting that of -7.4 kJ eq⁻¹, are significantly different from each other. On the basis of the minerals in which the site groups are found, the six groups can be associated with four main types of 2:1 layer silicate surface: micaceous, hydrous mica (illitic), vermiculitic, and montmorillonitic, as in Table II. Such an identification of exchange sites with specific mineral surfaces must be tentative in the absence of other supporting data, however. The present 'classification' also differs somewhat from that given earlier (16), because our ideas have changed as more data were collected.

If mica-type sites are truly indicative of mica surfaces, and so on for montmorillonite etc., then of the clays examined, only the mica could be regarded as pure. The Montana vermiculite contained much montmorillonite, the Fithian illite contained mica and montmorillonite, the Wyoming bentonite contained some mica, and the kaolinite a very small amount of vermiculite/ montmorillonite. Extending the investigation to the <0.2 μm fraction of montmorillonites from Wyoming, Redhill, Camp Berteau and New Mexico, produced the same types of site as already found, plus two more. These, being of intermediate differential enthalpy value between mica and montmorillonite, were, again tentatively, associated with hydrous mica surfaces (Table III). The differential enthalpy value of -11.7 found in the New Mexico sample and grouped under 'mica' is perhaps a little too negative for mica-type sites and may reflect some other type of surface. Again, if the various site groups do indicate particular mineral surfaces, then only the Wyoming bentonite can be regarded as a pure montmorillonite; the other samples contain mica, vermiculite, etc. (Table III and reference 11).

     Looking at a series of kaolinites from St. Austell, England, and Georgia, U.S.A., suggested that all of them, even two reckoned pure according to XRD and viscosity measurements, contained a small amount of 2:1 mineral surfaces that would account for all of their exchange characteristics (Table IV and reference 12). Using a characteristic exchange capacity, the amount of charge occupied by the various 'impurities' was converted into a percentage composition by weight (7, 11, 12). If such a procedure is correct, it offers a means of quantitatively assessing very small amounts of 'impurities' that have an exchange capacity in minerals. Certainly there was semi-quantitative agreement between the amount of 'impurity' in the kaolinites calculated from enthalpy measurements and by x-ray data (Table IV). Further verification of the technique was provided by Arkcoll et al (13) who used it to identify small amounts of mica and hydrous mica as the source of hitherto unexplained K reserves in some strongly weathered Brazilian soils. However, conclusive verification requires a detailed comparison of enthalpy and x-ray data on the same materials using the most modern XRD techniques.

Soils. The soils examined contained the same groups of exchange sites as those found in the clays (Table V, and references 14 and 15). For some soils the surfaces identified corresponded exactly to the clay mineralogy determined by X-ray diffraction (XRD). For example, a sample of the Worcester series contained 60-70% mica and 20-30% chlorite in its clay fraction according to XRD, and the measured enthalpy of Ca → 2K exchange showed only mica surfaces. However, the clay composition of other soils as determined by differential enthalpies, did not completely agree with that determined by XRD. Assuming that, for example, the mica-type surfaces identified by differential enthalpies correspond to a mica clay identified by XRD, then in these latter soils, surfaces have been altered by weathering and by coatings of organic material, oxides and hydroxides.

     When K fertilizers are added to soil, some of the K is used by crops but much is adsorbed on external exchange sites in the soil and even more on slowly exchangeable sites. From the discussion above on types of exchange sites, one would expect that these residues of K would absorb on K selective sites, and thus,

Table III. Groups of exchange sites in some <0.2 μm montmorillonites as distinguished by differential enthalpies of exchange. (From reference 11)

Material	Cation exchange capacity (Ca-form) ($\mu eq\ g^{-1}$)	Differential enthalpy values ($kJ\ eq^{-1}$) and, in brackets, μeq charge occupied by each type of site				
		Mica	Hydrous Mica	Montmorillonite		
Wyoming bentonite	1050	–	–	-7.6 (204)	-6.7 (346)	–
Redhill	1120	–	-9.4 (93)	-7.5 (309)	–	-5.8 (718)
Camp Berteau	1150	-10.3 (53)	-8.7 (207)	-7.4 (447)	–	-6.0 (443)
New Mexico	1350	-11.7 (246)	-8.7 (1104)	–	–	–
Mean (±SE)	–	-11.0 (0.8)	-8.7 (0.1)	-7.5 (0.2)	-6.7 (0.2)	-5.9 (0.4)

Table IV. Groups of exchange sites in some <5 μm kaolins as distinguished by differential enthalpies of Ca → 2K exchange. (From reference 12)

Material	Cation exchange capacity (Ca-form) (μeq g⁻¹)	Differential enthalpy values (kJ eq⁻¹) and, in brackets, eq charge occupied by each type of site				Content of 2:1 minerals (%w/w)	
		Vermic.	Mica	Hydrous Mica	Montmorillonite	XRD	'Enthalpies'
A	17	-13.7 (3)	-10.8 (4)	-8.7 (10)	–	0	0.9
B	25	–	–	-9.4 (9)	-7.2 (16)	Trace	1.7
C	30	–	-11.0 (12)	-8.2 (18)	–	3	1.5
D	30	-13.4 (10)	–	-9.4 (6); -8.0 (14)	–	5	1.5
E	30	–	–	-9.1 (7); -8.0 (13)	-7.2 (10)	11	1.9
F	145	–	–	-9.6 (15); -8.2 (20)	-7.1 (66); -5.1 (44)	15	10.4
Mean (SE)	–	-13.6 (0.2)	-10.9 (0.2)	-9.4 (0.2); -8.2 (0.2)	-7.2 (0.2); -5.1 (–)	–	–

Table V. Groups of exchange sites in some soils as distinguished by differential enthalpies of Ca → 2K exchange. (From references 14,15)

Soil		Differential enthalpy values (kJ eq^{-1})			
		Vermic.	Mica	Hydrous mica	Montmorillonite
Batcombe	Nil	-12.9	-	-	-
	PK	-	-	-9.7	-7.0
	NPK	-	-	-8.6	-6.7
	FYM	-17.3	-11.5	-	-7.6
Beccles	Nil	-12.3	-	-9.8	-6.7
	PK	-	-10.8	-8.1	-
	FYM	-	-10.0	-	-
Andover		-	-11.4	-7.9	-
Hanslope		-	-10.4	-8.0	-
Newport		-12.2	-10.1	-	-
Worcester		-	-10.8	-	-

soils treated with K fertilizer would show a decrease in or an absence of K selective sites in comparison with the untreated soil. There is a decrease in K selectivity (free energy) and K binding strength (enthalpy) (Table V and Figure 2) in the samples of soil treated with K (the PK, NPK and Farm Yard Manure (FYM) soils) over those not given K (the Nil soils), but no great difference in the order of the systems (entropy). The decrease in K preference and binding strength is over the whole of the exchange sites, however, and does not consist merely of the decrease in or disappearance of K selective sites. A simple collapse of frayed-edge or wedge type K-selective sites would also be expected to cause a change in entropies which is not seen, and a decrease in CEC which is not apparent. It appears that adsorption and fixation of K in field soil over a long time (> 100 years) causes changes to surfaces far more complex than those observed in laboratory experiments on clays and soils. Looking at the Nil (unfertilized) and PK or NPK (mineral fertilized) plots of the Batcombe and Beccles soils in Table V shows that K has changed vermiculite-like sites into mica-like and hydrous mica-like sites. Perhaps this is indicative of reverse weathering (cf 34). Another possibility is that adsorption of a monovalent cation, in this case K$^+$, causes some clay dispersion and some movement of clay (especially those particles with K selective sites) down the profile (35).

In contrast to the effect of K from mineral fertilizers, organic matter affects free energies, enthalpies and entropies, and sometimes in a very complex way. In the Beccles soil listed in Table V, organic residues decreased K preference by decreasing K binding strength and by increasing disorder. In the Batcombe soil, the large organic residues in the soil from the FYM (Farm Yard Manure) treated plot greatly reduced K preference and

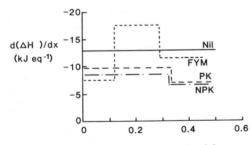

Figure 2. The differential enthalpy of Ca → 2K exchange
(d(ΔH_x/dx) as a function of fractional K saturation for the
Batcombe series soils from various plots of the Broadbalk
Classical Experiment at Rothamsted.

appeared to cause a reversal of the normal pattern of differential
enthalpy curves (Figure 2). Sites that bound K weakly appeared to
be filled before those that bound K more strongly. This sharply
contrasting behaviour has not yet been examined further. The
differences between the two soils could be because (i) exchange
first takes place on organic exchange sites which bind K only
weakly but which are favoured by entropy changes, (ii) the organic
material, while not directly involved in the exchange, coats
mineral surfaces blocking some of the exchange sites. There may
also be some effect of the decomposition of organic matter on clay
weathering.
        Entropy changes show an increase in order over all the
exchange when K is adsorbed, an increase so large that the
Batcombe soil treated with FYM prefers Ca. This may reflect the
characteristics of organic exchange sites, but it could also
reflect the blocking action of organic materials. Other workers
have suggested that organic material blocks exchange sites (36,
37), and the destabilization of clay-Ca-organic bridges by
monovalent cations has been noted before (38). Certainly the K
on the plots receiving FYM on some Rothamsted Classical
Experiments has been observed as reacting differently, namely
being 'more available', before (39).
        Generally, clay mineral content was the most important factor
determining K selectivity in soil, with weathering, fertilizers and
organic residues affecting selectivity through their modifying
effects on mineral surfaces.

Conclusions

Potassium-calcium exchange equilibrium studies, involving
particularly the measurement of differential enthalpies of
exchange, quantitatively identify types of cation exchange sites
in soils and phyllosilicate minerals which can be interpreted in
terms of mica, hydrous mica, vermiculite and montmorillonite
surfaces. Combined with XRD and other analytical techniques, such
surface measurements provide a means of more precisely determining
mineral composition, and of examining the effects of weathering
and of various coatings on the surfaces of all aluminosilicate
minerals with an exchange capacity. At this stage it is not

possible to identify conclusively the various types of sites with known physical characteristics of aluminosilicates, particularly because charge density (coulombic) effects will also be a strong force in determining selectivity. However, the mass of evidence points to the existence of (i) a few highly K selective, specific sites which bind K strongly at particle and step edges, at cracks, and around islands of 10 Å cores in weathered micas (illites); (ii) sites on the surfaces of micas, which are very K selective because of entropic effects but which do not bind K especially strongly; (iii) sites of intermediate K selectivity on illitic and vermiculitic surfaces; (iv) sites of low K selectivity on external planar sites on montmorillonitic surfaces.

It has often been argued that clay minerals cannot be used as models for soils in surface chemistry because soil clays are too heterogeneous and impure. The work reported here shows that, for both equilibrium and kinetic studies, standard aluminosilicates are useful models for soils.

Literature Cited

1. Arnold, P.W. Nature 1960, 182, 1594-5.
2. Arnold, P.W. Proc. Fert. Soc. 1962, No. 72, 25-43.
3. Arnold, P.W.; Close, B.M. J. Agric. Sci., Camb. 1961, 57, 381-6.
4. Talibudeen, O. In "Potassium in the Soil", Proc. 9th Colloq. Int. Potash Inst. 1972, 97-112.
5. Talibudeen, O. Rep. Prog. App. Chem. 1973, 58, 403-8.
6. Talibudeen, O.; Beasley, J.D.; Lane, P.; Rajendran, N. J. Soil Sci. 1978, 29, 207-18.
7. Goulding, K.W.T.; Talibudeen, O. J. Coll. Int. Sci. 1980, 78, 15-24.
8. Talibudeen, O.; Goulding, K.W.T.; Edwards, B.S.; Minter, B.A. Lab. Practice 1977, 26, 952-5.
9. Minter, B.A.; Talibudeen, O. Lab. Practice 1982, 31, 1094-6.
10. Talibudeen, O. Ads. Sci. Tech. 1984, 1, 235-46.
11. Talibudeen, O.; Goulding, K.W.T. Clays Clay Mins. 1983, 31, 37-42.
12. Talibudeen, O.; Goulding, K.W.T. Clays Clay Mins. 1983, 31, 137-42.
13 Arkcoll, D.B.; Goulding, K.W.T.; Hughes, J.C. J. Soil Sci. 1985, 36, 123-8.
14. Goulding, K.W.T.; Talibudeen, O. J. Soil Sci. 1984, 35, 397-408.
15. Goulding, K.W.T.; Talibudeen, O. J. Soil Sci. 1984, 35, 409-20.
16. Goulding, K.W.T.; Talibudeen, O. J. Soil Sci. 1979, 30, 291-302.
17. Ross, G.J.S. Proc. 40th Session Int. Stat. Inst., Warsaw 1975, Paper 81.
18. Gaines, G.L.; Thomas, H.C. J. Chem. Phys. 1953, 21, 714-8.
19. Assa, A.D. Cah. ORSTOM Ser. Pedol. 1976, 14, 219-26, 279-86.

20. Schouwenberg, J.Ch.van; Schuffelen, A.C. Neth. J. Agric. Sci. 1963, 11, 13-22.
21. Lee, R.; Rowell, D.L. J. Soil Sci. 1975, 26, 418-25.
22. Bolt, G.H.; Sumner, M.E.; Kamphorst, A. Soil Sci. Soc. Am. Proc. 1963, 27, 294-9.
23. Sposito, G. "The Surface Chemistry of Soils"; Oxford University Press: Oxford, U.K., 1984.
24. Norrish, K. Proc. Int. Clay Conf., 1972 1973, pp.417-32.
25. Farmer, V.C.; Russell, J.D. Trans. Far. Soc. 1971, 67, 2737-49.
26. Ross, D.K.; Hall, P.L. In "Advanced Chemical Methods for Soil and Clay Mineralogy Research"; Stucki, J.W.; Banwart, W.L., Eds., Reidel, Boston, 1980; p.93.
27. Brouwer, E.; Baeyens, B.; Maes, A.; Cremers, A. J. Phys. Chem. 1983, 87, 1213-9.
28. Plancon, A.; Besson, G.; Gaultier, J.P.; Mamy, J; Tchoubar, C. In "Developments in Sedimentology Vol. 27"; Mortland, M.M.; Farmer, V.C., Eds.; Elsevier, Oxford, 1979; pp. 45-54.
29. Pashley, R.M. Clays Clay Mins 1985, 33, 193-9.
30. Maes, A.; Verheyden, D.; Cremers, A. Clays Clay Mins. 1985, 33, 251-7.
31. Brown, G.; Newman, A.C.D.; Rayner, J.H.; Weir, A.H. In "The chemistry of soil constituents"; Greenland, D.J.; Hayes, M.H.B., Eds., John Wiley and Sons, Chichester, 1978, pp. 29-178.
32. Nadeau, P.H.; Wilson, M.J.; McHardy, W.J.; Tait, J.M. Science 1984, 225, 923-5.
33. Le Roux, J.; Rich, C.I.; Ribbe, P.H. Clays Clay Mins. 1970, 18, 333-8.
34. Ross, G.J.; Phillips, P.A.; Culley, J.L.P. Can. J. Soil Sci. 1985, 65, 599-603.
35. Chen, Y.; Banin, A.; Borochovitch, A. Geoderma 1983, 30, 135-47.
36. Beckett, P.H.T.; Nafady, M.H.M. J. Soil Sci. 1967, 18, 263-81.
37. Greenland, D.J.; Mott, C.J.B. In "The Chemistry of Soil Constituents"; Greenland, D.J.; Hayes, M.H.B., Eds., John Wiley and Sons, Chichester, 1978; pp 321-353.
38. Theng, B.K.G. "Formation and Properties of Clay-Polymer Complexes" (Developments in Soil Science, Vol. 9); Elsevier, Amsterdam, 1979; ch. 12, pp. 283-326.
39. Addiscott, T.M.; Johnston, A.E. J. Agric. Sci., Camb. 1971, 76, 553-61.

RECEIVED June 18, 1986

SURFACE SPECTROSCOPY

16

Adsorption of Metal Ions and Complexes on Aluminosilicate Minerals

B. A. Goodman

The Macaulay Institute for Soil Research, Craigiebuckler, Aberdeen AB9 2QJ, Scotland

Adsorption of metal ions and complexes on the various
types of aluminosilicate mineral is briefly reviewed
along with contributions made by several spectroscopic
methods to understanding the nature of adsorbed
species. A knowledge of the chemical forms of adsorbed
species is an important preliminary to any
understanding of their reactivities in either natural
or artificial situations and, although significant
progress has been made in some systems, there is
clearly still a great deal of work necessary in order
to characterize fully the environments of many
adsorbed species. Examples of the types of reaction
that may be carried out specifically by
metal-exchanged clays are given and serve as
illustrations of the importance of such species in
natural systems and of the tremendous potential that
such systems have in performing novel chemical
reactions.

The surfaces of many aluminosilicate minerals carry a charge and in
other cases reactive centres may exist. Any surface charge is
usually compensated by association with, often involving adsorption
of, equivalent amounts of ions of opposite charge and such
adsorption processes are of particular importance for clay minerals
(which have high surface areas) in natural systems. Clay minerals
play a major role in regulating the mobility of ions with both
beneficial and toxic properties in soils and waters and the nature
of the adsorbed species can have an influence on the physical and
chemical properties of the clay. Therefore, an understanding of
adsorption processes and the nature of the species adsorbed on
clays is of great importance in many areas of science ranging from
the chemical and agricultural industries, through environmental
science to the engineering and construction industries.
    Theoretical models of adsorption have been reviewed along with
the adsorption of metal ions, complex formation and the influence of
surface properties of the adsorbent and solution pH on adsorption

0097-6156/86/0323-0342$06.00/0
© 1986 American Chemical Society

processes in a previous ACS publication (1). Sorption processes can take a number of forms, with the broadest distinction being whether there is chemical reaction with the mineral surface, i.e. chemisorption, or not, in the case of physisorption. In addition precipitation may occur on the solid surface with the resultant 3-dimensional growth of a new molecular species. The problem of distinguishing between adsorption and surface precipitation is discussed by Sposito (this volume) and will not be considered in detail here. However, in order to improve our understanding of adsorption processes we need knowledge of the chemical nature of the sites at which adsorption occurs and their variations with external conditions, especially pH. Also, since solutions rarely have a simple composition, information is required on the chemical forms of species in solution and those adsorbed onto minerals in addition to the amounts of metals involved. This is a formidable task and one that cannot be achieved by conventional approaches. In a number of specific cases, however, a great deal of information can be obtained from a variety of spectroscopic techniques and this review will focus attention on some examples of their usage.

Spectroscopic techniques exist that cover virtually the complete energy range of electromagnetic radiation and most have some potential value in studies of surface adsorption. In order of increasing energy the various techniques mentioned in this review are nuclear magnetic resonance (NMR), electron spin or paramagnetic resonance (ESR or EPR), electron spin double resonance (ENDOR) and electron spin echo modulation (ESEM), infra-red (IR) and Raman spectroscopy, photoacoustic and ultra-violet (UV)-visible spectroscopy, photoelectron spectroscopy and Mössbauer spectroscopy. NMR is concerned with transitions between nuclear spin states (usually $\pm$ 1/2) in an external magnetic field and provides information on the chemical environment of the nucleus. Both solids and liquids can be studied, although in the former case it is usual to spin the sample rapidly at the "magic angle" (54.7°) to remove anisotropic effects and produce solution-like information. ESR is similar in priniple to NMR except that transitions are observed between electron spin states, the chemical environment(s) of unpaired electron(s) being probed. ENDOR is essentially a combination of ESR and NMR in that a nuclear resonance transition is observed whilst performing an electron resonance excitation. ESEM is also related to ESR spectroscopy and in this case the electron spin echo is modulated by a series of pulses to produce a decay curve whose shape is determined by the nature of neighbouring and next-nearest-neighbour atoms. All of these spin resonance techniques are amenable to the study of solid and liquid samples. IR and Raman spectroscopy are concerned with molecular vibrations and IR spectroscopy in particular has wide-ranging applications in the characterization of surface groups (particularly OH) and adsorbed molecules (particularly organic). Transitions involving valence electrons occur in the UV-visible and near IR regions of the spectrum and there are many instances where species have absorptions at these frequencies that are sensitive enough to be of value in sorption studies. Again liquid- and solid-state samples can be investigated, the latter also being amenable to study by photoacoustic spectroscopy. X-ray photoelectron spectrosopy (XPS) probes transitions involving inner-shell electrons. It has applications to a large number of elements and also provides information on their chemical nature, making it a valuable technique in sorption studies. It is, however, limited to solid-state investigations and the insertion of the sample into a high vacuum creates possible difficulties in relating results to practical situations. Mössbauer spectroscopy is concerned with transitions between nuclear ground and excited states, the energy occurring in the γ-ray region of the electromagnetic spectrum. In practice the exciting radiation is obtained by modulating by a Döppler velocity the energy of radiation emitted by a nucleus decaying from its excited to ground state, the excited state nucleus being generated by decay of a suitable radioactive precursor. Information is provided on the chemical environment of the nucleus but only solid materials contribute to a spectrum. In the conventional approach of using the unknown material as absorber Mossbauer spectroscopy is very limited in its applications to sorption processes. It is an insensitive method often requiring many hours (or even days) to produce a single spectrum and, for most practical purposes, Fe-57 is the only isotope suitable for this type of experiment. However, sorption studies involving the radioactive source nucleus can be performed with much greater sensitivity and examples of the use of Co-57 and Sb-119 are given in the chapter by Ambe et al. (this volume). Further details of the various methods are beyond the scope of this article and the interested reader is referred to some of the very many texts in the chemical literature.

The nature of clay surfaces

As far as adsorption properties are concerned, aluminosilicate minerals may be divided into 3 general groups:-
(i) expanding layer structures, such as smectites, which can greatly increase their surface area on solvation and which generally carry a negative charge over a very wide range of pHs,
(ii) cage structures, such as zeolites, which have internal surfaces accessible only to ions or molecules below a certain size, and
(iii) structures whose adsorption properties are determined solely by the chemical natures of their surfaces.
The surface charge of aluminosilicate minerals may arise either from isomorphous cation substitution within the structure, which is pH independent, or by protonation/deprotonation reactions at oxide/hydroxide surface groups which is pH dependent. The former mechanism is generally considered to dominate in smectite minerals, whereas the latter may also be important in zeolites. The determination of surface charge can sometimes present problems, especially with minerals that have a low net charge. The principal difficulties in these cases arise from either the presence of small amounts of undetected impurities of minerals with high surface charge or from partial decomposition of the mineral during the experiment. Both situations have been reported for kaolin minerals under different circumstances. Thus Bolland et al., (2) have shown that, unless account is taken of dissolution of structural Al, the negative charge on kaolinite surfaces can be mistakenly attributed to an oxide-like source. In contrast with a number of natural kaolins Lim et al., (3) concluded, on the basis of x-ray diffraction

(XRD) and chemical extraction procedures, that much of the observed cation exchange capacity (CEC) arises from smectite impurities, the external surface CEC of kaolinite ranging from 0-1 meq/100 g. The assessment of surface area may sometimes present difficulties, e.g. with smectites N_2 adsorption measurements grossly underestimate the area that is exposed in solution when the layers are fully expanded. In an attempt to overcome this problem a simple theoretical model has recently been developed (4) for deriving double-layer potentials for the clay-solution interface from co-ion exclusion measurements. The results of this work suggest that the surfaces of montmorillonite and illite have constant potentials and do not behave like constant-charge surfaces as is generally assumed.

The characterization of acidic sites on solid surfaces represents a fundamental problem in mineral characterization. Total concentrations can be determined by titration, but this does not distinguish between Brönsted and Lewis acid sites. By the adsorption of bases, such as pyridine, these sites can be distinguished by IR spectroscopy (5), but quantitative measurements are difficult, requiring the determination of molar absorptivities. Quantum mechanical calculations have recently been used (6) in an attempt to rationalize the interaction mechanisms of surface hydroxyl groups of the Lewis acid type with adsorbent molecules in zeolites and amorphous aluminosilicates, and some papers have been published which demonstrate the use of NMR spectroscopy in resolving this type of problem. For example, N-15 NMR under conditions of cross polarization and "magic angle" spinning can distinguish the Brönsted and Lewis acid sites by the systematic displacement of N-15 enriched pyridine by unenriched n-butylamine (7). Silylation followed by Si-29 NMR has been used as a probe of the surface hydroxyl groups of silica (8,9) and could presumably also be used with mineral hydroxyl surfaces. However, IR spectroscopy is the technique that has been used most frequently (e.g. 10,11) in the characterization of surface hydroxyl groups in minerals.

Experimental information on the origin and nature of sites for adsorption on the surfaces of minerals can be obtained from ESR investigations of adsorbed paramagnetic ions. For example, the results from Cu(II) and Mn(II) exchanged kaolinite, talc and pyrophyllite (12) demonstrate that most of the exchange sites arise from ionic substitutions in each of the minerals. The dipolar broadening of the Mn spectra was consistent with the divalent exchange ions being about 11-12 Å apart on the mineral surfaces. The Cu spectra at low relative humidity were orientation dependent with the principal axis perpendicular to the plane of the clay sheet, suggesting that the cation exchange was not associated with edge sites. Recent work on the adsorption of Cd(II) and Cu(II) on montmorillonite in the pH range 4-8.5 (13) has shown that at low levels of metals, however, the pH dependence associated with adsorption is reversible and much lower than for oxide minerals or for precipitation reactions. From these measurements it was argued that the adsorption occurs preferentially at the constant potential edge sites in this pH range.

Adsorption of metal ions on minerals

Selectivity of adsorption can vary greatly and is influenced by a variety of factors, the most significant being the charge and chemical form of the adsorbate, the nature of other ions in the external solution and the character of the surface of the mineral under investigation. Recognition of the various factors contributing to the adsorptivity is important in understanding and interpreting experimental results. In this section, selected examples from the literature will be presented as illustrations of the development of our knowledge on adsorption processes.

Exchange isotherms have been produced (14) for several divalent cations with Na montmorillonite and found to be virtually identical at constant pH (5.5). The reactions were stoichiometric and reversible when the equivalent fraction of the trace metal in the exchange phase was <0.7, and the selectivity coefficient was found to vary with the amount of metal in the exchange phase. Although this behaviour was described as non-ideal, it has been shown subsequently that, according to thermodynamic principles, the sodium-trace metal cation exchange produces an exchanger phase on montmorillonite that behaves as an ideal mixture (15). When exchanging ions are of unequal charge, it has been shown (16), from an analysis of ion exchange data, that the formation of tactoid structures may influence selectivity of adsorption. In this latter work the degree of deviation from ideal mass-action exchange was related to the dissimilarity of the ions undergoing exchange and data involving trivalent ion adsorption on smectites suggest that mass action is a poor approximation when the adsorbing and desorbing ions have different hydration energies and charge. No form of exchange equation successfully described ion exchange for a wide range of experimental conditions, although fluctuation of the selectivity coefficient followed consistent trends with changing experimental conditions. There are also cases where individual minerals show strong specific adsorption for a particular cation, e.g. halloysite has a high selectivity for Zn (17,18).

The nature and ionic strength of the medium can exert a significant influence on adsorption behaviour, e.g. adsorption of Ni by kaolinites decreased with increasing ionic strength, although greater adsorption occurred with Na- than with Ca-saturated minerals (19). When sulphate was the dominant solution anion, adsorption of Ni was depressed relative to that obtained with nitrate. This was explained by the formation of neutral ion pairs with sulphate and the consequent reduction in activity of the adsorbing species in solution. Adsorption of uranium in the form of uranyl ions by montmorillonite has been variously reported as roughly equal to the CEC of the clay for nitrate and acetate solutions (20,21), to approximately one half of the CEC for nitrate solutions and very much less for sulphate solutions (22). Isotherms for adsorption of uranyl ions on montmorillonite from nitrate solutions have been reported to follow Langmuir-type curves with increasing U concentrations, approaching the CEC of the clay at maximum adsorption (23). In competition with alkali and alkaline earth ions, U was preferentially adsorbed, there being a strong preference relative to Na and K but a smaller preference over Mg, Ca and Ba.

The pH dependence of the adsorption of Co on montmorillonite in the pH range 5-6 has been interpreted (14,24) as due to the behaviour of structural hydroxyl groups on the clay or to hydroxy-Al compounds. In other work, adsorption of Cd, Co and Sr on to montmorillonite from solutions with widely different salt concentrations has been investigated in the pH range 5-6.5 (25). Adsorption of Sr was consistent with a simple ion exchange process (26). At moderate to high salt concentrations (>0.01 M) the adsorbability of Cd was lower from chloride than from nitrate solutions as was the increase in distibution coefficent with increasing pH, possibly as a result of the formation in solution of complexes of the type $CdCl_n^{2-n}$, which have a lower affinity for the clay than the free solvated ion. Adsorbability of Cd and Co increased with increasing pH, particularly at high salt concentrations. Their distribution coefficents decreased with increasing salt concentrations, but less sharply than that of Sr. At very low loading levels there was a decrease in distribution coefficient with increased loading of Cd and Co compared to Sr, suggesting that 2 different types of site participate in the adsorption of both Cd and Co.

Complexing ligands often, but not always, affect the adsorption of trace metals on mineral surfaces. For example, Inskeep and Baham (27) have reported that addition of natural water soluble organic ligands from forest litter, sewage sludge or soil had little effect on Cd adsorption on montmorillonite (27), but Farah and Pickering (28) found a significant difference in the adsorption of Cd from a landfill leachate as compared to that from pure sodium nitrate solution. The presence of ligands causes the threshold pH, at which precipitation/sorption of hydroxy species occurs, to be shifted to higher values for Pb and Cd on kaolinite, illite and montmorillonite, the magnitude of the effect depending on the stability of the metal-ligand complex (29). Natural water-soluble organic ligands produced a marked decrease in Cu adsorption on montmorillonite with increased pH values, in contrast to the increase that was observed in the absence of organic ligands (27).

Extensive measurements have been made of the influence of organic ligands on the uptake of Cu(II) by kaolinite, illite and montmorillonite (30,31). With kaolinite in alkaline media the clay acts as a nucleation site for the formation of hydroxy-bridged copper species and the major role of many ligands is to "mask" this precipitation reaction, since uncharged and negatively-charged complexes are not adsorbed to any measurable extent. When kaolinite was allowed to come into contact with the metal prior to the addition of the ligand, the amount retained by the clay was greater than if the initial contact was with the ligand or the complex. The absorptive capacity of kaolinite for cations increased with pH up to a limiting value at about pH 7. Other divalent ions were found to compete with Cu for adsorption sites, e.g. with 5-fold excess of Mg or Ca the Cu adsorption fell to 30% and 18%, respectively, of its original level on Na montmorillonite. The behaviour of illite was similar to that of kaolinite with the controlling process apparently the formation of polymeric hydroxy species on particular surface sites of the clay. The development of hydroxy ions is controlled by the solution pH and inhibited by complexation with the ligands.

Montmorillonite, in contrast, appeared to behave as an ion-exchanger with results being interpreted in terms of competition between all positively charged solution species for the adsorption sites.

The retention of Cu by allophane is enhanced by phosphate regardless of the sequence of Cu and phosphate adsorption, although Cu has been found to have no effect on the simultaneous and subsequent adsorption of phosphate on surface bound Cu (32). ESR results suggest that the Cu binds to surface AlOH groups of the allophane irrespective of the presence of phosphate and it was proposed that the enhanced Cu retention was the result of the formation of a ternary complex by the binding of phosphate to the axial position of the surface-bound Cu ion.

The formation of polymeric metal ions on mineral surfaces readily occurs in nature and aluminosilicate clay minerals in their natural state are often associated with surface coatings of iron and/or aluminium oxides. Mössbauer spectroscopy has been used extensively in the identification and characterization of such iron oxide species in minerals (e.g. 33), but little work has been carried out on the similar aluminium oxide species. However, the production of hydroxy-aluminium interlayers in expanding layer minerals, possibly involving the $Al_{13}O_4(OH)_{24}^{7+}$ ion, has received considerable attention (34). The polymeric cations partially neutralize the charge of the aluminosilicate sheets, which, therefore, exhibit a reduced CEC, but at the same time the interlayer ions function as pillars which prevent the mineral from collapsing on heating. Thus structures resembling those of zeolites are formed and have attracted interest because of their molecular sieve and catalytic properties. Hydroxy-magnesium interlayers also occur extensively in phyllosilicates. Such interlayers can be prepared synthetically (35,36) by titrating MgCl and NaOH into a suspension of a layer silicate mineral, although the extent of interlayer formation is strongly dependent on pH and the nature of the mineral (37).

Characterization of chemical forms of adsorbed metal ions

Information on the nature of the chemical environment of trace metal ions adsorbed on clay minerals can be obtained by a number of spectroscopic methods, but the principal applications have used either XPS or ESR spectroscopy, or one of its related techniques, such as ENDOR and ESEM spectroscopy.

With XPS it is possible to obtain good analytical information on the amount of metal adsorbed and, in favourable cases, to identify the chemical form of that metal. Oxidation states are readily determined and it can be shown, for example, that adsorption of Co(II) on manganese oxides results in oxidation to Co(III) (38,39), whereas adsorption of Co(II) on zirconia and alumina leads to the formation of cobalt(II) hydroxide (40). With Y-type zeolites hexaaquacobalt(II) is adsorbed as Co(II), and cobalt(III) hexaammine is adsorbed as Co(III). The XPS spectrum of Co(II) adsorbed on chlorite was consistent with the presence of the hexaaquacobalt(II) ion for pH 3-7 and indicated that no cobalt(II) hydroxide was present (41). With kaolinite and illite, Co is adsorbed as Co(II) over the pH range 3-10 (39,42), it being bound as the aqua ion below pH 6 and as the hydroxide above pH 8. Measurements involving Pb have shown that adsorbed Pb(II) on montmorillonite remains as Pb(II) (43), whereas oxidation to Pb(IV) occurs on β- and δ-MnO_2 (39). Results from the adsorption of Cr(III) on chlorite, kaolinite and illite (44) indicate that the Cr remains as Cr(III) and that the adsorbed species is the Cr(III) aqua ion at pH values below 4 and chromium(III) hydroxide at pH values above 6.

XPS is able to distinguish between metal ions in a mineral structure and those adsorbed on the surface with the same oxidation state. As an example, the Mg 1s photoelectron- and Auger electron-spectra of Mg montmorillonite reveal considerable differences in the electronic states of the exchangeable and skeletal Mg (45), the former being similar to typical ionic compounds, such as magnesium fluoride, whilst the latter resembles magnesium oxide.

XPS has general applications in the study of trace metal species on mineral surfaces. However, because measurements have to be made under high vacuum, there is always a question as to the relationship of the sample being investigated to that which exists under aqueous conditions. ESR and related techniques use samples in their natural state and chemical information is obtained from both solution and solid phases. Information on the orientation of adsorbed species on the clay surfaces is provided in addition to probing their chemical environments more sensitively than other methods for the study of trace components. In these cases, though, only a small number of paramagnetic ions can be investigated and much of the published ESR work is concerned with adsorption of Cu(II), although there are reports of experiments with other transition metal ions.

With zeolites the nature of the Cu(II) ESR spectra varies with the degree of hydration of the mineral, and a range of parameters have been reported in the literature. When Na-Y zeolites were exchanged with low concentrations of Cu(II) at ambient temperature and short equilibrium times (46), hydrated Cu(II) ions were found in the supercages. Rapid dehydration leaves the Cu(II) ions in the supercages where they exhibit distinctive ESR parameters, but slow dehydration followed by prolonged evacuation allows the Cu(II) to migrate into the small cages, where they exhibit a different distinctive set of parameters. These latter ions are the last to undergo reduction at high temperature. Dipolar coupling between pairs of Cu(II) ions in Cu/Ce exchanged Y-zeolites has also been identified in ESR spectra (47). The magnitude of this dipolar interaction permits a calculation of a 4.2 Å separation of the ions and oxygen-broadening experiments reveal the presence of these pairs in the large cavities of the zeolite lattice.

Not all paramagnetic ions produce an ESR spectrum and, indeed, with many ions the ability to produce a spectrum is dependent upon the chemical environment of that ion. By performing double integrations of the 1st derivative ESR spectra from Cu(II) exchanged Y-zeolite with varying degrees of dehydration, Conesa and Soria (48) have shown that there is an ESR spectral intensity minimum at 100°C, while the magnetic susceptibility measurements show no change in paramagnetism under the same conditions. The effect was interpreted as arising from line broadening of Cu(II) in trigonal symmetry. Another decrease in intensity occurred at temperatures above 300°C, but in this case it was accompanied by a similar decrease in susceptibility and corresponded to a reduction of Cu(II) to Cu(I).

Small amounts of Cu(II) adsorbed on gibbsite at low pH gave ESR spectra that were consistent with the presence of free hexaaquacopper(II) and a rigid-limit component from monomeric Cu(II) oriented with its principal axis perpendicular to the (001) plane of the mineral (49). Raising the pH above 5 resulted in an increase in the amount of Cu adsorbed but a decrease in the ESR signal intensity as a result of the formation of copper hydroxide polymers on the mineral surface. The formation of this hydroxide phase on gibbsite occurs 1 pH unit lower than the commencement of precipitation from solution and indicates that the gibbsite surfaces promote the copper hydrolysis, lowering the apparent solubility of the metal. Because IR results indicated that there was no interaction between the copper and surface hydroxyls and no separate-phase copper hydroxide could be detected by electron microscopy, it was proposed that the adsorption occurs at the edges of crystal steps on the (001) faces where Al-OH and Al(OH)$_2$ groups are present.

With layer silicate structures, it has been shown (50) that, when a monolayer of water occupies the interlayer space, adsorbed Cu(II) has axial symmetry with the symmetry axis perpendicular to the silicate layers. Such behaviour is exhibited by a range of layer silicates and is independent of the origin of the layer charge, i.e. from octahedral substitutions in hectorite or tetrahedral substitutions in saponite. With two layers of water in the interlayer space, no appreciable changes with orientation in the spectra from films of the clays were observed, and this result was interpreted in terms of the principal axis of the adsorbed Cu being aligned at approximately 45° to the silicate. However, with such results it is not possible to distinguish between this type of preferential alignment and the situation where there is no preferred orientation relative to the clay structure. Doping Cu(II) into Mg smectites, where the air-dried samples retain an interlayer of three water molecules, produced ESR spectra that demonstrated that Cu once again adopted a preferential alignment with its principal axis perpendicular to the plane of the silicate sheets (51).

With Mn(II)-exchanged smectites, having significant amounts of structural iron, dipolar interactions between Fe and Mn and between neighbouring Mn ions result in large anisotropic line widths (52). In hectorite, which contains little structural Fe, dipolar effects can be minimized by doping small amounts of Mn into the Mg exchanged form. It was found that under fully hydrated conditions hexaaquamanganese(II) ions were present with somewhat broader lines than are found for fluid solutions and which were consistent with an approximately 30% increase in time between collisions with water molecules (52). Line widths increased significantly on drying in air, thus demonstrating that the mobility of the ion is greatly decreased when the interlayer is limited to two water molecules. Dehydration at 200°C caused the Mn ions to move into hexagonal positions in the silicate structure and a spectrum typical of Mn in crystalline matrices was observed.

Exchange of varying quantities of the oxovanadium(IV) ion on Mg hectorite resulted in hydrolysis of V at low levels of adsorption (53). The hydrolyzed product that was adsorbed on the clay surface was interpreted as having a ligand environment that was partially aqueous and partially hydroxide in nature. With increasing V adsorption on wetted hectorite the partially hydrolyzed product was obscured by a spectrum from the solvated oxovanadium(IV) ion which had a linewidth greater than in aqueous solution as a result of restricted mobility. Under strongly dehydrating conditions the adsorbed V was observed to align with the principal V=O axis perpendicular to the plane of the clay platelets.

ESR has also been used in the characterization of species adsorbed on pillared clays, i.e. smectites with hydroxy-aluminium interlayers. Adsorption of Cu(II) on hydroxy-aluminium hectorite produced mobile hexaaquacopper(II) and Cu(II) chemisorbed to discrete sites of the OH-Al interlayer(54). The ratio of chemisorbed to mobile Cu increased with increasing pH, but even at pH>7, when the solubility product of copper(II) hydroxide was exceeded, chemisorbed Cu(II) remained the dominant species. This is in complete contrast to the results with gibbsite (49), where precipitation of copper hydroxide was observed at pH>5. Spectra from air-dried films showed that the Cu(II) had axial symmetry with principal axis perpendicular to the OH-Al-hectorite ab plane. At higher pH, a spectrum similar to that of Cu(OH)$_4^{2-}$ on alumina was observed, suggesting a ligand exchange mechanism for Cu(II) adsorption on the complex. With hydroxy-aluminium montmorillonite there was an increasing capacity for Na adsorption with increasing pH (55) and Na was not displaced by adsorption of low levels of Cu, indicating the existence of Cu-specific sites. In contrast to the hectorite system, there was evidence of a hydroxy or hydroxy carbonate precipitate analogous to the situation with gibbsite, although there was no gibbsite detected in this system. ESR spectra showed the existence of a chemisorbed species as well as electrostatically bound hexaaquacopper(II), but rather unexpectedly the Cu could be solubilized with ammonia only after extraction with barium chloride.

The specific adsorption of Cu(II) and Co(II) by imogolite, an aluminosilicate mineral with tubular morphology of composition $Al_2SiO_3(OH)_4$, was found to be lower than for allophanes, which are related amorphous aluminosilicate materials covering a range of Al:Si ratios. Cu adsorption on synthetic allophanes was dependent on the Si:Al ratio (increasing with increasing Al) (56) but no consistent effect was found with natural allophanic clays or for Co(II) adsorption. ESR spectra were interpreted as indicating that adsorption of monomeric Cu(II) occurred on an alumina-like surface, where OH was coordinated to a single Al ion and at a second type of site, which was thought to be a single SiOH or AlOH group, with the distribuion of Cu(II) between the sites being dependent on the Si:Al ratio, pH and adsorbate concentrations. Exposure of the adsorbed Cu to ammonia resulted in ligand exchange and the formation of Cu(II)-NH$_3$ -surface complexes. With imogolite some Cu(II) was desorbed from the surfaces as tetraamminecopper(II) ions and desorption of both Cu(II) and Co(II) ions was readily effected by complexation with EDTA or by competition with Pb(II) or protons.

In the ESR spectra of adsorbed oxovanadium(IV) ions on minerals, information on the nature of the adsorbed species is obtained from the g-values and V hyperfine coupling constants, but ligand hyperfine structure is seldom, if ever, observed. With ENDOR much smaller hyperfine splittings can be observed than with ESR and it is possible to measure hyperfine coupling from nuclear spins in the neighbourhood of the paramagnetic cation. For example, proton splittings for both axial and equatorial water associated with oxovanadium(IV) adsorbed on Y-zeolite were found to be similar to those from the ion in solution, but in the zeolite there was no matrix ENDOR peak, indicating that the V was shielded from interaction with protons outside of its 1st coordination sphere(57). Because there was no close approach of bulk water molecules, the complex could not be in the centre of the supercage surrounded by one or more solvent layers. Also, although evacuation did not cause a significant change in the ESR spectrum (58), the ENDOR spectrum lost its proton signal, and exhibited a more complex Na ENDOR spectrum (57). This result showed that on dehydration V loses its axial water molecule and binds to four oxygens of the zeolite structure, thus illustrating the type of chemical change that can occur in adsorbed species simply as a result of dehydrating the sample. ENDOR spectra are more sensitive than ESR to changes in structure and there is great value in performing both types of measurement.

The ESEM technique provides an alternative method for measuring small hyperfine interactions and has the advantage of providing information on the number of spins responsible for a particular interaction in addition to distinguishing isotropic and dipolar contributions to the hyperfine coupling constants. Although analysis of the time domain results can sometimes present difficulties, several papers have been published on the application of ESEM to the adsorption of cations on inorganic solids. Ichikawa et al (59) have shown that Cu(II) ions exchanged into silica gel and interacting with ammonia or water, have only two ligand molecules in their coordination spheres. [It should be noted that the Cu(II) exchanged into silica is completely different to that adsorbed onto silica gel, where the hexaaquacopper(II) species is present.]

ESR and ESEM studies of Cu(II) in a series of alkali metal ion-exchanged Tl-X zeolites were able to demonstrate the influence of mixed co-cations on the coordination and location of Cu(II) (60). The presence of Tl(1) forces of Cu(II) into the α-cage to form a hexaaqua species, whereas Na and K result in the formation of triaqua or monoaqua species. In NaTl-X zeolite, both species are present with the same intensity, indicating that both cations can influence the location and coordination geometry of Cu(II). The Cu(II) species observed after dehydration of Tl-rich NaTl-X and KTl-X zeolites was able to interact with ethanol and DMSO adsorbates but no such interaction was observed with CsTl-X zeolites. This interaction with polar adsorbates was interpreted in terms of migrations of the copper from the β-cages.

ESR spectroscopy can be used with adsorbed paramagnetic ions to study the liquid associated with mineral surfaces. Cu(II) and Mn(II) have been used in his type of investigation, although difficulties are encountered with observing a resonance from Mn(II) in distorted environments. Measurements of Cu(II) on silica at room temperature and above have shown that adsorbed water behaves in the same manner as bulk water, but at lower temperatures it experiences a decreased mobility (61). On freezing two types of water are found; one which is freezable and undergoes crystallization and the other which is unfreezable, in which the ice structure cannot be formed because of the surface interaction. NMR, IR and differential thermal analysis (DTA) have also been employed in the investigation of the strucure of adsorbed water (62-65), which can have an effect on mineral propertes, e.g. the NMR line width, and hence the electric field gradient around Al, in ZSM-5 zeolites has been shown to be influenced by the degree of hydration (66).

Adsorption and characterization of complex species on minerals

The adsorption of transition metal complexes by minerals is often followed by reactions which change the coordination environment around the metal ion. Thus in the adsorption of hexaamminechromium(III) and tris(ethylenediamine) chromium(III) by chlorite, illite and kaolinite, XPS showed that hydrolysis reactions occurred, leading to the formation of aqua complexes (67). In a similar manner, dehydration of hexaamminecobalt(III) and chloropentaamminecobalt(III) adsorbed on montmorillonite led to the formation of cobalt(II) hydroxide and ammonium ions (68), the reaction being conveniently followed by the IR absorbance of the ammonium ions. Demetallation of complexes can also occur, as in the case of dehydration of tin tetra(4-pyridyl) porphyrin adsorbed on Na hectorite (69). The reaction, which was observed using UV-visible and luminescence spectroscopy, was reversible indicating that the Sn(IV) cation and porphyrin anion remained close to one another after destruction of the complex.

Inelastic electron tunneling spectroscopy(IETS) is a novel spectroscopic technique that produces vibrational spectra and gives the possibility of observing bands that are Raman- and/or IR-allowed as well as transitions that are forbidden in the photon spectroscopies. Hipps and Mazor (70) have used IETS to investigate Ni and Co glycinates adsorbed on alumina and have shown that the glycine acts as a bidentate ligand in the surface complexes in the pH range 4-9. It was not possible, however, to determine whether mono- or diglycinates were formed.

Adsorption of a number of different copper complexes on kaolinite and illite has been shown to involve the formation of polymeric hydroxy species at particular sites on the clays (30,31). In contrast, with montmorillonite adsorption appeared to proceed via an ion exchange process with all positively charged solution species competing for availabile sites on the mineral (32). ESR measurements of bisglycineCu(II) adsorbed on montmorillonite and imogolite at pH 6 showed the presence of two different adsorbed species, one resembling the solution complex, the other being different for the two minerals (71). Furthermore, with a large Cu humate complex, adsorption on montmorillonite appeared to involve a physical association of the Cu humate with the mineral surface, whereas imogolite extracted the metal from the organic matter to produce an ESR spectrum that was identical to that which was obtained by adsorbing the uncomplexed ion (71).

ESEM results on the interaction of silica-exchanged Cu(II) with a range of adsorbates showed that one or two adsorbate molecules were able to coordinate to the Cu depending on the chemical interaction, polarity and size (72). Differences in $A_{//}$ were observed for N- and O-coordinated ligands, but these seem to reflect a change in coordination symmetry and not a difference in adsorbate ligand number. N-coordinated ligands form approximately square planar

complexes with two adsorbates and two lattice oxygens whereas
0-coordinated ligands involve four lattice oxygens in distorted
octahedral complexes. Distorted 5-coordinated complexes, involving
four lattice oxygens, and one adsorbate molecule, were formed with
π-bond coordinated ligands.

Complexes of 3d transition metal ions with ammines have a high
affinity for montmorillonite-type clay minerals with overall
stability constants much higher than for the complexes in aqueous
solution (73), although molecular orbital calculations based on d-d
transition energies and ESR parameters for bis(ethylenediamine)
Cu(II) indicated similar values for the σ and in-plane π orbitals
for solution and adsorbed complexes (74). The crystal field
splitting parameter, 10Dq, and the crystal field stabilization
energy of bis(ethylenediamine)Cu(II) adsorbed on a number of
smectite minerals was observed to increase linearly with increasing
average negative charge density of the minerals and to decrease with
increased loading of the clay (75). These results could be explained
if the clay acts as a heterogeneous anionic solvent towards the
complex. XRD shows that tris(ethylenediamine) Cu(II) ions are too
large to fit between montmorillonite sheets at 50% humidity, and,
when large amounts of the complex are added to the clay, the square
planar bis complex is adsorbed (76). With diethylenetriamine and
tetraethylenepentamine complexes the situation is more complicated,
with several different complexes existing in solution as a function
of pH (77). Adsorption studies of Cu(II) and Ni(II) complexes of
these ligands on hectorite using ESR, IR and UV-visible absorption
spectroscopies have shown that the hectorite surface prefers
tetragonally-distorted complexes. Axially-coordinated water
molecules are readily lost and planar complexes are formed on the
interlamellar surface. The planar Ni(II) complexes are diamagnetic,
indicating that the clay surface functions as only a very weak axial
ligand.

Cu(II) and Fe(II) complexes with 1,10-phenanthroline (phen)
show a high affinity for a smectite (hectorite) surface (78). As was
the case with ethylenediamine, adsorbed Cu(II) exists as a bis
complex. Although Fe remains bound in a tris complex, an increase in
the oxidation potential of the $Fe(phen)_3^{2+}$ - $Fe(phen)_3^{3+}$ couple,
above that in pure solvent, was seen when the complexes were
associated with the mineral surface. Adsorption of enantiomeric
$Fe(phen)_3^{2+}$ on montmorillonite has been shown to produce $Fe(phen)_3^{2+}$
as the adsorbed species, whereas racemic $Fe(phen)_3^{2+}$ is adsorbed in
twice the amount as $Fe(phen)_3^{2+}$ X^- ion pairs (79). This property of
adsorption of a racemic mixture to twice the CEC has been exploited
in the antiracemization of a racemic mixture of a labile metal
complex (80). An adduct of montmorillonite with Λ-$Ru(phen)_3^{2+}$
accepted the adsorption of Δ-$M(phen)_3^{2+}$ (M=Ru,Fe,Ni), but not
Λ-$M(phen)_3^{2+}$, and racemic mixtures of either the Co or Fe complexes
were observed to antiracemize in the presence of Λ-$Ru(phen)_3^{2+}$
-montmorillonite. Adsorption of an achiral molecule, such as the
acridine orange cation by Δ-$Ni(phen)_3^{2+}$ -montmorillonite resulted in
optical activity being introduced in the electronic spectrum of the
adsorbate. The Δ-$Ni(phen)_3^{2+}$ -montmorillonite complex has recently
been used to resolve chromatographically enantiomers of organic
molecules having two aromatic or one aromatic and one aliphatic ring
(81).

Reactions involving metal ions adsorbed on minerals

The chemisorption or reaction of organic molecules with metal-exchanged clays is an area of science of great relevance to the chemical industry as well as being of importance in the understanding of natural processes and the control of environmental quality. Examples of this type of reaction are presented in the following sections.

Several papers have been published in which reactions of simple molecules with metal exchanged clays have been investigated. Lawless and Levi (82) have studied the effect of exchangeable cation on the condensation of glycine and alanine in bentonites during drying, warming and wetting cycles. Peptide bond formation was observed and the effectiveness of the metal to catalyze the condensation was $Cu^{2+} > Ni^{2+} = Zn^{2+} > Na^+$. Glycine showed 6% of the monomer incorporated into oligomers with the pentamer being the largest detected. Less peptide bond formation occurred with alanine and only the dimer was observed. Adsorption of ATP and ADP was found to be greater for Mg and Zn montmorillonite than for Na montmorillonite (83), presumably because of complex formation. Pure ATP decomposed on heating and the rate of decomposition was accelerated in the presence of glycine with the production of small yields of peptide. In the presence of Mg- or Zn-kaolinite or Mg-montmorillonite the rate of decomposition of ATP decreased, although the peptide formation still occurred. The 5'-AMP nucleotide was not adsorbed by alkali-metal exchanged clays, but with divalent metal ions the adsorption increased in the order $Mg < Co < Ni < Cu < Zn$ (84). Also purine nucleotides are adsorbed more strongly than pyrimidine nucleotides on Zn bentonite. In a competitive study between 2'-, 3'- and 5'-AMP for Zn bentonite there was a large preference for adsorption of the 5'-AMP over the 2'- and 3'-isomers.

Two types of complex are formed on reaction of benzene with Cu montmorillonite. In the Type 1 species the benzene retains its aromaticity and is considered to be edge bonded to the Cu(II), whereas in the Type 2 complex there is an absence of aromaticity (85,86). ESR spectra of the Type 2 complex consist of a narrow peak close to the free spin g-value and this result can be explained in terms of electron donation from the organic molecule to the Cu(II), to produce a complex of Cu(I) and an organic radical cation. Similar types of reaction occur with other aromatic molecules. However with phenol and alkyl-substituted benzenes only Type 1 complexes were observed (87), although both types of complex were seen on the adsorption of arene molecules on to Cu(II) montmorillonites (88) and anisole and some related aromatic ethers on to Cu(II) hectorite (89).

The reaction of benzene with Cu(II) and Fe(III)-exchanged hectorites at elevated temperatures produced a variety of organic radical products, depending on the concentration of water in the reaction medium and the reaction time (90). The formation of free radicals was accompanied by a reduction in oxidation state of the metals, a process that had a zero-order dependence on the metal ion concentration. Under anhydrous conditions the free radicals appeared to populate sites in the interlayer region, the activation energies under these conditions being lower than in the hydrated samples.

Activation energies were also lower for the Fe(III) than the Cu(II) systems and agreed well with numbers obtained from kinetic measurements. There was little temperature dependence of the ESR signals from the organic radicals suggesting that exchange processes involving organic free radicals were probably not important in determining the ESR line shapes.

Fe(III) and Cu(II) adsorbed on smectites (hectorite), either as hydrated or partially hydrolyzed ions, have a high degree of activity in oxidizing benzidine to form the benzidine blue semiquinone radical cation (91), whereas in the absence of clay further oxidation occurs. Small amounts of structural iron were also effective in initiating this reaction, but surface iron oxide coatings exhibited no significant activity. A similar type of reaction has been observed to occur with dioxins when refluxed with a Cu(II) smectite in n-hexane (92). Formation of the radical cation was established by IR, UV-visible and ESR spectroscopy. Mass spectroscopy showed some evidence for the formation of dioxin polymers, with molecular ions for both dimers and trimers being seen. It has been proposed that this type of reaction may provide a cheap and convenient means of detoxifying dioxins by reacting the dioxin radical cations with other organic species to yield products of a less undesirable nature.

In addition to stabilizing organic products by reaction with metal-exchanged clays, as indicated above, aluminosilicate minerals may enable the preparation of metal organic complexes that cannot be formed in solution. Thus a complex of Cu(II) with rubeanic acid (dithiooxamide) could be prepared by soaking Cu montmorillonite in an acetone solution of rubeanic acid (93). The intercalated complex was monomeric, aligned with its molecular plane parallel to the interlamellar surfaces, and had a metal:ligand ratio of 1:2 despite the tetradentate nature of the rubeanic acid.

Specific reactions may occur between organic molecules and particular metals exchanged onto clays, e.g. the reaction between flavomononucleotide (FMN) and Fe(III) montmorillonite. Adsorption isotherms, UV-visible and Mössbauer spectroscopic data showed that a 1:1 Fe:FMN ratio existed at maximum adsorption. Significantly lower levels of adsorption were exhibited by Cu, Zn and Ca clays, in which there was apparently less specific interaction (94). A completely different type of behaviour has been reported for the interaction of picloram (4-amino-, 5, 6-trichloropicolinic acid) with metal-containing montmorillonites (95). Interaction with Al or Fe-saturated clay or clay coated with Al or Fe hydrous oxides was concentration dependent, the monomeric acid being present on the mineral surface at low concentration (<1 meq/g clay) and the salt at higher levels of adsorption. The IR spectrum of picloram adsorbed on montmorillonite coated with a hydrous oxide of Cu(II) resembled that of Cu(II) picloram, indicating that coordination-type bonding occurred, but when picloram was adsorbed onto Cu(II)-exchanged clay a completely different type of spectrum was obtained. With both copper systems, however, the interactions were independent of concentration in the range 0.44-2.20 meq/g clay.

Interactions between adsorbed metal ions and complex organic molecules have been observed in other systems. For example, IR results indicated that complexation occurred between sulfolane ($C_4H_8SO_2$) and either Cu or Ni montmorillonites, with the result that desorption rates were very much lower than with Na or H clays (96). The adsorption of bitumen on montmorillonite, kaolinite, illite and chlorite was found to be influenced by the nature of the exchangeable cation and the solvent carrier (97). Ca clays adsorbed the organic material more strongly than Na clays and, even though adsorption occurred primarily on external surfaces, the clay organic complexes were sufficiently stable to resist powerful organic solvents (97).

The use of photoacoustic spectroscopy as a tool for monitoring surface species has been demonstrated (98) in investigations of the complexation of Cu(II) with an ethylenediamine analog immobilized on silica gel. Nearly consecutive formation of bis and mono complexes occurred with increasing Cu(II) concentration although there was no evidence for the conversion of bis to mono sites. A high level of agreement was found between these results and ESR results (99) thus demonstrating that photoacoustic spectroscopy may have potential uses in other systems where the adsorbed species has absorption at suitable wavelengths. Plots of the reciprocal of the amount of immobilized metal against the reciprocal of the mobile metal concentration offer a convenient way of describing the heterogeneous metal distribtion as a function of metal ion concentration and also provide a basis for comparison with the corresponding solution phase equilibria.

Conclusions

Chemical reactions of adsorbed species are of importance in vast areas of science, the involvement of adsorbed metal ions in catalysis being one example of great economic value. In addition reactions involving adsorbed species can sometimes produce products that may be either difficult or impossible to prepare away from the mineral surface. Therefore, an understanding of the chemical processes that occur in such systems is of potential economic benefit to industrial operations. Such knowledge is also of much wider significance, however, because the movement of ions in most environmental situations is controlled by sorption processes, and aluminosilicate minerals play a major role in many situations.

Sorption processes are influenced not just by the natures of the absorbate ion(s) and the mineral surface, but also by the solution pH and the concentrations of the various components in the solution. Even apparently simple absorption reactions may involve a series of chemical equilibria, especially in natural systems. Thus in only a comparatively small number of cases has an understanding been achieved of either the precise chemical form(s) of the adsorbed species or of the exact nature of the adsorption sites. The difficulties of such characterization arise from (i) the number of sites for adsorption on the mineral surface that are present because of the isomorphous substitutions and structural defects that commonly occur in aluminosilicate minerals, and (ii) the difference in the chemistry of solutions in contact with a solid surface as compound to bulk solution. Much of our present understanding is derived from experiments using spectroscopic techniques which are able to produce information at the molecular level. Although individual methods may often be applicable to only special situations, significant advances in our knowledge have been made using this type of approach. Even so, many systems are poorly understood and there is still considerable scope for advancement to be made in our understanding of sorption reactions by utilising fully the information that can be obtained from the various spectrscopic techniques.

Literature Cited

1. Tewari, P.H. "Adsorption from aqueous solutions"; Plenum Press: New York and London, 1981.
2. Bolland, M.D.A.; Posner, A.M.; Quirk, J.P. _Clays Clay Minerals_ 1980, 28, 412-418.
3. Lim, C.H.; Jackson, M.L.; Koons, R.D.; Helmke, P.A. _Clays Clay Minerals_ 1980, 28, 223-229.
4. Chan, D.Y.C.; Pashley, R.M.; Quirk, J.P. _Clays Clay Minerals_ 1984, 32, 131-138.
5. Parry, E.P. _J. Catalysis_ 1963, 2, 371-379.
6. Geerlings, P.; Tariel, N.; Botrel, A.; Lissillour, R.; Mortier, W.J. _J. Phys. Chem._ 1984, 88, 5752-5759.
7. Haw, J.F.; Chuang, I.-S.; Hawkins, B.L.; Maciel, G.E. _J. Am. Chem. Soc._ 1983, 105, 7206-7207.
8. Sindorf, D.W.; Maciel, G.E. _J. Am. Chem. Soc._ 1983, 105, 3767-3776.
9. Sindorf, D.W.; Maciel, G.E. _J. Phys. Chem._ 1983, 87, 5516-5521.
10. Little, L.H. "Infrared spectra of adsorbed species"; Academic Press: London and New York, 1966
11. Kiselev, A.V.; Lygin, V.I. "Infrared spectra of surface compounds"; Keter Publishing House; Jerusalem, 1975.
12. McBride, M.B. _Clays Clay Minerals_ 1976, 24, 88-92.
13. Inskeep, W.P.; Baham, J. _Soil Sci. Soc. Am. J._ 1983, 47, 660-665.
14. Maes, A.; Peigneur, P.; Cremers, A. _Proc. Int. Clay Conf._ Mexico City, 1975, p 319-329.
15. Sposito, G.; Mattigod, S.V. _Clays Clay Minerals_ 1979, 27, 125-128.
16. McBride, M.B. _Clays Clay Minerals_ 1980, 28, 255-261.
17. Wada, K.; Abd-Elfattah, A. _J. Soil Sci._ 1979, 30, 281-290.
18. Wada, K.; Kakuto, Y. _Clays Clay Minerals_ 1980, 28, 321-327.
19. Mattigod, S.V.; Gibali, A.S.; Page, A.L. _Clays Clay Minerals_ 1979, 27, 411-416.
20. Goldsztaub, S.; Wey, R. _Bull. Soc. Franç. Miner. Crist._ 1955, 78, 242-248.
21. Nuss, M.L.; Wey, R. _Bull. Gr. Franç. Argiles_ 1956, 7, 15-19.
22. Davey, P.T.; Scott, T.R. _Nature_ 1956, 178, 1195.
23. Tsunashima, A.; Brindley, G.W.; Bastovanov, M. _Clays Clay Minerals_ 1981, 29, 10-16.
24. Peigneur, P.; Maes, A.; Cremers, A. _Clays Clay Minerals_ 1975, 23, 71-75.
25. Egozy, Y. _Clays Clay Minerals_ 1980, 28, 311-318.
26. Shiao, S.-Y.; Rafferty, P.; Meyer, R.E.; Rogers, W.J. _ACS Symposium 100,_ 1979, p 297-324.
27. Inskeep, W.P.; Baham, J. _Soil Sci. Soc. Am. J._ 1983, 47, 1109-1115.

28. Farrah, H.; Pickering, W.F.   Aust. J. Chem. 1976, 29, 1167-1176.
29. Farrah, H.; Pickering, W.F.   Aust. J. Chem. 1976, 29, 1177-1184.
30. Farrah, H.; Pickering, W.F.   Aust. J. Chem. 1977, 30, 1417-1422.
31. Frost, R.R.; Griffin, R.A.   J. Environ. Sci. Health 1977, A12, 139-156.
32. Clark, C.J.; McBride, M.B.   Soil Sci. 1985, 139, 412-421.
33. Fysh, S.A.; Cashion, J.D.; Clark, P.E.   Clays Clay Minerals 1983, 31, 293-298.
34. Rich, C.I.   Clays Clay Minerals 1968, 16, 15-30.
35. Carstea, D.D.; Harward, M.E.; Knox, E.G.   Clays Clay Minerals 1970, 18, 213-222.
36. Slaughter, M.; Milne, I.H.   Clays Clay Minerals Proc. 7th Conf 1958 (Publ. 1960) 114-124.
37. Keren, R.   Clays Clay Minerals 1979, 27, 303-304.
38. Murray, J.W.; Dillard, J.G.   Geochim. Cosmochim. Acta 1979, 43, 781-787.
39. Dillard, J.G.; Koppelman, M.H.; Crowther, D.L.; Schenck, C.V.; Murray, J.W.; Balistrieri, L.   In "Adsorption from Aqueous Solutions"; Tewari, P.H., Ed.,; Plenum Press: New York and London, 1981, p 227-240.
40. Tewari, P.H.; Lee, W.   J. Colloid Interface Sci. 1975, 52, 77-88.
41. Koppelman, M.H.; Dillard, J.G.   J. Colloid Interface Sci. 1978, 66, 345-351.
42. Dillard, J.G.; Koppelman, M.H.   J. Colloid Interface Sci. 1982, 87, 46-55.
43. Counts, M.E.; Jen, J.S.C.; Wightman, J.P.   J. Phys. Chem. 1973, 77, 1924-1926.
44. Koppelman, M.H.; Emerson, A.B.; Dillard, J.G.   Clays Clay Minerals 1980, 28, 119-124.
45. Seyama, H.; Soma, M.   Chem. Letts. 1981, 1009-1012.
46. Herman, R.G.; Flentge, D.R.   J. Phys. Chem. 1978, 82, 720-729.
47. Conesa, J.C.; Soria, J.   J. Phys. Chem. 1978, 82, 1575-1578.
48. Conesa, J.C.; Soria, J.   J. Phys. Chem. 1978, 82, 1847-1850.
49. McBride, M.B.; Fraser, A.R.; McHardy, W.J.   Clays Clay Minerals 1984, 32, 12-18.
50. Clementz, D.M.; Pinnavaia, T.J.; Mortland, M.M.   J. Phys. Chem. 1973, 77, 196-200.
51. McBride, M.B.; Pinnavaia, T.J.; Mortland, M.M.   Am. Mineral. 1975, 60, 66-72.
52. McBride, M.B.; Pinnavaia, T.J.; Mortland, M.M.   J. Phys. Chem. 1975, 79, 2430-2435.
53. McBride, M.B.   Clays Clay Minerals 1979, 27, 91-96.
54. Harsh, J.B.; Doner, H.E.; McBride, M.B.   Clays Clay Minerals 1984, 32, 407-413.
55. Harsh, J.B.; Doner, H.E.   Soil Sci. Soc. Am. J. 1984, 48, 1034-1039.
56. Clark, C.J.; McBride, M.B.   Clays Clay Mineral 1984, 32, 300-310.
57. van Willigen, H.; Chandrashekar, T.K.   J. Am. Chem. Soc. 1983, 105, 4232-4235.

58. Martini, G; Ottaviani, M.F.; Seravalli, G.L. J. Phys. Chem. 1975, 79, 1716-1720.
59. Ichikawa, T.; Yoshida, H.; Kevan, L. J. Chem. Phys. 1981, 75, 2485-2488.
60. Lee, H.; Narayama, M.; Kevan, L. J. Phys. Chem. 1985, 89, 2419-2425.
61. Bassetti, V.; Burlamacchi, L.; Martini, G. J. Am. Chem. Soc. 1979, 101, 5471-5477.
62. Hougardy, J.; Stone, W.E.E.; Fripiat, J.J. J. Chem. Phys. 1976, 64, 3840-3851.
63. Fripiat, J.J. Bull. Gp. Franç. Argiles. 1971, 23, 1-8.
64. Tait, M.J.; Franks, F. Nature (London) 1971, 230, 91-94.
65. Hair, M.L. "Infrared spectroscopy in surface chemistry" Arnold: London 1967.
66. Kentgens, A.P.M.; Scholle, K.F.M.G.J.; Veeman, W.S. J. Phys. Chem. 1983, 87, 4357-4360.
67. Koppelman, M.H.; Dillard, J.G. Clays Clay Minerals 1980, 28, 211-216.
68. Fripiat, J.J.; Helsen, J. Clays Clay Minerals, Proc. 14th Conf. 1966, 163-179.
69. Abdo, S.; Cruz, M.I.; Fripiat, J.J. Clays Clay Minerals 1980, 28, 125-129.
70. Hipps, K.W.; Mazur, U. Inorg. Chem. 1981, 20, 1391-1395.
71. Goodman, B.A.; Green, H.L.; McPhail, D.B. Geochim. Cosmochim. Acta 1984, 48, 2143-2150.
72. Ichikawa, T.; Yoshida, H.; Kevan, L. J. Phys. Chem. 1982, 86, 881-883.
73. Maes, A.; Peigneur, P.; Cremers, A. J. Chem. Soc. Faraday Trans. 1 1978, 182-189.
74. Schoonheydt, R.A. J. Phys. Chem. 1978, 82, 497-498.
75. Maes, A.; Schoonheydt, R.A.; Cremers, A.; Uytterhoeven, J.B. J. Phys. Chem. 1980, 84, 2795-2799.
76. Burba, J.L.; McAtee, J.L. Clays Clay Minerals 1977, 25, 113-118.
77. Schoonheydt, R.A.; Velghe, F.; Baerts, R.; Uytterhoeven, J.B. Clays Clay Minerals 1979, 27, 269-278.
78. Berkheiser, V.E.; Mortland, M.M. Clays Clay Minerals 1977, 25, 105-112.
79. Yamagishi, A.; Tanaka, K.; Toyoshima, I. J. Chem. Soc., Chem. Commun. 1982, 343-344.
80. Yamagishi, A. Inorg. Chem. 1985, 24, 1689-1695.
81. Yamagishi, A. J. Am. Chem. Soc. 1985, 107, 732-734.
82. Lawless, J.G.; Levi, N. J. Mol. Evol. 1979, 13, 281-286.
83. Rishpon, J.; O'Hara, P.J.; Lahav, N.; Lawless, J.G. J. Mol. Evol. 1982, 18, 179-184.
84. Lawless, J.G.; Edelson, E.H. In "GOSPAR Life Sciences and Space Research"; Holmquist, R., Ed.; Pergamon Press: Oxford and New York, 1980; p.83-88.
85. Doner, H.E.; Mortland, M.M. Science 1969, 166, 1406-1407.
86. Mortland, M.M.; Pinnavaia, T.J. Nature (Phys. Sci.) 1971, 229 75-77.
87. Fenn, D.B. and Mortland, M.M. Proc., Int. Clay Conf. 1972, (Pub. 1973) Ed.; Serratosa, J.M. p591-603.
88. Rupert, J.P. J. Phys. Chem. 1973, 77, 784-790.
89. Fenn, D.B.; Mortland, M.M.; Pinnavaia, T.J. Clays Clay Minerals 1973, 21, 315-322.

90. Eastman, M.P.; Patterson, D.E.; Pannell, K.H. Clays Clay Minerals 1984, 32, 327-333.
91. McBride, M.B. Clays Clay Minerals 1979, 27, 224-230.
92. Boyd, S.A.; Mortland, M.M. Nature 1985, 316, 532-534.
93. Son, S.; Ueda, S.; Kanamura, F.; Koizumi, M. J. Phys. Chem. 1976, 80, 1780-1782.
94. Mortland, M.M.; Lawless, J.G.; Hartman, H.; Frankel, R. Clays Clay Minerals 1984, 32, 279-282.
95. Aochi, Y.; Farmer, W.J. Clays Clay Minerals 1981, 29, 191-197
96. Lorprayoon, V.; Condrate, R.A. Clays Clay Minerals 1981, 29, 71-72.
97. Czarnecka, E.; Gillott, J.E. Clays Clay Minerals 1980, 28, 197-203.
98. Burggraf, L.W.; Kendall, D.S.; Leyden, D.E.; Pern, F.-J. Anal. Chim. Acta 1981, 129, 19-27.
99. Pinnavaia, T.J.; Lee, J.G.; Abedeni, M. In "Silylated Surfaces", Leyden, D.E.; Collins, W.T., Eds.; Gordon and Breach: New York, 1980; p333-379.

RECEIVED September 15, 1986

17

Paramagnetic Probes of Layer Silicate Surfaces

Murray B. McBride

Department of Agronomy, Cornell University, Ithaca, NY 14853

Electron spin resonance (ESR) is a useful technique
for investigating the mobility and orientation
of exchange cations at the surface of layer silicate
clays in various states of hydration. Using Cu^{2+}
and the charged nitroxide spin probe, TEMPAMINE[+]
(4-amino-2,2,6,6-tetramethylpiperidine N-oxide),
adsorbed on a small fraction of the exchange sites
of several smectites (expanding layer silicates),
it is shown that interlayers with thicknesses equivalent to four molecular layers of water or more
allow a high degree of rotational mobility, while
less expanded interlayers impose a specific rigid
orientation on the probe cations. It is proposed
that quantity and location of structural charge
modify chemical interactions between probe cations
and surfaces via surface-H_2O hydrogen bonding.

The layer silicate clays contribute significantly to the chemical
properties of soils and sediments due to their very high specific
surface areas in aqueous media and their capacity for cation
exchange. Figure 1 depicts the structure of a single crystallite
(platelet) of those minerals termed 2:1 layer silicates. The
terminology derives from the fact that the unit cell consists
of a single gibbsite-like or brucite-like layer (with Al^{3+} or
Mg^{2+} in octahedral coordination) sandwiched between two planar
layers of corner-linked silica tetrahedra. If they have no structural defects, these silicate structures are classified under
the pyrophyllite-talc subgroup, and are essentially inert. However, isomorphous substitution of Al^{3+} for Si^{4+} in the tetrahedral
layer, and Mg^{2+} for Al^{3+} (or Li^+ for Mg^{2+}) in the octahedral
layer bestows internal negative charge on the crystallites, which
must then be compensated by exchangeable cations at the surface.
These cations are depicted in two possible states of hydration
at the surface in Figure 1. Depending upon the charge, hydration
energy, redox properties and acidity of these cations, the
chemical and physical properties of the layer silicates can be

0097-6156/86/0323-0362$07.75/0

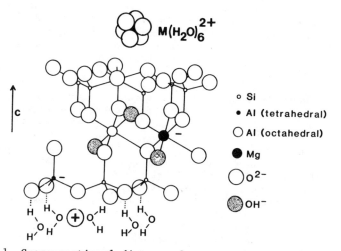

Figure 1. Cross-sectional diagram of an expanding 2:1 layer silicate showing the octahedral layer, tetrahedral layer, and hydrated exchange cations in the interlayer.

greatly modified. One example is the exchange of strongly hydrating cations such as Ca^{2+} by alkylammonium ions, resulting in increased hydrophobicity of the clay as well as a reduced ability of the interlayer (region between adjacent crystallites) to imbibe water (1).

Layer silicates, if initially in the dry state, can expand in water by hydration of the exchangeable cation and the surface. For layer silicates with low structural charge (i.e. smectites), this expansion is limited to about four molecular layers of water if the exchangeable cation has a charge of +2. Since the silicate platelet is about 0.96 nm thick, the repeat spacing along the c-axis is then approximately 0.96 + (4x.26) = 2.0 nm.

However, smectites with singly-charged exchangeable cations such as Li^+ and Na^+ demonstrate free-swelling behavior in water, with the c-axis spacing increasing to very large values. Removal of water by air-drying collapses the spacing to the equivalent of one molecular layer of water (in the case of cations with +1 charge) or two layers of water (cations with +2 charge) at most between the crystallites. One and two layers of water correspond to approximately 1.2 and 1.5 nm c-axis spacings, the result of strongly bonded hydration water in the inner hydration sphere of the cations.

Vermiculites have a 2:1 layer structure similar to smectites, but expand less freely in water, presumably because of the higher layer charge in the former minerals. Most of this structural charge resides in the tetrahedral layers of the vermiculite platelets. Even when fully wetted, vermiculites do not expand beyond the two water-layer stage (~1.5 nm c-spacing).

There is considerable evidence that, despite the structural similarity of vermiculites and smectites, their surfaces have quite different adsorptive properties. For instance, hydrogen bonding between adsorbed water and surface oxygen atoms is stronger for tetrahedrally-charged than for octahedrally-charged layer silicates (2), a result of the greater negative charge imparted to surface oxygens by the tetrahedral location of charge (Figure 1). Thus, vermiculites bond water at the surface more tenaciously than octahedrally-charged smectites such as montmorillonite (2). This fact may explain the superiority of montmorillonite over vermiculite as an adsorbent for organocations (3, 4). Complicating this description, however, is the fact that a sample of any particular layer silicate can have layer charge properties which vary widely from one platelet to another (5). By measuring the c-axis spacings, cation exchange capacity, water retention, and other properties of layer silicates, one obtains the "average" behavior of the mineral surfaces.

Spectroscopic techniques such as electron spin resonance (ESR) offer the possibility to "probe" the chemical environment of the interlayer regions. With the ESR technique, an appropriate paramagnetic ion or molecule is allowed to penetrate the interlayer, and chemical information is deduced from the ESR spectrum. Transition metal ions, such as Cu^{2+}, and nitroxide radical cations, such as TEMPAMINE (4-amino-2,2,6,6-tetramethylpiperidine N-oxide) have been used as probes in this manner (6-14). Since ESR is a sensitive and non-destructive method, investigations of small quantities of cations on layer silicate clays at various stages

of hydration is possible. Several other surface analysis methods
have been limited in value by the need to dry the clay or conduct
the analysis in a vacuum. This paper outlines the theory of
ESR as it applies to nitroxide spin probes and Cu^{2+}, and demonstrates the information that can be obtained by the analysis
of ESR spectra of paramagnetic probes on layer silicate clays
with a range of properties.

Materials and Methods

The Na^+- and Ca^{2+}-saturated montmorillonite, saponite and beidellite were obtained from Dr. J. L. McAtee, Chemistry Department,
Baylor University. These minerals had been purified and prepared
by methods described by Callaway and McAtee (15). Included in
this paper for purposes of comparison are relevant data from
previous studies in which hectorite and vermiculite had been
doped with paramagnetic probes (7, 10). The chemical formulae
of the low-Fe smectites and mermiculite are reported in Table I.

Table I.   Unit cell formulae and sources of the Na^+-saturated
           layer silicate clays.

Species	Unit cell formula	Source
Hectorite	$Na_{1.04}(Si_{8.00})(Mg_{4.70}Li_{1.15}Fe_{0.06})O_{20}(OH)_4$	Hector, CA
Saponite	$Na_{0.98}(Si_{6.85}Al_{1.15})(Mg_{5.79}Al_{0.12}Fe_{0.07})O_{20}(OH)_4$	Ballarat, CA
Montmorillonite	$Na_{0.73}(Si_{8.00})Al_{3.20}Mg_{0.72}Fe_{0.06})O_{20}(OH)_4$	Gonzales County, TX
Beidellite	$Na_{1.10}(Si_{6.95}Al_{1.05})(Al_{3.91}Mg_{0.02}Fe_{0.06})O_{20}(OH)_4$	Black Jack Mine, ID
Vermiculite	$Na_{2.00}(Si_{5.72}Al_{2.28})(Mg_{5.66}Al_{0.30}Fe_{0.02})O_{20}(OH)_4$	Llano, TX

A sufficient quantity of $10^{-3}\underline{M}$ $CuCl_2$ or acidified TEMPAMINE
solution was added to a 50 mg quantity of the clay powder to
attain an approximate adsorption level of 20% or 1% of the CEC,
respectively, assuming complete adsorption. Only the Na^+-saturated
clays were treated with TEMPAMINE. The suspensions were then
well mixed and allowed to form self-supporting films by air-drying.
Washing of the clays was avoided since hydrolysis of Cu^{2+} can
result from this pretreatment. The Cu^{2+} and $TEMPAMINE^+$-doped
clays were then analyzed by ESR, using a Varian E-104 (X-band)
spectrometer. The clay films were oriented in the sample cavity
with the plane of the film perpendicular ($\theta=90°$) and parallel
($\theta=0°$) to the magnetic field in order to determine the directional
dependence of the ESR spectrum. Spectra of the films were obtained
in the air-dry state, after a week of exposure to free water
(100% R.H.), after immersion in distilled water (TEMPAMINE-treated

clays only), and after one day of exposure over concentrated NH_4OH (Cu^{2+}-treated clays only).

Powder x-ray diffraction patterns of oriented clay films on glass slides were obtained for both the air-dry and fully wet Na^+- and Ca^{2+}-smectites using a Philips Norelco diffractometer. The measured c-axis spacings are reported in Table II.

Table II. C-axis spacings of layer silicate clays

Clay	Exchange cation	Spacing (nm)	
		air-dry	wet
Montmorillonite	Ca	1.40	1.84
	Na	1.21	>4.0
Saponite	Ca	1.47(1.29)	1.84
	Na	1.37	1.77
Beidellite	Ca	1.21	1.80
	Na	1.24	>4.0
Hectorite	Ca	1.54	2.10
	Na	1.26	>4.0

Results and Discussion

Nitroxide Spin Probes Adsorbed on Layer Silicates.

A brief review of the principles of ESR is necessary before the spectra of nitroxides adsorbed on clays are analyzed. An unpaired electron in a molecule, when placed in a magnetic field H, has its magnetic dipole aligned parallel or antiparallel to H. If the magnetic field direction is taken to be the z-axis, this alignment of dipoles corresponds to quantization of electron energy states, with the quantum number ($m_S = \pm 1/2$) describing the two allowed z-axis components of the electron spin. The energy of interaction between the magnetic field and the electron spin is:

$$E = g\beta Hm_S$$

where g is the electron g-factor and β is the electronic Bohr magneton. The allowed energies of the unpaired electron are then:

$$E = \pm 1/2 \ g\beta H$$

with the difference, $\Delta E = g\beta H$, representing the quantity of energy required to "flip" the electron from one allowed spin state to the other. If an electromagnetic field of constant frequency, ν, is applied perpendicular to H, transition between these two energy levels will occur at a particular magnetic field strength, H_o, determined by the resonance condition:

$$h\nu = \Delta E = g\beta H_o$$

The allowed transition in ESR is diagrammed in Figure 2. The ESR experiment is commonly conducted at a fixed frequency near 9.5×10^9 Hz by scanning through a magnetic field range until absorption of electromagnetic radiation is detected at H_o. The value of H_o can then be used to calculate the electron g-factor. The exact value of g depends on the magnitude of local magnetic fields induced in the atom by the applied external field, H. Nitroxide spin probes possess an unpaired electron which is largely localized on the $2p\pi$ orbital of the N atom (16), as is shown in the molecular structure of TEMPAMINE[+] below:

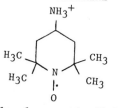

The main source of local magnetic fields causing g to deviate from the "free electron" value of 2.0023 is orbital magnetic moment generated from spin coupling to excited electronic states (spin-orbit coupling). For the electron in the p-orbital of TEMPAMINE, this coupling is orientation-dependent (i.e. anisotropic), and slightly different values of g are observed for different orientations of the nitroxide relative to the applied field, H. If the axis system is defined as shown below:

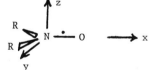

then typical g-values for the nitroxide with its x, y and z axes aligned parallel to H are $g_{xx} = 2.0089$, $g_{yy} = 2.0061$, and $g_{zz} = 2.0027$ (16).

Because the unpaired electron interacts with the local magnetic field generated by the nuclear spin (spin quantum number, $I = 1$) of the N atom, each of the electronic spin states depicted in Figure 2 is actually split into three energy levels by the allowed m_I quantum numbers (1, 0, -1) of the nucleus, and three electronic transitions are detected in the ESR experiment, as shown in Figure 3. The observed first-derivative spectrum of a nitroxide is, therefore, composed of three peaks separated by the nuclear hyperfine splitting, A. Like the g-value, the hyperfine splitting of the nitroxide is anisotropic, a result of the direction-dependent dipole-dipole interaction between the p-orbital electron and the nitrogen nucleus. Typical principal-axis A values for the nitroxide with its x, y and z axes oriented parallel to H are $A_{xx} = 6.5$, $A_{yy} = 6.7$, and $A_{zz} = 33$ gauss. Figures 4a, 4b and 4c depict the expected first-derivative spectrum of a nitroxide radical with the magnetic field, H, along the x, y and z axes of the molecule, respectively. For a randomly oriented collection of these radicals in the solid state, a relatively broad "glassy" or "powder" spectrum

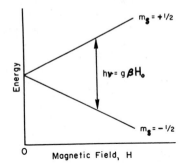

Figure 2. Electron spin levels in a magnetic field, showing the resonance conditions at $H = H_0$.

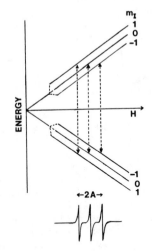

Figure 3. Spin levels for an electron interacting with the N atom (I=1) in the nitroxide radical. The three allowed transitions generate an ESR spectrum with hyperfine splitting, A.

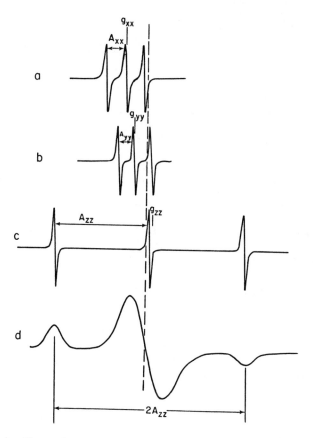

Figure 4. First-derivative ESR spectra of a nitroxide spin probe with the x-axis (a), y-axis (b), and z-axis (c) aligned parallel to H. The rigid-limit "powder" spectrum (d) is also shown. The dashed line marks the free electron field position (g = 2.0027). Reproduced from Ref. (8) with permission of D. Reidel Publishing Co.

is obtained (Figure 4d), the resultant of the spectra generated
by nitroxides in all orientations. Although less information
can be obtained from "powder" than from "single crystal" spectra,
Figure 4d demonstrates that the outer peaks of the "powder" spectrum
at least provide a measure of the A_{zz} hyperfine splitting. TEMPAMINE
in aqueous solution at room temperature has an almost perfectly
symmetrical three line ESR spectrum (Figure 3) because rapid rotational motion of the molecule completely averages the anisotropy
of the g-value and A value. The average values, g_o and A_o, are
calculated from the principal values:

$$g_o = 1/3 \; g_{xx} + 1/3 \; g_{yy} + 1/3 \; g_{zz}$$

$$A_o = 1/3 \; A_{xx} + 1/3 \; A_{yy} + 1/3 \; A_{zz}$$

However, when protonated TEMPAMINE adsorbs by cation exchange
on fully hydrated layer silicate clays (10, 11), the spectrum
becomes less symmetrical as shown in Figure 5. The beidellite
and montmorillonite spectra have line shapes typical for nitroxide
molecules with rotational frequencies on the order of 10^9 Hz (17).
The reciprocal of the frequency can be defined as the rotational
correlation time, τ_c, which represents the approximate time required
for reorientation by random thermal tumbling. For values of τ_c
$< 5 \times 10^{-9}$ s, an approximate theoretical solution for the linewidths
of the three nitroxide ESR lines is given by (17):

$$\frac{\Delta H(M_I)}{\Delta H(0)} = 1 - \frac{\tau_c}{[\sqrt{3}\pi\Delta\nu(0)]} \; [C_1 M_I + C_2 M_I^2]$$

where $\Delta H(M_I)$ is the peak-to-peak linewidth (in gauss units) of
the low ($M_I = +1$), center ($M_I = 0$) and high ($M_I = -1$) field resonance lines, $\Delta\nu(0)$ is the linewidth of the center resonance in
Hz units, and C_1 and C_2 are constants determined by the principal
A and g values. Estimates of τ_c from the beidellite and montmorillonite spectra based upon this equation are 0.4×10^{-9} and 1.2
$\times 10^{-9}$ s, respectively, values that are 10 to 25 times longer
than the 5×10^{-11} s estimate of τ_c for the nitroxide in aqueous
solution (10). Thus, adsorption reduced the mobility of the probe,
and the orientation-dependence of hyperfine splitting, most noticeable for hectorite and montmorillonite clay films (Figure 5),
can be attributed to some degree of preferred orientation of the
TEMPAMINE molecule on the wet clay surfaces. Since the prepared
clay films have the layer silicate sheets aligned in the plane
of the film, the effect of the angle between the plane of the
clay film and H on the hyperfine splitting, as shown in Figure
6, is a measure of ordering of the spin probe on the silicate
surfaces. For example, if the nitroxide were rigidly adsorbed
on the silicate with its z-axis at exactly 90° to the clay surfaces,
the 90° clay film orientation to H would generate an ESR spectrum
with hyperfine splitting equal to A_{zz} (see Figure 4), while the
0° film orientation to H would produce a spectrum with splitting
of $1/2 \; A_{xx} + 1/2 \; A_{yy}$. The nature of smectite films is such that,
while the c-axes of the individual platelets are oriented at $\sim$90°
to the plane of the film, the a and b-axes of the platelets are
randomly oriented within the plane of the film. Thus, only the

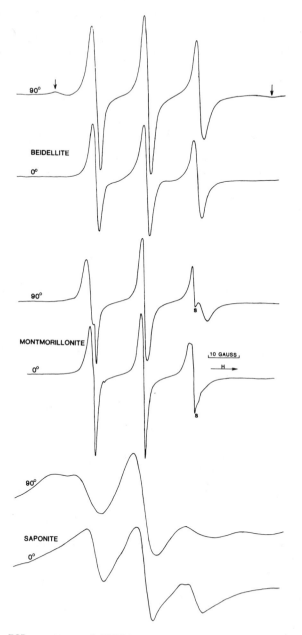

Figure 5. ESR spectra of TEMPAMINE adsorbed on wet Na$^+$-smectite films, oriented at 0° and 90° to the applied magnetic field. Weak rigid-limit spectra (labelled with arrows) and solution spectra (labelled with "s") are evident in beidellite and montmorillonite, respectively.

A_{zz} value and the <u>average</u> of the A_{xx} and A_{yy} values are theoretically accessible to measurement from oriented smectite films. Then,
A_{zz} - $(1/2\ A_{xx} + 1/2\ A_{yy})$ is the maximum anisotropy in hyperfine
splitting that is possible. The anisotropy observed in Figure
6, and reported in Table 3 below, is much smaller than the maximum.
One can quantify the degree of surface ordering by defining an
order parameters, s:

$$ s = \frac{A_\perp - A_\parallel}{A_{zz} - (1/2\ A_{xx} + 1/2\ A_{yy})} $$

$A_\perp$ and $A_\parallel$ are the experimentally determined hyperfine splittings
for the clay films at $90°(\perp)$ and $0°(\parallel)$ to the magnetic field.

Table III. Hyperfine splitting (A) values of TEMPAMINE[+] adsorbed
on fully wetted oriented clay films.

Clay	$A_\parallel$(gauss)	$A_\perp$(gauss)	s	Estimated rotational correlation time (ns)
Na^+-hectorite	15.1	19.2	.16	1.7
Na^+-montmorillonite	16.5	18.4	.08	1.2
Na^+-beidellite	16.7	17.2	.02	0.4

Since s in Table III is positive, there is a measurable tendency
for the nitroxides to orient with the molecular z-axis perpendicular
to the plane of the clay film. This can be deduced from the fact
that $A_{zz} \gg A_{xx} \sim A_{yy}$ and $A_\perp > A_\parallel$. It is concluded that thermal
tumbling of the probe on the clay surfaces is not isotropic since
there is incomplete averaging of the anisotropy of hyperfine splitting. However, the value of s very close to zero for Na^+-beidellite
indicates little tendency for preferred orientation of the probe
on beidellite surfaces. It is tentatively suggested that beidellite, with its tetrahedral charge and consequently stronger tendency
to hydrogen-bond water, may not interact as strongly with TEMPAMINE[+]
as hectorite or montmorillonite. Other studies have linked the
strength of organocation adsorption to the ease of displacement
of hydration water from the layer silicate surface ($\underline{3}, \underline{4}$).
    Saponite, as the spectra of Figure 5 depict, retains TEMPAMINE[+]
in a more rigid environment, a fact which may be related to the
inability of this particular smectite in the Na^+-saturated form
to fully expand in water (Table II). The saponite spectra of
Figure 5 cannot be analyzed by the approach outlined above for
nitroxides undergoing rapid motion, since these slow-motion spectral
lineshapes are determined by anisotropies in the A and g values
which are not completely time-averaged. Only if the rotational
frequency of the nitroxide is much greater than the anisotropy
of the A and g-values ($\underline{i.e.}$ $|A_{zz} - A_{xx}|$, $|g_{xx} - g_{zz}|$) expressed
in frequency units, is the anisotropy averaged by molecular motion.
The A-value anisotropy, $|A_{zz} - A_{xx}|$, corresponds to a frequency
of 73 MHz or a rotational correlation time of 1.4×10^{-8}s. The
general lineshape of the saponite spectra indicate a τ_c value

of $\sim 10^{-8}$s and a fairly high degree of orientation, with the nitroxide z-axis tending to be aligned at 90° to the plane of the saponite film. The c-spacing of wet Na^+-saponite (1.77 nm) provides an approximately 0.8 nm interlayer, probably imposing steric restrictions on the rotational motion of the TEMPAMINE molecule, which has a diameter of ~ 0.8 nm.

For the relatively rapid anisotropic motion ($\tau_c < 3 \times 10^{-9}$s) of spin probes observed on fully hydrated smectites such as montmorillonite, hectorite, and beidellite, an external fixed axis system (x', y', z') can be defined in addition to the molecule-fixed axes (x, y, z). The angle, θ, between the z and z' axes is given by:

$$\cos \theta = \vec{z} \cdot \vec{z}$$

If we select the z' axis to be normal to the plane of the clay film, the angle θ can be imagined to fluctuate rapidly as the probe tumbles in the interlayer. This time-averaged system is symmetric about the z' axis because of the random orientation of the a and b axes of the clay platelets in the x'y' plane. Also, the probe molecule is unlikely to prefer any particular orientation relative to the a and b axes of each individual platelet. This situation is diagrammed in Figure 7. Equations for the hyperfine splitting of the probe for orientations of the magnetic field perpendicular ($\perp$) and parallel ($\parallel$) to the x'y' plane (or ab plane) can then be derived (8):

$$A_\perp = \langle \cos^2\theta \rangle (A_{zz} - A_{xx}) + A_{xx}$$

$$A_\parallel = 1/2(1 - \langle \cos^2\theta \rangle)(A_{zz} - A_{xx}) + A_{xx}$$

where $\langle \cos^2\theta \rangle$ is the time-averaged value of $\cos^2\theta$. When these equations are solved for θ using the $A_\parallel$ and $A_\perp$ values from Table III, apparent average θ values of 45°, 52°, and 54° are estimated for hectorite, montmorillonite and beidellite, respectively. Completely isotropic motion of the probe in the interlayer would produce $A_\parallel = A_\perp$ and $\theta = 54.7°$, the "magic angle". Thus, hectorite has the greatest anisotropy of motion, with a tendency of the z-axis of the nitroxide to tilt toward the z' axis. This orientation may facilitate direct surface-methyl group interaction as well as close approach of the $-NH_3^+$ group to surface charge sites.

Factors found to influence the mobility and average orientation of TEMPAMINE$^+$ on smectites are also factors which modify the swelling behavior. For example, hectorites saturated with 2+ exchange ions (Mg^{2+}, Ca^{2+}, Ba^{2+}) reveal significantly reduced rotational mobility of TEMPAMINE$^+$ and less anisotropy of motion when compared with Na^+-hectorites (10). The restricted swelling of smectites saturated by divalent cations probably accounts for this difference in mobility. Organic solvents also alter the mobility and average orientation of the adsorbed probes (10), effects attributable to reduced interlayer expansion in these solvents, and a tendency for less polar solvents to be attracted to the TEMPAMINE$^+$ molecule. Several possible orientations of TEMPAMINE$^+$ are illustrated in Figure 8.

Heterogeneity of clay surface charge, which may result in

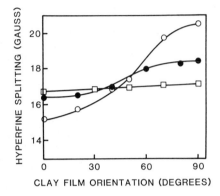

Figure 6. Plot of TEMPAMINE hyperfine splitting as a function
of orientation of the clay film (angle between plane of film
and the applied magnetic field) for hectorite (O), montmorillonite
(●) and beidellite (□).

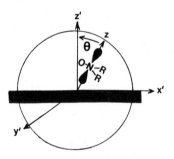

Figure 7. Average orientation of the xyz axis system of TEMPAMINE
in the fixed (x'y'z') axis system of clay.

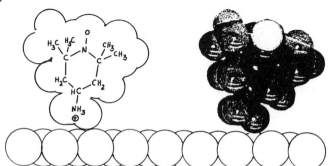

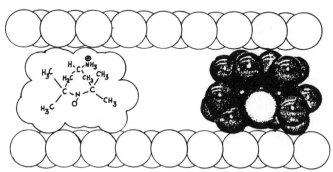

Figure 8. Several orientations of TEMPAMINE at the layer silicate surface with a. maximum CH_3-surface interaction ($\theta \sim 45°$), b. minimum CH_3-surface interaction ($\theta > 45°$), c. minimum mobility ($\theta \sim 0°$). θ is defined in Figure 7.

differential swelling behavior, is reflected in the mobility of
the adsorbed probe on wet clays. The beidellite spectra of Figure
5 possess weak resonances which arise from rigidly-bound highly-
oriented TEMPAMINE$^+$. The outer resonances of this rigid-limit
spectrum (denoted by arrows) are observed for the 90° orientation
only, a fact consistent with the "flat" alignment of the probe
shown in Figure 8c, with the N-O bond axis in the plane of the
clay film. The high tetrahedral charge of the beidellite may
prevent some interlayer regions from swelling in excess water,
thereby trapping TEMPAMINE$^+$ cations between platelets. More dramatic
evidence for heterogeneity of swelling is seen in the beidellite
spectra after equilibration at 100% relative humidity (Figure
9). The spectrum of mobile probe (denoted by "m") remains, but
the rigid-limit spectrum (denoted by "r") is much more intense
than for the beidellite in excess water. Clearly, two distinct
interlayer environments are detected by the probe. To a lesser
extent, montmorillonite also shows evidence of highly oriented,
rigidly-bound probes under conditions of 100% humidity (spectra
not shown). Saponite, because of its inability to expand beyond
1.77 nm, fails to reveal mobile TEMPAMINE$^+$ cations at any degree
of hydration.

Dehydration of the smectite interlayers by air-drying reduces
the c-axis spacing to the point where steric restrictions force
the TEMPAMINE molecules to align at the surfaces as shown in Figure
8c. The spectrum of air-dry beidellite, depicted in Figure 10,
is typical of the rigid-limit spectra observed for all the smectites, and demonstrates the high degree of ordering in the narrow
(0.2 - 0.3 nm) interlayers. The rigidity is defined in terms
of the minimum rotational correlation time of the nitroxide which
produces a spectrum with characteristic rigid-limit lineshape.
This value is approximately 10^{-6} s; that is, the rigid-limit spectrum of Figure 10 indicates that τ_c for TEMPAMINE$^+$ is $\geq 10^{-6}$.

Cu Adsorbed on Layer Silicates. While the basic principles of
ESR used to interpret the nitroxide spectra also apply to Cu^{2+},
several significant spectral differences need to be explained.
The ESR spectrum of $Cu(H_2O)_6^{2+}$ results from an unpaired electron
in the $d_{x^2-y^2}$ orbital, as illustrated in Figure 11. Elongation
of the axial (z-axis) Cu-H$_2$O bonds relative to the equatorial
(x, y-axis) bonds lowers the symmetry of the $Cu(H_2O)_6^{2+}$ complex
from octahedral to tetragonal. Termed the Jahn-Teller effect,
this distortion is favored by the energy gained in lowering the
energy level of the d_{z^2} orbital and raising the level of the $d_{x^2-y^2}$
orbital (Figure 11). There are four allowed electron transitions
for the unpaired d-electron, as Figure 12 reveals, because Cu^{2+}
has a nuclear spin of 3/2. Thus, the M_I quantum numbers of the
nucleus can be -3/2, -1/2, 1/2 or 3/2, splitting each electron
spin state into four energy levels. The highly orientation-dependent
spin-orbit coupling and electron spin-nuclear spin interaction
in the d-orbital result in a high degree of anisotropy in both
the g-value and the A value. As is the case for nitroxide probes,
Cu^{2+} ions tumble rapidly enough in aqueous solution at room temperature to average the anisotropy, producing a symmetrical spectrum
at g = 2.18 - 2.20 (18). However, the expected four hyperfine

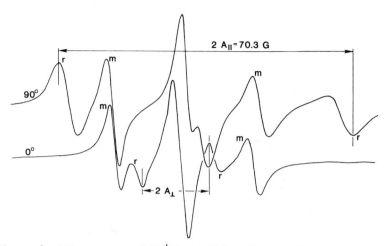

Figure 9. ESR spectra of Na⁺-beidellite films with adsorbed
TEMPAMINE equilibrated at 100% R.H., oriented at 0° and 90°
to the applied magnetic field. Resonances attributed to rigid
and mobile probes are denoted by "r" and "m", respectively.

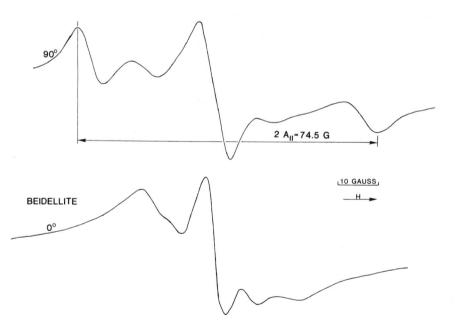

Figure 10. ESR spectra of air-dry Na⁺-beidellite films with
adsorbed TEMPAMINE, oriented at 0° and 90° to the applied
magnetic field.

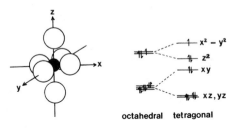

Figure 11. Stereochemistry and d-orbital energy diagram of the tetragonal $Cu(H_2O)_6{}^{2+}$ ion.

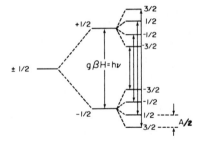

Figure 12. Energy level diagram for the spin states of Cu^{2+} ($S=1/2$, $I=3/2$).

lines are not resolved for aqueous solutions, a possible consequence of the rapid exchange of Jahn-Teller distortion axes which is possible in a symmetrically coordinated complex such as $Cu(H_2O)_6^{2+}$ (18).

Figure 13 depicts ESR spectra of Cu^{2+} in hydrated oriented beidellite films. The nearly symmetrical and orientation-independent nature of the spectra suggests that $Cu(H_2O)_6^{2+}$ ions in the interlayer have mobility near that of $Cu(H_2O)_6^{2+}$ ions in solution. In fact, the addition of excess water to these clays alters the Cu^{2+} spectra to fully symmetrical peaks at g = 2.18 - 2.19. ESR studies of other paramagnetic metal ions (Mn^{2+}, VO^{2+}) as surface probes have confirmed that cation rotational mobility on fully wetted smectites with interlayer spacings of 1.0 nm or more is not more than 50% reduced relative to aqueous solution (12, 14). Apparently, about four molecular layers of water in the interlayer provide sufficient space for solution-like rotation rates of hydrated metal ions.

A very different result is evident for Cu^{2+} in air-dry smectites, which generally have one to two layers of water between the platelets. Whether saturated by Na^+ or Ca^{2+} exchange ions, all of the Cu^{2+}-doped smectites produce rigid limit spectra, shown in Figures 14, 15 and 16. The dependence of the spectra on clay film orientation in the applied field can only be interpreted as a strong tendency for the Cu^{2+} ions to align on the silicate surfaces. For an axially symmetric species such as $Cu(H_2O)_6^{2+}$, the symmetry (z) axis generates a different g-value when aligned with the magnetic field than do the x and y axes; i. e.:

$$g_{xx} = g_{yy} \neq g_{zz}$$

To simplify terminology of axial systems, g_{zz} is defined to be $g_{||}$ (the g-value observed with the symmetry axis of Cu^{2+} parallel to the applied field), and g_{xx} (= g_{yy}) is defined to be $g_\perp$ (the g-value observed with the symmetry axis perpendicular to the applied field). An elongated z-axis (depicted in Figure 11 for $Cu(H_2O)_6^{2+}$) results in $g_{||} > g_\perp$. For axially symmetric Cu^{2+} rigidly bound in a crystal, the g-value can then vary between the minimum ($g_\perp$) and maximum ($g_{||}$), depending on orientation of the crystal within the magnetic field. However, for axial Cu^{2+} bound in a powdered clay sample, all possible orientations, and therefore all g-values between $g_\perp$ and $g_{||}$ are represented in the "powder" spectrum. Therefore, electron spin resonance occurs only for field values, H, between $H_{||}$ and $H_\perp$, where:

$$H_{||} = \frac{h\nu}{g_{||}\beta} \qquad H_\perp = \frac{h\nu}{g_\perp \beta}$$

As the spectrum is recorded, with H approaching $H_{||}$ from low field, absorption of energy suddenly begins at H = $H_{||}$, continues between $H_{||}$ and $H_\perp$, and suddenly ends at H = $H_\perp$. The shape of the absorption spectrum of an axially symmetric paramagnetic is shown in Figure 14a. The absorption of energy is much stronger near $H_\perp$ than $H_{||}$ because the probability that ions in the powder have their z-axis oriented nearly in the plane perpendicular to H is much higher than the probability that they have their z-axes almost parallel

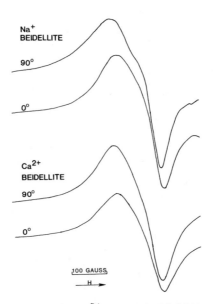

Figure 13. ESR spectra of Cu^{2+}-doped beidellite films after equilibration at 100% R.H., oriented at 0° and 90° to the applied magnetic field.

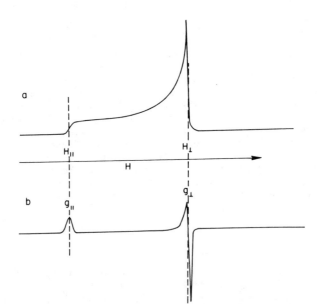

Figure 14. Absorption (a) and first-derivative (b) "powder" or "glass" ESR spectrum of an axially symmetric spin system.

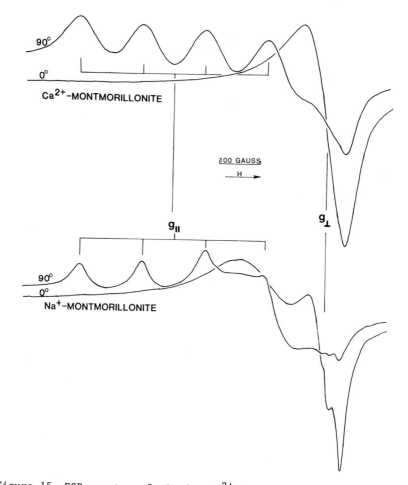

Figure 15. ESR spectra of air-dry Cu^{2+}-doped montmorillonite films oriented at 0° and 90° to the applied magnetic field. The vertical line in this and subsequent figures marks the g = 2.0023 field position.

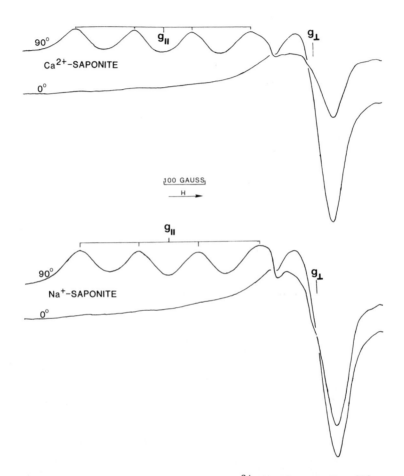

Figure 16. ESR spectra of air-dry Cu^{2+}-doped saponite films oriented at 0° and 90° to the applied magnetic field.

to H. Figure 14b demonstrates that the first derivative of this absorption curve (the usual mode of reporting ESR spectra) allows the values of $g_\parallel$ and $g_\perp$ to be determined in the powder sample.

ESR spectra of Cu^{2+} in dry smectite powders are similar to that shown in Figure 14b, but the 3/2 nuclear spin of Cu splits each $g_\parallel$ and $g_\perp$ resonance into four peaks ($\underline{6}$). The individual hyperfine lines of the $g_\parallel$ resonance are easily resolved because $A_\parallel$ is large, but the $g_\perp$ resonances are not always resolved because $A_\perp$ is small. Thus, $A_\parallel$, $g_\parallel$, and $g_\perp$ can be measured from the powder spectrum.

Figures 15, 16, and 17 depict Cu^{2+} spectra of air-dry smectites in oriented films at 0° and 90° to the applied magnetic field. In all smectites examined, the four well-separated resonances of $g_\parallel$ were observed in the 90° orientation, while the largely unresolved resonances of $g_\perp$ were observed in the 0° orientation. It is concluded that the symmetry axis of Cu^{2+} is oriented perpendicular to the layer silicate sheets, consistent with the orientations of hydrated Cu^{2+} revealed in Figure 18A and 18C. Since the air-dry clays have between one and two molecular layers of water between the platelets (Table II), all three stereochemical arrangements in Figure 18 have to be considered, but the tilted arrangement of $Cu(H_2O)_6^{2+}$ in Figure 18B is ruled out by the ~45° angle of the symmetry axis relative to the silicate sheets. Most air-dry smectites saturated with 2+ cations have c-axis spacings of 1.4 - 1.5 nm, suggesting that the exchange cations retain a complete inner-sphere hydration shell. Interestingly, Cu^{2+}-saturated hectorite has an air-dry spacing of 1.24 nm, leaving room for only a single monolayer of water ($\underline{9}$). This unique behavior for a 2+ cation is attributed to the Jahn-Teller distortion, allowing the loss of axial H_2O ligands to occur relatively easily upon drying. The stereochemistry of the resulting $Cu(H_2O)_4^{2+}$-clay complex is shown in Figure 18A. Similarly, the c-axis spacings of the air-dry Na^+-saturated smectites (Table II) require this same stereochemistry for Cu^{2+} introduced into the interlayer. An exception may be Na^+-saponite, with a 1.37 nm spacing, which is probably interstratified between the single and double layer of water.

It is believed that $Cu(H_2O)_6^{2+}$ orients in vermiculite interlayers (2 water layers thick) as shown in Figure 18B ($\underline{6}$). Yet, the smectites with more vermiculite-like properties (high tetrahedral charge, high total charge) showed no evidence of this orientation, even in cases where two layers of water were situated between the plates. It is necessary to conclude that $Cu(H_2O)_6^{2+}$ or $Cu(H_2O)_5^{2+}$ ions are found in the two-layer hydrates of the smectites, with the orientation shown in Figure 18C.

Only in the case of Na^+-beidellite were at least two different ligand environments of Cu^{2+} evident from the ESR spectra (Figure 17). This may indicate interstratification and the existence of interlayers with one and two layers of water. The g-values and hyperfine splittings for the rigid-limit spectra of Figures 15-17, as well as for previously studied minerals ($\underline{7}$), are given in Table IV.

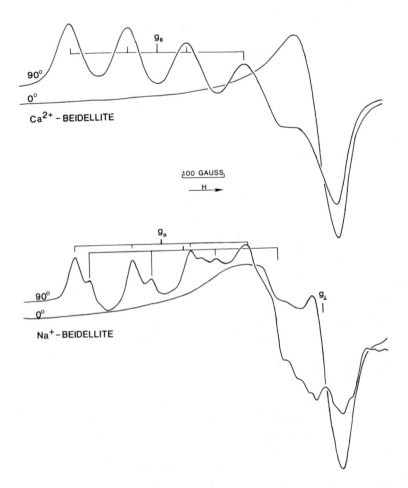

Figure 17. ESR spectra of air-dry Cu^{2+}-doped beidellite films oriented at 0° and 90° to the applied magnetic field.

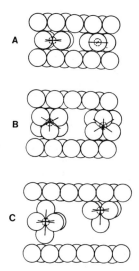

Figure 18. Diagrams of three possible stereochemical arrangements of Cu-aquo complexes in the interlayer of layer silicates: A. square planar; B. octahedral with symmetry axes ∿45° to the silicate layers; C. octahedral or square pyramidal with symmetry axis at 90° to the silicate layers.

Table IV. ESR parameters of Cu^{2+} in air-dry oriented layer silicate clay films.

Clay	Exchange cation	$A_{\parallel}(cm^{-1} \times 10^4)$	$g_{\parallel}$	$g_{\perp}$
Beidellite	Ca	155	2.37	2.08
	Na	155	2.37	2.08
		(163)	(2.33)	(2.08)
Saponite	Ca	159	2.36	2.08
	Na	159	2.35	2.08
Montmorillonite	Ca	165	2.35	2.08
	Na	167	2.35	2.07
Hectorite	Mg	156	2.34	2.07
Vermiculite (Llano)	Mg	123	2.40	2.09
	Na	138	2.39	2.07
Montmorillonite + NH_3	Na	202	2.22	2.06
Beidellite + NH_3	Na	196	2.22	2.05

Several general observations can be made from the spectra and the calculated ESR parameters. First, the Na^+-smectites possess narrower resonance line widths than the Ca^{2+}-smectites, with the exception of the saponite. Since Ca^{2+}-smectites, unlike Na^+-smectites, do not disperse into individual platelets in aqueous suspension, the Na^+-smectite films formed by drying suspensions onto a smooth flat surface have the silicate surfaces more perfectly oriented in the plane of the film. As a result, less angular variation of the z-axis of Cu^{2+} relative to the plane of the film would narrow the spectra. One can see evidence of hyperfine splitting in the $g_\perp$ component of the Na^+-smectite spectra, but not in the Ca^{2+}-smectite spectra. Saponite, unlike the other smectites, has very similar spectral linewidths for the Na^+ and Ca^{2+} form (Figure 16). Since this Na^+-saponite sample does not disperse completely in water (Table II), the alignment of Na^+-saponite platelets in the clay film may be no better than that of the Ca^{2+}-saponite.

The g-values and A values of Table IV reveal that the particular layer silicate has more effect on ESR parameters of adsorbed Cu^{2+} than saturation of exchange sites with different cations such as Na^+ and Ca^{2+}. Also, the smectites as a group have lower g and higher A values than vermiculite. From the perspective of molecular orbital theory, low g and high A values correspond to more covalent bonds between Cu^{2+} and the ligand (19). Thus, Cu^{2+} in the interlayer of smectites would seem to have a stronger covalent bond to H_2O than Cu^{2+} in vermiculites or aqueous solution. The ESR parameters for $Cu(H_2O)_6^{2+}$ in solution have been reported as $g_\parallel = 2.39$, $g_\perp = 2.07$, and $A_\parallel = 142 \times 10^{-4}$ cm^{-1} (18). One interpretation of this result is based on the different extent of hydrogen bonding between the silicate surface and the coordination water of Cu^{2+}. In smectites, the hydrogen bonding is weak (2), and would be unlikely to perturb the preferred $Cu-OH_2$ bonding geometry, which is pyramidal (i.e. tetrahedral) in Cu^{2+} hydrates (20). Such geometry presumably maximizes the covalency of $Cu-OH_2$ bonds by allowing optimal overlap between Cu orbitals and an sp^3 lone pair of H_2O. In vermiculite, the stronger hydrogen bonding of water to the surface may distort the pyramidal hydrate structure. This explanation cannot be verified without further investigation; other mechanisms could explain these ESR parameters. For example, monomeric Cu-hydroxy complexes have lower g and greater A values than Cu-water complexes (21). If Cu^{2+} were hydrolyzed to a greater extent in smectites than in vermiculite, the difference in ESR parameters could be explained. There is evidence that octahedrally-charged clays exchanged with acidic cations such as Cu^{2+} or Ca^{2+} are more acidic than their tetrahedrally-charged counterparts, based upon a comparison of their abilities to protonate adsorbed organic molecules (22, 23). Since this Bronsted acidity in clays arises from hydrolysis of the exchange cations, it is reasonable to expect a stronger tendency for hydrated Cu^{2+} to hydrolyze on smectite than on vermiculite surfaces. The reason for this is unclear, but may be related to the relative ability of the tetrahedral and octahedral sites of negative charge in the silicate structure to act as proton acceptors.

Adsorbed Cu^{2+} on layer silicate clays readily forms complexes with neutral ligands by displacement of water, as proven by the shifts in ESR parameters following exposure of the clay films to NH_3 vapor (Table IV). The larger hyperfine splitting in combination with lower g values (compared to Cu-water complexes) is evidence for the stronger ligand field of Cu-ammonia complexes. Orientation-dependence of the ESR spectra (not shown) revealed the symmetry axis of the Cu^{2+}-ammonia complexes to be aligned in the interlayers of the smectites in the same manner as the Cu^{2+}-water complexes. The ESR parameters and the orientation data point to the existence of $Cu(NH_3)_4^{2+}$ (planar) or $Cu(NH_3)_5^{2+}$ (square pyramidal) species in the interlayer; the $g_{\parallel}$ values were somewhat lower and the $A_{\parallel}$ values higher than those reported for $Cu(NH_3)_4^{2+}$ in solution (24). Other similarly oriented Cu-ligand complexes have been reported from ESR studies of dehydrated smectites, including $Cu(pyridine)_4^{2+}$ (25), $Cu(1,10-phenanthroline)_2^{2+}$ (26), and Cu^{2+}-amino acid complexes (27).

Summary

The orientation and mobility of organocations and hydrated metal cations at exchange sites of layer silicate clays is controlled to a large extent by steric restraints imposed by the narrow interlayer regions of layer silicate clays. Expansion of these interlayers by the introduction of water or other solvents allows rotational motion which appears to be isotropic for metal ions such as Cu^{2+}, but is somewhat anisotropic for the TEMPAMINE organocation. The latter effect cannot be attributed to steric restrictions, but rather to physical bonding forces which tend to favor certain orientations of the organocations at the silicate surface. Smectites with high tetrahedral charge have less tendency to adsorb the organocation in a preferred orientation than smectites with low tetrahedral charge, a property attributed to the more hydrophilic nature of tetrahedrally-charged smectites. There was also some evidence that hydrated Cu^{2+} in the interlayers of air-dry smectites has chemical properties different from Cu^{2+} in vermiculites.

The paramagnetic probes detected heterogeneity of swelling behavior in smectites that was not obvious from X-ray diffraction data, presumably caused by variation of charge density on different clay layers within the same smectite sample.

Literature Cited

1. Theng, B.K.G. "The Chemistry of Clay-Organic Reactions"; Wiley:New York, 1974; Chap. 5.
2. Farmer, V.C.; Russell, J.D. Trans. Faraday Soc. 1971, 67, 2737-49.
3. Hayes, M.H.B.; Pick, M.E.; Toms, B.A. J. Colloid Interface Sci. 1978, 65, 254-65.
4. Hayes, M.H.B.; Pick, M.E.; Toms, B.A. J. Colloid Interface Sci. 1978, 65, 266-75.
5. Lagaly, G. Clays and Clay Minerals 1982, 30, 215-22.
6. Clementz, D.M.; Pinnavaia, T.J.; Mortland, M.M. J. Phys. Chem. 1973, 77, 196-200.
7. McBride, M.B. Clays and Clay Minerals 1976, 24, 211-12.

8.   McBride, M.B. In "Advanced Chemical Methods for Soil and
     Clay Minerals Research"; Stucki, J.W.; Banwart, W.L., Eds.;
     NATO Advanced Study Institutes Series C, D. Reidel Publishing
     Co.: Dordrecht, Holland, 1980; Chap. 9.
9.   McBride, M.B. Soil Sci. Soc. Am. J. 1976, 40, 452-56.
10. McBride, M.B. Clays and Clay Minerals 1977, 25, 205-10.
11. McBride, M.B. Clays and Clay Minerals 1979, 27, 97-104.
12. McBride, M.B. Clays and Clay Minerals 1979, 27, 91-6.
13. McBride, M.B.; Pinnavaia, T.J.; Mortland, M.M. J. Phys.
     Chem. 1975, 79, 2430-5.
14. McBride, M.B.; Pinnavaia, T.J.; Mortland, M.M. American
     Mineralogist 1975, 60, 66-72.
15. Callaway, W.S.; McAtee, J.L. Manuscript in press.
16. Nordio, P.L. In "Spin Labeling. Theory and Applications";
     Berliner, L.J., Ed.; Academic:New York, 1976, Chap. 2.
17. Smith, I.C.P. In "Biological Applications of Electron Spin
     Resonance"; Swartz, H.M.; Bolton, J.R.; Borg, D.C.; Eds.;
     Wiley:New York, 1972; Chap. 11.
18. Poupko, R.; Luz, Z. J. Chem. Phys. 1972, 57, 3311-8.
19. Kivelson, D.; Neiman, R. J. Chem. Phys. 1961, 35, 149-55.
20. Friedman, H.L.; Lewis, L. J. Solution Chem. 1976, 5, 445-55.
21. Ottaviani, M.F.; Martini, G. J. Phys. Chem. 1980, 84, 2310-5.
22. Mortland, M.M.; Raman, K.V. J. Agr. Food Chem. 1967, 15,
     163-7.
23. Mortland, M.M.; Raman, K.V. Clays and Clay Minerals 1968,
     16, 393-8.
24. Vierke, G. Z. Naturforsch. 1971, A26, 554-60.
25. Berkheiser, V.; Mortland, M.M. Clays and Clay Minerals 1975,
     23, 404-10.
26. Berkheiser, V.E.; Mortland, M.M. Clays and Clay Minerals
     1977, 25, 105-12.
27. Nagai, S.; Ohnishi, S.; Nitta, I.; Tsunashima, A.; Kanamaru,
     F.; Koizumi, M. Chem. Phys. Letters 1974, 26, 517-20.

RECEIVED May 20, 1986

Applications of Surface Techniques to Chemical Bonding Studies of Minerals

Dale L. Perry

Lawrence Berkeley Laboratory, University of California, Berkeley, CA 94720

In the last several years, a number of new instrumental techniques have been developed that are quite effective in detecting changes in the surfaces of
minerals that have undergone chemically induced reactions or natural geologic alterations. These techniques are able to yield a large amount of information concerning the chemical bonding of both the
mineral itself and any surface species that are
present. The present paper addresses x-ray photoelectron, Auger, and combined x-ray photoelectron/
Auger spectroscopy (including the Auger parameter
concept), and discusses many of the important spectral parameters associated with them that are useful
for determining the chemical states of elements found
in mineral systems. Problems related to the vacuum-
oriented techniques - such as dehydration, charging,
and metal ion reduction - are discussed, along with
methods of studying samples as a function of depth.
Examples of applications of these concepts to studies
of bonding in minerals and related surface species
are discussed.

During the last several years, a number of new instrumental surface
techniques have been developed that are quite effective in detecting
changes in the surfaces of minerals that have undergone chemically
induced or natural geologic alteration. These techniques are quite
sensitive (approximately 0.1-0.5% atomic concentration for x-ray photoelectron and Auger spectroscopy, for example), and they make it
possible to monitor very small amounts of elements that may be
present in the near surface material. Any change in the surface with
respect to chemical composition may readily be measured qualitatively

0097-6156/86/0323-0389$06.00/0
© 1986 American Chemical Society

and, in some cases, quantitatively. As a result, this type of research in surface science is merely another application to reaction systems involving aqueous solution/solid interfaces: many other areas of research, such as aqueous corrosion (1) and electrochemistry (2), have used surface science almost since its inception.

Surface measurements as part of a geochemical study have several advantages. First, the techniques collectively are sensitive to all elements in the periodic table, although hydrogen and helium cannot be detected by x-ray photoelectron and Auger spectroscopy (two of the most widely used techniques). Second, several of the techniques, such as x-ray photoelectron spectroscopy, give important information about the chemical bonding environment of the element being studied. Third, surface studies involve only the first several layers of the substrate, where mineral-solution interactions have their greatest effect. Finally, many surface methods can be coupled with depth profiling to actually measure changes in elemental concentration as a function of depth below the surface layer of minerals.

Even though the vacuum-oriented surface techniques yield much useful information about the chemistry of a surface, their use is not totally without problems. Hydrated surfaces, for example, are susceptible to dehydration due to the vacuum and localized sample heating induced by x-ray and electron beams. Still, successful studies have been conducted on aquated inorganic salts (3), water on metals (3), and hydrated iron oxide minerals (4). Even aqueous solutions themselves have been studied by x-ray photoelectron spectroscopy (5). The reader should also remember that even dry samples can sometimes undergo deterioration under the proper circumstances. In most cases, however, alterations in the sample surface can be detected by monitoring the spectra as a function of time of x-ray or electron beam exposure and by a careful, visual inspection of the sample.

Another problem that can be encountered in x-ray photoelectron (and Auger) spectroscopy is that of charging. This problem arises from the fact that many samples under study are not in electrical equilibrium. This equilibrium can be attained only if an electron flow from ground can be achieved to neutralize the residual positive charge on the sample surface created by the photoemission process. For those minerals that are insulators, the conductivity may not be sufficient to prevent a positive charge from building on the surface; this results in a false increase in the binding energy. There are several ways of addressing charging, including the use of low energy electrons to neutralize the positive charge (6), gold decoration of the sample (7), and the contaminant carbon 1s line (8). This problem has been reviewed by several authors (9,10), as has been the Auger parameter concept discussed below, which circumvents this problem.

Metal ion reduction can also occur during surface studies. One of the best documented examples is that of the reduction of copper(II) to copper(I), a process that has been reviewed extensively in the research literature (11). Certain copper(II) minerals, such as CuO (tenorite), are quite susceptible to photoreduction (11), and care must be taken in conducting surface studies on them. Uranium(VI) salts chemisorbed on zeolites have also been observed (12) to undergo photoreduction to uranium(IV) compounds. Mechanisms for photoreduction of CuO, for example, have been widely reported in the chemical research literature (13-17).

Keeping these problems in mind, several of the more widely used techniques are discussed below, with an emphasis placed on their application to bonding studies of minerals and species adsorbed on them during mineral/water interface reactions. As with any experimental techniques, many of these problems can be minimized or even eliminated if due caution is taken in performing the studies.

X-Ray Photoelectron Spectroscopy (XPS)

X-ray photoelectron spectroscopy (also called electron spectroscopy for chemical analysis, or ESCA) is a surface technique that can be used to detect elements qualitatively (and quantitatively, in some cases) in the surface layers of solids, as well as the chemical states (species) of the elements. The basic experimental apparatus for performing XPS studies includes an x-ray source (most commonly, Mg Kα or Al Kα radiation), an electron energy analyzer, a specimen holder, and a vacuum chamber. As the x-rays impinge on the solid samples, electrons are emitted from the surface. The electron energy analyzer then measures the kinetic energy of the emitted electrons, which can be related directly to the binding energy of the electron in that particular atom. The binding energies of the individual atoms yield much information about their chemical states.

X-ray photoelectron spectroscopy can be performed on any sample that is vacuum compatible, i.e., does not decompose under vacuum in the range $10^{-7} - 10^{-10}$ torr. In most cases, the technique does little or no damage to the sample surface, although x-ray-induced chemical reactions such as those discussed above may occur. As a surface technique, it gives a signal that mirrors approximately the top 15 to 40Å of the surface (18).

The following aspects of x-ray photoelectron spectroscopy are important in terms of determining bonding both in chemical species on minerals and in the minerals themselves. Data obtained from these spectral parameters are both structural and electronic, and, when considered with crystallographic structural data where possible, give a comprehensive bonding picture. Of course, for general survey treatises of this technique, prior works (19-21) should be consulted.

Binding Energy. The most often cited spectral parameter in x-ray photoelectron spectroscopy is the binding energy. Not only do photoelectron lines indicate a particular element, but they also represent a chemical state (or species) of that element. The core photoelectron lines of many elements are quite sensitive to the electronic state of that element from two standpoints: oxidation state and the functional group of which the element is a component. In some cases, the differences in binding energies for various oxidation states are quite large, e.g., Cr(III) vs. Cr(VI), Fe(II) vs. Fe(III). In other cases, the binding energy separation between oxidation states is not so large as to differentiate unequivocally between them, e.g., Cu(I) vs. Cu(0), Co(II) vs. Co(III). The distinction between the members of the latter sets of ions must be made on the basis of other spectroscopic parameters, such as multiplet splitting, satellite structure, and spin-orbit splitting (discussed below). The

same is also true when one is trying to differentiate among members
of a series of similar compounds (e.g., different oxides of a metal
in the same oxidation state).

Figure 1, for example, shows a survey scan of natural galena,
PbS. Besides lead and sulfur, one also sees the oxygen inherent in
natural samples and the chlorine resulting from groundwater fluid
inclusions; all are nicely separated in terms of their binding energies. But if one looks at a high resolution spectrum of the fused
sulfur $2p_{3/2,1/2}$ line (Figure 2), it is seen that there are several
sulfur species present. The sulfide form, part of the galena itself,
is observed at 160.4 eV, while the elemental sulfur and sulfate forms
are found at 163.8 and 168.3 eV, respectively[22]. In some weathered galena samples, the thiosulfate ion ($S_2O_3^{2-}$) can also be detected
[23]; this ion has additional line components in both the sulfide and
sulfate region due to the presence of two states of sulfur in that
ion. In this case, the different forms of the sulfur can be quite
easily differentiated on the basis of the marked differences in the
binding energies and line shapes. Figure 3 shows a similar situation
for the sulfur 2p spectra of chalcocite and covellite, two naturally
occurring copper(I) minerals [24]. The sulfide region of the spectrum of covellite is more complex, since it contains both S^{2-} (sulfide) and S_2^{2-} (disulfide) ions; chalcocite contains only the simple
sulfide ion. In these cases, the line shapes (the contour of the
photoelectron peak of an element) are also quite different.

Satellite Structure. Another experimental parameter obtained from
x-ray photoelectron spectra is that of satellites. These lines,
which can be either weak or strong relative to the main photoelectron
line with which they are associated, appear to the high binding
energy side of that photoelectron line. They result from coupled
electronic processes brought about by the initial ionization of an
atom in the emission of the photoelectron. These satellites can
arise from several different types of these processes. When the
kinetic energy of the primary ejected electron is "shared" with
valence electrons to promote another electron to an excited state,
the resulting satellite is referred to as a "shake-up" satellite;
when this "sharing" results in another electron being promoted to a
continuum state, the result is a "shake-off" satellite. These satellites vary in their position relative to the main peak as a function
of the chemical state of the element being studied. For extensive
treatises on these satellites, the reader should consult more
comprehensive review articles [25-27].

Figure 4 is a good example of satellite structure associated
with a surface species. The chromium $2p_{3/2,1/2}$ spectrum results from
the reaction of the dichromate ion, $Cr_2O_7^{2-}$, with galena to yield both
chromium(VI) (indicated by an asymmetric, broadened shoulder on the
high binding energy side of the main chromium(III) peak) and a
reduced chromium(III) species. The satellite located 10.3 eV to the
high binding energy side of the chromium(III) line (the main $2p_{1/2}$
line at 586.4 eV) has been shown [28] to be a mixed chromium(III)
oxide-carbonate complex. The chromium(VI) species, which is diamagnetic and thus yields no significant satellite structure, was identified as $PbCrO_4$.

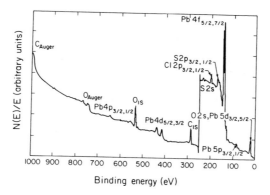

Figure 1. X-ray photoelectron survey scan of naturally weathered galena, PbS.

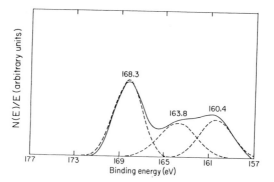

Figure 2. High-resolution X-ray photoelectron spectrum of naturally weathered galena, PbS, in the sulfur $2p_{3/2,1/2}$ region.

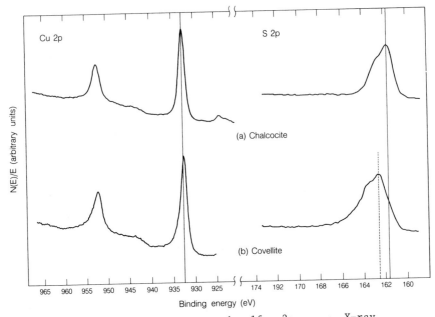

Figure 3. The copper $2p_{3/2,1/2}$ and sulfur $2p_{3/2,1/2}$ X-ray photoelectron spectra of natural chalcocite (Cu_2S) and covellite (CuS). (Reproduced with permission from Ref. 24. Copyright 1986 Chapman & Hall).

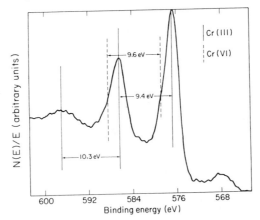

Figure 4. The chromium $2p_{3/2,1/2}$ photoelectron spectrum of galena that has been reacted with an aqueous solution of $Na_2Cr_2O_7$. The broken vertical lines through the main peaks indicate the photoelectron peaks associated with the chromium(VI) species, and the solid vertical lines indicate the photoelectron lines associated with the chromium(III) species. (Reproduced with permission from Ref. 28. Copyright 1984 Elsevier Sequoia.)

Spin-Orbit Splitting. When ionization of a p, d, or f orbital occurs to produce a photoelectron, one of the results is a phenomenon referred to as spin-orbit splitting. This ionization (19-21) yields two ionic states and thus two major peaks. If one again looks at Figure 3, he can see the copper 2p photoelectron line split into the $2p_{3/2}$ and $2p_{1/2}$ components; the same doublet also is observed for the chromium 2p line in Figure 4, where the splitting is 9.6 eV for the chromium(VI) species and 9.4 eV for the chromium(III) species. The magnitude of the splitting is many times a function of the oxidation state of the element being studied, such as in the case of chromium. It is sometimes associated with the differences in ligands or the coordination sphere of the central element, although the difference in splitting is rather minimal for these two latter cases. It is usually greatest for elements that exhibit differences in oxidation state along with differences in electronic spin state; for example, diamagnetic cobalt(III) exhibits a spin-orbit splitting much different from that of paramagnetic, "high-spin" cobalt(II) (15).

Multiplet Splitting. Photoionization in a core shell coupled with one or more valence shell electrons results in multiplet splitting, a phenomenon (25) that can be utilized to study paramagnetic systems. The effects of multiplet splitting may manifest themselves in a variety of ways. In the case of 3d ions, such as those found in the first row transition elements, the 2p photoelectron lines undergo broadening. The copper $2p_{3/2,1/2}$ lines shown in Figure 3 would be considerably broader if the copper species being observed were the paramagnetic, $3d^9$ copper(II) system. Also, the 3s photoelectron lines of the 3d block of transition elements actually undergo a splitting into a doublet for paramagnetic ions. By careful study of the 3s photoelectron level, in conjunction with good chemical standards, one can learn much about the bonding of species on a reacted mineral surface or about the substrate itself.

Taken as a whole, then, the above spectroscopic parameters – binding energy, satellite structure, spin-orbit and multiplet splitting, and line shapes – give a more comprehensive picture of the chemistry of a surface. If one looks at the entire set of data (Table I) pertaining to the spectrum shown in Figure 4, he can see that the chromium(III) species on the galena surface is hydrated chromium(III) oxide with a reaction layer film of CO_2 that has formed a mixed hydrated oxide/carbonate complex (28). Both "$Cr(OH)_3$" and "$Cr_2(CO_3)_3 \cdot nH_2O$" are reaction products of the reaction of hydrated chromium(III) oxide and CO_2; the chief difference is that the "$Cr(OH)_3$" will contain less carbonate material, resulting in a smaller spin-orbit splitting than "$Cr_2(CO_3)_3 \cdot nH_2O$" (28).

Auger Electron Spectroscopy (AES)

Auger electron spectroscopy (29) is a type of electron spectroscopy that is used for determining solid surface elemental and electronic composition. An experiment is conducted by bombarding a solid surface with an electron beam of energy ranging from 1 keV to 10 keV. Alternatively, an x-ray source can be used. The Auger electrons, emitted from an atom by means of a radiationless transition, are

Table I. X-Ray Photoelectron Data for the Galena/$Cr_2O_7^{2-}$ Reaction and Related Cr(III) Compounds[a,b]

Sample	Cr $2p_{3/2}$	Cr $2p_{1/2}$	O 1s	C 1s[c]
Galena/$Cr_2O_7^{2-}$	577.0–Cr(III)	586.4–Cr(III)	530.9	288.0
	578.8–Cr(VI)	588.4–Cr(VI)		
Cr_2O_3	576.3	586.1	529.9	
$Cr(OH)_3$[e]	576.6	586.4	530.8	288.0
$Cr_2(CO_3)_3 \cdot nH_2O$[f]	577.0	587.1	531.2	288.0

Sample	Cr ($2p_{1/2}$–$2p_{3/2}$)	Cr 3s Splitting	Cr $2p_{1/2}$ Satellite[d]
Galena/$Cr_2O_7^{2-}$	9.4-Cr(III)	4.1–4.3	10.3
	9.6-Cr(VI)		
Cr_2O_3	9.8	4.3	10.3
$Cr(OH)_3$[e]	9.8	4.1	10.3
$Cr_2(CO_3)_3 \cdot nH_2O$[f]	10.1	4.1	10.3

[a]All values in electron volts, eV.
[b]Adventitious (contaminant) carbon 1s = 284.6 eV.
[c]Carbon line attributable to the carbonate species.
[d]Distance (to the high binding energy side) from the main $2p_{1/2}$ line.
[e]Also formulated as $Cr_2O_3 \cdot nH_2O \cdot xCO_2$; please see text.
[f]Also formulated as $Cr_2O_3 \cdot nH_2O \cdot xCO_2$; please see text.

detected by an energy analyzer. The sensitivity of this technique is a function of the probability of the Auger transition peculiar to a particular element, the current and energy of the impinging electron or x-ray beam, and the efficiency of the electron analyzer.

The Auger effect is based on the following processes. Upon ionization of a core atomic level in a solid sample, the atom may undergo a decay to a lower energy level. This leaves the atom in a doubly ionized state, and the energy difference between the two states is transmitted to the ejected Auger electron. Those Auger transitions that occur near the surface of the solid result in ejected Auger electrons that do not undergo electron energy loss. The shape and energy of the resulting Auger peaks are thus useful in identifying the elemental composition of the sample surface as well as obtaining useful chemical information.

While historically many Auger data have been obtained using an electron beam to generate the Auger spectra in the form of derivative (or dN/dE vs. electron binding energy) spectra, the spectra can be recorded also in the counting mode (or N(E)/E vs. electron binding energy, the format used in Figure 1). This spectral representation allows the researcher to study spectral line shapes for elements that many times have much more detailed features than they would exhibit

in the derivative mode. Two additional advantages of x-ray-induced
Auger electron spectroscopy are that surface charging is much less of
a problem for geologic materials (most of which are insulators), and
surface damage is generally less for samples irradiated under x-ray
beams than for the same samples irradiated by an electron beam.
Auger spectroscopy can also be used in a scanning mode (scanning
Auger microscopy, or SAM) to yield surface topographical and elemental distribution data (30).

Figure 5 gives a good illustration of high resolution x-ray-
induced Auger electron spectra for lead oxide minerals and related
intermediates and products arising from their oxidation and thermal
decomposition reactions (31). One can readily see, for example, that
the oxygen KVV spectrum of the yellow form of PbO (the mineral massicot), shown in Figure 5a, is quite different from that of PbO_2
(plattnerite), shown in Figure 5d; the kinetic energies of the oxygen
KVV Auger lines are also quite different.

Combined X-Ray Photoelectron/X-Ray-Induced Auger Spectroscopy
(XPS/XAES)

As mentioned above, an Auger spectrum also can be generated by an
appropriate x-ray source such as Al Kα or Mg Kα, which are used to
generate x-ray photoelectron spectra; one result is shown in Figure
1. Not only are photoelectron lines such as the lead 4f, 4d, and 5d
lines present, but one also sees oxygen and carbon Auger lines. High
resolution studies of these Auger lines, coupled with the x-ray photoelectron lines in the same spectrum, allow a more complete and
detailed study of the chemical and electronic states of the elements
present.

One chief advantage of having both the x-ray photoelectron and
Auger spectra combined is that a researcher can make use of Auger
parameters. Auger parameters (32–34) are a concept by which use is
made of both x-ray photoelectron lines and the x-ray-induced Auger
lines to characterize a chemical species. The approach has two major
advantages. First, chemical shifts in x-ray-excited Auger lines are
usually larger and very different from those observed for the photoelectron lines; this is quite important in terms of being better
able to identify a particular compound. Secondly, since the positions of the Auger line and photoelectron line are relative to one
another in the same spectrum, the uncertainty in spectral line position due to charging no longer exists. In effect, the Auger parameter provides an internal reference value for each compound of a particular element. In the ideal case, this value will be a unique one
and will be sufficiently different from those of other compounds to
differentiate it unequivocally. Unfortunately, Auger parameters of
several compounds or techniques may be very close to one another, and
the use of other parameters may be necessary to identify them spectroscopically.

While the Auger parameter can be expressed in several ways, one
of the most commonly used definitions is that shown in Equation 1

$$\alpha' = BE_{photo} + KE_{Auger} \tag{1}$$

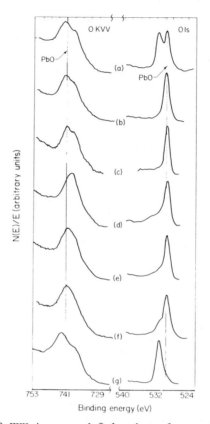

Figure 5. The O KVV Auger and O 1s photoelectron lines for
(a) powdered yellow PbO (massicot) as received, (b) same sample
heated in situ in O_2, (c) clean metallic lead exposed to 500 L
of O_2 at 150 °C, (d) powdered PbO_2 (plattnerite) as received,
(e) same sample heated to 320 °C in vacuo, (f) powdered Pb_3O_4
(minium), and (g) powdered $2PbCO_3 \cdot Pb(OH)_2$ (hydrocerussite).
(Reproduced with permission from Ref. 31. Copyright 1984
American Institute of Physics.)

where α' is the modified Auger parameter, BE_{photo} is the charge-
corrected binding energy of the photoelectron line for the element
studied, and KE_{Auger} is the charge-corrected kinetic energy of an
appropriate Auger line for the same element. In choosing which photoelectron and Auger lines are used to compute the Auger parameter,
the deciding factors are the sharpness and intensity of the lines as
well as their sensitivities to changes in the differences in the
bonding environment of the element studied.

Table II shows representative Auger parameters for several silicon and aluminum species. Both mineral and non-mineral species are
shown to give some idea of the range of values. For these calculations, the fused silicon and aluminum $2p_{3/2,1/2}$ photoelectron lines
are used in conjunction with the KLL Auger lines.

After one considers the experimental aspects of combined x-ray
photoelectron/Auger spectroscopy discussed above, he can then use the

data in Table III to differentiate between the minerals chalcocite (Cu_2S) and covellite (CuS) (Fig. 3). Differences are in the sulfur $2p_{3/2,1/2}$ binding energies and the kinetic energies of the copper LVV observed Auger lines. Additionally, the line shape and peak width differences of the sulfur $2p_{3/2,1/2}$ spectra, coupled with the differences between the copper $2p_{3/2}$ and sulfur $2p_{3/2}$ binding energies (a full electron volt), allow the investigator to differentiate between the two minerals spectroscopically. Again, all of the spectral parameters as a whole must be taken together to obtain the best results in distinguishing between the two sulfides.

In addition to studying an "as received" mineral sample surface, the researcher can also study samples as a function of the depth into the bulk by means of depth profiling studies. This can be performed in two ways: ion sputtering and electron grazing angle analysis. In ion sputtering, rare gas (usually argon) ions are used to remove layers of the sample surface. This approach is a destructive technique and can totally alter the chemistry of the surface that existed before sputtering; functional groups such as carbonates, sulfates, and nitrates are usually destroyed by sputtering. Also, sputtering affects the contaminant carbon 1s peak, thus causing some uncertainty in using this line for charge compensation. Another problem observed for this technique is that of differential elemental sputtering: different elements sputter at different rates. All of these problems have been discussed extensively in review articles (35,36).

Electron grazing angle analysis takes advantage of the increase of surface sensitivity in x-ray photoelectron spectroscopy by using variations in grazing angle of electron emission from the surface sample. For optimum results, the sample surface should be without irregularity or roughness, since changes in the surface contour can severely affect the surface sensitivity as a function of the electron emission angle. Still, this technique has the chief advantage of being non-destructive; this allows depth profiling without disturbing the initial surface chemistry of the sample surface. This approach has been reviewed by several investigators (37).

The above techniques have a wide array of applications, including those that are both analytical and physicochemical (such as bonding) in nature. Typical examples of research include the surface chemistry of ferrite minerals (38) and the valence states of copper in a wide array of copper (39) minerals. Other areas of bonding that have been studied include the oxidation state of vanadium (40) in vanadium-bearing aegirines (also using x-ray photoelectron spectroscopy) and the surface features of titanium perovskites (41).

Other areas of geologic research that can be explored using a combined instrumentation approach are many. Surface reactions of solids that have been suspended in aqueous solutions can be studied; this type of work has important applications in mineral processing research. The analysis of precious and strategic metal ores can be studied in order to monitor their inherent material and chemical properties and their surface characteristics before and after reaction.

In summary, any mineral solid surface that is involved in a geologic process (either natural or laboratory induced) or reaction can be studied from a standpoint of the bonding and chemistry of the surface as well as from an analytical standpoint. As a result of this approach, more valuable information can be obtained about the mineral surface than if the mere analytical approach is used exclusively.

Table II. Representative Auger Parameters
for Silicon and Aluminum[a,b]

Silicon	$2p_{3/2,1/2}$	$KL_{23}L_{23}$	α'
α-cristobalite, SiO_2	103.2	1608.6	1711.8
α'-quartz, SiO_2	103.6	1608.6	1712.2
Vycor, SiO_2	103.5	1608.5	1712.0
Si_3N_4	101.9	1612.2	1714.1
Si	99.6	1616.3	1715.9

Aluminum	$2p$	$KL_{23}L_{23}$	α'
Boehmite, AlO(OH)	74.2	1387.6	1461.8
Bayerite, $Al(OH)_3$	74.3	1387.7	1462.0
Gibbsite, $Al(OH)_3$	74.0	1387.4	1461.4
AlAs	73.6	1391.2	1464.8
AlN	74.0	1388.9	1462.9
Al	72.9	1393.2	1466.1

[a]Abstracted from Wagner, C.D., In "Practical Surface Analysis by
Auger and X-Ray Photoelectron Spectroscopy"; Briggs, D.
and Seah, M.P., Eds.; John Wiley and Sons, Ltd.: New York, 1983;
Appendix 4, p. 477.
[b]In electron volts, eV.

Table III. X-Ray Photoelectron and Auger Data for
Chalcocite and Covellite[a,b]

	S $2p_{3/2}$	Cu $2p_{3/2}$	Cu $2p_{1/2}$	Cu L_3VV[c]	$\alpha*$[d]	Δ[e]
Cu_2S	161.7(2.2)	932.6(1.9)	952.3(2.3)	917.3	1849.9	770.9
CuS	162.5(2.8)	932.4(1.9)	952.2(2.3)	917.9	1850.3	769.9

[a]All spectra referenced to C 1s at 285.0 eV.
[b]Numbers in parentheses next to the copper and sulfur 2p binding
energies represent the full width at half maximum (FWHM) of those
photoelectron lines.
[c]Kinetic energies for the Auger lines were determined from their
apparent binding energies (KE = $h\nu$-BE).
[d]Modified Auger parameter for copper.
[e]The differences in binding energies between the Cu $2p_{3/2}$ and
S $2p_{3/2}$ photoelectron lines.

The additional information regarding the surface is useful for predicting reaction stoichiometries, mechanisms, and rates that are pertinent to aqueous geochemical processes.

Acknowledgments

This research was partially supported by the U.S. Department of Energy, Office of Basic Energy Sciences, Division of Engineering and Geosciences, under Contract No. DE-AC03-76SF00098, and the U.S. Department of the Interior, Bureau of Mines, under Contract No. J0145057.

Literature Cited

1. Baer, D.R.; Thomas, M.T. In "ACS Symposium Series, No. 199, Industrial Applications of Surface Analysis," L.A. Casper and C.J. Power, Eds., American Chemical Society: Washington, D.C., 1982.

2. Sherwood, P.M.A. Chem. Soc. Rev. 1985, 14, 1–44.
3. Hirokawa, K.; Danzaki, Y. Surf. Interface Anal. 1982, 4, 63–7.
4. McIntyre, N.S.; Zetaruk, D.G. Anal. Chem. 1977, 19, 1521–29.
5. Avanzino, S.C.; Jolly, W.L. Inorg. Chem. 1978, 100, 2228–30.
6. Lewis, R.T.; Kelly, M.A. J. Electron Spectrosc. Relat. Phenom. 1980, 20, 105–115.
7. Hnatowich, D.J.; Hudis, J.; Perlman, M.L.; Ragaini, R.C. J. Appl. Phys. 1971, 42, 4883.
8. Swift, P. Surf. Interface Anal. 1982, 4, 47–51.
9. Broughton, J.Q.; Perry, D.L. Surf. Sci. 1978, 74, 307–17.
10. Reference 8 and references therein.
11. Klein, J.C.; Li, C.P.; Hercules, D.M.; Black, J.F. Appl. Spectrosc. 1984, 38, 729–34.
12. Perry, D.L.; Suib, S., unpublished data.
13. Wallbank, B.; Johnson, C.E.; Main, I.G. J. Electron Spectrosc. Relat. Phenom. 1974, 4, 263–69.
14. Rosencwaig, A.; Wertheim, G.K. J. Electron Spectrosc. Relat. Phenom. 1972/73, 1, 493–96.
15. Frost, D.C.; Ishitani, A.; McDowell, C.A. Mol. Phys. 1972, 24, 861–77.
16. Hirokawa, K.; Honda, F.; Oku, M. J. Electron Spectrosc. Relat. Phenom. 1975, 6, 333–45.
17. Sarma, D.D. Indian J. Chem. 1980, 19, 1046–49.
18. Klasson, M.; Hedman, J.; Berndtsson, A.; Nilsson, R.; Nordling, C. Physica Scripta 1972, 5, 93–5.
19. Siegbahn, K.; Nordling, C.N.; Fahlman, A.; Nordberg, R.; Hamrin, K.; Hedman, J.; Johansson, G.; Bergmark, T.; Karlsson, S.E.; Lindgren, I.; and Lindberg, B. "ESCA, Atomic, Molecular, and Solid State Structure Studied by Means of Electron Spectroscopy"; Almqvist and Wiksells: Upsala, 1967.
20. Siegbahn, K.; Nordling, C.; Johansson, G.; Hedman, J.; Heden, P.F.; Hamrin, K.; Gelius, U.; Bergmark, T.; Werme, L.O.; Manne, R.; and Baer, Y. "ESCA Applied to Free Molecules"; North-Holland: Amsterdam, 1969.
21. Jolly, W.L. Coord. Chem. Rev. 1974, 13, 47–81.

22. Lindberg, B.J. Int. J. sulfur. Chem. C 1972, 7, 33–53.

23. Lindberg, B.J.; Hamrin, K.; Johansson, G.; Gelius, U.; Fahlmann,
     A.; Nordling, C.; Siegbahn, K. Phys. Scr. 1970, 1, 297–309.

24. Perry, D.L.; Taylor, J.A. J. Mat. Sci. Lett., in press.

25. Vernon, G.A.; Stucky, G.; Carlson, T.A. Inorg. Chem. 1976, 15,
     278–84.

26. Brisk, M.A.; Baker, A.D. J. Electron Spectrosc. Related Phenom.
     1975, 7, 197–213.

27. Aarons, L.J.; Guest, M.F.; Hillier, I.H. J. Chem. Soc., Faraday
     Trans. II 1972, 68, 1866–74.

28. Perry, D.L.; Tsao, L.; Taylor, J.A. Inorg. Chim. Acta 1984, 85,
     L57–L60.

29. Thompson, M.; Baker, M.D.; Cristie, A.; Tyson, J.F. "Auger
     Electron Spectroscopy"; Wiley-Interscience: New York, 1985.

30. Prutton, M. Scanning Electron Microsc. 1982, 1, 83–91.

31. Taylor, J.A.; Perry, D.L. J. Vac. Sci. Technol. A 1984, 2,
     771–4.

32. Wagner, C.D.; Gale, L.H.; Raymond, R.H. Anal. Chem. 1979, 51,
     466–82.

33. Wagner, C.D. In "Practical Surface Analysis by Auger and X-Ray
     Photoelectron Spectroscopy"; D. Briggs, and M.P. Seah Eds.; John
     Wiley and Sons, Ltd.: New York, 1983; Appendix 4, p. 477.

34. Wagner, C.D.; Passoja, D.E.; Hillery, H.F.; Kinisky, T.G.; Six,
     H.A.; Jansen, W.T.; Taylor, J.A. J. Vac. Sci. Technol. 1982,
     21, 933–44.

35. Coyle, G.J.; Tsang, T.; Adler, I.; Ben-Zvi, N. J. Electron
     Spectrosc. Related Phenom. 1981, 24, 221–36.

36. Hofman, S. Surf. Interface Anal. 1980, 2, 148–60.

37. Ebel, M.F. J. Electron Spectrosc. Relat. Phenom. 1978, 14,
     287–322.

38. Perry, D.L.; Bonnell, D.W.; Parks, G.D.; Margrave, J.L. High
     Temp. Sci. 1977, 9, 85–98.

39. Nakai, I.; Sugitani, Y.; Nagasimi, K.; Niwa, Y. J. Inorg. Nucl.
     Chem. 1978, 40, 789–91.

40. Nakai, I.; Ogawa, H.; Sugitani, Y.; Niwa, Y.; Nagashima, K.
     Mineral. J. 1976, 8, 129–34.

41. Myhra, S.; Bishop, H.E.; Riviere, J.C. Surf. Technol. 1983, 19,
     161–72.

RECEIVED August 4, 1986

In Situ Mössbauer Studies of Metal Oxide–Aqueous Solution Interfaces with Adsorbed Cobalt-57 and Antimony-119 Ions

F. Ambe, S. Ambe, T. Okada, and H. Sekizawa

Institute of Physical and Chemical Research (Riken), Wako-shi, Saitama 351-01, Japan

In situ emission Mossbauer spectroscopy provides
valuable information on the chemical structure of
dilute metal ions at the metal oxide/aqueous solution
interface . The principles of the method are described with some experimental results on divalent
Co-57 and pentavalent Sb-119 adsorbed on hematite.
The chemical structure of the adsorbed ions was found
to be dependent on pH of the aqueous phase. Most of
the divalent Co-57 and pentavalent Sb-119 ions form
strongly bonded surface complexes under alkaline and
acidic conditions, respectively.

In spite of the development of physicochemical techniques for
surface analysis, spectroscopic methods applicable to the study of
bonding between adsorbed metal ion species and substrate are limited, especially those applicable to in situ measurement at interfaces between solid and aqueous phases (1,2). In previous papers,
we showed that emission Mossbauer measurement is useful in
clarifying the chemical bonding environment of dilute metal ions
adsorbed on magnetic metal oxide surfaces (3,4).
     We now extend the work to in situ measurements on metal ions
adsorbed at the metal oxide/aqueous solution interface. In this
report, our previous results are combined with new measurements to
yield specific information on the chemical structure of adsorbed
species at the solid/aqueous solution interface. Here, we
describe the principles of emission Mossbauer spectroscopy, experimental techniques, and some results on divalent Co-57 and pentavalent Sb-119 ions adsorbed at the interface between hematite
(α-Fe$_2$O$_3$) and aqueous solutions.

Principles

Although radioactive isotopes have been widely utilized as tracers
in the study of adsorption equilibrium and kinetics, in these types
of studies they provide no direct information on chemical structure

0097-6156/86/0323-0403$06.50/0
© 1986 American Chemical Society

of the adsorbed species. However, when a Mossbauer source nuclide
is adsorbed on a magnetic oxide, structural information about
adsorbed ions can be obtained from their emission Mossbauer
spectra. The principles of the method are described below with the
examples Co-57 and Sb-119 (Figure 1), which are the sources of the
most popular Mossbauer nuclides, Fe-57 and Sn-119. The two source
nuclides may be regarded as representatives of a transition and
non-transition element, respectively.

The radioactive nuclide Co-57 decays through the 136 keV second
nuclear excited level to the first excited level of Fe-57, which
then emits the 14.4 keV Mossbauer gamma-ray with a half-life of 98
ns. If Co-57 ions are adsorbed on a certain surface firmly enough
to provide an appreciable recoilless fraction on the emission of the
gamma-rays, the chemical environment of the Fe-57 in the first
excited level can be determined by analyzing the resonant gamma-rays
with a standard absorber (emission Mossbauer spectroscopy). Similarly, the chemical structure of Sn-119 arising from adsorbed
pentavalent Sb-119 can be determined through emission Mossbauer
spectra.

Since the observation of Mossbauer spectra on Fe-57 and Sn-119
is made immediately after the EC (electron capture) decays of Co-57
and Sb-119 (after 141 ns and 25.7 ns on average), the chemical
structure of Co-57 and Sb-119 can be regarded as essentially that
resulting from Fe-57 and Sn-119 emission Mossbauer spectra. The
Auger cascade following the EC decay results in multiple ionization
of the decaying atom. In the case of Co-57, charge states up to 7+
are theoretically predicted for the daughter nuclide Fe-57 (5).
Such highly ionized species have been detected for Cl-37 produced
by the EC decay of Ar-37 in gaseous phase (6). In solids, however,
such anomalous states are not realized or their life time is much
shorter than the half-life of the Mossbauer level (Fe-57: 98 ns and
Sn-119: 17.8 ns) because of fast electron transfer, and usually
species in ordinary valence states (2+, 3+ for Fe-57 and 2+, 4+ for
Sn-119) are observed in emission Mossbauer spectra (7,8). The
distribution of Fe-57 and Sn-119 between the two valence states
depends on the physical and chemical environments of the decaying
atom in a very complicated way, and detection of the counterparts of
the redox reaction is generally very difficult. The recoil energy
associated with the EC decays of Co-57 and Sb-119 is estimated to be
insufficient to induce displacement of the atom in solids.

In absorption Mossbauer spectroscopy, a source nuclide in a
standard form (usually in a metallic matrix) is coupled with a
sample to be investigated. This method requires at least 100 µg of
Fe or Sn in the usual experimental setup even if a Mossbauer sensitive enriched stable isotope Fe-57 or Sn-119 is employed. In
emission Mossbauer spectroscopy, however, 1 mCi of Co-57 or Sb-119,
which corresponds nominally to 120 ng of Co-57 or 1.4 ng of Sb-119,
is sufficient to permit measurement. This technique enables study
of very dilute systems, especially those with ions directly bound to
the substrate.

Mossbauer isomer shift and quadrupole splitting are commonly
used to obtain information about the bonding environment around
source nuclides. The isomer shift arises from the electric monopole
interaction of the nucleus with the electrons and depends on the

s-electron density at the nucleus. It is observed as the shift of the line in the Mossbauer spectrum. If the electronic environment of Fe-57 or Sn-119 nucleus is not spherically symmetric, its first excited level (I = 3/2) splits into two sublevels (m = ±1/2 and ±3/2) as a result of the electric quadrupole interaction with the electrons. This is observed as the quadrupole splitting of the Mossbauer line into a doublet in the absence of magnetic field. These parameters vary with coordination number of metal by oxygen and with distortion of oxygen environment, respectively. However, they are not informative enough in determining whether or not chemical bonds are formed between the hydrolytically adsorbed metal ions and an oxide substrate. In this study another Mossbauer parameter, magnetic splitting, was used to investigate the chemical structure of the adsorbed metal ions. When Fe-57 or Sn-119 nucleus is in a magnetic field, their ground and first excited levels (I = 1/2 and 3/2) split into two and four sublevels with m = -1/2, 1/2 and m = -3/2, -1/2, 1/2, 3/2, respectively. The transitions with Δm = 0, ±1 are allowed and a symmetric sextet is observed in the absence of electric quadrupole field. If an electric quadrupole field overlaps a large magnetic field, the Mossbauer spectrum becomes an asymmetric sextet. When the magnetic field has a distribution with relatively large fraction in the region of low values, the sextet is not resolved and a broad band is observed.

When Fe-57 or Sn-119 ions arising from Co-57 or Sb-119 are bonded through oxide ions to the surface metal ions of a magnetic oxide, the Fe-57 or Sn-119 nuclei feel the hyperfine magnetic fields induced by interaction with the ordered magnetic metal ions of the substrate. The hyperfine magnetic fields are observed as broadening or splitting in the Mossbauer spectra. Substantial magnetic interaction occurs only on adsorbed ions that are bonded through oxide ions to the ordered magnetic ions of the substrate. For those ions bonded to the surface magnetic ions through two or more ions in sequence (for example, species hydrogen-bonded to oxide surfaces), the interaction is much weaker. Therefore, the chemical structure of Co-57 or Sb-119 ions adsorbed on the surfaces of the magnetic oxide can be estimated by analyzing the broadening or splitting of the emission lines.

The present method is applicable not only to ferromagnetic and ferrimagnetic oxides but also to antiferromagnetic oxides, which are macroscopically not magnets.

Experimental Section

The ferric oxide, hematite, used in the present work was a high purity powder reagent with a BET surface area of 27 m^2/g; 30 mg was employed in each run. Some measurements were made on hematite calcined in air to see the effects of sintering the surface on the chemical structure of the adsorbed metal ions. The hematite samples were checked by Mossbauer absorption and powder X-ray diffraction measurements. The Mossbauer absorption spectra consisted of a magnetic sextet with no superparamagnetic component due to fine particles (9).

The Mossbauer source nuclide Co-57 was obtained commercially; Sb-119 was produced by the 160 cm RIKEN cyclotron at our institute.

The procedure for separating Sb-119 from an alpha-irradiated tin target has been described elsewhere (10,11). The amounts of cobalt and antimony coexisting with the nuclides are estimated to have been about 400 ng/mCi and 300 ng/mCi, respectively, i.e., to have been much smaller than that required for monolayer coverage of 30 mg of the hematite sample. About 10 cm^3 of an aqueous solution containing 1 - 2 mCi of divalent Co-57 or 0.1 - 1 mCi of pentavalent Sb-119 was adjusted to an appropriate pH value in a Teflon vessel with a 0.5 mm-thick Teflon window at the bottom, and about 30 mg of hematite powder was added to the solution. The suspension was shaken for 30 min at room temperature. After settling of the powder at the bottom of the vessel, the pH was remeasured.

The hematite with adsorbed Co-57 or Sb-119 along with the solution was subjected to emission Mossbauer measurement at 24±1°C with the experimental setup shown in Figure 2. The absorber, Fe-57-enriched potassium ferrocyanide (0.5 mg Fe-57/cm^2) or barium stannate (0.9 mg Sn-119/cm^2), was driven by a Ranger 700-series Mossbauer spectrometer connected to a Tracor-Northern TN-7200 multi-channel analyzer. The Mosssbauer gamma-rays of Co-57 and Sb-119 were detected respectively with a Kr(+3% carbon dioxide)-filled proportional counter and with a 2 mm-thick NaI(Tl) scintillation counter through 65 μm-thick Pd critical absorber for Sn K X-rays. The integral errors in the relative velocity were estimated to be of the order of 0.05 mm/s by repeated calibration measurements using standard absorbers.

In parallel with the emission measurements, in situ Mossbauer absorption measurements on hematite suspensions treated in a similar manner as in the emission measurements were performed to check the effects of aqueous phase pH on the substrate. The absorption spectra obtained in the pH region 5 - 12 consisted of the same well-defined sextet as dry hematite powder, indicating that no appreciable change occurred in the state of dispersion and particle size of hematite in the studied pH range.

Since the substrate hematite contains Fe-57, a certain Mossbauer self-absorption is inevitable in the case of measurements on Co-57. But, the effect of the absorption is considered to be not important as far as the pH dependence of the spectra is concerned, since the amount of hematite was kept constant and the adsorbed divalent Co-57 was dispersed uniformly in it.

The data obtained were analyzed with a FACOM M380 computer.

Experimental Results

In Situ Mossbauer Measurement on Hematite/Divalent Co-57. The adsorption behavior of cobaltous ions on hematite surfaces was essentially the same as that on silica reported by James and Healy (12). Appreciable adsorption begins at about pH 4 followed by an abrupt increase in adsorption between pH 6 and 8. Beyond pH 9, adsorption is practically complete. Emission Mossbauer spectra of Fe-57 arising from the divalent Co-57 ions at the interface between hematite particles and the 0.1 mol/dm^3 NaCl solutions of different pH at room temperature are shown in Figure 3. The emission spectra show a marked dependence on the pH of the aqueous phase. No emission lines ascribable to paramagnetic iron species are recognized in

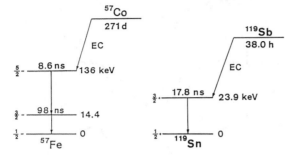

Figure 1. Simplified decay schemes of the Mossbauer source nuclides.

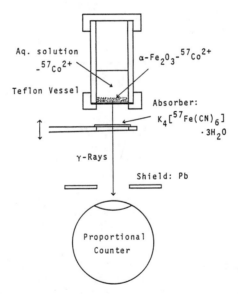

Figure 2. Experimental setup for in situ emission Mossbauer measurement. Setup for Co-57 is shown.

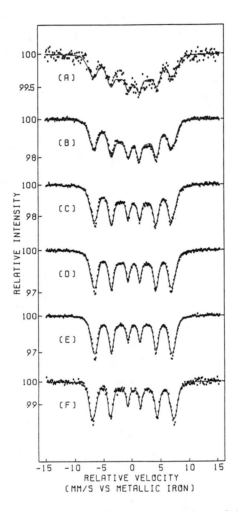

Figure 3. In situ emission Mossbauer spectra of Fe-57 arising
from divalent Co-57 at the hematite/0.1 mol dm^{-3} NaCl solution
interface for various pH values of the aqueous phase
(measurement at room temperature): (A) pH 5.7, (B) pH 7.4, (C)
pH 9.6, (D) pH 11.0, (E) pH 12.7, (F) pH 3.0. (F) was measured
after readjustment of pH from 12.7 to 3.0. The isomer shift is
given relative to metallic iron and the sign of relative velocity is defined as in ordinary absorption spectra. The curves
are composed from the results of Hesse-Rubartsch analysis given
in Figure 10.

the spectra within the experimental uncertainties. The isomer shift indicates that Fe-57 ions arising from divalent Co-57 at the interface are predominantly in the trivalent state. As described above, it is extremely difficult to identify the oxidants of divalent Co-57 ions to trivalent Fe-57. It is because the trivalent Fe-57 ions are detected with high sensitivity by their gamma-rays, while the species reduced are not radioactive. At pH 5.7, the spectrum consists of a partly resolved sextet suggesting a hyperfine magnetic field distribution extending down to low values of magnetic field (Figure 3(A)). With increase in pH of the aqueous phase the splitting of the sextet increases and simultaneously the width of each line diminishes (Figure 3(B)-(E)).

When the pH is readjusted to a higher value, the splitting increases and a spectrum which is essentially the same as the one obtained originally at the higher pH is observed. However, when the pH of a sample solution is lowered from 12.7 down to 3.0, slow desorption of Co-57 ions was observed, and the Mossbauer spectrum virtually remained unchanged (Figure 3(F)). These observations suggest that the chemical form of divalent Co-57 adsorbed from an alkaline solution is retained when the pH of the solution is lowered down to an acidic pH value.

Adsorption of Pentavalent Sb Ions on Hematite. So far as we know, there are no experimental data on the adsorption equilibrium of dilute pentavalent Sb ions on metal oxides. Therefore, the pH dependence of the adsorption of pentavalent Sb ions on hematite was measured. Carrier-free pentavalent Sb-119 ions were adsorbed on 30 mg of hematite (prefired at 900°C for 2 hours) from 10 cm3 of 0.25 mol/dm3 LiCl solutions at $24\pm1^\circ$C. The amount of antimony employed in each run is estimated to be about 50 ng. The adsorption proceeds with a measurable rate and attains an apparent equilibrium after shaking for several hours. The reaction is second order with respect to the concentration of pentavalent Sb ions in the solution (13). The values given in Figure 4 are those obtained after 22 hours equilibration. As seen in Figure 4, strong adsorption of pentavalent Sb ions is observed below pH 7, while the percent adsorbed diminishes abruptly above that. Most of the Sb ions adsorbed on hematite from solutions of pH 2-5 are not desorbed by subsequent adjustment to alkaline conditions. Results on desorption of Sb ions pre-adsorbed at pH 4 are shown in Figure 4.

In dilute concentration of pentavalent Sb ions, the $[Sb(OH)_6]^-$ form of the complex is reported to predominate over the pH range studied in the present work (14). The zero point of charge (ZPC) of hematite is reported to be pH 6.5-8.6 (15). The surface of hematite particles are positively charged due to excess surface protons below the ZPC, while they are charged negatively above the ZPC. Therefore, the observed pH dependence in adsorption of pentavalent Sb-119 on hematite is apparently interpreted in terms of electrostatic attraction and repulsion between the negatively charged Sb complex and the positively or negatively charged surface.

In Situ Mossbauer Measurement on Hematite/Pentavalent Sb-119. The Mossbauer measurement on Sb-119 was continued for 1 - 3 days after shaking the suspension for 30 min and waiting for settling of the

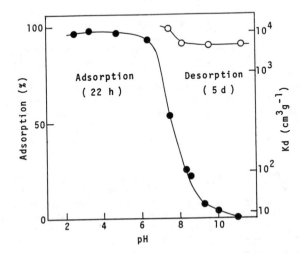

Figure 4. pH dependence of the adsorption and desorption of carrier-free pentavalent Sb-119 on hematite at room temperature (30 mg of hematite prefired at 900°C in 10 cm3 of 0.25 mol/dm3 LiCl solutions). Desorption was measured on pentavalent Sb-119 pre-adsorbed at pH 4. Shaking time was 22 hours for the adsorption and was 5 days for the desorption.

hematite powder at the bottom. Within the experimental uncertainties, no change in the spectrum was observed during each run. The in situ emission Mossbauer spectra of Sn-119 arising from pentavalent Sb-119 ions adsorbed on non-pretreated hematite particles in 0.25 mol/dm^3 LiCl solutions of different pH values at room temperature are shown in Figure 5. Because of the absence of the Mossbauer effect for chemical species in the solution, the spectra represent only information on Sn-119 arising from pentavalent Sb-119 at the interface. As seen in Figure 5, the emission spectra are essentially symmetric lines centered at the zero relative velocity against the barium stannate absorber. This demonstrates that Sn-119 at the interface is exclusively in the tetravalent state. The expected magnetic sextet appears as a single broad band due to overlapping of components distributed in low magnetic field region. The line width shows a strong pH dependence, suggesting that the chemical structure of adsorbed pentavalent Sb-119 is considerably dependent on pH of the solution. At pH 8.5, where a poor quality spectrum is obtained because of low adsorption of Sb-119, the full width at the half maximum amounts to 2.5 mm/s (Figure 5(A)). This value is about twice as large as that for Sn-119 arising from pentavalent Sb-119 on corundum-type oxide surface without magnetic interactions. (The pentavalent Sb-119 ions on chromic oxide give a single line with a full width at the half maximum of 1.3 mm/s above the Neel temperature (303 K) against the same absorber as employed in the present work (3)). Therefore, we conclude that most of the adsorbed pentavalent Sb-119 ions on hematite particles have a certain magnetic interaction with the ordered ferric ions of the substrate at the pH value. With the decrease in pH of the aqueous phase (Figure 5(B)-(E)), the line width increases remarkably and the full width at the half maximum attains a value of about 7 mm/s at pH 2.5, which suggests stronger magnetic interactions of Sn-119 with ferric ions of the substrate in the weakly acidic region.

In contrast to the case of divalent Co-57 ions described above, the spectra show no hysteresis against the lowering of pH. Conversely, the spectrum of a sample previously adjusted to a pH of 2.5 was found to remain broadened after the pH had been raised to 8.6 (Figure 5(F)). Thus, the chemical structure of pentavalent Sb-119 adsorbed from an acidic solution is considered to be retained when the pH of the solution is raised above 7.

In Figure 6 are shown the emission Mossbauer spectra of the hematite/pentavalent Sb-119/0.25 mol dm^{-3} LiCl solution systems of three different pH values before heating ((A1)-(C1)) and after heating at 98°C for 30 min ((A2)-(C2)). The experiment was performed on the hematite sample prefired at 900°C, which gives spectra with larger line width than the non-pretreated sample over the whole pH region studied. A considerable change in pH was observed after heating the system of pH 6.6. At each pH value studied, the emission spectra show broadening or even immature splitting to a sextet after the heating, suggesting incorporation of more Sb-119 ions in the surface metal ion sites. Further heating up to 100 min brought about no further appreciable change in the spectra.

In order to study the effect of the amount of pentavalent Sb ions on adsorption state, in situ emission Mossbauer measurement was made on Sb-119 adsorbed on hematite with non-radioactive pentavalent

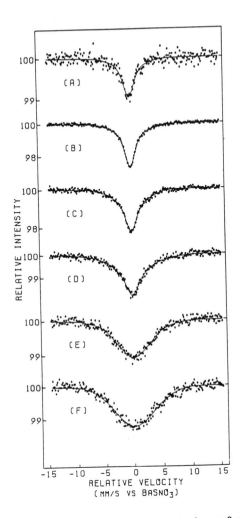

Figure 5.   In situ emission Mossbauer spectra of Sn-119 arising from pentavalent Sb-119 at the hematite/0.25 mol dm-3 LiCl solution interface for various pH values of the aqueous phase (measurement at room temperature): (A) pH 8.5, (B) pH 6.6, (C) pH 4.6, (D) pH 3.4, (E) pH 2.5, (F) pH 8.6.   (F) was measured after readjustment of pH from 2.5 to 8.6.   The isomer shift is given relative to barium stannate and the sign of relative velocity is defined as in ordinary absorption spectra.   The curves are composed from the results of Hesse-Rubartsch analysis given in Figure 11.

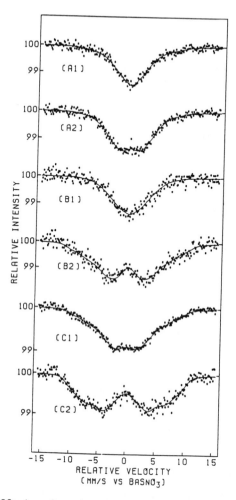

Figure 6.   Effects of preheating of sample suspensions at 98°C for 30 min on the in situ emission Mossbauer spectra of Sn-119 arising from pentavalent Sb-119 at the hematite/0.25 mol dm^{-3} LiCl solution interface (measurement at room temperature): (A1) Before heating, pH 6.6 and (A2) after heating, pH 7.9; (B1) Before heating, pH 4.4 and (B2) after heating, pH 4.3; (C1) Before heating, pH 2.6 and (C2) after heating, pH 2.6.   The curves are composed from the results of Hesse-Rubartsch analysis given in Figure 12.

Sb carrier ions. The Sb-119 ions were adsorbed on 30 mg of hematite
from 10 cm3 of a 0.25 mol/dm3 KCl solution containing about 1 mg of
pentavalent Sb ions. About 0.3 mg of Sb was adsorbed at pH 2.5 and
4.0. The amounts of Sb adsorbed are less than that required to
cover all the hematite surfaces as a monolayer. The emission
Mossbauer spectra obtained are shown in Figure 7. It is seen from
Figure 7 that the width of the emission Mossbauer spectrum at pH 2.5
is much smaller than that of the carrier-free one, while essentially
no effect of carrier Sb ions is observed at pH 4.0.

Effects of Pentavalent Sb Ions on the Adsorption of Divalent Co-57
on Hematite. Benjamin and Bloom reported that arsenate ions enhance
the adsorption of cobaltous ions on amorphous iron oxyhydroxide
(16). Similarly, when divalent Co-57 ions were adsorbed on hematite
together with pentavalent Sb ions, an increase of adsorption in the
weakly acidic region was observed. For example, when 30 mg of
hematite was shaken with 10 cm3 of 0.1 mol/dm3 KCl solution at pH
5.5 containing carrier-free Co-57 and about 1 mg of pentavalent Sb
ions, 95 % of Co-57 and about 30 % of Sb ions were adsorbed. The
emission spectra of the divalent Co-57 ions adsorbed under these
conditions are shown in Figure 8 together with the results obtained
under different conditions. As seen in Figure 8, the spectra of
divalent Co-57 co-adsorbed with pentavalent Sb ions are much different from those of Co-57 adsorbed alone (Figure 3). These observations show a marked effect of the co-adsorbed pentavalent Sb ions on
the chemical structure of adsorbed Co-57.

Analysis of the Mossbauer Data

Magnetic Interactions on Hematite Surfaces. In magnetic metal
oxides, the localized spin densities on the metal ions interact with
each other through the superexchange interaction (17-19). The main
component of the hyperfine magnetic field on trivalent Fe-57 arising
from adsorbed Co-57 ions originates in their own d-electrons ordered
by the superexchange interaction with the neighboring ferric ions.
In the simplest case in which the trivalent Fe-57 ions are completely incorporated into the cooperative antiferromagnetic system
of the bulk substrate, the Fe-57 ions are expected to align parallel
or antiparallel to the magnetic ions of the substrate in a similar
manner as the ferric ions of the substrate. When the trivalent
Fe-57 ions are on the surface, however, their magnetization is
considered to be reduced to some extent due to reduction in the
number of neighboring magnetic metal ions interacting with them.
     In the following discussion, we treat the surface effect on the
basis of the Weiss field (molecular field) approximation (17-19),
assuming no relaxation (fluctuation of the electron spins). In the
treatment, the reduced magnetization m (magnetization at a certain
temperature divided by that at 0 K) of the surface ferric ions at
temperature T K is described by

$$m(\text{surface Fe}^{3+}) = B_S(g\beta S f_r H_W/kT)$$

Here, B_S is the Brillouin function for a paramagnetic ion with spin
quantum number S (= 5/2 for ferric ion), while g, β, and k are the

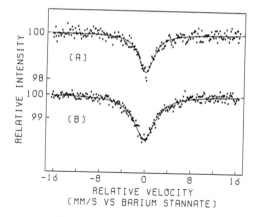

Figure 7. In situ emission Mossbauer spectra of pentavalent Sb-119 adsorbed on hematite with Sb carrier from 0.25 mol/dm3 KCl solution: (A) pH 4.0 and (B) 2.5.

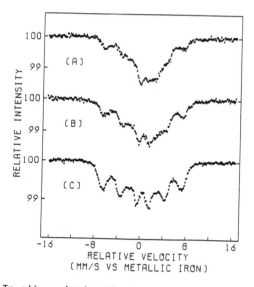

Figure 8. In situ emission Mossbauer spectra of divalent Co-57 adsorbed on hematite with pentavalent Sb ions from 0.1 mol/dm3 KCl solution: (A) pH 5.5, (B) pH 9.2. (C) From 0.3 M KOH.

g-factor(~ 2 for ferric ion), the Bohr magneton, and the Boltzmann constant, respectively. The reduction factor f_r for the Weiss field H_W due to the surface effect is given by

$$f_r = \sum_{\text{surface}}' n_i G_i \; / \sum_{\text{bulk}} n_j G_j$$

where G_i is a microscopic Weiss field constant for each ordered ferric ion acting on the surface ferric ion in question, n being the number of ions with the same G.

The nature of magnetic hyperfine interaction on tetravalent Sn-119 ions is different from that on paramagnetic Fe-57. Although tetravalent Sn-119 ions are diamagnetic, i.e., they don't have their own unpaired electron spins, spin densities from the ordered magnetic ions can be "supertransferred" to the Sn-119 ions through chemical bonds. Substantial spin transfer occurs only to such diamagnetic cations that are bound through oxide ions to the ordered magnetic metal ions. We assume that the supertransferred hyperfine (STHF) magnetic field H_{hf} acting on the surface Sn-119 ions is proportional to the algebraic sum of the magnetization m_i of the ferric ions interacting with them. Namely,

$$H_{hf}(\text{surface } ^{119}\text{Sn}^{4+})$$
$$= \left[\sum_{\text{surface}} n_i m_i (\text{Fe}^{3+}) / \sum_{\text{bulk}} n_j m_j (\text{Fe}^{3+}) \right] \cdot H_{hf}(\text{bulk } ^{119}\text{Sn}^{4+})$$

Corundum-type Magnetic Oxide Surfaces. The substrate hematite with the corundum-type crystal structure is an antiferromagnet below 963 K. In the corundum-type structure of hematite, pairs of ferric ions are in a row spaced by single vacant sites along the $\langle 111 \rangle$ direction. The positions of ferric ions in each pair are shifted slightly upward or downward in the $\langle 111 \rangle$ direction. We denote these lattice positions as up and down sites (A^u and A^d), respectively. In our simplified model for the dominant (111) surfaces of hematite, four kinds of metal ion sites are distinguished, that is, up and down sites in the zeroth and first metal ion layers (A_0^u, A_1^u, A_0^d and A_1^d in Figure 9. In the model, the second metal ion layer is not distinguishable from the bulk.

The results of Weiss field calculation on ferric ions at the surface metal ion sites are given in Figure 6 of ref [4], and the values for room temperature are shown in Figure 10. Since both ferric and pentavalent Sb ions can occupy octahedral or distorted octahedral sites with six ligand oxide ions and bulk hematite is considered to accommodate pentavalent Sb-119 ions in the metal ion sites ([3]), we can estimate STHF interactions on tetravalent Sn-119 ions at the surface metal ion sites of hematite. Using the magnetization of surface ferric ions at room temperature, the STHF magnetic fields on tetravalent Sn-119 ions at the surface sites are calculated to be

$$H_{hf}(A_0^u) = (3 \times 0.76/9) H_{hf}(\text{bulk})$$

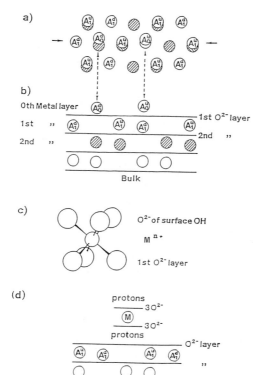

Figure 9. Simplified model of the (111) surface of the corundum-type structure. (a) A view of the surface from a direction slightly shifted from <111>. Only metal ions of the zeroth, first, and second layers are shown. (b) A section of the surface along the arrows depicted in part a. Hexagonally close-packed oxide ion layers are shown with lines. Surface protons are not shown. (c) A divalent Co-57 or pentavalent Sb-119 ion on the zeroth metal ion layer. (d) Aquo or hydroxyl complex of divalent Co-57 or pentavalent Sb-119 hydrogen-bonded to the surface oxide ion layers of hematite.

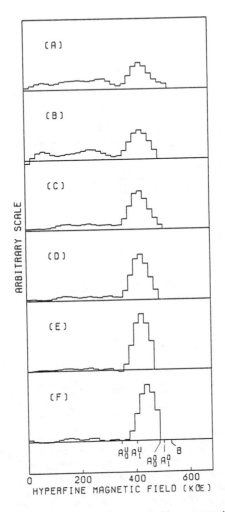

Figure 10. Distribution of the hyperfine magnetic fields on trivalent Fe-57 obtained by the Hesse-Rubartsch analysis of the spectra shown in Figure 3.

$$H_{hf}(A_0^d) = ((3\times0.95+3\times0.76)/9)H_{hf}(\text{bulk})$$

$$H_{hf}(A_1^u) = (3/9)H_{hf}(\text{bulk})$$

$$H_{hf}(A_1^d) = (6/9)H_{hf}(\text{bulk})$$

A bulk hematite/pentavalent Sb-119 sample prepared by coprecipitation of ferric and pentavalent Sb-119 ions and by calcination gave a STHF field of 122 kOe on Sn-119 ions arising from Sb-119 at room temperature. Using this value for $H_{hf}(\text{bulk})$, we obtain the STHF fields on the surface Sn-119 ions shown at the bottom of Figure 11. In our earlier work on ex situ measurements of pentavalent Sb-119 (3), we did not apply this treatment. But, the fields for A^u's, A^d's and the bulk (B) correspond roughly to the upper limits of the Regions I, II, III of ref 3, respectively. For tetravalent Sn-119 arising from Sb-119 species adsorbed on the surfaces by hydrogen bonding, the hyperfine field is estimated to be much smaller than the above values and a broad distribution near the zero magnetic field is expected to appear. The Sn-119 ions having no STHF interactions with the substrate give rise to a delta-function-like sharp peak at zero field.

Because of the simplifications described above as well as the limit of fitting, the error associated with the results of the analysis is estimated to be by one order larger than the experimental error, namely about 15 kOe for Fe-57 and 3 kOe for Sn-119.

Discussion

Divalent Co-57 Ions on Hematite. Figure 10 shows the distribution of hyperfine magnetic fields acting on the Fe-57 nuclei calculated from the emission Mossbauer spectra given in Figure 3 by the method of Hesse and Rubartsch (20) assuming no relaxation. It can be seen from Figure 10 that the adsorbed divalent Co-57 ions are at least in two chemical forms: one giving a peak in the region corresponding to the calculated values for the surface sites A's (Figure 9) and another giving the broad distribution in the lower fields. The former, whose fraction increases with increase in pH, is attributed to the divalent Co-57 ions in surface sites A's. One of the probable forms for the latter is $[Co(H_2O)_6]^{2+}$ or some hydrolyzed form $[Co(H_2O)_n(OH)_{6-n}]^{(n-4)+}$ hydrogen-bonded to the hematite surface as shown in Figure 9(d). Other chemical forms with one or two Co-O-Fe(substrate) bonds are also possible. At pH 5.7 (Figure 10(A)), about half of the adsorbed Co-57 ions are in these weakly bound states. With increase in pH, the fraction of Co-57 ions in the surface sites A's increases and becomes dominant in the strongly alkaline region. When the pH of the solution was reajusted below 5 after adsorption in the alkaline region, slow but steady desorption of divalent Co-57 ions was observed by measuring the radioactivity of the aqueous phase. For the spectrum on which Figure 10(F) is based, desorption was continuing during the Mossbauer measurement due to the pH lowered from 12.7 to 3.0. Figure 10(F) shows, therefore, that desorption of the divalent Co-57 ions from the surface sites A's occurs from strongly bound surface complexes.

Our discussion has so far ignored the effect of relaxation, namely spin flip of electrons in a time comparable to the Larmor precession period (21). The apparent low field (weak bonding) component in the analysis above might be the result of relaxation. However, the decrease in relaxation time is also considered to reflect the decrease in the strength of chemical bonds between the adsorbed metal species and the substrate. Therefore, the conclusion described above is considered to remain valid.

Pentavalent Sb-119 Ions on Hematite.   In Figure 11 are shown the results of Hesse-Rubartsch analysis of the emission Mossbauer spectra given in Figure 5 on the distribution of STHF magnetic fields acting on the nuclei of Sn-119 arising from pentavalent Sb-119 ions adsorbed on hematite surface.   In contrast to trivalent Fe-57 described above the magnetic interaction between the substrate and the tetravalent Sn-119 ions increases gradually with decreasing pH of the aqueous phase.   At neutral and slightly acidic pH values, the STHF fields show dominantly a broad distribution near zero magnetic field.   The observation demonstrates that tetravalent Sn-119 species not in the surface metal sites but having a certain STHF interaction with the substrate are overwhelming at the interface.   As the most probable dominant chemical form of the adsorbed Sb-119 ions in this pH region (above pH 4), we propose the $[Sb(OH)_6]^-$ complex attached to the oxide surfaces by hydrogen bonding like the structure shown in Figure 9(d).   In the neutral and slightly acidic region, the surfaces are not negatively charged and few excess protons exist on them.   This situation is considered favorable for the adsorption of the hydroxyl complex by hydrogen-bonding.

With the decrease in pH, the broad peak near the zero magnetic field diminishes and the distribution extends to higher fields (Figure 11).   This is interpreted to show that formation of Sb-O-Fe bonds occurs on the surface.   The distribution has a relatively large fraction in the region corresponding to the magnetic fields for surface sites.   Therefore, most of the adsorbed pentavalent Sb-119 ions are considered to be in the zeroth or in the first metal ion layer of the hematite surface (A's in Figure 9), when the aqueous phase is acidic.

After heating the hematite/pentavalent Sb-119 suspension at pH 4.3 (B2) and 2.6 (C2), a considerable amount of Sb-119 ions are in the bulk (the second or deeper layers in Figure 9), based on the results of calculation shown in Figure 12.   In the dried hematite/pentavalent Sb-119 sample reported previously, no apprecia-ble diffusion of the surface Sb-119 ions into the second or deeper layers was observed on heating the sample at 200° C for 2 hours (Figure 3 of ref 3).   Therefore, the change of the field distribu-tion observed after heating the suspension of hematite with adsorbed Sb-119 ions at 98°C is not likely to be the result of simple diffu-sion of surface Sb-119 ions into the second or deeper layers. Moreover, the fact that no further change is observed after 30 min's heating suggests a different mechanism.   A plausible explanation is that a chemical rearrangement of the surface occurs when the hema-tite particles are exposed to aqueous solutions at 98°C.   For example, some ferric ions are released from the surfaces of hematite

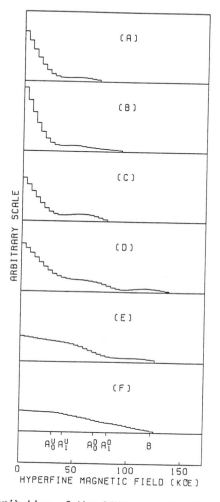

Figure 11.   Distribution of the STHF magnetic fields on tetravalent Sn-119 nuclei obtained by the Hesse-Rubartsch analysis of the spectra shown in Figure 5.

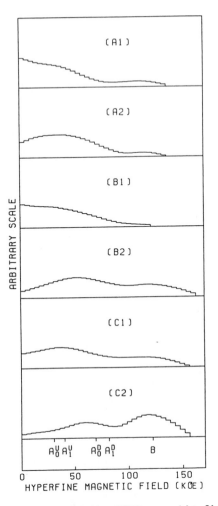

Figure 12. Distribution of the STHF magnetic fields on tetravalent Sn-119 nuclei obtained by the Hesse-Rubartsch analysis of the spectra shown in Figure 6.

particles and are again attached to them so as to incorporate a part of the pentavalent Sb-119 ions into the metal ion sites of the second or deeper layers. Since the change does not proceed further after 30 min, it is concluded that the rearrangement is limited only to a few layers of the surfaces and the Sb-119 ions are distributed among them.

The observation that the line width of the emission spectrum of pentavalent Sb-119 ions adsorbed with Sb carrier at pH 2.5 is much smaller than that of carrier-free Sb-119 is indicative of higher energy bonding at lower coverages, i.e., in the absence of carrier Sb, the sites forming stronger chemical bonds are occupied preferentially by Sb-119. These interactions of Sb-119 with strong bonding site are diluted out in the presence of carrier Sb leading to the smaller line width, weaker average bonding.

Effects of Pentavalent Sb on the Adsorption of Divalent Co-57.

The emission Mossbauer spectra of divalent Co-57 adsorbed on hematite with pentavalent Sb ions (Figure 8) are complex and we have not yet succeeded in their analysis. It is certain, however, from the spectra that trivalent Fe-57 ions produced by the EC decay of Co-57 are interacting magnetically with the ferric ions of the substrate. This means that the divalent Co-57 are not adsorbed on the pentavalent Sb ions, but on hematite directly. The $[Sb(OH)_6]^-$ anions are considered to facilitate direct adsorption of divalent Co-57 ions on the positively charged surfaces of hematite in the acidic region.

Conclusion

In situ emission Mossbauer spectroscopic measurement of the hyperfine magnetic fields on trivalent Fe-57 and tetravalent Sn-119 arising from divalent Co-57 and pentavalent Sb-119, respectively, yields valuable information on the chemical structure of adsorbed metal ions at the interface between hematite and an aqueous solution.

In the slightly acidic region, adsorbed Co-57 ions are distributed between at least two chemical forms: one attributable to the Co-57 ions coordinatively bound to surface sites and the other to Co-57 weakly bound to the hematite surfaces. In the alkaline region, most of the adsorbed Co-57 ions are in the zeroth or first metal-ion layers of the substrate forming Co-O-Fe bonds. Desorption of divalent Co-57 from strongly coordinated surface complexes occurs when the pH is lowered from alkaline to acidic values.

In contrast, the pentavalent Sb-119 ions at the interfaces are weakly bonded to the oxide ion layer of the hematite surfaces in neutral and slightly acidic region, while in the acidic region most of the adsorbed Sb-119 ions are in the zeroth or first metal ion layers of the substrate forming Sb-O-Fe bonds. The pentavalent Sb-119 ions having once been incorporated into the surface metal ion sites retain their chemical form, even when the pH of the aqueous phase is raised above 7. Heating of suspensions at $98°C$ results in chemical rearrangement of the hematite surfaces to yield pentavalent Sb-119 ions in the second or deeper metal ion layers.

Future Prospect

The present method is still in its early stage of application. Both ex situ and in situ type measurements are applicable to a variety of mineral/aqueous solution interfaces. For example, the mechanism of selective adsorption of cobaltous ions on manganese minerals can be studied by this method. In addition to the two Mossbauer source nuclides described in the present article, there are a number of other nuclides which can be studied. We have recently started a series of experiments using Gd-151 which is a source nuclide of Eu-151 Mossbauer spectroscopy. Development of theory on surface magnetism, especially one including relaxation is desirable. Such a theory would facilitate the interpretation of the experimental results.

Acknowledgments

Cooperation of the staff of RIKEN cyclotron in many irradiation runs
for the production of Sb-119 is gratefully acknowledged.

Literature Cited

1. Jenne, E. A. "Chemical Modeling in Aqueous System"; ACS SYMPO-
    SIUM SERIES No. 93, American Chemical Society: Washington,
    D.C., 1979.
2. Tewari, P. H. "Adsorption from Aqueous Solutions"; Plenum
    Press: New York, 1981.
3. Okada, T.; Ambe, S.; Ambe, F.; Sekizawa, H. J. Phys. Chem.
    1982, 86, 4726.
4. Ambe, F.; Okada, T.; Ambe, S.; Sekizawa, H. J. Phys. Chem.
    1984, 88, 3015.
5. Pollak, H. Phys. Status Solids 1962, 2, 720.
6. Snell, A. H.; Pleasonton, F. Phys. Rev. 1955, 100, 1396.
7. Wertheim, G. K. In "The Electronic Structure of Point Defects";
    Amelinckx, S.; Gevers, R.; Nihoul, J., Eds.; North-Holland:
    Amsterdam, 1971; Part 1.
8. Friedt, J. M.; Danon, J. Radiochim. Acta 1972, 17, 173.
9. Kundig, W.; Bommel, H.; Constabaris, G.; Lindquist, R. H. Phys.
    Rev. 1966, 142, 327.
10. Ambe, F.; Ambe, S.; Shoji, H.; Saito, N. J. Chem. Phys. 1974,
    60, 3773.
11. Ambe, S. J. Radioanal. Nucl. Chem., Articles 1984, 81, 77.
12. James, R. O.; Healy, T. W. J. Colloid Interface Sci. 1972, 40,
    42.
13. Ambe, S., to be published.
14. Jander, G.; Ostmann, H. J. Z. Anorg. Allg. Chem. 1962, 315,
    241.
15. Parks, G. A.; de Bruyn, P. L. J. Phys. Chem. 1962, 66, 967.
16. Benjamin, M. M.; Bloom, N. S. In "Adsorption from Aqueous
    Solutions"; Tewari, P. H. Ed.; Plenum Press: New York, 1981;
    pp. 41-60.
17. Goodenough, J. B. "Magnetism and the Chemical Bond";
    Interscience-Wiley: New York, 1963.
18. Morrish, A. H. "The Physical Principles of Magnetism"; Wiley:
    New York, 1965.
19. Kittel, C. "Introduction to Solid State Physics", 5th ed.;
    Wiley: New York, 1976.
20. Hesse, J.; Rubartsch, A. J. Phys. E 1974, 7, 526.
21. Hoy, G. R. In "Mossbauer Spectroscopy Applied to Inorganic
    Chemistry"; Long, G. J., Ed; Plenum Press: New York, 1984; Vol.
    I, p. 195.

RECEIVED June 18, 1986

TRANSFORMATION REACTIONS
AT THE MINERAL-WATER INTERFACE

20

Photoredox Chemistry of Colloidal Metal Oxides

T. David Waite[1]

School of Chemistry, University of Melbourne, Parkville, Victoria 3052, Australia

Redox reactions involving particulate metal oxides that
may be affected by absorption of light are considered
in this review with attention focussed on the possible
geochemical implications of such reactions. Of
particular significance in the natural (particularly
aquatic) environment are redox processes induced or
enhanced on absorption of light by chromophores at the
metal oxide surface in which the metal of the oxide
lattice constitutes the cationic partner. Light induced
electron transfer within such a chromophore may result
in significant rates of particle dissolution. In
addition, the effect of metal oxides in inducing
spectral shifts in sorbed molecules, and the
stabilization of photoproducts against degradative
recombination through the influence of particle surface
charge may be significant in some instances.

Redox reactions of metal oxides is a subject of considerable
interest not only in the field of aquatic geochemistry but also in
the fields of metal corrosion, plant nutrition, solar energy
research and hydrometallurgy. Electron transfer at the metal oxide
- solution interface frequently (though not always) results in
dissolution of the solid phase and for this reason has been studied
extensively (often in a relatively empirical way) with a view to
technological application. The large majority of articles published
to date deal with the thermally induced reductive dissolution of
metal oxides by either metal ion reductants (such as Cr^{2+}, Fe^{2+} or
V^{2+}), other inorganic reductants (particularly hydrazine,
hydroxylamine and dithionite), organic reductants (such as oxalic
and thioglycollic acids) or free-radical reductants produced by
γ-radiolysis (particularly the α-hydroxy radical of 2-propanol). A
number of excellent reviews of such redox reactions at solid-liquid
interfaces are available ([1-4]). Considerable progress is now being
made in elucidating the kinetic and mechanistic details of the

[1]Current address: Environmental Sciences Division, Australian Atomic Energy
Commission, Sutherland, N.S.W. 2232, Australia

thermal dissolution of metal oxides by reducing agents with most
attention being focussed on the interaction of colloidal manganese
and iron oxides with organic acids (5-8) and on the dissolution of
ferrites, common deposits in water-cooled nuclear reactor pipes, by
low oxidation-state metal ions (such as V^{2+}) and organic reducing
agents (9,10).

Essentially all the above examples of redox reactions of metal
oxides involve the thermal redistribution of electron density at the
metal oxide - solution interface typically resulting in a change in
oxidation state of the lattice metal with resultant dissolution of
the metal oxide. Not surprisingly, the absorption of additional
energy in the form of light may induce or enhance electron transfer
in some cases with a resultant initiation or rate enhancement of
certain redox reactions. In this paper, redox reactions involving
metal oxides that may be affected by absorption of light are
considered. This overview is restricted to the reactions of
particulate metal oxides with discussion of effects due to light
absorption both by the bulk metal oxide and by chromophores located
at the solid-solution interface. In addition, the effects of metal
oxide surface properties on photoinduced redox reactions are also
briefly considered.

Light Absorption by the Metal Oxide Bulk

While most metal oxides are poorly conducting, a variety of metal
oxides - some of which occur widely naturally (for example hematite,
α-Fe_2O_3, rutile, TiO_2 and cassiterite, SnO_2) - exhibit semiconducting
properties. Semiconducting materials are characterized by a full
valence band separated by an energy gap from a vacant conduction
band. Electrical conduction in these materials is due to the motion
of free charge carriers (electrons in the normally empty conduction
band or holes in the normally full valence band) arising from
deviation from stoichiometry, trace element impurities or excitation
of electrons across the band gap on input of energy (11,12). The
latter source of charge carriers is of major interest here.

When a semiconductor is immersed in a solution, charge transfer
occurs at the interface because of the difference in the tendency of
the two phases to gain or lose electrons. The net result is the
formation of an electrical field at the surface of the semiconductor.
The direction of this field depends on the relative electron
affinities of the semiconductor and solution. For an n-type
semiconductor, the field typically forms in the direction from the
bulk of the semiconductor toward the interface. Thus, if an e^--h^+
pair forms in the interface region of the semiconductor due to light
absorption, the electron moves towards the bulk of the semiconductor
and the hole moves towards the surface. The charge separation so
induced prevents recombination of the e^--h^+ pair and generates holes
at the particle surface capable of oxidizing adsorbed or
solution-phase species. The charged aggregate generated by electron
movement may build up and act as a reducing centre effecting
solution-phase reductions (13). Under certain conditions, the
photoproduced electrons or holes may induce semiconductor dissolution
(14).

Much of the early work on redox reactions of colloidal

semiconducting metal oxides dealt with the production of H_2O_2 by
reduction of the O_2 present in solution (15,16) but in recent years
the range of redox processes investigated has expanded considerably.
The decomposition of water by light-induced electron transfer
reactions at the surface of colloidal semiconductors has been
investigated extensively (17) as has the oxidation of a variety of
inorganic and organic compounds. The photocatalyzed oxidation of
CN^- to OCN^- and SO_3^{2-} to SO_4^{2-} by a range of semiconducting powders
including TiO_2, ZnO and α-Fe_2O_3 have been reported (18) with trapped
(tr) valence band (VB) holes and conduction band (CB) electrons
initiating oxidation and reduction reactions respectively at the
particle surface; i.e.

$$(TiO_2) + h\upsilon \rightarrow e_{CB}^- + h_{VB}^+$$
$$e_{CB}^- \rightarrow e_{tr}^-$$
$$h_{VB}^+ \rightarrow h_{tr}^-$$
$$CN^- + 2OH^- + 2h_{tr}^+ \rightarrow OCN^- + H_2O$$
$$O_2 + 2e_{tr}^- + 2H_2O \rightarrow H_2O_2 + 2OH^-$$

Other inorganic reactions shown to be photo-induced at colloidal
semiconducting metal oxide surfaces include the synthesis of ammonia
from water and nitrogen (19) and the oxidation of halide ions I^-,
Br^- and Cl^- to the respective radical anions I_2^-, Br_2^- and Cl_2^-
(20-22).

A variety of photocatalyzed decarboxylation reactions on TiO_2
powder including the decomposition of acetate to methane and carbon
dioxide and the breakdown of benzoic acid yielding predominantly CO_2
have been reported by Bard and coworkers (23,24). Evidence for the
occurrence of these "photo-Kolbe" reactions has stimulated the search
for other organic reactions that might be photochemically initiated
by excitation of semiconductors and extensive work in this area is in
progress (25).

Despite the widespread interest by geochemists in the
possibility of semiconducting mineral ores and clays inducing the
types of reactions outlined above in natural systems, very little
substantial evidence has yet been presented to indicate that such
reactions occur to any significant extent. Oliver, Cosgrove and
Carey (26) investigated the effect of 1% slurries of the
titanium-containing ores rutile and ilmenite on the
photodecomposition (at 300nm) of a saturated aqueous solution of
p-dichlorobenzene (p-DCB) and found no decay despite the fact that a
1% anatase slurry rapidly degraded the p-DCB under the same
conditions with a half-life of approximately 5 minutes. A wide range
of other clays (montmorillonite, bentonite, illite, kaolinite and
dolomite) were also shown to have no effect. There have been
preliminary reports that solar photolysis of beach sand coated with
natural organic matter releases carbon dioxide (27) and a
semiconducting mechanism has been suggested but significantly more
confirmation is needed. In addition, sands from various geographic
locations have been shown to abiotically reduce N_2 from the air to
NH_3 and traces of N_2H_4 on exposure to sunlight (28). This
abiological N_2 photofixation has been associated with the presence of
semiconducting titanium oxide in the sands; i.e.

$$\overset{TiO_2}{N_2 + 3H_2O + nh\nu \rightarrow 2NH_3 + 3/2O_2}$$

$$\overset{TiO_2}{N_2 + 2H_2O + mh\nu \rightarrow N_2H_4 + O_2}$$

and is suggested to be a significant component of the nitrogen cycle in arid and semiarid regions. In general, however, the complex nature of mineral ores (possessing, at best, only a few percent of semiconducting material) and the low product yields to be expected from semiconducting mechanisms operating under conditions that are far from optimal render isolation and observation of such specific processes difficult.

Light Absorption by Surface-Located Chromophores Incorporating the Metal of the Oxice Lattice

Of more apparent significance in the aquatic environment are redox processes induced or enhanced on absorbance of light by chromophores at metal oxide surfaces in which the metal of the oxide lattice constitutes the cationic partner. Light induced electron transfer within such a chromophore often results in disruption of the oxide lattice. The photoredox-induced dissolution of iron and manganese oxides by such a mechanism has been proposed as a possible means of supply of essential trace-metal nutrients to plants and aquatic organisms (29-31).

Photodissolution of Metal Oxides in Additive-Free Medium. Photolysis of colloidal suspensions of both the relatively crystalline iron oxide lepidocrocite (γ-FeOOH) and an amorphous iron oxide (am-FeOOH) by simulated solar radiation in the absence of any reducing agents has been shown (32) to induce particle dissolution to varying degrees (Table I, Figure 1). This dissolution has been associated with light-induced charge transfer within surface-located ferric hydroxy groups - a process suggested to be similar to the photoreduction of solution-phase Fe(III) hydroxy species (33):

$$FeOH^{2+} + h\nu \rightarrow Fe^{2+} + {}^{\bullet}OH$$

Both more highly hydrolyzed (e.g. $Fe(OH)_2^+$) and less hydrolyzed (Fe_{aq}^{3+}) species have been reported to be photochemically less active than $FeOH^{2+}$ (34). In the homogeneous case, only light of relatively high energy would be expected to induce significant charge transfer as the $O^{2-} \rightarrow Fe^{3+}$ band of $FeOH^{2+}$ exhibits an absorption maximum at approximately 300nm (33). Some debate exists concerning the location of the $O^{2-} \rightarrow Fe^{3+}$ charge transfer bands of solid iron oxides and hydroxides. Marusak et al. (35) report a relatively low energy charge transfer band for hematite at 375nm but more recent molecular orbital calculations indicate that the lowest energy ligand to metal charge transfer transition is nearer 270nm with lower energy bands attributable to Fe^{3+} ligand field transitions (36).

The amorphous iron oxide is observed to be considerably more photoactive than the crystalline oxide - presumably as a result of the greater number of surface-located ferric hydroxy chromophores (the BET surface area of the synthesized γ-FeOOH is only 34 m^2/g

Table I.   Initial Rates of Photodissolution of γ-FeOOH and am-FeOOH
           Suspended in 0.01M NaCl Using a Simulated Solar
              Spectrum of Total Radiation Output 300
                    µEinsteins cm^{-2} min^{-1}.

Conditions	Initial Rate of Oxide Dissolution (nmol Fe solubilized (µmol FeOOH)$^{-1}$ min^{-1})	
	pH 4.0	pH 6.5
γ-FeOOH in 0.1M NaCl	0.26	<0.05
am-FeOOH in 0.1M NaCl	40	1.4

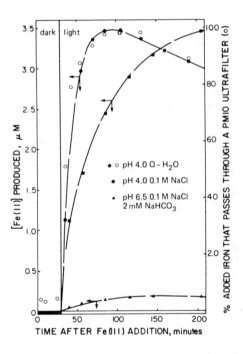

Figure 1. Production of Fe(II) on irradiation of pH 4.0
solutions of distilled, deionized water and 0.1M NaCl and a
pH 6.5 solution of 0.1M NaCl/2mM NaHCO$_3$ containing 3.5µM
am-FeOOH. The ultrafilterability of the iron through a membrane
exhibiting a nominal molecular weight cutoff of 10,000 (Amicon
PM10) in the pH 4.0 study is also shown. Light source:
simulated solar spectrum of total intensity 300 µEinsteins cm^{-2}
min^{-1}. (Reproduced from Ref. 32. Copyright 1984, American
Chemical Society.)

compared to a value of $\approx 200 m^2/g$ for the am-FeOOH) and as a consequence of the less ordered nature of its surface (higher proportion of kinks and edges). The decreasing rate of photodissolution on increasing pH is to be expected in view of i) the increasing proportion of more hydrolyzed (less photoactive) surface groups, and ii) the increasing tendency of photoproduced Fe^{2+} to readsorb to the colloid surface (37) and reoxidize to am-FeOOH.

Photoredox Reactions of Metal Oxides in the Presence of Organic Radical Scavengers. In additive free homogeneous systems, the yield of Fe^{2+} on photolysis of ferric hydroxy species is very low due particularly to product recombination (38). This recombination can be prevented by performing the photolysis in the presence of organic radical scavengers such as alcohols. Thus, Carey *et al.* (39) photolyzed $Fe(H_2O)_6^{3+}$ in the presence of ethylene glycol and observed Fe^{2+} formation as well as the production of the organic degradation products formaldehyde and glycolaldehyde. These workers concluded that the organic molecule did not enter the inner coordination sphere, but reacted as an encounter partner in the Fe(III) solvent cage. The formation of formaldehyde on photolysis of a goethite-ethylene glycol mixture (40) has been interpreted in terms of a mechanism similar to that proposed for the homogeneous case. Direct photoexcitation of an outer-sphere surface complex leading to electron transfer from ethylene glycol to Fe^{3+} has been postulated; *i.e.*

$$FeOOH(s) + HOCH_2CH_2OH = FeOOH, HOCH_2CH_2OH \text{ (sorbed layer)}$$
$$FeOOH, HOCH_2CH_2OH + h\nu \rightarrow Fe^{2+}(lattice) + [HOCH_2CH_2OH]^{+\bullet}$$
$$[HOCH_2CH_2OH]^{+\bullet} \rightarrow {}^{\bullet}CH_2OH + CH_2O + H^+$$
$$FeOOH(s) + {}^{\bullet}CH_2OH \rightarrow Fe^{2+} \text{ (lattice)} + CH_2O + H_2O$$

Reduced iron so formed would be expected to migrate rapidly out of the oxide lattice but, under the pH conditions of this study (pH 6.5) will be readily readsorbed to the oxide surface and, predictably, is not a major solution phase product.

Direct Photolysis of Surface-Located Inner Coordination Sphere Complexes. In the presence of a strong metal binding ligand, the underlying central metal ion in the surface layer of a metal oxide can exchange its structural OH^- ions for the ligand. Thus, the association of citrate with an iron oxyhydroxide surface may be represented:

$$>FeOH^0 + H_3Cit = >FeCit^{2-} + 2H^+ + H_2O$$

(where > denotes a surface grouping). Solution phase ferric hydroxycarboxylate complexes exhibit active photochemical properties principally because of strong ligand to metal charge transfer transitions in the UV and near UV spectral region (38). Surface-located ferric hydroxycarboxylates appear to exhibit similar photochemical activity to their homogeneous counterparts resulting, at low pH, in rapid dissolution of the iron oxide with concomitant oxidation of the adsorbed ligand (41). The results of studies into the simulated solar radiation induced dissolution of dilute

suspensions of γ-FeOOH in the presence of a large excess of citrate
are shown in Figure 2. At low pH (pH 4), a constant rate of
dissolution is obtained as the photoproduced Fe^{2+} rapidly detaches
from the solid phase and experiences little tendency to resorb or
oxidize. While surface ferric citrate groups are irreversibly
decomposed in the dissolution process, a constant rate of dissolution
is maintained on continued photolysis by rapid adsorption of fresh
citrate from solution. At higher pH, not only do we observe a
lowered initial rate of dissolution (due to the decreased tendency
for citrate to form surface complexes) but we also observe a
decreasing rate of dissolution over time as photoproduced Fe(II) is
resorbed and possibly oxidized at the oxide surface.

Similar photo-induced reductive dissolution to that reported for
lepidocrocite in the presence of citric acid has been observed for
hematite (α-Fe$_2$O$_3$) in the presence of S(IV) oxyanions (42) (see
Figure 3). As shown in the conceptual model of Faust and Hoffmann
(42) in Figure 4, two major pathways may lead to the production of
$Fe(II)_{aq}$: i) surface redox reactions, both photochemical and thermal
(dark), involving Fe(III)-S(IV) surface complexes (reactions 3 and 4
in Figure 4), and ii) aqueous phase photochemical and thermal redox
reactions (reactions 11 and 12 in Figure 4). However, the rate of
hematite dissolution (reaction 5) limits the rate at which $Fe(II)_{aq}$
may be produced by aqueous phase pathways (reactions 11 and 12) by
limiting the availability of $Fe(III)_{aq}$ for such reactions. The rate
of total aqueous iron production ($d[Fe(aq)]_T/dt = d\{[Fe(III)_{aq}] +$
$[Fe(II)_{aq}]\}/dt$) in the dark represents an upper bound to the rate of
$Fe(II)_{aq}$ generation from all redox reactions other than the
photo-induced surface redox reaction (reactions 4+11+12). The rate
of aqueous iron production in the dark (1.0 nM/min) is far slower
than the observed rates of $Fe(II)_{aq}$ production in illuminated
suspensions of hematite containing S(IV). The production of $Fe(II)_{aq}$
in irradiated suspensions of hematite containing S(IV) is therefore
due to a surface photo-induced reductive dissolution reaction. Faust
and Hoffmann (42) note that the >Fe(III)-S(IV) surface complex is
likely to exhibit a ligand to metal charge transfer band at a similar
wavelength to that of the corresponding aqueous phase complex
($\lambda_{max} = 367nm$) and observed a significant increase in quantum yield
for $Fe(II)_{aq}$ production in the vicinity of this wavelength.
Photo-induced charge transfer within the surface-located
>Fe(III)-S(IV) complex provides the simplest explanation consistent
with experimental results but reductive dissolution as a consequence
of direct excitation of the bulk solid cannot be discounted.
Excitation of the $O^{2-} \rightarrow Fe^{3+}$ lattice charge transfer band in hematite
may create "holes" localized on lattice oxygen (O^-) and electrons
localized on lattice iron (Fe^{2+}). Provided sufficient charge carrier
mobility exists, the holes will migrate to the surface where electron
transfer from adsorbed S(IV) occurs. It should be noted however
that, in this proposed reaction scheme, the lattice recombination
reaction ($O^- + Fe^{2+}$) is likely to limit the yield of the overall
reaction (42).
      At low pH, where resorption reactions are minimal, the
photodissolution process may be represented as a two-step process
involving adsorption of ligand L to metal oxide surface sites
followed by detachment of reduced metal ions; that is, for an
iron oxyhydroxide:

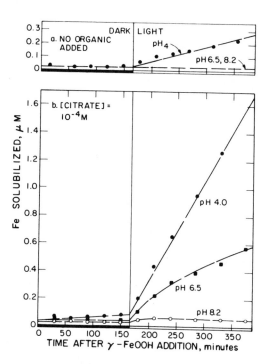

Figure 2.   Dissolution of 5µM γ-FeOOH under dark and light
conditions in a) the absence, and b) the presence of 10^{-1}M
citrate.   Light source: simulated solar spectrum of total
intensity 300 µEinsteins cm^{-2} min^{-1}.   (Reproduced with permission
from Ref. 41.   Copyright 1984, Academic Press, Inc.)

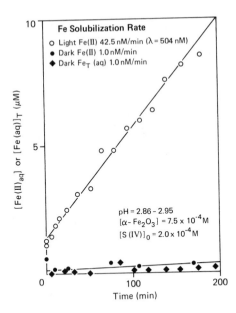

Figure 3. Photochemical and thermal solubilization of hematite in deoxygenated hematite suspensions containing S(IV). (Reproduced from Ref. 42. Copyright 1985, American Chemical Society.)

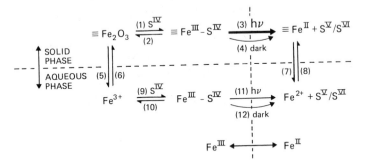

Figure 4. Conceptual model of iron redox chemistry in hematite suspensions containing S(IV). (Reproduced from Ref. 42. Copyright 1985, American Chemical Society.)

$$>FeOH^0 + L = >FeL \qquad \text{step 1}$$
$$>FeL + h\nu \rightarrow Fe(II)_{aq} \qquad \text{step 2}$$

For monochromatic illumination of constant intensity, the rate of $Fe(II)_{aq}$ production can be formulated as:

$$\frac{d[Fe(II)_{aq}]}{dt} = k[>FeL]$$

where k is the first-order rate constant for the reduction/detachment step (step 2). The rate of adsorption of ligand L to the oxide surface is typically considerably faster than the rate of detachment of reduced metal ions, thus step 1 may be viewed as a pre-equilibrium step. If this step has equilibrium constant K and a steady-state condition is applied to the concentration of surface-complex sites, then the rate of $Fe(II)_{aq}$ production may be expressed as (43):

$$\frac{d[Fe(II)_{aq}]}{dt} = \frac{kKS[L]}{1 + K[L]}$$

where S is the total "concentration" of surface sites (S ≈ [>FeOH⁰] + [>FeL]). Under these conditions the rate of reaction is dependent upon the adsorption isotherm that defines the equilibrium adsorption of L on the oxide surface. When the adsorbate concentration is low the reaction is first-order but, as the concentration is increased the order becomes less than one and finally zero when the surface is saturated with adsorbate. The ligand concentration dependencies of the rate of photoinduced dissolution of γ-FeOOH at pH 4 in the presence of tartaric and salicylic acids (Figures 5a and b) clearly exhibit this saturation effect. Equilibrium and first-order rate constants (K and k respectively) obtained by fitting rectangular hyperbolae to γ-FeOOH dissolution rate *versus* total organic concentration data for citrate, tartrate, oxalate, and salicylate ligands are given in Table II (columns 3 and 4). The actual rates of dissolution of γ-FeOOH obtained under both dark and light conditions in the presence of excess (100μM) ligand are also given (columns 5 and 6). All studies were performed in pH 4, 0.01M NaCl solutions containing 5μM total iron and were photolyzed using a Hg arc lamp source with 365nm band pass filtering (44). The oxide used exhibited a BET surface area of 144 m^2/g giving a total surface site concentration of 0.84μM (assuming an OH site density of 8 OH groups/nm^2). Insight into the likely photoactivity of the surface-located ferric complexes has been obtained by examination of the concentration of Fe(II) produced on photolysis of the solution phase complexes (44). This concentration represents a steady-state between production of Fe(II) by photodegradation of the complexes and Fe(II) oxidation by powerful photoproduced oxidants (particularly H_2O_2). Photoreduction rate constants obtained by assuming both oxidation and reduction processes to be first order are given in Table II (column 2) as are the molar absorptivities of the ferric complexes at 365nm (column 1). Note that the organic ligands used exhibit essentially no absorption over the irradiation wavelengths

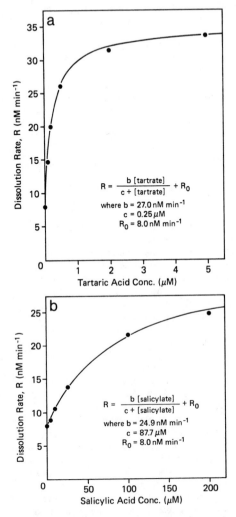

Figure 5. Dependence of rate of dissolution of 5μM γ-FeOOH in pH 4.0, 0.01M NaCl on concentration of a) tartaric acid, and b) salicylic acid. Fitted parameters obtained for rectangular hyperbolic model are given. Light source: mercury arc lamp with 365nm band-pass filtering.

Table II. Measured and Fitted Parameters Obtained in Studies of the Photodissolution of γ-FeOOH Suspended in pH 4, 0.01M NaCl Containing Various Organic Acids. Illumination provided by a Hg arc lamp source and 365 nm band-pass filter (total radiation output 85 μE cm^{-2} min^{-1}).

Organic Acid	(1) $^2\varepsilon$	(2) $^3k_{soln}$	(3) ^4k	(4) ^5K	(5) $^6R_{light}$	(6) $^7R_{dark}$
	M^{-1}cm^{-1}	min^{-1}	min^{-1}	M^{-1}	nM min^{-1}	nM min^{-1}
Tartaric	320	0.49	0.032	$10^{6.6}$	37.8	1.2
Citric	208	0.27	0.028	$10^{5.7}$	33.1	1.1
Oxalic	230	0.63	0.034	$10^{5.3}$	51.5	9.8
Salicylic	140	0.67	0.030	$10^{4.1}$	21.5	0.6
Phthalic	120	0.14	–	–	2.3	0.2
No organic					8.0	<0.2

[1]all studies performed in pH 4.0, 0.01M NaCl solutions using mercury arc lamp illumination with 365nm band-pass filtering (total radiation input 85 μEinsteins cm^{-2} min^{-1});
[2]molar absorptivity of soluble ferric complexes at 365nm;
[3]first-order rate constant for photoreduction of soluble ferric complexes;
[4]first-order rate constant for reduction/detachment step in photodissolution of γ-FeOOH (obtained from rectangular hyperbolic fit to dissolution data);
[5]formation constant for pre-equilibrium step in photodissolution of γ-FeOOH (obtained from rectangular hyperbolic fit to dissolution data);
[6]rate of dissolution of 5μM γ-FeOOH in presence of 100μM organic ligand on continuous photolysis;
[7]rate of dissolution of 5μM γ-FeOOH in presence of 100μM organic ligand under dark conditions.

used (300 $\leq \lambda \leq$ 400nm) implying that any photoeffects observed arise
as a result of absorption by the ferric complexes (or by the iron
oxide).

Except for phthalic acid, all other carboxylic acids studied
induce considerable increases in the light compared to the dark
values (the relatively high rate of iron oxide dissolution induced by
oxalic acid has been extensively studied (5,8). Phthalic acid
actually appears to stabilize the iron oxide against photodissolution
despite the solution phase complex exhibiting some photoactivity.
The first-order rate constants for the reduction/detachment step are
surprisingly constant compared to the photoreducibility of the
solution phase complexes suggesting that detachment (i.e. the
movement of Fe^{2+} from the lattice to solution) rather than reduction
is rate-limiting. The extent of dissolution is controlled
principally by the pre-equilibrium step (i.e. by the concentration
of surface-located ferric complexes). This concentration will be
determined by the strength of Fe(III)-ligand binding and by the
nature of the electrostatic interactions between the charged colloid
surface and the organic ligand. The calculated pre-equilibrium
constant, K, is a measure of the tendency of the various ligands to
form photoactive surface-located Fe(III) complexes. Note that the
light induced generation of powerful Fe(II) oxidizing agents such as
H_2O_2, possibly resulting in the formation of fresh iron oxide
precipitates, may also be an important factor controlling the extent
of photodissolution.

There have been preliminary reports that the dissolution of iron
and manganese oxides by naturally occurring humic and fulvic acids is
enhanced on photolysis (30,32,45). As can be seen from Figure 6,
addition of 10 mg/L aquatic fulvic acids to a pH 4, 5μM suspension of
γ-FeOOH results in an approximately 4-fold increase in solar
radiation-induced dissolution (the extent of dissolution in both the
presence and absence of fulvic acids under dark conditions was
negligible over the time scale of interest). At pH 6.5, the increase
observed on addition of fulvic acids is small but measurable. The
effect of near-UV light (300 $\leq \lambda \leq$ 400nm; λ_{max} = 365nm) in enhancing
the dissolution of 5μM of the synthetic manganese oxide vernadite,
δ-MnO_2 in the presence of 10 mg/L Suwanee River fulvic acid (an
International Humic Substances Society standard material) are
somewhat more dramatic with ≈90% solubilization after three hours of
continuous photolysis compared with only ≈20% dissolution after
suspension in the same medium for three hours with no illumination
(Figure 7). Interestingly, the presence or absence of oxygen appears
to have little effect on the photodissolution process - a somewhat
surprising result in view of reports that photogenerated H_2O_2 may
solubilize δ-MnO_2 (46). Presumably, the extent of H_2O_2 production
due to photodegradation of fulvic acid is too small to exert any
significant effect on the dissolution process under these conditions.

The actual excitation centre in these fulvic acid containing
systems is difficult to identify as the organic molecule itself
absorbs relatively strongly in the near-UV region. Solution phase
studies of iron-fulvic acid systems have indicated the presence of
salicylate-like binding sites exhibiting charge transfer
characteristics (47); however excitation of n-π* or π-π* transitions
is also likely to enhance the electron donating ability of these
molecules and hence their ability to solubilize iron and manganese

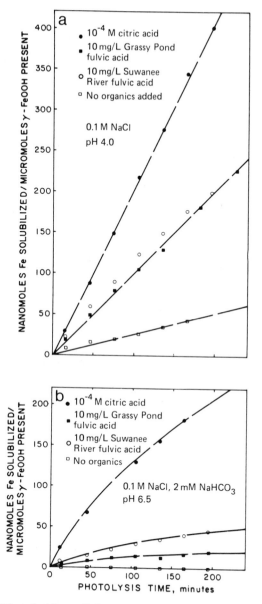

Figure 6.   Dissolution of 5μM γ-FeOOH on photolysis of a) pH 4.0, and b) pH 6.5 solutions containing either 10^{-4}M citrate, 10 mg/L aquatic fulvic acids, or no added organic agent.   Light source: simulated solar spectrum of total intensity 300 μEinsteins cm^{-2} min^{-1}.   (Reproduced from Ref. 32.   Copyright 1985, American Chemical Society.)

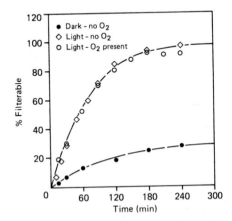

Figure 7.   Dissolution of 5μM δ-MnO$_2$ in pH 7.1, 0.01M NaCl/2mM NaHCO$_3$ solutions containing 10 mg/L Suwanee River fulvic acid under dark and light conditions in the presence and absence of oxygen.   Light source: mercury arc lamp with 365nm band-pass filtering.   (Reproduced from Ref. 45.).

oxides. Such an effect has been suggested in studies of the dissolution of hematite and lepidocrocite by a range of α-mercaptocarboxylic acids (48,49) in which high energy UV light (unfiltered Xenon arc lamp source) was found to significantly enhance the rate of colloid dissolution (Figure 8).

As discussed earlier for the additive-free system, the form of the metal oxide is critical in determining the dissolution characteristics. As shown for the dissolution of 25μM α-Fe_2O_3 and γ-FeOOH by 10^{-2} M mercaptoacetic acid in Figure 9, the same is true when a strong Fe(III) ligand is present (49). Analysis of dissolution rate *versus* ligand concentration data for the hematite (BET surface area 65 m^2/g) and lepidocrocite (BET surface area 144 m^2/g) systems yields first-order rate constants for the rate-limiting detachment step of 0.012 and 0.84 min^{-1} respectively - a result that is not particularly surprising given the relative solubilities of these two phases (the solubility products for γ-FeOOH and α-Fe_2O_3 are on the order of 10^{-39} and 10^{-41} respectively).

Effect of Metal Oxide Surface Properties on Photoinduced Redox Reactions

In addition to the primary photoredox properties of metal oxides in which either the bulk oxide or surface-located groups incorporating the oxide cation absorb light directly, secondary photoredox effects in which the presence of the oxide alters the photochemical characteristics of molecules or groups of molecules interacting with the oxide surface should also be mentioned. Of particular interest are i) the effect of metal oxides in inducing spectral shifts in sorbed molecules, and ii) the stabilization of photoproducts against degradative recombination through the influence of colloid surface charge.

i) Significant spectral shifts in molecules have been shown to occur on adsorption to a polar surface with the direction of the shift dependent on the transition involved (50). A blue shift is to be expected for n-π* transitions because of stabilization of the ground state by the polar surface while the reverse is expected for π-π* transitions. Extensive results confirming these expectations have been reported (50). For example, strong, photochemically active bands of xenobiotics such as anthracene, polychlorinated aromatic compounds and certain pesticides are red-shifted into the near-UV region on adsorption to sands and silica gel (51). Photolysis of these compounds in this spectral region has been shown to result in significant mineralization to CO_2 and H_2O (52).

ii) In photosensitized electron-transfer reactions, a photoinduced electron transfer from a donor, D, to an acceptor, A, results in the reduced and oxidized products, *i.e.*

$$D + A + h\nu \rightarrow (D^+...A^-) \rightarrow D^+ + A^-.$$

These photoproducts are initially in an "encounter cage complex" and may either recombine within this cage structure or

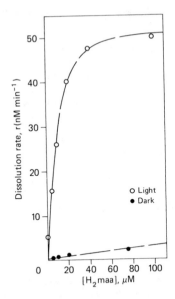

Figure 8. Rate of dissolution of 5μM γ-FeOOH suspended in pH 3.0, 0.01M NaCl containing various concentrations of mercaptoacetic acid under dark and light conditions. Light source: unfiltered 150W Xenon arc lamp. (Reproduced from Ref. 49.)

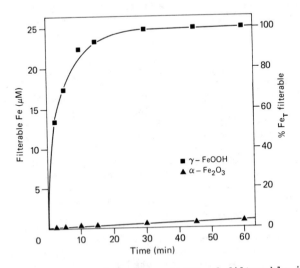

Figure 9. Comparison of the percentage of filterable iron obtained on photolysis of 25μM suspensions of γ-FeOOH and α-Fe$_2$O$_3$ in pH 3.0, 0.01M NaCl containing 10mM mercaptoacetic acid. Light source: unfiltered 150W Xenon arc lamp.

dissociate to separated ions. The separated ions A^- and D^+ can then recombine in a diffusion recombination process or be utilized in subsequent oxidation and reduction reactions. Willner and coworkers (53-55) have examined a variety of electron transfer reactions on SiO_2 colloids in aqueous medium using ruthenium tris(bipyridine) as photosensitizer and anthraquinone sulfonates as electron acceptors and have shown that the electrical potential of the particles assists the separation of the products from the encounter cage complex and also retards the back-reaction of the separated ions resulting in significant enhancements in rate and extent of product formation. There has been speculation that the silica frustules of diatoms may similarly affect redox processes occurring at the organism/water interface though no concrete evidence of such an effect has yet been presented.

Conclusions

The area of colloidal metal oxide photoredox chemistry has been one of particularly intensive investigation over the last ten years. The possibility of using colloidal semiconducting oxides as catalysts in the degradation of water has stimulated much of this interest. Of possibly more significance to the field of aquatic geochemistry is the ability of light to initiate or enhance the dissolution of colloidal metal oxides through excitation of surface-located chromophores incorporating the cation of the metal oxide as an active redox partner. While preliminary investigations in this area have confirmed the surface-controlled nature of this process and delineated the major controlling factors, extensive mechanistic work is still needed in order to clarify the details of electron transfer and molecule detachment. The effects of the charged metal oxide surface on the spectral properties of sorbed chromophores and on the recombination rate of photogenerated species at the colloid surface are also areas of possible considerable implication in the aquatic environment.

Literature Cited

1. Diggle, J.W. In "Oxides and Oxide Films"; Diggle, J.W., Ed.; Marcel Dekker: New York, 1973; Vol. 2, p. 281.
2. Valverde, N.; Wagner, C. Ber. Bunsenges. Phys. Chem. 1976, 80, 330.
3. Segal, M.G.; Sellers, R.M. Adv. Inorg. Bioinorg. Mech. 1984, 3, 97.
4. Gorichev, I.G.; Kipriyanov, N.A. Russ. Chem. Rev. 1984, 53, 1039.
5. Baumgartner, E.; Blesa, M.A.; Marinovich, H.A.; Maroto, A.J.G. Inorg. Chem. 1983, 22, 2224.
6. Stone, A.T.; Morgan, J.J. Environ. Sci. Technol. 1984, 18, 450.
7. Stone, A.T.; Morgan, J.J. Environ. Sci. Technol. 1984, 18, 617.
8. Zhang, Y.; Kallay, N.; Matijevic, E. Langmuir 1985, 1, 201.
9. Segal, M.G.; Sellers, R.M. J. Chem. Soc., Faraday Trans. 1 1982, 78, 1149.
10. Sellers, R.M.; Williams, W.J. Faraday Discuss. Chem. Soc. 1984, 77, 265.

11. Shuey, R.T. "Semiconducting Ore Minerals"; Developments in Economic Geology, 4; Elsevier: New York, 1975.

12. Morrison, S.R. "Electrochemistry at Semiconductor and Oxidized Metal Electrodes"; Plenum Press: New York, 1980.

13. Bard, A.J. Science 1980, 207, 139.

14. Gerischer, H. Faraday Discuss. Chem. Soc. 1980, 70, 137.

15. Freund, T.; Gomes, W.P. Cat. Rev. 1969, 3, 1.

16. Dixon, D.R.; Healy, T.W. Aust. J. Chem. 1971, 24, 1193.

17. Kiwi, J.; Kalyanasundaram, K.; Gratzel, M. Struct. Bond. 1982, 49, 37.

18. Frank, S.N.; Bard, A.J. J. Phys. Chem. 1977, 81, 1484.

19. Miyama, H.; Fujii, N.; Nagae, Y. Chem. Phys. Lett. 1980, 74, 523.

20. Henglein, A. Ber. Bunsenges. Phys. Chem. 1982, 86, 241.

21. Moser, J.; Gratzel, M. Helv. Chim. Acvta 1982, 65, 1436.

22. Moser, J.; Gratzel, M. J. Am. Chem. Soc. 1983, 105, 6547.

23. Kraeutler, B.; Bard, A.J. J. Am. Chem. Soc. 1978, 100, 5985.

24. Izumi, I.; Fan, F-R.F.; Bard, A.J. J. Phys. Chem. 1981, 85, 218.

25. Fox, M.A. Acc. Chem. Res. 1983, 16, 314.

26. Oliver, B.G.; Cosgrove, E.G.; Carey, J.H. Environ. Sci. Technol. 1979, 13, 1075.

27. Sancier, K.M.; Wise, H. Atmos. Env. 1981, 15, 639.

28. Schrauzer, G.N.; Strampach, N.; Hui, L.N.; Palmer, M.R.; Salehi, J. Proc. Natl. Acad. Sci. USA 1983, 80, 3873.

29. Krizek, D.T.; Bennett, J.H.; Brown, J.C.; Zaharieva, T.; Norris, K.H. J. Plant Nut. 1982, 5, 323.

30. Sunda, W.G.; Huntsman, S.A.; Harvey, G.R. Nature 1983, 301, 234.

31. Finden, D.A.S.; Tipping, E.; Jaworski, G.H.M.; Reynolds, C.S. Nature 1984, 309, 783.

32. Waite, T.D.; Morel, F.M.M. Environ. Sci. Technol. 1984, 18, 860.

33. David, F.; David, P.G. J. Phys. Chem. 1976, 80, 579.

34. Mulay, L.N.; Selwood, P.W. J. Am. Chem. Soc. 1955, 77, 2693.

35. Marusak, L.A.; Messier, R.; White, W.B. J. Phys. Chem. Solids 1980, 41, 981.

36. Sherman, D.M.; Waite, T.D. Am. Mineral. (in press).

37. Sigg, L.; Stumm, W. Colloids and Surfaces 1981, 2, 101.

38. Balzani, V.; Carassiti, V. "Photochemistry of Coordination Compounds"; Academic Press: New York, 1970.

39. Carey, J.H.; Cosgrove, E.G.; Oliver, B.G. Can. J. Chem. 1977, 55, 625.

40. Cunningham, K.M.; Goldberg, M.C.; Weiner, E.R. Photochem. Photobiol. 1985, 41, 409.

41. Waite, T.D.; Morel, F.M.M. J. Colloid Interface Sci. 1984, 102, 121.

42. Faust, B.C.; Hoffmann, M.R. Presented at American Chemical Society 189th National Meeting, Miami, Florida, April 1985.

43. Castellan, G.W. In "Physical Chemistry", 2nd ed.; Addison-Wesley: Reading, MA, 1971; Chap. 30.

44. Waite, T.D., unpublished data.

45. Waite, T.D.; Wrigley, I.C., unpublished data.

46. Waite, T.D.; Wrigley, I.C.; Smith, J.D., unpublished data.

47. Waite, T.D.; Morel, F.M.M. Anal Chim. Acta 1984, 162, 263.

48. Waite, T.D.; Torikov, A.; Smith, J.D. J. Colloid Interface Sci. (in press).

49. Waite, T.D.; Torikov, A.; Smith, J.D., unpublished data.

50. Nicholls, C.H.; Leermakers, P.A. In "Advances in Photochemistry"; Pitts, J.N.; Hammond, G.S.; Noyes, W.A., Eds.; Wiley: New York, 1971; Vol. 8.
51. Leermakers, P.A.; Thomas, H.T. J. Am. Chem. Soc. 1965, 87, 1620.
52. Gab, S.; Schmitzer, J.; Thamm, H.W.; Parlar, H.; Nature 1979, 270, 331.
53. Willner, I.; Yang, J-M.; Laane, C.; Otvos, J.W.; Calvin, M. J. Phys. Chem. 1981, 85, 3277.
54. Willner, I.; Degani, Y. Isr. J. Chem. 1982, 22, 163.
55. Degani, Y.; Willner, I. J. Am. Chem. Soc. 1983, 105, 6228.

RECEIVED June 18, 1986

21

Adsorption of Organic Reductants and Subsequent Electron Transfer on Metal Oxide Surfaces

Alan T. Stone

Department of Geography and Environmental Engineering, Johns Hopkins University, Baltimore, MD 21218

Rates of reductive dissolution of transition metal oxide/hydroxide minerals are controlled by rates of surface chemical reactions under most conditions of environmental and geochemical interest. This paper examines the mechanisms of reductive dissolution through a discussion of relevant elementary reaction processes. Reductive dissolution occurs via (i) surface precursor complex formation between reductant molecules and oxide surface sites, (ii) electron transfer within this surface complex, and (iii) breakdown of the successor complex and release of dissolved metal ions. Surface speciation is an important determinant of rates of individual surface chemical reactions and overall rates of reductive dissolution.

Oxide/hydroxide minerals of Mn(III,IV), Fe(III), Co(III), and Pb(IV) are thermodynamically stable in oxygenated solutions at neutral pH, but are reduced to divalent metal ions under anoxic conditions in the presence of reducing agents. Changes in oxidation state dramatically alter their solubility. Reduction of Fe(III) to Fe(II), for example, increases iron solubility with respect to oxide/hydroxide phases by as much as eight orders of magnitude ($\underline{1}$).

Organic reductants are present in higher concentrations and found in a wider variety of aquatic environments than inorganic reductants, and therefore have the most significant impact on transition metal geochemistry. In homogeneous solution, reduction of metal ion complexes by organic compounds is relatively well understood. Catechol, for example, reduces $Fe(H_2O)_6^{3+}$ in 0.1M $HClO_4$ in two, one-equivalent steps ($\underline{2}$):

$$Fe(H_2O)_6^{3+} + QH_2 \rightleftharpoons Fe(H_2O)_6^{2+} + QH\cdot + H^+ \qquad (1)$$

$$Fe(H_2O)_6^{3+} + QH\cdot \rightleftharpoons Fe(H_2O)_6^{2+} + Q + H^+ \qquad (2)$$

0097–6156/86/0323–0446$06.00/0

Within the pH range of natural waters, ter- and tetra-valent transition metals are incorporated into oxide minerals. Reduction by organic compounds is, therefore, a surface chemical reaction. The following reaction steps can be postulated for the reduction of iron oxide surface sites by catechol:

$$>Fe^{III}OH + QH_2 \rightleftharpoons >Fe^{II}OH^- + QH\cdot + H^+ \tag{3}$$

$$>Fe^{III}OH + QH\cdot \rightleftharpoons >Fe^{II}OH^- + Q + H^+ \tag{4}$$

$$>Fe^{II}OH^- \xrightarrow{\text{dissolution}} Fe(H_2O)_6^{2+} \tag{5}$$

where the symbol > denotes bonds to the surface metal center in the oxide lattice. Equations 3-5 represent surface reaction steps associated with precursor complex formation, electron transfer, and dissolution, respectively. Comprehensive reviews by Diggle (3) and Gorichev and Kipriyanov (4) outline in general terms how bulk and surface properties influence rates of metal oxide dissolution. Much recent interest has focussed upon the impact of surface speciation and surface chemical reactions on dissolution rates (5-7), and upon possible analogies with ligand substitution and electron transfer reactions in homogeneous solution (5). This paper discusses reductive dissolution by examining the mechanisms of pertinent surface chemical reactions using theories of ligand substitution and electron transfer reactions developed from kinetic studies in homogeneous solution. Differences between reactions in homogeneous solution and on metal oxide/hydroxide surfaces arising from surface structure, chemical speciation, and interfacial electrical properties are discussed. Molecular descriptions of surface chemical reactions will prove useful in interpreting results from actual reductive dissolution experiments and in predicting the kinetic behavior of unexplored reactions.

Reductive Dissolution: Overall Reaction Scheme

Reductive dissolution of transition metal oxides by organic reductants can be described as occurring in the following sequence of steps: (i) diffusion of reductant molecules to the oxide surface, (ii) surface chemical reaction, and (iii) diffusion of reaction products away from the oxide surface. In situations where transport steps (i) and (iii) are rate-controlling, fluxes of charged reactant and product species are influenced by the combined effects of the interfacial concentration gradient and electrical potential gradient arising from the net charge on the oxide surface (8). Many reductive dissolution reactions of environmental and geochemical interest are controlled by rates of surface chemical reactions, which are the focus of this discussion.

Transition metal oxide/hydroxides differ in their ability to oxidize organic compounds. Table I lists reduction potentials $E°$ (for $[H^+]=[Me^{2+}]=1.0M$) and E' (for $[H^+]=10^{-7}M$ and $[Me^{2+}]=10^{-6}M$) for several first-row transition metals.

Table I. Reduction Potentials of Selected Transition Metal Oxide/
Hydroxides

Half-Reaction	$E°$	E'	Ref.
$\frac{1}{2} \, \delta\text{-MnO}_2(s) + 2H^+ + e^- = \frac{1}{2}Mn^{2+} + H_2O$	+1.29v	+0.64	(9)
$\gamma\text{-MnOOH}(s) + 3H^+ + e^- = Mn^{2+} + 2H_2O$	+1.50	+0.61	(9)
$\alpha\text{-FeOOH}(s) + 3H^+ + e^- = Fe^{2+} + 2H_2O$	+0.67	-0.22	(1)
$CoOOH(s) + 3H^+ + e^- = Co^{2+} + 2H_2O$	+1.48	+0.59	(10)
$\frac{1}{2}Ni_3O_4(s) + 4H^+ + e^- = \frac{3}{2} Ni^{2+} + 2H_2O$	+1.98	+0.85	(11)

Based upon thermodynamic data given in Table I, oxidant strength
decreases in the order: $Ni_3O_4 > MnO_2 > MnOOH > CoOOH > FeOOH$. Rates
of reductive dissolution in natural waters and sediments appear to
follow a similar trend. When the reductant flux is increased and
conditions turn anoxic, manganese oxides are reduced and dissolved
earlier and more quickly than iron oxides (12, 13). No comparable
information is available on release of dissolved cobalt and nickel.

Surface Chemical Reaction Mechanism

Reductive dissolution of metal oxide surfaces occurs through a
series of ligand substitution and electron transfer reactions which
resemble in many respects analogous reactions in homogeneous solution. Two general mechanisms for electron transfer between metal
ion complexes and organic compounds have been developed, based upon
studies of reactions in homogeneous solution, and are reviewed by
Littler (14). Figure 1 illustrates both the inner-sphere mechanism
and outer-sphere mechanism for the reduction of a tervalent metal
ion $(Me(H_2O)_6{}^{3+})$ by phenol (HA) in homogeneous solution. Both
mechanisms have in common the formation of a precursor complex,
electron transfer within this complex, and subsequent breakdown of
the successor complex. The free radical A· produced by one-equivalent electron transfer is quickly consumed by further reactions with
metal ion oxidant complexes and by radical coupling reactions. In
the inner-sphere mechanism, the reductant enters the inner coordination sphere via ligand substitution and bonds directly to the metal
center prior to electron transfer (14). The highest electron transfer rate by the inner-sphere mechanism is limited by the rate of
ligand substitution. Reaction via an outer-sphere mechanism, in
contrast, leaves the inner coordination sphere intact; electron
transfer is facilitated by an outer-sphere precursor complex (14).
These mechanisms can operate in parallel, with the overall reaction
dominated by the fastest pathway. Substitution-inert metal ion
complexes necessarily react via an outer-sphere mechanism.

Similarly, inner-sphere and outer-sphere mechanisms can be postulated for the reductive dissolution of metal oxide surface sites,
as shown in Figure 2. Precursor complex formation, electron transfer, and breakdown of the successor complex can still be distinguished. The surface chemical reaction is unique, however, in that
participating metal centers are bound within an oxide/hydroxide

	Inner Sphere	Outer Sphere
(A) Precursor Complex Formation	$Me^{III}(H_2O)_6^{3+} + HA \underset{k_{-1}}{\overset{k_1}{\rightleftharpoons}} Me^{III}(A)(H_2O)_5^{2+} + H_3O^+$	$Me^{III}(H_2O)_6^{3+} + HA \underset{k_{-1}}{\overset{k_1}{\rightleftharpoons}} Me^{III}(H_2O)_6^{3+}, HA$
(B) Electron Transfer	$Me^{III}(A)(H_2O)_5^{2+} \underset{k_{-2}}{\overset{k_2}{\rightleftharpoons}} Me^{II}(\cdot A)(H_2O)_5^{2+}$	$Me^{III}(H_2O)_6^{3+}, HA \underset{k_{-2}}{\overset{k_2}{\rightleftharpoons}} Me^{II}(H_2O)_6^{2+}, A\cdot + H^+$
(C) Breakdown of Successor Complex	$Me^{II}(\cdot A)(H_2O)_5^{2+} + H_2O \underset{k_{-3}}{\overset{k_3}{\rightleftharpoons}} Me^{II}(H_2O)_6^{2+} + A\cdot$	$Me^{II}(H_2O)_6^{2+}, A\cdot \underset{k_{-3}}{\overset{k_3}{\rightleftharpoons}} Me^{II}(H_2O)_6^{2+} + A\cdot$

Figure 1.   Reduction of $Me(H_2O)_6^{3+}$ by phenol (HA) in homogeneous solution.

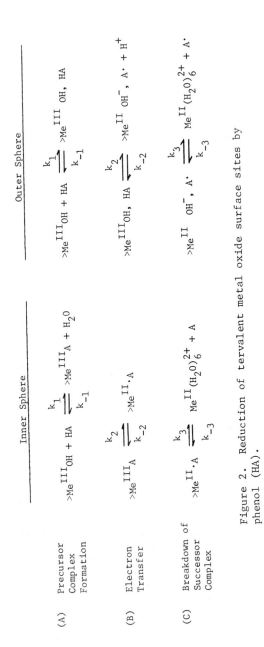

Figure 2. Reduction of tervalent metal oxide surface sites by phenol (HA).

phase; this coordinative environment may differ from that of mono-nuclear aquated species in several important respects.

The kinetic behavior of the reductive dissolution mechanisms given in Figure 2 can be found by applying the Principle of Mass Action to the elementary reaction steps. The rate expression for precursor complex formation via an inner-sphere mechanism is given by:

$$\frac{d[>Me^{III}A]}{dt} = k_1 \ [>Me^{III}OH][HA] - (k_{-1}+k_2)[>Me^{III}A] + k_{-2} \ [>Me^{II} \cdot A]$$

(6)

For the special case of steady-state precursor complex concentration (and assuming negligible back reactions for B and C) the rate of Me^{2+}(aq.) formation is given by:

$$\frac{d[Me^{2+}]}{dt} = \frac{k_1 k_2}{(k_{-1} + k_2)} \ [>Me^{III}OH][HA](1-e^{-k_3 t})$$

(7)

A similar rate expression can be written for the outer-sphere mech-anism. Based upon Equation 7, high rates of reductive dissolution are favored by (i) high rates of precursor complex formation (large k_1), (ii) low desorption rates (small k_{-1}), (iii) high electron transfer rates (large k_2), and (iv) high rates of product release (high k_3). The next section in this paper examines how chemical characteristics of metal oxide surface sites and reductant molecules influence the rate of each of these contributing reaction steps.

Description of Reductive Dissolution on a Molecular Level

Electron transfer reactions of metal ion complexes in homogeneous solution are understood in considerable detail, in part because spectroscopic methods and other techniques can be used to monitor reactant, intermediate, and product concentrations. Unfavorable characteristics of oxide/water interfaces often restrict or compli-cate the application of these techniques; as a result, fewer direct measurements have been made at oxide/water interfaces. Available evidence indicates that metal ion complexes and metal oxide surface sites share many chemical characteristics, but differ in several important respects. These similarities and differences are used in the following discussions to construct a molecular description of reductive dissolution reactions.

Coordinative Environment. The coordinative environment of transi-tion metal ions affects the thermodynamic driving force and reaction rate of ligand substitution and electron transfer reactions. $Fe^{III}OH^{2+}$(aq) and hematite (α-Fe_2O_3) surface structures are shown in Figure 3 for the sake of comparison. Within the lattice of oxide/hydroxide minerals, the inner coordination spheres of metal centers are fully occupied by a regular array of O^{2-} and/or OH^- donor groups. At the mineral surface, however, one or more coordi-native positions of each metal center are vacant (15). When oxide surfaces are introduced into aqueous solution, H_2O and OH^- molecules

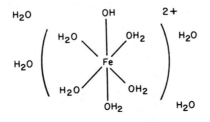

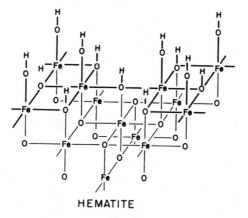

HEMATITE

Figure 3. Fe(III) coordinative environment in $Fe^{III}OH^{2+}$(aq.) and on a hematite ($\alpha-Fe_2O_3$) surface (adapted from 48).

bind to vacant coordinative positions on metal centers, and H^+ binds to oxygen donor atoms. Thus, a fully hydrated surface is covered by a layer of H_2O and OH^- molecules, which must be displaced in order for solute molecules to form complexes with surface sites. All coordinative positions of metal ion complexes in homogeneous solution are available for interactions with incoming solvent and solute molecules. At metal oxide surface sites, however, one or more coordinative positions are shielded from interaction by the underlying oxide.

Hydrolysis of metal ion complexes in homogeneous solutions of increasing pH is an important phenomenon:

$$Me(H_2O)_6^{3+} \rightleftharpoons Me(H_2O)_5(OH)^{2+} + H^+ \tag{8}$$

$$Me(H_2O)_5(OH)^{2+} \rightleftharpoons Me(H_2O)_4(OH)_2^+ + H^+ \tag{9}$$

Hydrated metal oxide surface sites (>MeOH) undergo analogous protonation/deprotonation reactions, creating a distribution of positive, negative, and neutral surface sites (16):

$$>MeOH + H^+ \rightleftharpoons >MeOH_2^+ \tag{10}$$

$$>MeOH \rightleftharpoons >MeO^- + H^+ \tag{11}$$

An excess of either $>MeOH_2^+$ or $>MeO^-$ imparts a net charge to the surface. This net electrical charge creates a potential gradient in the interfacial region which, in turn, disturbs the distribution of charged species near the interface (17). Near a negatively charged oxide surface, cation concentrations are substantially higher than bulk solution concentrations, and anion concentrations are substantially lower. The dielectric constant and other medium characteristics are also perturbed near charged interfaces (18). Electrostatics, therefore, has a substantial impact on reactions of metal oxide surface sites.

Precursor Complex Formation. Inner-sphere precursor complex formation and breakdown of the successor complex are ligand substitution reactions, and are therefore limited by the lability, or ligand exchange rate, of the metal center (6). Dissolved complexes and oxide surface sites of different transition metals have different labilities towards ligand substitution, reflecting differences in metal-ligand bond strengths. Water exchange rates of hexaquo complexes in homogeneous solution are one possible measure of relative lability, which show substantial variations. Characteristic times for water exchange at $Cr(H_2O)_6^{3+}$ (2.0×10^6 seconds (19)) and at $Fe(H_2O)_6^{3+}$ (6.3×10^{-3} seconds (20)), for example, differ by eight orders of magnitude. To a first approximation, ligand exchange at Cr(III) oxide surface sites can be expected to occur more slowly than exchange at Fe(III) oxide surface sites. Substantially slower specific adsorption of phosphate onto α-Cr_2O_3 has, in fact, been observed (21). Unfortunately, the number of reliable measurements of ligand exchange rates at metal oxides surfaces is small (22-25) and so there are few opportunities for direct comparison.

If the reductant is a metal ion, adsorption may also precede electron transfer:

$$>Me^{III}OH + Me'(H_2O)_6^{2+} \underset{k_{-1}}{\overset{k_1}{\rightleftharpoons}} >Me^{III}OMe'(H_2O)_5^+ + H_3O^+ \tag{12}$$

In this case, precursor complex formation depends upon the lability of the incoming metal ion, rather than that of the oxide surface site, since the inner coordination ligands of the surface site are not exchanged (26).

Rates of ligand exchange depend quite strongly on the coordinative environment of the metal center. The water exchange rate of $Fe(H_2O)_5(OH)^{2+}$ is almost three orders of magnitude higher than that of $Fe(H_2O)_6^{3+}$, and follows a dissociative, rather than an associative exchange mechanism (20). $Fe(H_2O)_5(OH)^{2+}$ has also been shown to form inner-sphere complexes with phenols (27), catechols (28), and α-hydroxycarboxylic acids (29) much more quickly than $Fe(H_2O)_6^{3+}$. The mechanism for complex formation with phenolate anion (A^-) is shown below (27):

$$Fe(H_2O)_6^{3+} \overset{fast}{\rightleftharpoons} Fe(H_2O)_5OH^{2+} + H^+ \qquad K_{OH} = 2.75 \times 10^{-3} M \tag{13}$$

$$Fe(H_2O)_6^{3+} + A^- \underset{k_{-a}}{\overset{k_a}{\rightleftharpoons}} Fe(H_2O)_5A^{2+} + H_2O \tag{14}$$

$$Fe(H_2O)_5OH^{2+} + A^- \underset{k_{-b}}{\overset{k_b}{\rightleftharpoons}} Fe(H_2O)_4(OH)A^+ + H_2O \tag{15}$$

$$Fe(H_2O)_4(OH)A^+ + H^+ \overset{fast}{\rightleftharpoons} Fe(H_2O)_5A^{2+} \tag{16}$$

The contribution of reaction 14 relative to reaction 15 is insignificant. Rate constants for a variety of substituted phenols are quite similar, indicating that a dissociative ligand exchange mechanism is operative (27).

For reactions involving metal oxide surface sites, the situation is considerably more complex. Note in reaction 15 that H_2O, rather than OH^-, is the leaving ligand. In the surface ligand exchange reaction, only outfacing coordinated ligands are available for exchange, since other coordinated ligands are shielded by surrounding oxide. Thus, rates of ligand substitution at $>MeOH_2^+$, $>MeOH$, and $>MeO^-$ sites depend upon the ability of H_2O, OH^-, and 0^{2-} to act as leaving ligands. Metal-ligand bond lengths decrease in the order $H_2O>OH^->0^{2-}$, and therefore the ability to act as a leaving group may decrease following the same trend.

Oxide composition and lattice structure influences the coordinative environment of surface sites, and should have an impact on rates of ligand substitution. Hematite (Fe_2O_3), goethite (α-FeOOH), and lepidocrocite (γ-FeOOH), for example, are all Fe(III) oxide/hydroxides, but may exhibit different rates of surface chemical reactions because of variations in surface site bond lengths and geometry.

An additional complicating factor is the nonuniformity of surface sites, discussed at length by Benjamin and Leckie (30). Lattice imperfections, surface inhomogeneities, and neighboring group effects broaden the distribution of site energies, such that some surface sites have higher affinities for adsorbate molecules than for others. Most likely, rates of surface complex formation are influenced by these differences in site energies. The Elovich Equation has been used to model rates of adsorption on non-uniform surfaces (31,32). It is based on the assumption that activation energies for adsorption increase linearly with increasing surface coverage.

Surface speciation can be expected to have a tremendous impact on rates of precursor complex formation. R_1, the rate of precursor complex formation, may depend upon the extent of surface protonation, since ligand exchange rates of $>MeOH_2^+$, $>MeOH$, and $>MeO^-$ may vary substantially:

$$R_1 = k_1[HA][>MeOH_2^+] + k_1'[HA][>MeOH] + k_1''[HA][>MeO^-] \qquad (17)$$

Additional surface species form when specifically-adsorbing ions such as calcium and phosphate are added: $>MeOCa^+$, $>MePO_4H_2$, $>MePO_4H^-$, $>MePO_4^{2-}$, $>Me_2PO_4H$, and $>Me_2PO_4^-$. Non-labile surface groups are not available for reaction, while other, slowly-exchanging sites can gradually be replaced by incoming reductant molecules.

Experiments examining the influence of calcium and phosphate on the reductive dissolution of manganese oxides by hydroquinone have, in fact, shown inhibition by adsorbed ions (33). As the total phosphate in solution is increased, the rate of Mn^{2+} release diminished in proportion to the phosphate surface coverage.

Surface Coverage and Reaction Rate. If precursor complex formation is fast relative to electron transfer and product release, it can be treated as a quasi-equilibrium step:

$$>Me^{III}OH + HA \rightleftharpoons >Me^{III}A + H_2O \qquad (18)$$

$$K_s = k_1/k_{-1} = \frac{[>Me^{III}A]}{[>Me^{III}OH][HA]} \qquad (19)$$

Production of reduced surface sites is directly proportional to $[>Me^{III}A]$. Thus, the distribution of reduced surface sites and rates of subsequent metal ion release both increase as reductant surface coverage is increased.

Relatively little is known about the speciation of organic compounds, and organic reductants in particular, when adsorbed to metal oxides. It is known that surface coverage is higher for bidentate organic ligands, such as catechol and salicylate, than for monodentate ligands, such as benzoate and phenol (34, 35). It has been shown that surface coverage generally increases as the pH is lowered, and that stability constants for surface complexes and analogous metal ion complexes in homogeneous solution coordinate with one another (35).

Adsorbed organic compounds do not necessarily occupy surface sites in a random fashion. Favorable hydrophobic interactions, for example, may cause some organic compounds to adsorb in groups, enhancing overall surface coverage. Electrostatic repulsion between charged adsorbate molecules, in contrast, may space molecules on the surface apart from one another.

When free radicals or other reactive organic species are generated in the electron transfer step, the pattern of surface coverage may have a dramatic impact on the yields of various possible organic oxidation products. Consider the oxidation of mercaptans (RSH) by a metal oxide surface. One-equivalent oxidation forms mercapto radicals (RS·) which quickly couple with neighboring radicals to form a disulfide product (36).

$$RSH + Me^{III} \rightarrow RS\cdot + H^+ + Me^{II} \tag{20}$$

$$RS\cdot + \cdot SR \rightarrow RSSR \tag{21}$$

When two or more mercaptans are simultaneously oxidized, a mixture of several disulfide products can be formed. Oxidation of mercaptoacetate and n-butyl mercaptan (BSH) yields three possible disulfides: ASSA, ASSB, and BSSB.

The Electron Transfer Step. Inner-sphere and outer-sphere mechanisms of reductive dissolution are, in practice, difficult to distinguish. Rates of ligand substitution at tervalent and tetravalent metal oxide surface sites, which could be used to estimate upward limits on rates of inner-sphere reaction, are not known to any level of certainty.

The most direct evidence for surface precursor complex formation prior to electron transfer comes from a study of photoreductive dissolution of iron oxide particles by citrate (37). Citrate adsorbs to iron oxide surface sites under dark conditions, but reduces surface sites at an appreciable rate only under illumination. Thus, citrate surface coverage can be measured in the dark, then correlated with rates of reductive dissolution under illumination. Results show that initial dissolution rates are directly related to the amount of surface bound citrate (37). Adsorption of calcium and phosphate has been found to inhibit reductive dissolution of manganese oxide by hydroquinone (33). The most likely explanation is that adsorbed calcium or phosphate molecules block inner-sphere complex formation between metal oxide surface sites and hydroquinone.

The stoichiometry of surface complexes as well as the total surface coverage is important in determining electron transfer rate.

Consider, for example, an organic reductant that forms protonated (>MeAH), neutral (>MeA) and outer-sphere (>MeOH, HA) surface complexes. R_2, the rate of electron transfer, is determined by how quickly electron transfer occurs within each type of complex.

$$R_2 = k_2[>Me^{III}AH^+] + k_2'[>Me^{III}A] + k_2''[>Me^{III}OH, HA] \qquad (22)$$

The pH dependence of the reaction is determined, in part, by the relative magnitude of k_2, k_2', and k_2''.

Few studies have systematically examined how chemical characteristics of organic reductants influence rates of reductive dissolution. Oxidation of aliphatic alcohols and amines by iron, cobalt, and nickel oxide-coated electrodes was examined by Fleischman et al. (38). Experiments revealed that reductant molecules adsorb to the oxide surface, and that electron transfer within the surface complex is the rate-limiting step. It was also found that (i) amines are oxidized more quickly than corresponding alcohols, (ii) primary alcohols and amines are oxidized more quickly than secondary and tertiary analogs, and (iii) increased chain length and branching inhibit the reaction (38). The three different transition metal oxide surfaces exhibited different behavior as well. Rates of amine oxidation by the oxides considered decreased in the order Ni > Co > Fe; the thermodynamic driving force for reaction also decreases in the same order.

More recently, reductive dissolution of Mn(III,IV) oxide suspensions by 27 aromatic and nonaromatic compounds was examined by Stone and Morgan (39). At pH 7.2, saturated alcohols, aldehydes, ketones, and carboxylic acids showed no reactivity, except for pyruvic and oxalic acids. Catechols, hydroquinones, methoxyphenols, and resorcinols, as well as ascorbate, dissolved manganese oxides at appreciable rates. Rates of precursor complex formation and intramolecular electron transfer could not be distinguished, since only overall rates of reductive dissolution were measured. It was observed, however, that electron-withdrawing substituents on aromatic substrates lower the reaction rate with manganese oxides, while electron-donating groups increase the reaction rate (39). Recent, unpublished research on reductive dissolution by substituted phenols supports these findings.

Dissolution of Reduced Surface Sites. Following electron transfer, reduced metal ions are released into overlying solution. This process is similar in many respects to dissolution reactions not involving changes in oxidation state. This latter class of reactions has been extensively studied (5, 7). The rate of metal ion detachment, replacing lattice bonds with bonds to solvent molecules, is believed to be the rate-limiting step. Release via progressive cleavage of metal-lattice bonds has been postulated by Valverde and Wagner (5). Protonation of surface sites weakens Me-O lattice bonds, enhancing detachment (7). Fractional orders with respect to [H+] are commonly observed in detachment-controlled dissolution reactions. To explain this observation, it has been postulated that a required number (n) of neighboring surface sites must be protonated before detachment can occur (7).

$$R_{\text{dissolution}} = k_H \{>MeOH_2^+\}^n \tag{23}$$

Specific adsorption of ligands can enhance or inhibit dissolution rates by altering the strength and lability of Me-O lattice bonds. Salicylate, oxalate, and citrate promote the dissolution of alumina (40). In the presence of ligand (L) the dissolution rate becomes (7):

$$R_{\text{dissolution}} = k_H \{>MeOH_2^+\}^n + k_L \{>MeL\} \tag{24}$$

Organic ligands without redox reactivity that coordinate metal oxide surface sites have been found to enhance rates of both reductive and non-reductive dissolution reactions (7).

Reduction lowers the charge to radius ratio of transition metal ions, promoting higher rates of ligand substitution. Reduced, divalent oxidation states of manganese, iron, cobalt, and nickel are also quite soluble (Table II).

Table II. Solubility Product Constants for Divalent Metal Ion Hydroxides (41)

Reaction	logK	$[Me^{2+}]_{sat.}$ at pH 7
$Mn(OH)_2(s) + 2H^+ = Mn^{2+} + 2H_2O$	15.2	15.9 M
$Fe(OH)_2(s) + 2H^+ = Fe^{2+} + 2H_2O$	12.9	7.9×10^{-2}
$Co(OH)_2(s) + 2H^+ = Co^{2+} + 2H_2O$	12.3	2.0×10^{-2}
$Ni(OH)_2(s) + 2H^+ = Ni^{2+} + 2H_2O$	10.8	6.3×10^{-4}

The high solubility and lability of reduced metal centers should make metal ion release fast relative to preceding steps in many reductive dissolution reactions.

Effect of Oxide Mineralogy on Reductive Dissolution. Oxide/hydroxide surface structures and the coordinative environment of metal centers may change substantially throughout the course of a reductive dissolution reaction. Nonstoichiometric and mixed oxidation state surfaces produced during surface redox reactions may exhibit dissolution behavior that is quite different from that observed with more uniform oxide and hydroxide minerals.

Oxide mineralogy may influence rates of reductive dissolution in several ways. Hematite (α-Fe_2O_3) and maghemite (γ-Fe_2O_3), for example, have the same stoichiometry but contain Fe(III) in quite different coordinative environments. Fe(III) in hematite occupies trigonally-distorted octahedral sites, while Fe(III) in maghemite is found in both octahedral and tetrahedral sites (42). Differences

in coordinative geometry and Me-O bond lengths may influence rates
of (i) precursor complex formation, (ii) surface electron transfer,
and (iii) rates of reduced metal ion release from the oxide lattice.

Few comparative studies have been made on the reductive dissolution of different mineral phases. In one such study, the order
of reaction with seven organic and transition metal reductants was
found to be the same: hematite (α-Fe_2O_3)>magnetite (Fe_3O_4)>nickel
ferrite ($NiFe_2O_4$) (43). Magnetite is an interesting case, since
both Fe(III) and Fe(II) are present in the lattice prior to reaction. Evidence indicates that Fe(III) sites reduced to Fe(II) sites
by redox reaction dissolve more quickly than Fe(II) sites originally
present in the mineral lattice (6).

Semiconducting properties of transition metal oxides are also
potentially important in reductive dissolution reactions. Adsorption of thermal or photochemical energy exceeding the band gap
energy makes crystalline oxides conductive; facile electron transfer
from one metal center to another occurs (49). Band gap energies
range between 0.26 and 0.7 eV for manganese oxides, and between 1.1
and 2.3 eV for iron oxides (44). The average vibrational energy
of molecules is approximately equal to kT, which has a value of
0.026 eV at 25°C (45). Band gap energies of manganese oxides are
sufficiently low that thermal excitation of electrons into the
conduction band may occur to a significant extent at room temperature. The degree of crystallinity may determine whether or not
semiconducting properties are significant. In poorly ordered oxides,
electron transfer may be restricted by gaps and imperfections that
reduce the amount of orbital overlap between adjacent metal centers
(M. Fox, pers. commun.).

Electron transfer between metal centers can alter the course
of reaction in several ways (46). Thermal excitation may create
especially reactive electron holes on the oxide surface, causing
reductant molecules to be consumed at the surface at a higher rate.
More importantly, electrons deposited on surface sites by organic
reductants may be transferred to metal centers within the bulk
oxide (47). This returns the surface site to its original oxidation
state, allowing further reaction with reductant molecules to occur
without release of reduced metal ions. Electron transfer between
metal centers may therefore cause changes in bulk oxide composition
and delay the onset of dissolution.

Conclusions

Reductive dissolution occurs via (i) surface precursor complex
formation between reductant molecules and oxide surface sites, (ii)
electron transfer within this surface complex, and (iii) breakdown
of the successor complex and release of dissolved metal ions. Surface speciation is important in determining rates of each of these
contributing steps. Limited available evidence concerning rates
and mechanism of surface chemical reactions and analogy to similar
reactions in homogeneous solution both support this conclusion.
Rate expressions for homogeneous reactions are written as functions

of species concentrations, determined by spectroscopic or alternative methods. Determination of surface speciation in heterogeneous
reactions is considerably more difficult, and therefore rate expressions are often written as functions of composite terms (such as
total surface coverage) or as functions of overlying solution composition. Comprehensive understanding of reductive dissolution
reactions requires that much more be learned about surface speciation and its impact on rate.

Acknowledgments

Support for this research was provided by National Science Foundation Grant CEE-04076. The useful input by James J. Morgan and
Marye A. Fox is acknowledged.

Literature Cited

1. Stumm, W.; Morgan, J.J. "Aquatic Chemistry," 2nd ed.; Wiley-
    Interscience: New York, 1981.
2. Mentasti, E.; Pelizzetti, E.; Saini, G. J. Chem. Soc. Dalton
    Trans. 1973, 2609.
3. Diggle, J.W. In "Oxides and Oxide Films"; Diggle, J.W. Ed.;
    Marcel Dekker: New York; 1973; Vol. 2, pp. 281-386.
4. Gorichev, I.G.; Kipriyanov, N.A. Uspekhi Khimii 1984, 53, 1790.
5. Valverde, N.; Wagner, C. Ber. Bunsenges. Phys. Chem. 1976, 80,
    330.
6. Bruyere, V.I.; Blesa, M.A. J. Electroanal. Interfacial Chem.
    1985, 182, 141.
7. Stumm, W.; Furrer, G.; Wieland, E.; Zinder, B. In "The Chemistry
    of Weathering"; Drever, J. Ed.; NATO ASI Series. Series C.
    Vol. 149; 1985; pp. 51-74.
8. Dukhin, S.S. In "Surface and Colloid Science," Matijevic, E. Ed.;
    Wiley-Interscience: New York, 1974; Vol. 7.
9. Bricker, O. Am. Mineral. 1965, 50, 1296.
10. Hem, J.D.; Roberson, C.E.; Lind, C.J. Geochim. et Cosmochim.
    Acta 1985, 49, 801.
11. Pourbaix, M. "Atlas of Electrochemical Equilibria"; Pergamon:
    Oxford, 1966.
12. Balzer, W. Geochim. et Cosmochim. Acta 1982, 46, 1153.
13. Froelich, P.N.; Klinkhammer, G.P.; Bender, M.L.; Luedke, N.A.;
    Heath, G.R.; Cullen, C.; Dauphin, P.; Hammond, D.; Hartmann, B.;
    Maynard, V. Geochim. et Cosmochim. Acta 1979, 43, 1075.
14. Littler, J.S. Spec. Publ. - Chem. Soc. 1970, No. 24.
15. Schindler, P.W. In: "Adsorption of Inorganics at Solid-Liquid
    Interfaces"; Anderson, M. Ed.; Ann Arbor Science: Ann Arbor,
    1981; pp. 1-49.
16. James, R.O.; Healy, T.W. J. Colloid Interface Sci. 1972, 40,
    42.
17. Westall, J.; Hohl, H. Adv. Colloid Interface Sci. 1980, 12, 265.
18. Nurnberg, H.W. In "Membrane Transport in Plants"; Zimmerman, V.;
    Dainty, J. Eds.; Springer-Verlag: New York, 1974.
19. Burgess, J. "Metal Ions in Solution"; Ellis Horword: Sussex,
    England, 1978.
20. Grant, M.; Jordan, R.B. Inorg. Chem. 1981, 20, 55.

21. Yates, D.E.; Healy, T.W. J. Colloid Interface Sci. 1975, 52, 222.
22. Hiraishi, M.; Harada, S.; Uchida, Y.; Kuo, H.L.; Yasanuga, T. Int. J. Chem. Kin. 1982, 12, 387.
23. Hachiya, K.; Ashida, M.; Sasaki, M.; Karasuda, M.; Yasunaga, T. J. Phys. Chem. 1980, 84, 2292.
24. Ikeda, T.; Sasaki, M.; Hachiya, K.; Astumian, R.; Yasanuga, T.; Schelly, Z. J. Phys. Chem. 1982, 86, 3861.
25. Mikami, N.; Sasaki, M.; Hachiya, K.; Astumian, R.; Ikeda, T.; Yasanuga, T. J. Phys. Chem. 1983, 87, 1454.
26. Hachiya, K.; Susaki, M.; Ikeda, T.; Mikami, N.; Yasanuga, T. J. Phys. Chem. 1984, 88, 27.
27. Cavasino, F.P.; DiDio, E. J. Chem. Soc. A 1970, 1151.
28. Mentasti, E.; Pelizzetti, E. J. Chem. Soc. Dalton Trans. 1973, 2605.
29. Mentasti, E. Inorg. Chem. 1979, 18, 1512.
30. Benjamin, M.M.; Leckie, J.O. J. Colloid Interface Sci. 1981, 79, 209.
31. Hayward, D.O.; Trapnell, B.M.W. "Chemisorption"; Butterworths: London, 1964.
32. Hingston, F.J. In "Adsorption of Inorganics at Solid-Liquid Interfaces"; Anderson, M. Ed.; Ann Arbor Science: Ann Arbor, 1981; pp. 51-90.
33. Stone, A.T.; Morgan, J.J. Environ. Sci. Technol. 1984, 18, 450.
34. Davis, J.A.; Leckie, J.O. Environ. Sci. Technol. 1978, 12, 1309.
35. Kummert, R.; Stumm, W. J. Colloid Interface Sci. 1980, 75, 373.
36. Panpalia, S.S.; Mehrota, R.N.; Kapoor, R.C. Indian J. Chem. 1974, 12, 1166.
37. Waite, T.D.; Morel, F.M.M. J. Colloid Interface Sci. 1984, 102, 121.
38. Fleischman, M.; Korinek, K.; Pletcher, D. J. Chem. Soc. Perkin II 1972, 1396.
39. Stone, A.T.; Morgan, J.J. Environ. Sci. Technol. 1984, 18, 617.
40. Furrer, G.; Stumm, W. Chimia 1983, 37, 338.
41. Morel, F.M.M. "Principles of Aquatic Chemistry"; Wiley: New York, 1983.
42. Sherman, D.M. Phys. Chem. Minerals 1985, 12, 161.
43. Segal, M.G.; Sellers, R.M. J. Chem. Soc. Chem. Commun. 1980, 991.
44. Strehlow, W.H.; Cook. E.L. J. Phys. Chem. Ref. Data 1973, 2, 163.
45. Bard, A.J.; Faulkner, L.R. "Electrochemical Methods"; Wiley: New York, 1980.
46. Morrison, S.R. "Electrochemistry at Semiconductor and Oxidized Metal Electrodes," Plenum: New York, 1980.
47. Zabin, B.A.; Taube, H. Inorg. Chem. 1964, 3, 963.
48. Kawakami, H.; Yoshida, S. J. Chem. Soc., Farad. Trans. 2 1985, 81, 1117.

RECEIVED June 18, 1986

22

Abiotic Organic Reactions at Mineral Surfaces

Evangelos A. Voudrias and Martin Reinhard

Environmental Engineering and Science Group, Department of Civil Engineering, Stanford University, Stanford, CA 94305-4020

Abiotic organic reactions, such as hydrolysis, elimination, substitution, redox, and polymerization reactions,
can be influenced by surfaces of clay and primary minerals, and of metal oxides. This influence is due to
adsorption of the reactants to surface Lewis and Brønsted
sites. Temperature and moisture content are the most
important environmental variables. Under ambient environmental temperatures, some reactions are extremely
slow. However, even extremely slow transformation reactions may be important from environmental and geochemical
viewpoints.

Abiotic organic reactions that may be influenced by mineral surfaces
include hydrolysis, elimination, substitution, redox, and polymerization. The effect of the surface may be either to promote
(increase the rate of) or to inhibit (decrease the rate of) reactions that may occur in homogeneous solution. In addition, mineral
surfaces may promote reactions that do not occur in homogenous solution by selectively concentrating molecules at the mineral surface
(1), by stabilizing intermediates (2), and by activating components
that would not otherwise be reactive (3).

Generally, the reactions of interest involve a substrate A at
the surface site S to form product B (Equation 1):

$$A + S \longrightarrow B \qquad (1)$$

If S is consumed, the reaction will proceed until either S or A is
depleted. However, when S is regenerated (i.e., S is acting as a
catalyst), the reaction will proceed until A is consumed. In the
latter case, the surface reaction may be viewed as a sequence of the
following elementary processes: (1) reactant A adsorbs to site S,
(2) A forms the surface precursor complex (A*S) which then forms the
successor complex (B*S) after overcoming the activation energy of
the transition state denoted (A*S)#, and (3) B desorbs from the surface (4):

$$A + S \rightleftarrows A{\ldots}S \qquad (2)$$

$$A{\ldots}S \rightleftarrows A{*}S \qquad (3)$$

$$A{*}S \rightleftarrows (A{*}S){\#} \rightleftarrows B{*}S \qquad (4)$$

$$B{*}S \rightleftarrows B + S \qquad (5)$$

0097–6156/86/0323–0462$07.25/0
© 1986 American Chemical Society

This scheme disregards mass transfer limitations and represents only a simplified model. Formation of A*S may involve specific interactions, such as hydrogen bonds, coordination, or π-complex formation, or non-specific interactions, such as van der Waals or hydrophobic bonds. Non-specific interactions are insignificant for small polar molecules, but may contribute significantly to the surface complex formation if the hydrophobic moiety is large (5, 6).

Depending upon the relative rates of reactions 2 to 5, several different limiting cases can be evaluated. For example, (1) if A is not sorbing, no precursor surface complex will be formed and transformation will not occur; (2) if surface complexation does not lower the activation energy of the rate-controlling step of the intermediate formation, no rate acceleration relative to the bulk solution case will occur; (3) if precursor or successor surface complexes are stabilized too strongly, product formation will be slowed by a high energy barrier; and (4) if B is sorbed too strongly, the reaction sites are blocked and the reaction cannot proceed at appreciable rates. Sabbatier's rule of catalysis states that "a catalytic effect is observed if the potential energy of the reaction intermediate is stable enough that it is formed readily, but not too stable so that it can be decomposed easily" (7). Thus, catalysis represents a special case for which restrictive conditions apply: moderately sorbing substrates form moderately stabilized activated surface complexes, which in turn react to easily desorbing products.

Obtaining mechanistic information at the level of detail described by Equations 2 to 5 for reactions catalyzed by mineral surfaces is quite difficult for several reasons: (1) characterization and quantification of active surface sites is complicated because the sediment surface is difficult to characterize; (2) many substrates and products with a range of properties may be competing for reaction sites; (3) several products may be formed from a single substrate; (4) products may be formed that are difficult to recover from the mineral matrix; (5) unknown active phases or components may be present in the sediment; and (6) the reactions may be mass transfer limited. Thus, mechanistic investigations of organic reactions in sediment systems have been mostly restricted to relatively well-defined systems, such as homoionic clays, which are accessible to spectroscopic investigation (review by Theng (8) and recent works by Soma et al. (9-12)), purified oxides, where mass limitations are insignificant (13-18), or primary minerals (19).

The interest in mineral-promoted organic reactions stems from the need to understand the fate of pesticides in soils and pollutants in sedimentary environments (8), petrogenesis (20-27), humification (19, 28, 29), the origin and evolution of life (1, 30), the use of clays as catalysts in industrial processes (31-37), in pharmaceutical applications (3), and as pigments and fillers in paper, plastic, and rubber (37).

The mineral surface may be considered as a solid source of Lewis and/or Brønsted acidity and the reactive sites S as localized acidic or basic functional groups. Reactions involving such sites may be understood in terms of Lewis acid/base or Brønsted acid/base interactions (1, 5, 6, 8, 38). As the acidity of the reactive sites increases, increasingly weak bases are neutralized and reactive surface complexes (A*S) may be formed. The term "acidity" is often used in the broad sense of the word, including both Brønsted and

Lewis acidity. Brønsted acidity refers to the ability of the mineral surface to donate a proton and Lewis acidity to the ability to accept electrons, i.e., to act as an oxidizing agent (39). We will discuss examples of heterogeneous processes involving mineral surfaces and organic or inorganic species. The review has been restricted to systems which can be related to sedimentary conditions or which have mechanistic significance.

Reactions Promoted by the Lewis and Brønsted Acidity of Clay Minerals

Clay Minerals as Lewis Acids. Lewis acid sites in a clay mineral are exchangeable (2) or structural (40) transition metal cations in the higher valence state, such as Fe^{3+} and Cu^{2+}, and octahedrally coordinated aluminum exposed at the crystal edges (38). Reduction of both exchanged and structural (octahedral) transition metal cations in the upper oxidation state is a reversible process (12, 41). Thus, transition metal cations in the lower valence state may also act as Lewis bases. Factors that affect the reactions promoted by Lewis acidity are listed in Table I. Lewis acid sites reversibly adsorb water (6, 9, 42), which may thus strongly compete with organic compounds that have weaker Lewis base properties, such as aromatic hydrocarbons. Lewis acidity depends on the degree of hydration and is strongest under desiccating conditions. Examples of reactions that are promoted by Lewis acidity are summarized in Table II. Other examples have been reviewed by Solomon and Howthorne (37).

Charge Transfer Complex Formation and Oxidation Reactions. Insight into heterogeneous reaction mechanisms involving Lewis acidity has been obtained by studying the formation of aromatic charge transfer complexes with various smectites using a combination of spectroscopic techniques, including electron spin resonance (ESR), infrared (IR), ultraviolet/visible absorption (UV/VIS) (9, 30, 42-48) and, most recently, resonance Raman spectroscopy (9-12).
      Generally, the potential for charge transfer becomes more favorable when the aromatic ring has electron donating substituents (e.g., OH, OMe, alkyl), the moisture content decreases, and the valence state and the reduction potential of the transition metal cation is high. ESR and IR spectra have indicated that interlamellar complexes formed between aromatic molecules and Cu^{2+}-montmorillonite may be of two distinct types: type I, a π-complex in which the aromatic ring remains intact (i.e., planar), and type II in which the aromatic ring is distorted and the aromaticity is disturbed (44, 45, 47). Recent work (10-12) has demonstrated that both type I and type II complexes may be monomers, dimers, or polymers, depending on the substitution of the parent compound.
      Soma et al. (9) studied the interaction between a Ru^{3+}- and Cu^{2+}-exchanged montmorillonite and p-dimethoxybenzene (DMBO) under desiccating conditions at room temperature. On the basis of Raman, IR, and VIS absorption spectra, they demonstrated that DMOB is stably adsorbed on Cu^{2+} and Ru^{3+}-exchanged montmorillonites as a radical cation. Oxidation by the metal cation was found to be reversible in the presence of water vapor:

Table I.   Parameters Controlling Reactions Promoted by Lewis Sites
            of Clay Minerals

Type of Lewis Site

    1.Lewis acid sites:
- edge sites of metal ions with unsatisfied coordination
- transition metal ions in the upper oxidation state exchanged in
the interlayer region or incorporated into the clay structure

    2.Lewis base sites:
- ditrigonal cavities in the interlayer region
- transition metal cations in the lower oxidation state exchanged
in the interlayer region or incorporated into the clay structure.

Clay Properties
- charge density
- clay structure
- interlayer spacing, swelling
- redox potential
- polarization power

Substrate Properties
- size
- ionization potential
- stability of radical, radical cation intermediates
- Lewis basicity/acidity
- complexation ability
- redox potential of substrate

Reaction Conditions
- moisture content
- temperature
- oxygen
- organic co-solutes
- humic substances

Table II.   Organic Reactions Affected by Surface Lewis-Acidity of
Clay Minerals

Reaction	Examples	References
1. Charge Transfer Complexation of		
N-heterocycles	riboflavine	(30)
2. Formation of Monomeric Cations from		
p-substituted ben- zenes	p-dimethoxybenzene	(55)
4,4'-substituted biphenyls	4,4'-dimethoxybiphenyl	(12)
3. Dimerization of		
monosubstituted ben- zenes, symm.	toluene, mesitylene, xylenes phenol with Cu(II)-S	(44)
arenes	biphenyl, naphthalenene, anthracene with Cu(II)-M	(46)
aromatic ethers	anisole with Cu(II)-H, butylphenyl ether	(45)
4. Oligo- and Polymerization of		
aromatic hydro- carbons	benzene, toluene	(42)
phenols	phenol, phenol mixtures, phenol-amino acid mixtures	(49, 55, 61, 29, 14)
symm. arenes	benzene, biphenyl, anthracene	(46)
aromatic ethers with Cu(II)-H	anisole, butylphenyl ether	(45, 55)
aromatic amines	4,4'-diaminobiphenyl (benzidine) by M aniline by Fe(III), Cu(II)-M	(43) (2, 40, 51) (28, 64)
5. Ligand Exchange Reactions		
aromatic amines	H_2O against aniline in Cu(II)-H	(64)
6. Hydrogen Exchange of		
alkyl hydrogens	^{3}H-cumene detritiation	(24)
7. Redox Disproportionation of		
alkenes	limonene to p-menthene and p-cumene	(56, 90)
8. Oxidation of		
hydroquinone	by S to benzoquinone	(54)
phenols	phenol with Fe(III), Cu(II)-S	(49)
	anisole by Cu(II)-S	(45)
	2,6-dimethylphenol	(55)
	methyl and Cl-phenol by S	(57)

S: smectite; M: montmorillonite; H: hectorite.

$$Cu^{2+}(H_2O)_n[clay] + DMOB \rightleftarrows Cu^+(H_2O)_{n-m}[clay] + DMOB^+ + mH_2O \quad (6)$$

In subsequent work, Soma et al. (12) studied the reactions of 4,4'-dimethoxybiphenyl (4,4'-DMOBP), anisole, toluene, chloro- and fluorobenzene, and phenol with Cu^{2+}-, Fe^{3+}-, and Ru^{3+}-exchanged montmorillonite under dessicating conditions and resonance Raman, ESR, and VIS absorption spectroscopy. They found that 4,4'-DMOBP forms two types of complexes, both corresponding to 4,4'-DMOBP radical cation. One complex is susceptible to attack by water and is readily reduced to 4,4'-DMBP. The other interacts strongly with its surroundings (e.g., 4,4'-DMOBP molecule, exchange cation, and silicate layer) which protects it against reaction with water.

In the presence of the metal-exchanged montmorillonite, anisole is converted to its radical cation form, which reacts further with neutral anisole to form the 4,4'-DMOBP. Formation of the latter appears quantitative in a clay where only one-eighth of the exchangeable cations are Cu^{2+} and the remaining are Na^+ ions. However, in fully exchanged Cu^{2+}-montmorillonite, a significant fraction of anisole cations are stable, perhaps because the high density of Cu^{2+}-anisole complexes restricts access of neutral anisole molecules (12). Phenol reacts to 4,4'-dihydroxybiphenyl (12, 49), but other, not yet identified, reactions appear to occur in parallel. Toluene was suspected to react to a methylphenylphenylmethane and was the only monosubstituted benzene investigated that does not react to a biphenyl-type product (12).

Another well-studied electron transfer reaction is the oxidation of aqueous benzidine in the presence of various clays (2, 40, 43, 50, 51). An electron from the colorless benzidine molecule is abstracted by the clay with formation of a blue monovalent radical cation. Upon drying of the blue clay-benzidine system, a yellow color is produced. There is disagreement in the literature with respect to the chemical identity of the yellow product (2, 40, 52); however, in the case of hectorite, the yellow product has been suggested to be the protonated form of the radical cation (divalent radical cation) (2, 52). There is also disagreement about whether the electron-accepting sites of the clay are ferric iron at the planar surfaces, aluminum ions at the edges, or exchangeable cations (2, 8).

In the case of hectorite, a fast faint-blue coloring reaction and a slow intense-blue coloring reaction were observed (2). The fast faint-blue coloring reaction, which occurred also in a N_2 atmosphere, was attributed to oxidation of extremely small quantities of benzidine by structural Fe(III) (2). The intense-blue coloring reaction was attributed to oxidation by O_2, because it did not occur under N_2. The much slower rate of this reaction was explained by slow diffusion of O_2. Color formation was prevented by treatment of the clay with hydrazine for reduction of Fe(III) to Fe(II), and by treatment of the clay with polyphosphate for deactivation of the crystal edges (43). However, McBride (2) interpreted the action of the polyphosphate differently. He suggested that inhibition of the reaction with O_2 is due to the increased pH of the solution as opposed to blocking of the Al ions at the edges. Oxidation of benzidine by electron transfer to exchangeable cations such as Cu^{2+} and Fe^{3+} is also possible but, interestingly, different forms of surface-adsorbed Fe(III) impurities were found inactive (2).

The role of structural Fe(III) in reactions with benzidine was also demonstrated by Mossbauer spectroscopy at very high adsorption levels (i.e., intercalation conditions) on montmorillonite ($\underline{40}$). Upon dehydration of the clay, a yellow color and regeneration of Fe^{2+} to Fe^{3+} was observed. This was explained by the equilibria

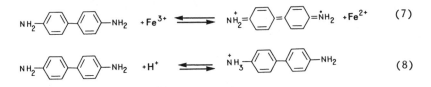

$$(7)$$

$$(8)$$

The source of H^+ is the dissociation of coordinated water (see section on Brønsted acid). The protonation of benzidine (Equation 8) shifts the oxidation reaction (Equation 7) to the left.

The presence or absence of oxygen is a critical factor in a number of other organic reactions. When the clays were saturated with alkali or alkaline earth cations, oxidation of aniline in the presence of smectites ($\underline{50}$) and dimethylaniline in the presence of Laponite ($\underline{53}$) took place only in the presence of oxygen. Oxygen was not required, however, when the exchange cations were Fe^{3+} or Cu^{2+}. Similar behavior was observed in the oxidation of hydroquinone to p-benzoquinone in smectite slurries ($\underline{54}$). The formation of biphenyldiols and diphenoquinones from 2,6-xylenol in the interlayer of Cu^{2+}-montmorillonites is retarded in a nitrogen atmosphere ($\underline{55}$). Oxygen serves as the oxydant to reoxidize the Cu^+ cation to the reactive Cu^{2+}. Oxygen has also been proposed to oxidize the phenol radical cation to the phenoxy radical, which may react with phenol molecules, oligomers, or polymers ($\underline{55}$).

Clays may also promote electron transfer between adsorbed organic reactants. This process is termed redox disproportionation if the electron transfer occurs between two identical species. The formation of p-cymene and p-menthene from p-menthene is an example ($\underline{56}$). In the presence of O_2, polymers were the main products. Redox disproportionation increased with the acidity of the clay and may be Brønsted acid-catalyzed. The mechanism was not established, however.

Oxidative Polymerization Reactions. Clays can initiate polymerization of unsaturated compounds through free radical mechanisms. A free radical R·, which may be formed by loss of a proton and electron transfer from the organic compound to the Lewis acid site of the clay or, alternatively, a free radical cation, $R^{+·}$, which may be formed by electron transfer of an electron from the organic compound to the Lewis acid site of the clay, can attack a double bond or an aromatic ring in the same manner as an electrophile. The intermediate formed is relatively stable because of resonance, but can react with another aromatic ring to form a larger, but chemically very similar, species. Repetition of the process can produce oligomers (dimers, trimers) and, eventually, polymers.

The mechanism indicated in Figure 1 is postulated for the clay-catalyzed polymerization of phenols through radical cations. According to this scheme, an acceptor for one electron, such as Cu(II) or Fe(III), must be present for the radical cation to form. In the presence of O_2 the reduced metal cation may be reoxidized which would result in the scheme shown in Figure 2. This scheme rationalizes the apparent stability of phenols in reduced systems of clays (54, 57), and anaerobic aquifers (58), where a suitable oxidant is lacking. The schemes shown in Figures 1 and 2 indicate the role of clays as hosts for metal cations and their redox properties. However, similar phenol coupling reactions may be promoted in the presence of a number of oxidants (such as O_2, H_2O_2, MnO_2, enzymes, and cell-free extracts) (59).

When phenol was adsorbed on Fe^{3+}- or Cu^{2+}-smectite under dehydration conditions, higher molecular weight products were formed (49, 55). ESR spectra of the fresh Cu^{2+}- smectite-phenol complex indicated formation of radical cations. Analysis by mass spectrometry and gel permeation chromatography indicated the presence of dimers, trimers, and tetramers, but higher molecular weight polymers (non-analyzable by mass spectrometry) were also present. The experiments were conducted by exposing freeze-dried clay to crystals of phenol from a different container, inside a P_2O_5 dessicator. Soma et al. (55) proposed that polymerization can be initiated by the phenoxy radical. The phenoxy radical may attack at either the para or the ortho position. Polymerization was also observed in benzene, and benzene-anisole and phenol systems, but these experiments were conducted by refluxing the clay with the organic substrate (49). Pinnavaia et al. (47) in their spectroscopic study on type I and type II complex formation of benzene, toluene, and anisole on Fe^{3+} and VO^{2+} exchange forms of hectorite under dehydration conditions also noted the formation of polymeric product. When benzene was subjected to the same reaction conditions, poly(p-phenylene) cations were identified in the interlayers of the clay mineral on the basis of resonance Raman spectra. When the dry clay-polymer complex was exposed to water vapor, the interlayer cation was re-oxidized and the poly(p-phenylene) cations were reduced to poly(p-phenylene). The reaction of Cu^{2+}- and Ru^{3+}-montmorillonites with biphenyl and p-terphenyl gave the products of different chain length (10, 11).

Isaacson and Sawhney (60) studied the reactions of a number of phenols and smectite with transition metal (Cu^{2+}, Fe^{3+}) and non-transition metal exchangeable cations. IR spectra of the clay-phenol complexes showed that all the clays studied transformed the sorbed phenols. The transformation occurred to a much greater extent in clays with transition metal cations than in those with the non-transition metal cations. In a subsequent study, Sawhney et al. (61) studied the polymerization of 2,6-dimethylphenol on air-dried homoionic Na-, Ca-, Al-, and Fe-smectite at 50°C. A portion of the adsorbed 2,6-dimethylphenol was transformed into dimers, trimers, tetramers, and quinone-type compounds. The nature of the exchange cations had an effect on both sorption and transformation and decreased in the order Fe >> Al > Ca > Na.

Wang et al. (62) reported the oxidative polymerization of a mixture of phenolic compounds in aqueous solution containing montmorillonite, illite, and kaolinite, each of which had been mixed with quartz in a 3:7 ratio, and by quartz alone. The mixture of phenolic compounds contained gallic acid, pyrogallol, protocatechuic acid, caffeic acid, orcinol, ferulic acid, p-coumaric acid, syringic acid, vanillic acid, and p-hydroxybenzoic acid. The oxidative

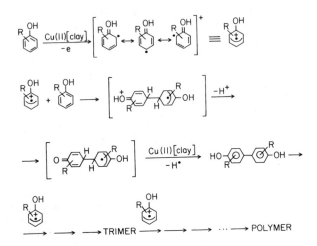

Figure 1. Postulated mechanism for the clay-catalyzed polymerization of phenols through radical cations.

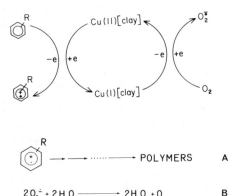

Figure 2. Schematic representation of electron transfer from an aromatic compound to O_2 with a Cu-exchanged clay as the catalyst and the formation of polymers (Reaction A) and hydrogen peroxide (Reaction B).

polymerization took place immediately and resulted in a dark mixture of "model" humic and fulvic acids, the infrared spectra of which resembled those of natural products. The results showed that all the clay minerals and the quartz catalyzed the oxidative polymerization, with quartz being the weakest catalyst.

Reactions of aniline with exchangeable transition metal cations may be through either interaction with the π-electrons of the aromatic ring or the nonbonding electron pair of the amino-group. Fe^{3+} was found to interact with the π-electrons, whereas Cu^{2+} was found to coordinate the amino group (63). Triphenyl amine reacted to N,N,N'N'-tetraphenylbenzidine when heated with Wyoming montmorillonite, possibly via the formation of N-N bonds (64). Cloos et al. (28) showed that when an aqueous aniline solution was percolated through a sand column containing some Fe^{3+}-montmorillonite, colored stripes and spots developed along the column, indicating type II complex formation between aniline and the clay. The presence of the type II complex was confirmed by IR and ESR spectrometry. The final product was a soil-humin or kerogen-like polymer. Such reactions through type II smectite complexes in aqueous media could have important implications for the transformations of organic compounds in natural systems, where, under aerobic conditions, the presence of structural Fe^{3+} is ubiquitous.

Soma et al. (12) have generalized the trends for aromatic compound polymerization as follows: (1) aromatic compounds with ionization potentials lower than approximately 9.7 eV formg radical cations upon adsorption in the interlayer of transition-metal ion-exchanged montmorillonites, (2) parasubstituted benzenes and biphenyls are sorbed as the radical cations and prevented from coupling reactions due to blockage of the para position, (3) monosubstituted benzenes react to 4,4'-substituted biphenyls which are stably sorbed, (4) benzene, biphenyl, and p-terphenyl polymerized, and (5) biphenyl methane, naphthalene, and anthracene are nonreactive due to hindered access to reaction sites. However, they observed a number of exceptions that did not fit this scheme and these were not explained.

Br∅nsted Acidity of Clay Minerals. The Br∅nsted acidity of clays primarily arises from the dissociation of water coordinated to exchangeable cations (6, 36, 65):

$$[M(H_2O)_x]^{n+} \rightleftharpoons [M(OH)(H_2O)_{x-1}]^{(n-1)+} + H^+ \qquad (9)$$

This Br∅nsted acidity depends on the size and charge of the cations, the moisture content, and the type and pretreatment of the clay system. Generally, the higher the charge and the smaller the radius of the exchange cation, the higher its polarizing power and, therefore, the higher the degree of dissociation of the adsorbed water. Thus, the surface acidity of homoionic montmorillonites saturated with Na, Mg, and Al decreases in the order Al > Mg > Na, but the reactive acidity and the order of acidity may also be influenced by the moisture content and the clay structure (65, 66). Dissociation of water coordinated to cations in the interlayers tends to be stronger than in bulk water (65).

Both the strength and the number of the Br∅nsted sites depend on the structure of the clay and its pretreatment. The acid strength is indicated by the pK_a range of the strongest sites. The number of sites, i.e., the total acidity, is indicated by the equivalent amount of base (amine) used to neutralize the sites. On certain dry clay minerals, such as kaolin and attapulgite, the pK_a of the strongest acid sites are below −8.2, with total acidities of 0.04 and > 0.2 mequiv/g, respectively, whereas for talc the pK_a range is only 4.4 to 3.3 with a total acidity of 0.005 mequiv/g (39). As the moisture content is increased, the surface acidity decreases rapidly. For a kaolinite system, it asymptotically approaches a pK_a of 4 at 15% (w/w) moisture content as indicated by an adsorbed Hammet indicator (Figure 3) (67). For near zero (< 1%) water content, the pK_a of the Hammet indicator is below −8 (equivalent to > 90% w/w sulfuric acid). By further removing the residual water, the surface acidity decreases, as it was shown by evacuating a synthetic Ni(II)−montmorillonite at high temperature (100–500°C) (67). IR spectra of pyridine surface complexes indicate that at 1% free moisture content a clay surface exhibits only Brønsted acidity. If the moisture content is further reduced, Lewis acidity is exhibited. Lewis sites appear to increase the acidity of adjacent Brønsted sites (67), and "dry" clays may exhibit properties of a superacid, such as HF/SbF_5 (22).

Brønsted acidity in the absence of water has been attributed to protons of structural hydroxyl groups (6, 69). The acid strength of such hydroxyl groups appears to be enhanced by the presence of adjacent electrons withdrawing Al ions (22). In 2:1 clays with no isomorphous substitution, i.e., with no ion-exchange capacity, one proton is coordinated by an oxygen of the octahedral layer. When isomorphous substitution with cations of a lower charge takes place in the octahedral layer, i.e., Me^{2+} for Al^{3+}, two protons are associated with this oxygen. When isomorphous substitution takes place in the tetrahedral layer (i.e., Me^{3+} for Si^{4+}), a proton is associated with an oxygen of the tetrahedral layer. Davitz (69) has shown on the basis of kinetic data that the catalytic acid activity is associated with charge of the tetrahedral layer. In contrast, octahedral sites were found to be inactive, perhaps due to shielding of the protons (69).

Table III summarizes the parameters that affect Brønsted acid-catalyzed surface reactions. The range of reaction conditions investigated varies widely, from extreme dehydration at high temperatures in studies on the use of clay minerals as industrial catalysts, to fully saturated at ambient temperatures. Table IV lists reactions that have been shown or suggested to be promoted by Brønsted acidity of clay mineral surfaces along with representative examples. Studies have been concerned with the hydrolysis of organophosphate pesticides (70–72), triazines (73), or chemicals which specifically probe neutral, acid-, and base-catalyzed hydrolysis (74). Other reactions have been studied in the context of diagenesis or catagenesis of biological markers (22–24) or of chemical synthesis using clays as the catalysts (34, 36). Mechanistic interpretations of such reactions can be found in the comprehensive review by Solomon and Hawthorne (37).

Reactions may be initiated by protonation of functional groups of a substrate to form a reactive intermediate. The tendency for protonation decreases with decreasing basicity, i.e., with decreasing pK_a of the conjugated acid. Qualitatively, the tendency to

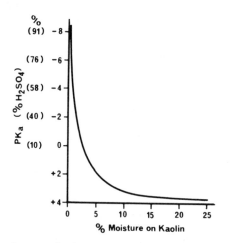

Figure 3. Dependence of the surface acidity of kaolinite on its moisture content. (Reproduced with permission from Ref. 67. Copyright 1971 Marcel Dekker.)

Table III.   Parameters Controlling Reactions Promoted by Brønsted
             Sites of Clay Minerals

Type of Brønsted Site
  - structural hydroxyl groups
  - water coordinated to exchange transition metal cations
  - surface-bound water
  - free water in the interlamellar region

Clay Properties
  - exchange cations, type, degree of exchange
  - charge density
  - interlayer spacing
  - clay structure

Substrate Properties
  - size
  - pK_a, pK_b
  - stability of carbonium ion intermediate

Reaction Conditions
  - moisture
  - polarity of organic co-solutes
  - humic acid, organic matter
  - inorganic co-solutes
  - temperature
  - ionic strength of reaction medium

---

accept a proton decreases in the order amines > C_1-C_2 alcohols,
water > ketones > higher alcohols, esters, ethers > hydrocarbons
(39). Under extreme conditions protonation of double bonds may
occur (75). Protonation of hydroxyl groups may lead to dehydration
of the protonated species and to the formation of a carbonium ion.
Examples are the reversible transformation of unsubstituted and
substituted triphenylcarbinol to the corresponding triphenylcar-
bonium by montmorillonite (76, 77). The reaction can be reversed by
adding water to the system (Equation 10):

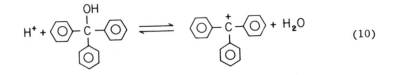

$$(10)$$

Hydrolysis Reactions. Hydrolysis reactions involve cleavage of a
single bond by reaction with water, a hydronium, or a hydroxide ion
(78). The bond is typically polarized between an electron-deficient
atom (C in carbonyl, P in organophosphates) and an electron-rich
atom (O, Cl, Br). The reaction may be neutral, base-, or acid-
promoted, depending on the substrate properties and the reaction
conditions, such as pH, temperature, and ionic strength (78, 79).

Table IV. Organic Reactions Affected by Surface Brønsted-Acidity of
Clay Minerals

Reaction	Examples	References
1. Demetallation of organometallic compounds	tin-tetrapyridyl-porphyrin by H, M, B	(81)
2. Hydrolysis of carboxyl acid esters	ethyl acetate by M and K	(74)
epoxides	cyclohexene oxide and substituted 2,3-epoxypropane by M, K to glycols	(74)
carbamates	N-methyl-p-tolyl carbamate by M, K to cresol and methyl-amine (73)	
alkyl halides	isopropyl bromide by M, K	(74)
N-heterocycles	s-triazines by M	(73)
organophosphates	parathion, methylparathion phosmet by K and M (69, 71)	(71)
3. Hydrogen Exchange of aromomatic hydrocarbons	detritiation of naphthalene by B	(23)
4. Condensation of NH₄ salts of carboxylic acids	ammonium acetate by B to acetamide	(1)
5. Alkene Formation by dehydration of alcohols	cholestanol by K and M to cholestene	(22, 34, 35)
6. Addition to Double Bonds of water	to ethylene by M to ethyl alcohol (81)	
acetic acid	to 2-methylpropene by Al-B to 2-methyl-propyl-2-acetate	(75)
7. Double Bond Isomerization in alkenes	limonene by M	(20)
sterenes		(26)
8. Ketal Formation from cyclic ketones	cyclohexanone and methanol by Al-B to dimethyl ketal	(75)
9. Rearrangments of sterols	diamerization of cholestanol	(22)
10. Dimerization of unsaturated fatty acids	dicarboxylic acid formation from unsaturated fatty acids	(1)
alkenes	propenylbenzenes	(33)
11. Polymerization of 2-butene styrene		(38)
12. Inversion of sugars	sucrose by H^+-B	(80)

M: montmorillonite; K: kaolinite; H: hectorite; B: bentonite.

Both acid- and base-promoted reactions may be affected by acidic
surfaces and, hence, by the factors which influence the surface
acidity. Kinetic evidence for increased Brønsted acidity at clay
surfaces has been presented by McAuliffe and Coleman (80) who studied the hydrolysis of ethylacetate and the inversion of sucrose.
They noted that potentionmetric pH measurements did not explain the
catalytically effective H^+-concentration at the clay surface.

Rate data for hydrolysis reactions in homogeneous aqueous solutions have been reviewed (79), but application of these data to
environmental conditions involving mineral surfaces remains difficult due to the unknown effects sorption may have. Several studies
have demonstrated that acid-catalyzed reactions are promoted if the
substrate is sorbed at clay surfaces (70-74; and other works
reviewed by Theng, 8), but inhibition may also occur if substrate
hydrolysis is base-promoted (74).

El-Amamy and Mill (74) determined the hydrolysis rates at 25°C
of several substrates with different hydrolysis characteristics
(neutral, acid-, and base-promotion), including 1-(4-methoxyphenyl)-
2,3-epoxypropane, ethyl acetate, cyclohexene oxide, and isopropyl
bromide in suspensions of montmorillonite and/or kaolinite. For
cyclohexene oxide and isopropyl bromide, they found no significant
difference between the rates determined at 4.5% (wt/wt) clay-
suspension and those in homogeneous buffers. For ethyl acetate,
only a slight increase was found. The small effects were probably
due to limited interaction with the clay surface because a significant promoting effect was observed for the less polar and more
strongly sorbing 1-(4-methoxyphenyl)-2,3-epoxypropane.

For N-methyl-p-toluyl carbamate, a chemical susceptible to
neutral and base-promoted decomposition, the rates of hydrolysis on
Na-montmorillonite and kaolinite were slower than in aqueous solution of the same bulk pH. A possible explanation is the lower pH at
the clay surface (74). The following hydrolysis mechanism in
aqueous solution was proposed:

$$CH_3NHCOOC_6H_4CH_3 \xrightarrow{OH^-} CH_3\bar{N}COOC_6H_4CH_3 \longrightarrow CH_3N{=}C{=}O + {}^-OC_6H_4CH_3 \quad (11)$$

$$CH_3N{=}C{=}O + H_2O \xrightarrow{H^+} CO_2 + CH_3NH_2 \quad (12)$$

The H^+ and NH_4^+ forms of homoionic montmorillonite promote the
hydrolysis of chloro-s-triazines to the hydroxy analogs (hydroxy-s-
triazines) (73). Apparently, the surface acidity of these clays was
extremely high, since no degradation was observed in control experiments conducted at pH 3.5 in homogeneous aqueous solution. Russell
et al. (73) suggested that the hydroxy-s-triazine products were
stabilized in the protonated form at the silicate surface. The IR
spectra of these surface complexes agreed with the spectra obtained
in 6N HCl, and it was inferred that the pH at the clay surface was 3
to 4 units lower than that measured in suspension.

Several studies have presented kinetic evidence that the Brønsted surface acidity increases with decreasing moisture content.
The extreme acidity at the surface of air-dried montmorillonite at
room temperature is demonstrated by the demetallation of Sn(IV)-
tetrapyridylporphyrin complexes. In contrast, in homogeneous

solutions demetallation of such complexes is not possible, even in
100% sulfuric acid. If too much water is present at the clay surface, the Sn(IV)-tetrapyridylporphyrin is not demetallated. Weaker
complexes, however, such as Fe(III)-tetraphenylporphyrin, are irreversibly demetallated and protonated under similar conditions (81).

Decreasing moisture content increases acid-catalyzed hydrolysis
rates, but if moisture is totally absent, the reactivity due to
Brønsted acidity of the surface may disappear (70, 71, 74). The
hydrolysis rates of the organophosphates parathion and methylpara-
thion on Na-, Ca-, and Al-kaolinite at room temperature were determined by Saltzman et al. (71). Figure 4 is a plot of the percent
parathion remaining on kaolinite after 15 days versus the kaolinite
water content. The figure shows that, for Na-kaolinite, the degradation rate was very slow up to a water content of about 0.8%.
Increasing the moisture content further, up to 2% water, resulted in
a very sharp increase in the degradation rate. Little change in the
rate occurred between 2 and 11.2% moisture. Moisture content of 2%
corresponds to saturation with water of the exchange cation's first
hydration shell, whereas 11% corresponds to the upper limit of chemically bound water. A slight increase in the water content above
11% resulted in a steep decrease of the hydrolysis rate. The presence of free water hindered almost completely the catalytic effect
of the clay surface. Similar trends were reported by El-Amamy and
Mill (74) for the hydrolysis of 1-(4-methoxyphenyl)-2,3-epoxypropane
on Na-montmorillonite.

A qualitatively similar relationship was observed for the
hydrolysis of parathion on Ca- and Al-kaolinite (Figure 4). In
order to explain the lower hydrolysis rate of parathion at Al-
kaolinite than at Na-kaolinite surfaces, it was suggested (70) that
steric hindrance may force the parathion molecule into a position or
conformation less favorable to hydrolysis. On the other hand, the
hydrolysis rate for methyl parathion on the Al-clay was higher than
the one on the Na-clay (71), probably due to the smaller size of the
methoxy group as compared to that of the ethoxy group of parathion
(70).

The experimental findings of Saltzman et al. (71) suggest that
sorbed water molecules serve as adsorption and degradation sites for
parathion. Sorbed water may be associated with surface oxygen atoms
or hydroxyl groups, edge sites of the lattice, or as a ligand with
the exchange cations. Work by Mingelgrin et al. (70) showed that
the hydrolysis of parathion, and other organophosphorous esters, in
general, occurs through the attack of a ligand water molecule of an
exchange cation at the P-O bond. The adsorbed phosphate moiety can
inhibit the reaction by blocking the reactive sites. The attack of
the ligand water molecule is enhanced by the cation-ligand interactions, which polarize the water and weaken the HO-H bond (70).

Sanchez Camazano and Sanchez Martin (72) studied the reaction
mechanism and the factors that influence the catalytic hydrolysis of
phosmet (o,o-dimethyl-S-(N-phthalimidomethyl)dithiophosphate) by
montmorillonite at 30°C. The montmorillonite was saturated with the
following cations: Ca^{2+}, Ba^{2+}, Cu^{2+}, Mg^{2+}, and Ni^{2+}. It was found
that after 24 hours at pH 6, the hydrolysis of phosmet in aqueous
suspension of Ca-montmorillonite was more than 60% complete, as
compared to 3% in homogeneous aqueous solution. The extent of
hydrolysis depended on the exchange cation and decreased in the

order Ca^{2+} (60%), Ba^{2+} (40%), Cu^{2+} (23%), Mg^{2+} (14%), and Ni^{2+} (3.2%).

Experiments with a series of minerals of the montmorillonite group (montmorillonite, nontronite, hectorite) all in the Ca-form, but of different layer charge, showed that the highest rate of hydrolysis corresponded to clays with a low total charge or with a particularly low tetrahedral charge. It appears that the type of the exchange cation and the interlayer charge (total or tetrahedral) control the accessibility of the reaction sites to the organic molecule (72). It was postulated that the low reactivity of the Cu^{2+}, Ni^{2+}, and the Mg^{2+} exchanged clays was due to strong attraction of the clay interlayers, which may hinder the intercalation of the organic compounds to the reaction sites.

The hydrolysis of phosmet in an aqueous solution takes place by nucleophilic attack of an OH^- group on the phosphorous atom, causing fission of the P-S bond.

$$(13)$$

The catalytic hydrolysis by montmorillonite is the result of the formation of a bidentate complex, in which interaction with the exchange cation takes place simultaneously through the sulfur atom of the P=S group and the oxygen of the C=O group. This interaction increases the electrophilic character of phosphorous, thus facilitating the nucleophilic attack by the OH^- group and producing hydrolysis of the P-S bond.

Hydrolysis of phosmet in the presence of montmorillonite occurs in two stages, both with apparent first-order kinetics. The first stage is of short duration and has a high hydrolysis rate constant. The rate varies as a function of the layer charge and exchangeable cations of the clay. The second stage, which is also a function of the same parameters, has a much slower rate constant. In some cases, its magnitude is very close to the hydrolysis constant of phosmet in aqueous solution. At the end of the first stage, most of the accessible sites in the clay have been reached (72).

Addition to Double Bonds and Elimination Reactions. Both Lewis and Brønsted acidity of mineral surfaces can promote addition and elimination reactions (37). Equation 14 shows an example for an addition/elimination equilibrium catalyzed by Brønsted acidity:

$$2R-CH_2-CH=CH_2 + 2R'-OH \underset{}{\overset{H^+}{\rightleftarrows}} R-CH_2-CHOR'-CH_3 + R-CHOR'-CH_2-CH_3 \quad (14)$$

where R' may be a proton, an alkyl or an acyl group. Elimination of water from alcohols (dehydration) catalyzed by surface acidity may be accompanied by other reactions, such as rearrangement of the intermediate carbocation to form mixtures of isomeric alkenes, intra-molecular and inter-molecular dehydration to form oligomeric and cyclic products, respectively (75), and polymerization (37). Alkylamines and alkyl thiols may undergo similar elimination and

condensation reactions; NH_3 and H_2S can react in a similar fashion as H_2O with alkenes to form alkylamines and alkyl thiols (37, 75).
   Cation-exchanged bentonites have been shown to be effective catalysts for addition of water, alcohols, and carboxylic acids to carbon-carbon double bonds under relatively mild conditions (75, 82). For instance, 2-methylpropene reacted with acetic acid at 18°C in the presence of Al-bentonite to form the ester product (75). Ion-exchanged bentonites are also efficient catalysts for formation of ketals from aldehydes or ketones. Cyclohexanone reacted with methanol in the presence of Al-bentonite at room temperature to give 33% yield of dimethyl ketal after 30 min of reaction time. On addition of the same clay to the mixture of cyclohexanone and trimethyl orthoformate at room-temperature, the exothermic reaction caused the liquid to boil and resulted in an almost quantitative yield of the dimethyl ketal in 5 min. When Na- instead of Al-bentonite is used, the same reaction did not take place (75). Solomon and Hawthorne (37) suggest that elimination reactions may have been involved in the geochemical transformation of lipid and other organic sediments into petroleum deposits.

Hydrogen-Exchange Reactions. Brønsted or Lewis acidity may promote hydrogen-exchange reactions (37). Both aromatic (23) and aliphatic (24) hydrogen exchange has been reported. Hydrogen-exchange reactions between the acidic clay surface and naphthalene derivatives were observed in aqueous slurries of homoionic bentonites at low temperatures (70°C). In contrast, hydrogen exchange (measured as detritiation or deuteration) between alkyl groups and acidic clay surfaces were observed at high temperatures (160°C) and long reaction times (670 h). A mechanism similar to electrophilic aromatic substitution was invoked to rationalize the aromatic H-exchange reaction.
   The mechanism of alkyl hydrogen exchange was not clarified, but a possible mechanism was postulated. Partial hydride abstraction by a Lewis acid site may have occured forming a carbocation-like species followed by exchange of a proton at a β-carbon. Such a mechanism predicts exchange to occur preferentially at methyl groups adjacent to the most stable carbocations (benzylic > 3° > 2° > 1°). This is consistent with the observed relative rates of epimerization of steranes during thermal maturation of sediments (83).

The Role of Soil Organic Matter. Humic compounds present in the sediment or soil may compete for reaction sites or block them altogether. In soils with moderate organic matter content, catalytic effects on hydrolysis may be smaller than in soils with low organic matter (74). The addition of humic acid to clay reduced hydrolysis rates of 1-(4-methoxyphenyl)-2,3-epoxypropane by coating the surface and, perhaps, by sorbing the organic into the lipophilic phase of the soil organic matter. Yaron (84) showed that parathion degradation decreased with an increase in the soil organic matter content. This was attributed to inactivation of the reactive sites available for the parathion decomposition.

Oxidation of Organic Compounds by Oxides and Primary Minerals

Certain Mn(IV)- and Fe(III)-oxides can promote the oxidation of organic compounds (14, 16). Since both Fe(III)- and Mn(IV)-oxides are ubiquitous in earth materials (85), their role in abiotically transforming organic compounds may be important. The mode of action for some metal oxides is to accept electrons from organic molecules (i.e., act as Lewis acids). For example, Mn(IV)-oxides may induce the oxidative polymerization of phenolic compounds to a humic acid-like product (15, 86), an effect similar to that observed with certain enzymes (87-89). When glycine was added to the reaction mixture, nitrogenous polymers and ammonia were formed (16). The various Mn(IV)-oxides studied were effective at promoting polymerization (measured as increased absorption at 600 nm), much more than Fe(III)-oxides. The rate of polymerization depended on the structure of the phenol and decreased in the order hydroquinone > resorcinol > catechol, for 7 days reaction time at pH 6 (15). Al- and Si-oxides did not significantly promote polymerization, although a previous study has indicated that metal impurities of freshly ground silica may promote hydroxylation and, presumably, polymerization of aromatics (13).

Similarly to Mn(IV)- and Fe(III)-oxides, some primary minerals were shown to promote polymerization of hydroquinone (19). Olivines, pyroxenes, and amphiboles accelerated the polymerization reaction to a greater extent than micas and feldspars. Microcline and quartz were ineffective. The effect was greatest for tephroite, a manganese-bearing silicate with the ideal chemical formula Mn_2SiO_4. Fayalite, the corresponding Fe(II) analog (Fe_2SiO_4), was effective, but to a lesser extent.

Similar results were reported previously by Larson and Hufnal (14) who studied the oxidative polymerization to humic-like materials of model organic compounds, such as catechol, pyrogallol, and their derivatives, in the presence of sediments, clays, transition metal oxides (MnO_2, ZnO, CuO) and cations (Mn^{2+}, Fe^{3+}). In stream water, the rate of transformation of catechol to brownish-green polymer products was increased when compared to a dilute phosphate buffer of comparable pH, probably because of the presence of transition metal cations which can promote polymerization. Addition of sediments or clays increased the rate of polymerization, whereas addition of EDTA or lowering of the pH reduced it. The effect of EDTA was explained by complexation of soluble transition metal ions, such as Fe^{3+} and Mn^{2+}, which promoted polymerization. In general, MnO_2 was the most effective catalyst, but it was found incapable of promoting polymerization of phenols without adjacent hydroxyl groups. The superior catalytic activity of MnO_2, as compared to other metal oxides, was also reported by Shindo and Huang (15), but no explanation was given. The polymerization reactions were rationalized by a mechanism involving hydroperoxyl radicals, $\cdot OOH$. These radicals can be formed from dissolved oxygen and Fe^{2+} sites in natural silicates, as shown in Figure 5.

The hydroperoxyl radical formed, $\cdot OOH$, has a pK_a of 4.75 and exists in water as a superoxide radical anion O_2^-. Either species in water dismutates to produce O_2 and hydrogen peroxide, H_2O_2, or, possibly, react with dissolved or adsorbed phenol. Dilute H_2O_2 does

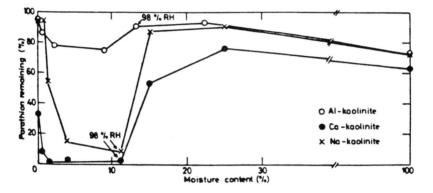

Figure 4. Degradation of parathion on kaolinite as a function of the clay moisture content. (Reproduced from Ref. 71. Copyright 1976 American Chemical Society.)

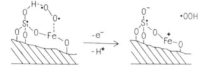

Figure 5. Formation of hydroperoxyl radicals by electron transfer from a clay surface to dissolved oxygen. (Reproduced from Ref. 14. Copyright American Chemical Society.)

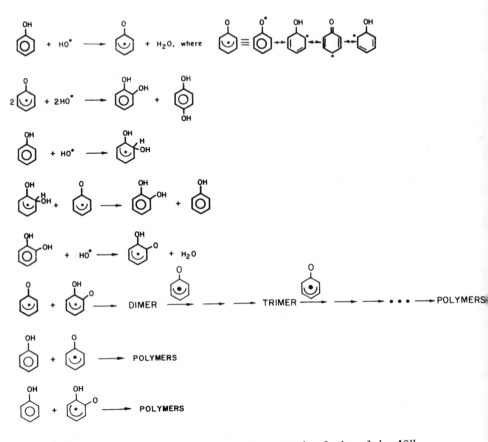

Figure 6. Reaction scheme for the attack of phenol by ·OH radicals (after Ref. 90).

not attack catechol rapidly in the absence of oxidizable or reducible cations, but in their presence, it is catalytically decomposed to form hydroxyl or hydroperoxyl radicals (Fenton reaction):

$$H_2O_2 + M^{n+} \longrightarrow \cdot OH + OH^- + M^{(n+1)+} \tag{15}$$

$$H_2O_2 + M^{(n+1)+} \longrightarrow \cdot OOH + H^+ + M^{n+} \tag{16}$$

$$2H_2O_2 \longrightarrow \cdot OH + \cdot OOH + H_2O \tag{17}$$

The $\cdot OH$ can react with catechol, by hydrogen abstraction or addition to the aromatic ring, to produce the resonance-stabilized radical. The latter could couple with other catechol molecules or oxygen to eventually form polymerized, highly colored materials, according to the scheme proposed for phenol by Voudrias (90) (Figure 6).

Summary and Conclusions

Abiotic organic reactions such as hydrolysis, elimination, substitution, redox and polymerization which are influenced by the surfaces of clay minerals, primary minerals, and metal oxides have been reviewed. Structural or exchangeable transition metal cations on clays can act as Lewis acid sites by accepting electrons from organic compounds. This electron transfer may lead to formation of charge transfer complexes or radical cations. Aromatic radical cations can react further with other aromatic ring-containing compounds to form dimers, trimers, or polymers. If oxygen is present, the reduced clay can be reoxidized and act as a catalyst.

The dissociation of water coordinated to exchangeable cations of clays results in Brønsted acidity. At low moisture content, the Brønsted sites may produce extreme acidities at the clay surface. As a result, acid-catalyzed reactions, such as hydrolysis, addition, elimination, and hydrogen exchange, are promoted. Base-catalyzed reactions are inhibited and neutral reactions are not influenced. Metal oxides and primary minerals can promote the oxidative polymerization of some substituted phenols to humic acid-like products, probably through $\cdot OH$ radicals formed from the reaction between dissolved oxygen and Fe^{2+} sites in silicates. In general, clay minerals promote many of the reactions that also occur in homogenous acid or oxidant solutions. However, rates and selectivity may be different and difficult to predict under environmental conditions. This problem merits further study.

Acknowledgments

This work was supported by the U.S. Environmental Protection Agency through the R.S. Kerr Environmental Research Laboratory, CR812462-01-0. However, this publication has not been subjected to the Agency's peer and administrative review and therefore does not necessarily reflect the views of the Agency and no official endorsement should be inferred.

Literature Cited

1.   Weiss, A. Angew. Chem. Int. Ed. Engl. 1981, 20, 850-60.
2.   McBride, M.B. Clays and Clay Minerals 1979, 27, 224-30.
3.   Cornejo, J; Hermosin, M.S.; White, J.L.; Barnes, J.R.; Hem,
     S.L. Clays and Clay Minerals 1983, 31, 109-112.
4.   Lasaga, A.C. Reviews in Mineralogy 1981, 8, 1-68.
5.   Mortland, M.M. In "Agrochemicals in Soil"; Banin; A.; Kafkafi,
     U., Eds.; Pergamon Press: Oxford, 1980, pp. 67-72.
6.   Yariv, S.; Cross, H. "Geochemistry of Colloid Systems";
     Springer Verlag: Berlin, 1979.
7.   Boudart, M.; Djega-Mariadassou, G. "Kinetics of Heterogeneous
     Reactions"; Princeton University Press: Princeton, NJ; 1984.
8.   Theng, B.K.G. In "International Clay Conf."; Van Olphen, H.;
     Vaniale, F., Eds.; Elsevier: Amsterdam, 1982, pp. 197-228.
9.   Soma, Y.; Soma, M.; Harada, I. Chem. Phys. Letters 1983, 94,
     475-8.
10. Soma, Y.; Soma, M.; Harada, I. Chem. Phys. Letters 1983, 99,
     153-6.
11. Soma, Y.; Soma, M.; Harada, I. J. Phys. Chem. 1984, 88, 3034-8.
12. Soma, Y.; Soma, M.; Harada, I. J. Phys. Chem. 1985, 89, 738-42.
13. Schofield, P.J.; Ralph, B.J.; Green, J.H. J. Phys. Chem. 1964,
     68, 472-6.
14. Larson, R.A.; Hufnal, J.M. Limnol. Oceanogr. 1980, 25, 505-12.
15. Shindo, H.; Huang, P.M. Soil Sci. Soc. Am. J. 1984, 48,
     927-34.
16. Shindo, H.; Huang, P.M. Nature (London) 1984, 308, 57-8.
17. Stone, A.T.; Morgan, J.J. Environ. Sci. Technol. 1984, 18,
     450-6.
18. Stone, A.T.; Morgan, J.J. Environ. Sci. Technol. 1984, 18,
     617-24.
19. Shindo, H.; Huang, P.M. Soil Science 1985, 139, 505-11.
20. Frenkel, M.; Heller-Kallai, L. Org. Geochem. 1977, 1, 3-5.
21. Johns, W.D. Ann. Rev. Earth Planet. Sci. 1979, 7, 183-98.
22. Sieskind, O.; Joly, G.; Albrecht, P. Geochim. Cosmochim. Acta
     1979, 43, 1675-9.
23. Alexander, R.; Kagi, R.I.; Larcher, A.V. Geochim. Cosmochim.
     Acta 1982, 46, 219-22.
24. Alexander, R.; Kagi, R.I.; Larcher, A.V. Org. Geochem. 1984,
     6, 755-60.
25. Aizenshtat, Z.; Miloslavsky, I.; Heller-Kallai, L. Org.
     Geochem. 1984, 7, 85-90.
26. Brassell, S.C.; McEvoy, J.; Hoffman, C.F.; Lamb, N.A.; Peakman,
     T.M.; Maxwell, J.R. Org. Geo. Chem. 1984, 6, 11-23.
27. Tannenbaum, E.; Kaplan, I.R. Geochim. Cosmochim. Acta 1985,
     49, 2589-2604.
28. Cloos, P.; Badot, C.; Herbillon, A. Nature 1981, 289, 391-3.
29. Wang, T.S-C.; Chen, J.-H.; Hsiang, W.-M. Soil Science 1985,
     140, 3-10.
30. Mortland, M.M.; Lawless, J.G. Clays and Clay Minerals 1983,
     31, 435-9.
31. Adams, J.M.; Ballantine, J.A.; Graham, S.H.; Laub, R.J.;
     Purnell, J.H. Reid, P.I.; Shaman, W.Y.M.; Thomas, J.M.
     J. Catal. 1979, 58, 239-52.

32. Adams, J.M.; Clement, D.E.; Graham, S.H. J. Chem. Res. 1981, 254-5.

33. Adams, J.M.; Bylina, A.; Graham, S.H. J. Catal. 1982, 75, 190-5.

34. Adams, J.M.; Clapp, T.V.; Clement, D.E. Clay Minerals 1983, 18, 411-21.

35. Adams, J.M.; Clement, D.E.; Graham, S.H. Clays and Clay Minerals 1983, 30, 129-34.

36. Pinnavaia, T.J. Science 1983, 220, 365-71.

37. Solomon, D.H.; Hawthorne, G.H. "Chemistry of Pigments and Fillers"; John Wiley & Sons: New York, 1983.

38. Solomon, D.H. Clays and Clay Minerals 1968, 16, 31-9.

39. Solomon, D.H.; Murray, H.H. Clays and Clay Minerals 1972, 20, 135-41.

40. Tennakoon, D.T.B.; Thomas, J.M.; Tricker, M.S. J. Chem. Soc. Dalton Trans. 1974, 20, 2211-5.

41. Rozenson, I.; Heller-Kallai, L. Clays and Clay Minerals 1978, 26, 88-92.

42. Rooney, J.J.; Pink, R.C. Trans. Farad. Soc. 1962, 58, 1632-41.

43. Solomon, D.H.; Loft, B.C.; Swift, J.D. Clay Minerals 1968, 7, 389-97.

44. Mortland, M.M.; Pinnavaia, T.J. Nature, Phys. Sci. 1971, 229, 75-7.

45. Fenn, D.B.; Mortland, M.M.; Pinnavaia, T.J. Clays and Clay Minerals 1973, 21, 315-22.

46. Rupert, J.P. J. Phys. Chem. 1973, 77, 784-90.

47. Pinnavaia, T.J.; Hall, P.L.; Cady, S.S.; Mortland, M.M. J. Phys. Chem. 1974, 78, 994-9.

48. Eastman, M.P.; Patterson, D.E.; Pannell, K.H. Clays and Clay Minerals 1984, 32, 327-33.

49. Mortland, M.M.; Halloran, L.J. Soil Sci. Soc. Amer. J. 1976, 40, 367-70.

50. Furukawa, T.; Brindley, G.W. Clays and Clay Minerals 1973, 21, 271-80.

51. Slade, P.G.; Raupach, M. Clays and Clay Minerals 1982, 30, 297-305.

52. Dodd, C.G.; Ray, S. Clays and Clay Minerals 1960, 8, 237-51.

53. Vansant, E.F.; Yariv, S. J. Chem. Soc. Faraday Trans. I 1977, 73, 1815-24.

54. Thompson, T. D.; Moll, W.F. Clays and Clay Minerals 1973, 21, 337-60.

55. Soma, Y.; Soma, H.; Harada, I. J. Contaminant Hydrology 1986, 1, 95-106.

56. Frenkel, M.; Heller-Kallai, L. Clays and Clay Minerals 1983, 31, 92-6.

57. Sawhney, B.L. Clays and Clay Minerals 1985, 33, 123-7.

58. Sawhney, B.L.; Kozloski, R.P. J. Environ. Qual. 1984, 13, 349-52.

59. Taylor, W.I.; Battersby, A.R. "Coupling of Phenols"; Marcel Dekker: New York; 1967.

60. Isaacson, P.J.; Sawhney, B.L. Clay Minerals 1983, 18, 253-65.

61. Sawhney, B.L.; Kosloki, R.K.; Isaacson, P.J.; Gent, M.P.N. Clays and Clay Minerals 1984, 32, 108-14.

62. Wang, T.S.C.; Li, S.W.; Ferng, Y.L. Soil Science 1978, 126, 15-21.

63. Moreale, A.; Cloos, P.; Badot, C. Clay Minerals 1985, 20,
    29-37.
64. Tricker, M.J.; Tennakoon D.T.B.; Thomas, J.M.; Heald, J. Clays
    and Clay Minerals 1975, 23, 77-82.
65. Mortland, M.M.; Raman, K.V. Clays and Clay Minerals 1968, 16,
    393-8.
66. Frenkel, M. Clays and Clay Minerals 1974, 22, 435-41.
67. Solomon, D.H.; Swift, J.D.; Murphy, A.J. J. Macromol. Sci.-
    Chem. 1971, A5, 587-601.
68. Sohn, J.R.; Ozaki, A. J. Catalysis 1980, 61, 29-38.
69. Davitz, J.C. J. Catalysis 1976, 43, 260-3.
70. Mingelgrin, U.; Saltzman, S.; Yaron, B. Soil Sci. Soc. Am. J.
    1977, 41, 519-23.
71. Saltzman, S.; Mingelrin, U.; Yaron, B. J. Agric. Food Chem.
    1976, 24, 739-43.
72. Sanchez Camazano, M.S.; Sanchez Martin, M.J. Soil Science
    1983, 136, 89-93.
73. Russell, J.D.; Cruz, M.; White, J.L.; Bailey, G.W.; Payne,
    W.R.; Pope, J.D.; Teasley, J.I. Science 1968, 160, 1340-2.
74. El-Amamy, M.M.; Mill, T. Clays and Clay Minerals 1984, 32,
    67-73.
75. Ballantine, J.A.; Purnell, J.H.; Thomas, J.M. Clay Minerals
    1983, 18, 347-56.
76. Fripiat, J.J.; Cruz, M.; Bohor, B.F; Thomas, J. Jr. Clays and
    Clay Miner. 1974, 22, 23-30.
77. Fusi, P.; Ristori, G.G.; Cecconi, S.;Franci, M. Clays and Clay
    Minerals 1983, 31, 312-4.
78. March, J. "Advanced Organic Chemistry, Reactions, Mechanisms,
    and Structure," 3rd. Ed.; John Wiley & Sons: New York; 1985.
79. Mabey, W.; Mill, T. J.Phys. Chem. Ref. Data 1978, 7, 383-415.
80. McAuliffe, C.; Coleman, N.T. Soil Sci. Soc. Proc. 1955, pp.
    156-60.
81. van Damme, H.; Crespin, M.; Obrecht, F.; Cruz, M.I.; Fripiat,
    J.J. J. Colloid Interface Sci. 1978, 66, 43-54.
82. Atkins, M.P.; Smith,D.J.H.; Westlake, D.J. Clay Minerals 1983,
    18, 423-9.
83. Mackenzie, A.S. Clay Minerals 1984, 19, 271-86.
84. Yaron, B. Soil Sci. Soc. Amer. Proc. 1975, 39, 639-43.
85. Jenne, E.A. In "Trace Inorganics in Water"; Baker, R.A., Ed.;
    Adv. Chem. Ser. 73; Amer. Chem. Soc.; 1968; pp. 337-87.
86. Shindo, H.; Huang, P.M. Nature (London) 1982, 298, 363-5.
87. Suflita, J.M.; Loll, M.J.; Snipes, W.C.; Bollag, J.-M. Soil
    Science 1981, 131, 145-50.
88. Flaig, W.; Beutelsbacher, H.; Rietz, E. In "Soil Components,
    Vol. 1, Organic Components"; Gieseking, J.E., Ed.; Springer:
    Berlin, 1975, pp. 1-211.
89. Haider, K.; Martin, J.P.; Filip, Z. In "Soil Biochemistry,"
    Vol. 4; Paul E.A.; McLaren, A.D., Eds.; Marcel Decker: New
    York, 1975, pp. 195-244.
90. Voudrias, E.A. "Effects of Activated Carbon on Free and Com-
    bined Chlorine with Phenols"; Ph.D. Thesis; University of Illi-
    nois, Urbana, 1985.

RECEIVED June 18, 1986

Mn(II) Oxidation in the Presence of Lepidocrocite: The Influence of Other Ions

Simon H. R. Davies[1]

Environmental Engineering Science, California Institute of Technology, Pasadena, CA 91125

Mn(II) oxidation is enhanced in the presence of lepido-crocite (γ-FeOOH). The oxidation of Mn(II) on γ-FeOOH can be understood in terms of the coupling of surface coordination processes and redox reactions on the surface. Ca^{2+}, Mg^{2+}, Cl^-, SO_4^{2-}, phosphate, silicate, salicylate, and phthalate affect Mn(II) oxidation in the presence of γ-FeOOH. These effects can be explained in terms of the influence these ions have on the binding of Mn(II) species to the surface. Extrapolation of the laboratory results to the conditions prevailing in natural waters predicts that the factors which most influence Mn(II) oxidation rates are pH, temperature, the amount of surface, ionic strength, and Mg^{2+} and Cl^- concentrations.

Manganese is an important element in the aquatic environment. It is an essential micronutrient (1,2) and is the subject of much interest because its oxides scavenge other heavy metals (3). Of particular interest are ferromanganese nodules, which are abundant in the aquatic environment. These nodules contain high concentrations of cobalt, nickel, copper and other heavy metals (4).

Manganese can exist in a number of oxidation states. The II, III, and IV oxidation states are found in the aquatic environment. Mn(III) and Mn(IV) are essentially insoluble in water, whereas, Mn(II) is relatively soluble. Because of the widely differing sol-ubilities of Mn(II) and the higher oxidation states the behaviour of manganese in the aquatic environment is strongly influenced by redox conditions. The relative stability of the different oxidation states depends upon pH, the oxidation potential and the concentration of complexing ligands present (5). While the III oxidation state is found in manganese minerals (6, 7), aqueous Mn(III) probably does

[1]Current address: Swiss Federal Institute for Water Resources and Water Pollution (EAWAG), CH-8600 Dübendorf, Switzerland.

not occur in natural waters, as it is thermodynamically unstable and disproportionates ($\underline{8}$):

$$2Mn^{3+} + 2H_2O = Mn^{2+} + MnO_2 + 4H^+ \quad K= 10^9 \quad (1)$$

At the pHs found in natural waters Mn(II) is thermodynamically stable in reducing environments. In oxic environments, Mn(IV) is the stable form, yet Mn(II) is found in these environments ($\underline{9}$, $\underline{10}$). The persistence of Mn(II) can be explained by the reduction of $\overline{Mn}$(III) and Mn(IV) oxides by naturally occurring organics ($\underline{2,11}$) and by the slow oxidation of Mn(II) ($\underline{8}$).

The rates of Mn(II) oxidation in natural waters, although slow, are typically orders of magnitude faster than the rate of oxidation of Mn(II) in solution ($\underline{8,12}$). It has been suggested that the enhanced rate of Mn(II) oxidation in natural waters is due either to bacterial oxidation ($\underline{13\text{-}16}$) or to the "catalytic" effects of surfaces such as metal oxides ($\underline{8}$, $\underline{17\text{-}19}$). The existing evidence suggests that in certain environments bacterial mediation of the reaction is important ($\underline{13\text{-}15}$). But in many cases the relative importance of bacterial and abiotic "catalysis" in natural waters has not been clearly defined.

It has been suggested that many surfaces including Ti(IV), Si(IV), Sn(IV), Fe(III), Mn(III), and Mn(IV) oxides, calcite, clay minerals, and feldspars accelerate the oxidation of Mn(II) ($\underline{8}$, $\underline{20\text{-}25}$). The interpretation of a number of these investigations is difficult because Mn(II) oxidation was studied under conditions where the concentration of manganese was greater than the concenentration of surface sites. Where the surface is not in considerable excess, it is difficult to determine whether the surface itself or oxidized manganese deposited on the surface accelerates the rate of Mn(II) oxidation. Another difficulty with the interpretation of some of these investigations is that the systems studied were greatly oversaturated with respect to pyrochroite, $Mn(OH)_2$ or rhodocrocite, $MnCO_3$. If these solids precipitate, the system may behave very differently from an undersaturated system.

Objectives

This paper discusses the oxidation of Mn(II) in the presence of lepidocrocite, γ-FeOOH. This solid was chosen because earlier work ($\underline{18}$, $\underline{26}$) had shown that it significantly enhanced the rate of Mn(II) oxidation. The influence of Ca^{2+}, Mg^{2+}, Cl^-, SO_4^{2-}, phosphate, silicate, salicylate, and phthalate on the kinetics of this reaction is also considered. These ions are either important constituents in natural waters or simple models for naturally occurring organics. To try to identify the factors that influence the rate of Mn(II) oxidation in natural waters the surface equilibrium and kinetic models developed using the laboratory results have been used to predict the

time scales for Mn(II) oxidation in natural waters. These predictions are compared with the rates of manganese removal observed in
natural waters.

Experimental

General Remarks. The details of the experimental procedures are
described in Davies (26). Deionized distilled water was used to
prepare all solutions. All reagents were analytical grade unless
otherwise noted. The gases used were filtered and scrubbed in a
Dreschel bottle. The glassware and plasticware were cleaned in a
strong detergent (Alconox), soaked in 4M nitric acid, and rinsed
with deionized distilled water. The solutions used in the adsorption
and oxidation experiments, other than those prepared from strong
acids, were filtered.

Preparation and Characterization of γ-FeOOH. The preparation and
characterization of the bulk and surface properties of the γ-FeOOH
studied is described in Davies (26). The surface properties of the
oxide are given in Table I.

Table I.   Surface Properties of γ-FeOOH.

	Batch L1	Batch L2
BET Surface Area, m^2/g	142	ND[1]
F⁻ Exchange Capacity, mmoles/g	2.27	2.58
Surface Acidity Constants [2]		
pKa_1	6.4 , 6.05 [3]	5.65
pKa_2	8.32, 8.14 [3]	8.13

[1] ND Not Determined
[2] In 0.1M $NaClO_4$, 25⁰C
[3] Duplicate experiments

pH Measurements. The pH measurements were made using a glass electrode and a double junction reference electrode; the outer compart-
ment of the reference electrode was filled with 0.1 M $NaClO_4$. The
electrode was calibrated using NBS buffers (pHydrion or Radiometer).
Accurate pH measurements in the suspensions were difficult to make
because the electrode potential drifted slowly with time apparently
due to the coating of the electrodes with an oxide film. To obtain
self consistent results, potential readings were taken after a set
time (5 minutes).

Analytical Procedures. Mn was determined by atomic absorption spectrophotometry (AAS) or the formaldioxime method (27). Ca, Mg and Fe were determined by AAS. Silicate, phosphate, sulphate and chloride were determined using techniques described in Standard Methods (28). The molybdosilicate method was used for silicate. Phosphate was determined using the vanadomolybdophosphoric acid method. Sulphate was determined by $BaSO_4$ gravimetry. Chloride was determined by the mercuric chloride method. Salicylate and phthalate were determined by UV spectrophotometry.

Adsorption Experiments. The adsorption experiments were performed at 25±0.2°C in 0.1 M $NaClO_4$. Mn adsorption experiments were performed under a nitrogen atmosphere at a total Mn(II) concentration of 50μM. Filterable concentrations were determined by analyzing 0.22 micron (Millipore Type GWSP) filtered samples. The first 5 mL of the filtrate was discarded. Adsorbed concentrations were determined by difference (i.e. adsorbed = total- filterable).

Oxidation Studies. These studies were performed at 25±0.2°C in a carbonate buffered 0.1 M $NaClO_4$ solution. The buffer was equilibrated with air or an O_2/CO_2 gas mixture for at least 24 hours. The pH was adjusted to 8.3 using $HClO_4$, NaOH, or Na_2CO_3. Where γ-FeOOH was present the suspension was deaerated by bubbling N_2/CO_2 for 1 hour, the Mn(II) was added and was allowed to equilibrate with the solid for 30 minutes and then the oxidation was commenced by switching to O_2/CO_2 or air bubbling. Where no solid was present the solution was not deaerated. The rate of oxidation was monitored by following the loss of filterable Mn.

Equilibrium Calculations. The computer program SURFEQL (29) was used to calculate the equilibrium distrubution of chemical species. The constant capacitance model (30, 31) was used for the surface equilibria calculations. The equilibrium constants used in these calculations are given in Davies (26).

Results and Discussion

Elsewhere (26) I have shown that the oxidation of Mn(II) in the presence of four metal oxides (α-FeOOH, γ-FeOOH, amorphous silica, and δ-Al_2O_3) can be understood in terms of the coupling of surface coordination processes and redox reactions on the metal oxide surface. These results are not discussed here, but the conclusions with regard to γ-FeOOH are summarized.

Mn(II) adsorption on metal oxide surfaces. The binding of Mn(II) on γ-FeOOH can be understood in a surface coordination chemical framework. The surface groups on a metal oxide are amphoteric and the hydrolysis reactions can be written:

$$\equiv FeOH_2^+ = \equiv FeOH + H^+ \qquad K_{a1} \qquad (2)$$

$$\equiv FeOH = \equiv FeO^- + H^+ \qquad K_{a2} \qquad (3)$$

where $\equiv FeOH$ is a surface hydroxyl group. The adsorption of Mn^{2+} can be described in terms of the competition with protons for surface sites. Equations 4 and 5 describe examples of such reactions.

$$\equiv FeOH + Mn^{2+} = \equiv FeOMn^+ + H^+ \qquad {}^*K_1 \qquad (4)$$

$$\begin{array}{c} =FeOH \\ | \\ =FeOH \end{array} + Mn^{2+} = \begin{array}{c} =FeO \\ | \quad \diagdown \\ \quad \quad Mn \\ | \quad \diagup \\ =FeO \end{array} + 2H^+ \qquad {}^*\beta_2 \qquad (5)$$

As shown in Figure 1, the adsorption of Mn(II) on γ-FeOOH can be successfully described using a constant capacitance model. In these calculations the hydrolysed surface complex $\equiv FeO-Mn-OH$ was not considered. The reason for not considering both the bidentate $(\equiv SO)_2 Mn$ and hydrolysed surface species is that both have virtually the same pH dependence, so it is impossible using the available data to make anything other than an arbitrary choice about the relative proportions of these two species. Based on the model calculations, in the pH range 8-9, the predominant Mn(II) species on the γ-FeOOH surface is the bidentate surface complex or the hydrolysed surface complex.

Mn(II) oxidation on metal oxides. The rate of oxidation of Mn(II) in the presence of γ-FeOOH can be described by the following equation (26):

$$- \frac{d[Mn(II)]}{dt} = k \frac{\{\equiv FeOH\} [Mn^{2+}]}{[H^+]^2} a\, pO_2 \qquad (6)$$

where $\{\equiv FeOH\}$ is the concentration of surface hydroxyl groups in moles/g of oxide, a is the concentration of oxide in g/L, and pO_2 is the partial pressure of oxygen in atmospheres. The pH dependence of the reaction suggests the reactive surface species is the bidentate surface complex or the hydrolysed surface complex. Alternative formulations of the rate expression are:

$$- \frac{d[Mn(II)]}{dt} = k'' \{(\equiv FeO)_2 Mn\}\, a\, pO_2 \qquad \underline{or} \qquad (7)$$

$$- \frac{d[Mn(II)]}{dt} = k'' \{\equiv FeOMnOH\}\, a\, pO_2 \qquad (8)$$

One possible mechanism for the oxidation of Mn(II) on an oxide surface is shown in Figure 2. The binding of Mn(II) to the surface may facilitate the electron transfer from Mn(II) to O_2 ($\underline{32}$). The surface groups on the metal oxide, if appropriately cooordinated, will exert a repulsive effect on the electron in the manganese d_z orbital, making it easier for the O_2 bound to the manganese atom to remove this electron. The nature of the products was not characterized so the overall reaction stiochiometry is unknown.

The effect of other ions on Mn(II) oxidation in the presence of γ-FeOOH. This section discusses the effect of Ca^{2+}, Mg^{2+}, Cl^-, phosphate, SO_4^{2-}, silicate, phthalate, and salicylate on Mn(II) oxidation in the presence of γ-FeOOH. Ca^{2+}, Mg^{2+}, Cl^-, and SO_4^{2-} are major components of natural waters. Previous studies have shown that these ions bind to iron oxide surfaces ($\underline{33}$, $\underline{34}$). Phosphate and silicate both interact strongly with iron oxide surfaces ($\underline{33}$). Salicylate and phthalate were chosen as simple models for naturally occurring organics. Polar organic compounds ($\underline{35\text{-}37}$) and naturally occurring organics ($\underline{38\text{-}41}$) are adsorbed on oxide surfaces. The effect of these ions on the manganese oxidation rate in homogeneous solution, at the concentrations used in the oxidation studies where γ-FeOOH is present (see Table IV), is not significant. At pH 8.3, the pseudo-first order rate constant k_1 for reaction in homogeneous solution is, in all cases studied, less than 4×10^{-5} min^{-1}. Since the rate of Mn(II) oxidation in solution is slow, reactions which displace Mn(II) from the surface will inhibit its oxidation. The surface coordination model presented earlier may be generalized to consider the interactions of other ions with the surface and the formation of solution complexes. Some examples of the possible reactions are given below:

Formation of surface complexes

$$\equiv FeOH + Mg^{2+} = \equiv FeOMg^+ + H^+ \qquad (9)$$

Ligand exchange

$$\equiv FeOH + SO_4^{2-} + H^+ = \equiv Fe\text{-}SO_4^- + H_2O \qquad (10)$$

Formation of solution complexes

$$Mn^{2+} + Cl^- = MnCl^+ \qquad (11)$$

The results found in the adsorption studies are given in Table II. From these results the surface equilibrium constants for the reactions given in Table III were determined using the SURFEQL program. In these calculations it is assumed that the Ca^{2+}, Mg^{2+}, Cl^-, SO_4^{2-}, phosphate and silicate surface species are those found previously to occur on the α-FeOOH surface ($\underline{33}$, $\underline{34}$). The phthalate and salicylate complexes $\equiv FeA^-$ and $\equiv FeAH$ (where $\overline{A^{2-}}$ is phthalate or salicylate) were considered in these calculations.

If, where the other ions are present, the Mn(II) oxidation rate is still described by equation 7 (or 8) then the psuedo-first order

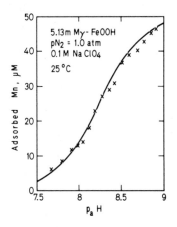

Figure 1. Mn(II) adsorption as a function of pH. The solid lines are calculated using the constant capacitance model.

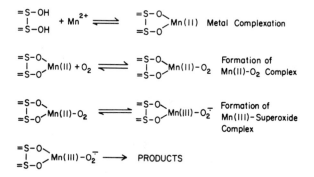

Figure 2. A possible mechanism for the oxidation of Mn(II) on a metal oxide surface.

Table II.   Adsorption of anions, Ca^{2+}, and Mg^{2+} on the γ-FeOOH
surface.

Ionic Medium[1]	pH	% Adsorbed
$10^{-3}M$ $Ca(ClO_4)_2$	8.30	11.3
$10^{-3}M$ $Mg(ClO_4)_2$	8.30	39.5
$10^{-3}M$ salicylate	8.29	15.2
$10^{-3}M$ salicylate	7.02	7.0
$10^{-3}M$ phthalate	8.30	<2
$10^{-3}M$ phthalate	7.00	75
$10^{-3}M$ Na_2SO_4	8.31	5
$10^{-3}M$ Na_2HPO_4	8.32	80
$10^{-3}M$ silicate	8.3	35
$10^{-3}M$ silicate	7.0	57

[1] All experiments in 10 mM γ-FeOOH, 0.1M $NaClO_4$ at 25^0C.

Table III.   Interactions of anions, Ca^{2+}, and Mg^{2+} with the γ-FeOOH
surface.

			$\log_{10}K$[1]
$\equiv FeOH + Ca^{2+} + H_2O$	=	$\equiv FeOCaOH + 2H^+$	-14.5
$\equiv FeOH + Mg^{2+} + H_2O$	=	$\equiv FeOMgOH + 2H^+$	-13.7
$\equiv FeOH + sal^{2-} + 2H^+$	=	$\equiv FesalH + H_2O$	23.7
$\equiv FeOH + sal^{2-} + H^+$	=	$\equiv Fesal^- + H_2O$	15.6
$\equiv FeOH + phth^{2-} + 2H^+$	=	$\equiv FephthH + H_2O$	17.2
$\equiv FeOH + phth^{2-} + H^+$	=	$\equiv Fephth^- + H_2O$	<9.5[2]
$\equiv FeOH + SO_4^{2-} + H^+$	=	$\equiv FeSO_4^- + H_2O$	8.4
$\equiv FeOH + PO_4^{3-} + H^+$	=	$\equiv FePO_4H^- + H_2O$	25.7
$\equiv FeOH + H_2SiO_4^{2-} + H^+$	=	$\equiv FeSiO_4H_2^- + H_2O$	18.2
$\equiv FeOH + H_2SiO_4^{2-} + 2H^+$	=	$\equiv FeSiO_4H_3 + H_2O$	25.1

sal^{2-} salicylate
$phth^{2-}$ phthalate
[1] K is the intrinsic constant at 25^0C in 0.1M $NaClO_4$.
[2] This species does not exist at significant concentrations in the
pH range studied.

rate constant k_1 for Mn(II) oxidation can be calculated using the expression:

$$k_1 = \frac{k'' \{(\equiv FeO)_2 Mn\}_0 \ a \ pO_2}{[Mn(II)]_0} \qquad (12)$$

(or the appropriate expression obtained from equation 8) where $[Mn(II)]_0$ is the initial concentration of Mn(II) and $\{(\equiv FeO)_2 Mn\}_0$ the initial concentration of the bidentate surface complex. The concentration of the bidentate complex present initially can be calculated using the SURFEQL program. In Table IV the ratio $k_1/(k_1,$ control) for the experimental results and those calculated using equation 12 are compared. The control experiments were in 0.1M $NaClO_4$, at pH 8.3, in a suspension of the same γ-FeOOH concentration.

Table IV. Comparison of the experimental and calculated values of the ratio k_1/k_1, control.

Ionic Medium [1]	k_1 / k_1, control [2]	
	Experimental	Calculated
10^{-2}M $Ca(ClO_4)_2$	0.35	0.73
10^{-2}M $Mg(ClO_4)_2$	0.07	0.28
10^{-3}M salicylate	0.36	0.80
5x 10^{-3}M salicylate	0.13	0.64
10^{-3}M phthalate	1.11	1.00
10^{-4}M phosphate	0.77	0.60
10^{-3}M phosphate	0.32	0.17
10^{-3}M silicate	0.12	0.56
0.33M Na_2SO_4	0.68	0.73

[1] All in 0.1M $NaClO_4$
[2] Phosphate and silicate experiments in 0.98 mM γ-FeOOH. Sulphate experiments in 2.5 mM γ-FeOOH. All other experiments in 1.0mM γ-FeOOH.

All the ions studied, except phthalate, inhibit the oxidation of Mn(II) to some degree. The relative extent to which these ions (at the concentrations indicated) affect the rate of Mn(II) oxidation is as follows:

$$10^{-2}M \ Mg^{2+} > 10^{-3}M \ silicate \ \sim \ 5x10^{-3}M \ salicylate >$$
$$10^{-3}M \ phosphate \ \sim 10^{-3}M \ salicylate \ \sim 0.7M \ NaCl \ \sim 10^{-2}M \ Ca^{2+}>$$
$$.033M \ Na_2SO_4 > 10^{-4}M \ phosphate > 10^{-3}M \ phthalate.$$

As shown in Table IV, generally the agreement between the experimentally obtained and calculated ratios is not quantitative. However
the model predicts that the relative effect of the added ions should
be:

$$10^{-3}M \ phosphate > 10^{-2}M \ Mg^{2+} >$$
$$10^{-3}M \ silicate \ \sim 5x10^{-3}M \ salicylate \ \sim 10^{-4}M \ phosphate >$$
$$10^{-2}M \ Ca^{2+} \ \sim \ .033M \ Na_2SO_4 \ \sim 10^{-3}M \ salicylate > 10^{-3}M \ phthalate.$$

which, except for the results obtained in the phosphate solutions,
is virtually the same order as found by experiment. Realizing the
limitations of the model calculations (30, 42), namely that (i) the
stoichiometry under the experimental conditions has been assumed ,
(ii) the equilibrium constants are not accurately determined
(estimated from single adsorption points using the assumed stoichiometry), (iii) the elucidation of the effects of the electric
properties of solid-water interface is difficult, and (iv) the
possible formation of ternary complexes is not considered, it is not
surprising that only qualitative agreement is obtained between the
model calculations and experimental results. That the model
calculations and experimental results agree qualitatively suggests
that the conceptual basis for the model is realistic, that is, the
effect that these ions have on the rate of Mn(II) oxidation can be
explained in terms of the effect they have on Mn(II) binding to the
surface. It is possible that the precipitation of iron phosphate on
the oxide surface occurs when phosphate is present in solution (43,
44). Should this occur then the number of sites on the oxide surface
available for Mn(II) binding would be greater than that predicted
using the competitive binding model. This could explain the fact
that in the experiments where phosphate is present the rate of
oxidation of Mn(II) is faster than that predicted using the
competitive binding model (and equation 12). Whatever the explanation for the results in phosphate solutions, it is likely that at
the concentrations found in natural waters ($\sim 10^{-6}M$) the effect of
phosphate on Mn(II) oxidation is minimal, because in $10^{-4}M$ phosphate
the Mn(II) oxidation rate is only 20% less than when no phosphate is
present.
       As shown in Table V the rate of Mn(II) oxidation in 0.7M NaCl
is about 3 times slower than in 0.7M NaClO$_4$, but the amount of
Mn(II) adsorbed is only about a third less in the chloride solution.
In the absence of surface complexation constants in these electrolyte matrices no model calculations can be made.

Table V. Mn(II) Adsorption and Oxidation in 0.7M NaClO$_4$ and 0.7M NaCl.

Ionic Medium	Mn$_{ads}$, μM	k_1, min^{-1}
0.7M NaClO$_4$	29.2	.0098
0.7M NaCl	18.3	.0035

All experiments pH 8.30, 10mM FeOOH, 25^{0}C

Mn(II) Oxidation in Natural Waters: Implications of Experimental Studies

In the studies described above the experimental conditions were chosen for experimental convenience, so they may differ greatly from those found in natural waters. To try to identify the factors that might influence Mn(II) oxidation on metal oxides surfaces in natural waters, the surface equilibria and kinetic models developed above can been used to predict the time scales for Mn(II) oxidation in these waters.

It has been shown elsewhere (26) that in natural waters the degree of enhancement of Mn(II) oxidation predicted on the basis of model calculations is as follows : γ-FeOOH > α-FeOOH > silica > alumina. It has also been shown that the rate of Mn(II) oxidation is strongly influenced by pH, γ-FeOOH concentration, temperature and ionic strength. Depending on the conditions, the predicted half-life ($\tau_{1/2}$ = ln 2/k_1) for Mn(II) oxidation may vary from a few days to thousands of years. By way of example, at pH 8, pO$_2$ 0.21 atm, 25^{0}C in waters containing 4μM γ-FeOOH and 0.2μM Mn(II), the half-life for oxidation is about 30 days.

Influence of other ions. Two examples are considered to illustrate the importance of other ions in natural waters on Mn(II) oxidation kinetics. In the first example the ionic composition of the solution is typical of that found in freshwaters and in the second example the composition of the solution is typical of that found in the low salinity region of an estuary (I ~0.1M). The composition of these solutions is given in Davies (26). The calculated oxidation half-lives based on the model given above for these systems are shown in Table VI.

In freshwater, Mn(II) oxidation is slightly slower than in 0.1M NaClO$_4$. The difference between the Mn(II) oxidation rate in freshwater and 0.1M NaClO$_4$ is greatest at pH 8.5, at this pH the rate of Mn(II) oxidation is only 40% lower in the freshwater than in 0.1M NaClO$_4$. In the estuarine-water at pH 8.5 the rate of Mn(II) oxidation is 20 times slower than in 0.1M NaClO$_4$. The speciation calculations indicate why the model predicts the oxidation is slower than in natural waters (see, for example Table VII).

Table VI. Predicted half-lives for Mn(II) oxidation in freshwater, estuarine-water and 0.1M NaClO$_4$. $\equiv$FeOH$_T$ = 2x10^{-6}M, 25^{0}C.

	Half-life, days		
pH	Freshwater	Estuarine-water	0.1M NaClO$_4$
7.5	290	470	240
8.0	37	94	25
8.5	5	57	3

Table VII. Calculated speciation in freshwater, estuarine-water and 0.1M NaClO$_4$. pH 8.5, 25^{0}C.

	% Total $\equiv$FeOH or Mn(II)		
	Freshwater	Estuarine-water	0.1M NaClO$_4$
Surface species			
$\equiv$FeOH	54	6	85
$\equiv$FeO$^-$	10	4	15
$\equiv$FeOMgOH	21	86	0
$\equiv$FeOCaOH	10	3	0
$\equiv$FeH$_3$SiO$_4$$^-$	3	3	0
Major Mn species			
Mn^{2+}	89	66	97
MnOH$^+$	1	<1	1
MnHCO$_3$$^+$	3	2	0
MnSO$_4$	<1	6	0
MnCl$^+$	4	25	0
MnCl$_2$	<1	1	0
($\equiv$FeO)$_2$Mn	1	<1	2

In the freshwater Ca^{2+} and Mg^{2+} have displaced Mn(II) from the surface, while in the estuarine-water the inhibition of Mn(II) oxidation is largely due to Mg^{2+} adsorption. The complexation of Mn^{2+} by Cl$^-$ also displaces some Mn(II) from the surface. According to the model calculations about 25% of the Mn(II) is present as Cl$^-$ complexes.
       In seawater the effect of the other ions on Mn(II) oxidation is

significant, but it is difficult to quantify this effect. A calculation, neglecting any effects that ionic strength may have on the
surface equilibria, indicates that in seawater the oxidation of
Mn(II) on γ-FeOOH is about 100 times slower than in 0.1M NaClO$_4$.

Mn(II) Oxidation in Natural Waters: A Comparison with Predictions
Based on Laboratory Data. The rates of manganese removal observed
in some natural waters are summarized in Table VIII. These estimates
for the rate of manganese removal have been obtained by various
means and thus it is difficult to make a direct comparison between
these data. For example, the data given are in some cases based on
the rates of manganese removal from the water column (see, e.g.
reference 49) and in other cases the data given are based on observations of manganese removal from solution using samples taken from
natural water bodies (see, e.g. reference 15). Despite these problems the data given in this table do serve to indicate the time
scales for the net removal of manganese in these waters. It is also
difficult to compare these data with the results from the laboratory
oxidation studies, for in general the surface properties of particles in natural waters are not known. Also, in natural waters the
relative importance of oxidative removal, adsorption, biological
uptake, and reductive dissolution is unknown.

Table VIII. Removal times for manganese in natural waters.

Source of Water	Temperature °C	pH	Half-life days	Reference
Groundwater	30	6.9-8.25	10-40	(45)
Lake Mendota	22	8.5	12	(46)
Griefensee (filtered)	20	8.8	100	(47)
Black Sea	8.5	7.7	2000	(48)
Open Ocean	Ambient	8.1	10000	(49)
Scheldt estuary	20	~7.5	3-8	(50)
Tamar estuary	20	~7.5	5-30	(19)
Onieda Lake	18	8.0-8.4	1-4	(15)
Saanich Inlet	9	7.4	2	(14)
MERL microcosm	Ambient	~8	4	(51)
Esthwaite Water	10	6.5-7.5	1-2	(16)

The rates of manganese removal observed in some natural waters
(e.g. Lake Mendota) compare favourably with the rates for Mn(II)
oxidation predicted on the basis of laboratory studies. In other
cases the predictions from the laboratory studies are difficult to

reconcile with the field observations. For example, in a sample
taken from Esthwaite Water, at pH 6.5 and 10^0C virtually all the
manganese was removed (from solution) within 2 days (16). This
removal rate is several orders of magnitude faster than would be
expected on the basis of the results presented in this paper.
Probably, manganese oxidizing bacteria play an important role in the
geochemistry of manganese in this lake (16).

The importance of bacteria in mediating Mn(II) oxidation in
certain environments is evident. But, the mechanisms whereby
bacteria oxidize Mn(II) are poorly understood. Some bacteria
synthesize proteins or other materials that enhance the rate of
Mn(II) oxidation (52). Other strains of bacteria require oxidized
manganese to oxidize Mn(II) (53), suggesting that they may catalyse
the oxidation of Mn(II) on the manganese oxide surface. Other
bacteria may catalyse the oxidation of Mn(II) on iron oxide
surfaces, as iron is associated with manganese deposits on bacteria
collected in the eastern subtropical North Pacific (54).

Summary and Conclusions

The oxidation of Mn(II) in the presence of γ-FeOOH can be described
by the equation:

$$- \frac{d[Mn(II)]}{dt} = k \frac{\{\equiv FeOH\} [Mn^{2+}]}{[H^+]^2} a \qquad (13)$$

The pH dependence of this reaction suggests that Mn(II) is
readily oxidizable if it is bound to the surface as a bidentate or
hydrolysed complex.

The influence of Ca^{2+}, Mg^{2+}, Cl^-, SO_4^{2-}, phosphate, silicate,
phthlate, and salicylate on the oxidation of Mn(II) in the presence
of γ-FeOOH can be explained, at least qualitatively, in terms of the
effect these ions have on the binding of Mn(II) to the γ-FeOOH
surface. The factors that most strongly influence Mn(II) oxidation
on γ-FeOOH are pH, temperature, γ-FeOOH concentration, ionic
strength, $[Mg^{2+}]$ and $[Cl^-]$.

The rates of Mn(II) removal in some natural waters are similar
to the Mn(II) oxidation rates predicted on the basis of these
laboratory studies. However, in other cases the rate of manganese
removal in natural waters is much faster than that expected on the
basis of this work. In these systems significant manganese removal
may occur as the result of adsorption, bacterially mediated
oxidation, or biological uptake.

Acknowledgments

I would like to thank Dr. J.J. Morgan for his advice and contributions to this work. I also thank the Jessie Smith Noyes Foundation
and the Office of Ocean Assessment, NOAA for financial support.

Literature Cited

1. Bowen, H.J.M. "Trace Metals in Biochemistry" Academic Press: New York, 1966.
2. Sunda, W.G.; Huntsman, S.A.; Harvey, G.R. Nature (London) 1983, 301, 234.
3. Singh, S.K.; Subramanian, V. Crit. Rev. in Environ. Control 1984 14, 33.
4. Roy, S. "Manganese Deposits", Academic Press: London, 1981.
5. Stumm, W.; Morgan, J.J. "Aquatic Chemistry", 2nd Ed., Wiley-Interscience: New York; 1981.
6. Giovanoli, R. In "Geology and Geochemistry of Manganese" Varentsov, I.M.; Grasseelly, Gy. Ed., Schweizerbart'sche Verlagbuchhandlung: Stuttgart, 1980; Vol. 1. pp 159- 202.
7. Murray, J.W.; Dillard, J.G.; Giovanoli, R.; Moers, H.; Stumm, W. Geochim. Cosmochim. Acta 1985, 49, 463.
8. Morgan, J.J. In "Principles and Applications of Water Chemistry" Faust, S.D. Ed.; Wiley and Sons Inc.: New York, 1967. pp 561-624.
9. Brewer, P.G. In "Chemical Oceanography" Riley, J.P.; Skirrow, G. Ed.; Academic: New York; 1975; Vol. 1. pp 415-496.
10. Carpenter, R. Geochim. Cosmochim. Acta 1983, 47, 875.
11. Stone, A.T.; Morgan, J.J. Environ. Sci. Technol. 1984, 18, 617.
12. Diem, D.; Stumm, W. Geochim. Cosmochim. Acta 1984, 48, 1571.
13. Emerson, S.; Cranston, R.E.; Liss, P.S. Deep-sea Res. 1979, 26A, 859.
14. Emerson, S.; Kalhorn, S.; Jacobs, B.M.; Tebo, K.H.; Nealson, K.H.; Rosson, R.A. Geochim. Cosmochim. Acta 1982, 46, 1073.
15. Chapnick, S.D.; Moore, W.S.; Nealson, K.H. Limnol. Oceanogr. 1982, 27, 1004.
16. Tipping, E.; Thompson, D.W.; Davison, W. Chem. Geol. 1984, 44, 359.
17. Wilson, D.E. Geochim. Cosmochim. Acta 1980, 44, 1311.
18. Sung, W.; Morgan, J.J. Geochim. Cosmochim. Acta 1981, 45, 2377.
19. Morris, A.W.; Bale, A.J.; Howland, R.J.M. Estuar. Coastal Shelf Sci. 1982, 14, 175.
20. Nichols, A.R.; Walton, J.H. J. Amer. Chem. Soc. 1942, 64, 1866.
21. Hem, J.D. "Chemical Equilibria and the Rates of Manganese Oxidation" U.S. Geol. Survey- Water Supply Paper 1667A, 1963.
22. Hem, J.D. Geochim. Cosmochim. Acta 1981, 45, 1369.
23. Morgan, J.J. Ph.D. Thesis, Harvard University, Cambridge, 1964.
24. Michard, G. Trans. Am. Geohys. Union 1969, 50, 349.
25 Coughlin, R.W.; Matsui, I. J. Catalysis 1976, 41, 108.
26. Davies, S.H.R. Ph.D. Thesis, California Institute of Technology, Pasadena, 1985.
27. Morgan, J.J.; Stumm, W. J. Am. Water Works Assoc. 1965, 57, 107.

28. Standard Methods for the Examination of Water and Wastewater, 15th Ed.; APHA, AWWA, WPCF: New York, 1980.

29. Faughnan, J. "The SURFEQL/MINEQL Manual", Environmental Engineering Science, California Institute of Technology: Pasadena, California, 1981.

30. Stumm, W.; Kummert, R.; Sigg, L. Croat. Chem. Acta 1980, 53, 291.

31. Schindler, P.W. In "Adsorption of Inorganics at Solid-Liquid Interfaces" Anderson, M.A.; Rubin, A.J. Eds.; Ann Arbor Science: Ann Arbor, 1981. pp 1-49.

32. Morgan, J.J.; Sung, W.; Stone, A.T. In "Environmental Inorganic Chemistry" Irgolic, K.J.; Martell, A.E., Eds.; VCH Publishers: Deerfield Park, FA; 1985.

33. Balistrieri, L.S.; Murray, J.W. In "Chemical Modeling in Aqueous Systems" Jenne, E.A. Ed.; ACS SYMPOSIUM SERIES No. 93; American Chemical Society: Wasington, DC, 1979. pp 275-298.

34. Sigg, L.; Stumm, W. Colloids and Surfaces 1981, 2, 101.

35. Parfitt, R.L.; Farmer, V.C.; Russell, J.D. J. Soil Sci. 1977, 28, 29.

36. Parfitt, R.L.; Fraser, A.R.; Farmer, V.C. J. Soil Sci 1977, 28, 289.

37. Davis, J.A.; Leckie, J.O. Environ. Sci. Technol. 1978, 12, 1309.

38. Davis, J.A. In "Contaminants and Sediments" Baker, R.A., Ed.; Ann Arbor Science: Ann Arbor, 1980; Vol. 2 . pp 279-304.

39. Davis, J.A.; Gloor, R. Environ. Sci. Technol. 1981, 15, 1223.

40. Tipping, E. Geochim. Cosmochim. Acta 1981, 45, 191.

41. Tipping, E. Chem. Geol. 1981, 33, 81.

42 Benjamin, M.M.; Leckie, J.O. Environ. Sci. Technol. 1981, 15, 1050.

43. Corey, R.B. In "Adsorption of Inorganics at Solid-Liquid Interfaces" Anderson, M.A.; Rubin, A.J., Eds.; Ann Arbor Science: Ann Arbor, 1981. pp 161-182.

44. Miesel, W.; Guttmann, H.-J.; Gütlich, P. Corrosion Sci. 1983, 23, 1373.

45. Handa, B.K. Chem. Geol. 1969/1970, 5, 161.

46. Delfino, J.J.; Lee, G.F. Water Res. 1971, 5, 1207.

47. Stumm, W.; Giovanoli, R. Chimia 1976, 30, 423.

48. Murray, J.W. and Brewer, P.G. In "Marine Manganese Deposits"; Glasby, G.P., Ed.; Elsevier: Amsterdam, 1977. pp 291-325.

49. Weiss, R.F. Earth Planet Sci. Lett. 1977, 37, 257.

50. Wollast, R.; Billen, G.; Duinker, J.C. Estuar. Coastal Mar. Sci. 1979, 9, 161.

51. Hunt, C.D. Limnol. Oceanogr. 1983, 28, 913.

52. Jung, W.K.; Schweisfurth, R. Zeit. für Allg. Mikrobiol. 1979, 9, 107.

53. Ehrlich, H.L. Appl. Microbiol. 1968, 16, 197.

54. Cowen, J.P.;   Silver, M.W. Science, Wash. 1984, 224, 1340.

RECEIVED June 25, 1986

Interaction of Co(II) and Co(III) Complexes on Synthetic Birnessite: Surface Characterization

John G. Dillard and Catherine V. Schenck

Department of Chemistry, Virginia Polytechnic Institute and State University, Blacksburg, VA 24061

The interaction of $Co(H_2O)_6^{2+}$ and Co(III) complexes; $Co(NH_3)_6^{3+}$, $Co(en)_3^{3+}$, (en = $NH_2CH_2CH_2NH_2$) with synthetic birnessite has been studied as a function of pH. The chemical nature of the birnessite sample surface was characterized using X-ray photoelectron spectroscopy (XPS) and secondary ion mass spectrometry (SIMS). Sorption of $Co(H_2O)_6^{2+}$ on birnessite in the pH range 4-7 results in the oxidation of Co(II) to Co(III) as shown by XPS results. Interaction of the Co(III) complexes with the birnessite surface occurs by loss of coordinated ligand, yielding Co(II) and Co(III) species from $Co(NH_3)_6^{3+}$, while Co(III) is the dominant cobalt state following the reaction of $Co(en)_3^{3+}$ with birnessite.

The interaction and sorption of metal ions with metal oxide and clay surfaces has occupied the attention of chemists, soil scientists, and geochemists for decades ([1-4]). Transition metal oxides receiving particular emphasis have included various oxides of manganese and iron ([5]). Interest in sorption phenomena is promoted by the desire to better understand incorporation of metals into minerals, especially marine deposits ([5]), the removal of trace metal pollutants and radionuclides from rivers and streams, via sorption and/or precipitation phenomena ([1,6]), and the deposition of metals on solid substrates in the preparation of catalysts ([7,8]). An important and significant task in sorption studies is the effort to identify the chemical form(s) of the sorbed metal ion.

0097-6156/86/0323-0503$06.00/0

Among the questions that arise are the following, at least, is
the metal present as a sorbed/hydrated species?, were the
conditions of metal incorporation such that precipitation of
metal-containing species occurred?, if a metal-organic compound
was present, what is the chemical form of the sorbed complex?
Answers to such questions have been provided in part from
thermodynamic studies (1-4,9) and from a variety of
spectroscopic techniques (10-12).

Manganese oxides play a significant role in concentrating
trace metals that have potential economic and strategic
importance (13). Studies of the formation processes of
manganese oxides under a variety of conditions have been
published (14,15). It was noted that the initial solid state
species formed by the oxidation of Mn(II) was Mn_3O_4 (hausmanite)
and that disproportionation or further oxidation processes
occurred depending on the conditions of the experiments. In the
reaction of Fe(II) with birnessite as a function of pH (16), it
was found that at pH<4 the rate of oxidation was controlled by
surface reactions at vacancies in the MnO_6 octahedral sheets.
Reactions at pH>4 proceeded at a slower rate presumably by the
formation of a FeOOH precipitate on surface active sites. The
reaction of As(III) with manganese(IV) oxide was found to yield
As(V) with about 19% of the Mn(IV) reduced to Mn(II) (17).
Under similar conditions As(III) was not oxidized on Fe(III)
oxide. Although such an oxidation is favored thermodynamically,
it was reasoned (17) that the kinetics of the process on the
Fe(III) oxide are too slow. The role of manganese dioxide in
the methylation of tin(II) and methyl tin(IV) ions has been
studied (18). It was found that Sn(II) and Sn(IV) form methyl
tin ions by CH_3 transfer. Of particular importance was the
result that manganese dioxide diminished the yield of all methyl
tin species. The decreased production of these ions was
attributed to the adsorption of Sn(II) on manganese dioxide
which prevented the formation of methyl tin intermediates. In
an investigation of the adsorption of Co(II) on birnessite the
role of oxygen dissolved in solution and the nature of the
manganese surface chemistry on the surface oxidation state of
cobalt were studied (19). Surface analysis results indicated
that cobalt(II) was oxidized to cobalt(III) under anoxic
conditions and surface Mn(III) was present on the birnessite
surface following the reaction. These studies illustrate the
importance of mineral surfaces in influencing the rates or
thermodynamic stability for reactions of geochemical importance.

In the present study the surface chemistry of birnessite and
of birnessite following the interaction with aqueous solutions
of cobalt(II) and cobalt(III) amine complexes as a function of
pH has been investigated using two surface sensitive
spectroscopic techniques, X-ray photoelectron spectroscopy (XPS)
and secondary ion mass spectrometry (SIMS). The significant
contribution that such an investigation can provide rests in the
information obtained regarding the chemical nature of the neat
metal oxide and of the metal oxide/metal ion adsorbate surfaces,
within about the top 50 Å of the material surface. The chemical

nature of the surface includes information on the oxidation state(s) of the metal ion, the extent of protonation of amine ligands, metal-ligand stoichiometry, and surface concentration of the constituent elements. This study is the first attempt to obtain and compare the results gathered using XPS and SIMS for chemical analysis and elemental speciation for sorption of metal ions on manganese-containing oxides. It is hoped that such investigations can provide procedures and methods for analysis, and generate information on the chemical nature of metal oxide surfaces and of adsorbates on substrates of geochemical importance.

Methods and Materials

The solid substrate, sodium birnessite, was prepared from sodium-buserite. Na-buserite was prepared according to the method outlined by Giovanoli et al. (20). Conversion of Na-buserite to Na-birnessite was accomplished by drying the solid at 120°C for 24 hrs in air. X-ray diffraction analysis of Na-birnessite exhibited peaks at 7.3, 3.5, 2.4 and 1.4 Å. The synthetic birnessite had a N_2 B.E.T. surface area of 27±2 m^2/gm and an ion exchange capacity (Ba^{2+} exchange at pH 7 on Na^+ saturated birnessite) of 62±3 meq/100 gm Na-birnessite. Metal complexes were purchased from commercial sources or were prepared according to methods in the literature (21-23).

The sorption studies were carried out at 25°C by suspending 0.10 g Na-birnessite in 20 mls of solution containing the adsorbate at pH's 4.0, 6.0, 7.0, 8.0, and 10.0. The solute solution concentrations were 10^{-2} M for $Co(H_2O)_6^{2+}$, $Co(NH_3)_6^{3+}$ and $Co(en)_3^{3+}$. The solution pH was maintained at ±0.2 pH units by addition of 0.1 M HNO_3, 0.1 M NaOH or 0.5 M NaOH. pH measurements were made using a glass electrode. Sorption reactions were allowed to continue for approximately one day at which time the solution and solid were separated by centrifugation. Following centrifugation, the solid was washed at least three times with doubly distilled deionized water. The wash water pH was equal to that at which the sorption reaction was carried out. The solid was re-suspended in the cobalt complex solution to effect additional sorption. The sorption cycle was repeated three times. Samples were dried in a desiccator over P_2O_5.

Surface characterization studies by X-ray photoelectron spectroscopy (XPS) were conducted using DuPont 650 and Perkin Elmer 5300 instruments. Samples were prepared by placing solid material on double stick adhesive tape, or by allowing solvent to evaporate from an acetone dispersion of a suspension placed on a stainless steel probe. A magnesium anode was used as the X-ray source (hν = 1253.6 eV). The temperature of samples during the analysis was approximately 30-40°C and the vacuum in the analysis chamber was about 10^{-7} torr. Potential

decomposition processes induced by X-ray exposure in the vacuum
were investigated by measuring the XPS spectra for the pure
complexes and for the cobalt-treated birnessite samples as a
function of time of exposure to the X-ray beam and as a function
of X-ray power. In these measurements no evidence for changes
in the chemical state of cobalt or of the elemental ratios,
particularly N/Co, Co/Mn, or O/Co, was found. This evidence was
obtained by examining the core level cobalt, nitrogen, oxygen
and manganese binding energies and photopeak shapes as a
function of time at a given X-ray power and at a fixed time for
a series of X-ray power levels (150-300 watts). In these
experiments it was found that X-ray induced decomposition
processes were not significant and did not affect the spectra at
power levels below about 225 watts. For these studies spectra
were obtained at 200 watts. The time required to obtain
reasonable core level spectra was not longer than one hour.

The spectrometer binding energy scale was calibrated for
each measurement using background carbon, where the C 1s binding
energy was taken as 284.6 eV. The carbon 1s energy was
calibrated against gold as a reference standard (24-26). Thus,
any charging effects were taken into account for each
measurement. The precision of the binding energy measurements
is ±0.1 eV, while the accuracy is approximately ±0.2 eV.

The relative instrumental sensitivity factors for cobalt and
nitrogen were determined by measuring core level (Co 2p and N
1s) XPS spectra for a series of pure cobalt amine complexes of
established stoichiometry. To evaluate the core level photopeak
intensities, peak areas, including shake-up satellite intensity
were used. The precision for the measurements of the nitrogen
to cobalt atomic ratio is ±10% while the accuracy is
approximately ±15%. Additional details of the XPS measurements
are contained in the literature (24,25).

Secondary ion mass spectra were measured using a Perkin-
Elmer PHI 3500 instrument. Experiments were carried out with 4
kV Ar$^+$ ions at beam currents of 3 and 300 nanoamps. Spectra
were measured to at least 500 daltons (d). Samples were
prepared in the manner used for the XPS studies. For
measurements on the pure complexes, sample charging occurred, as
evidenced by the inability to record secondary ion mass spectra.
To reduce charging, a low energy electron beam (50-400 eV) was
rastered across the sample during SIMS analysis. Positive and
negative ion SIMS spectra were recorded; however, only positive
ion spectra are of interest for this discussion. In the spectra
only unipositive ions were detected, so that the mass numbers
detected correspond to combinations of the various isotopes of
the elements in the ion. Thus an ion at m/z 17 d is assigned to
the $^{16}O^1H^+$ ion.

X-ray diffraction, surface area, and cation exchange
determinations were accomplished in the manner reported
previously (24,25). The bulk cobalt to manganese atomic ratios
were determined by dissolving the birnessite-treated samples in
acidic solutions, typically 1.0 M HNO$_3$, and measuring the
solution concentration for cobalt and manganese by atomic

absorption spectroscopy ($\underline{24,25}$). The atomic absorption results are taken as bulk analysis whereas the XPS atomic ratios correspond to surface analysis, i.e., within about the outer 50 Å of the solid surface.

Results and Discussion

XPS Studies/Characterization

The sorption of $Co(H_2O)_6^{2+}$, $Co(NH_3)_6^{3+}$ and $Co(en)_3^{3+}$ on Na-birnessite was studied as a function of pH. Surface analysis by XPS was used to determine the elemental content at the surface, to establish the oxidation state or chemical environment of cobalt, manganese and nitrogen, and to determine the surface concentration as indicated by adsorbate element to substrate element ratios. For $Co(H_2O)_6^{2+}$ sorption on Na-birnessite (Table I), results similar to those previously reported were obtained ($\underline{19}$). In Table I the Co $2p_{3/2}$ and Mn $2p_{3/2}$ binding energies are summarized and the energy separation between the Co $2p_{1/2}$ and Co $2p_{3/2}$ photopeaks is given. This information is used to establish the chemical nature of cobalt and manganese. The Co/Mn atomic ratios presented in Table I represent the value at the surface. At low pH values, 4.0, 6.0, and 7.0, Co(II) is oxidized to Co(III). Evidence for the presence of Co(III) is contained in the Co 2p spectra by the absence of shake-up satellite features on the high binding side of the main 2p photopeaks and by the 15.0 ±0.1 (eV) separation of the Co $2p_{1/2}$ and Co $2p_{3/2}$ main peaks ($\underline{19,24,26}$). At pH 8 and 10 the XPS spectra for cobalt include intense shake-up satellite features and a difference in the Co $2p_{1/2}$ - Co $2p_{3/2}$ photopeaks near 15.9±0.1 eV. These spectral features are consistent with the presence of cobalt as Co(II) ($\underline{19,24,26}$). The XPS results for the manganese core level indicate that Mn(IV) is the dominant surface species, in that the Mn $2p_{3/2}$ binding energy is equal within experimental error to the value for untreated birnessite ($\underline{19}$). The very large Co/Mn atomic ratios at pH 7, 8, and 10 suggest that a non-uniform deposition of cobalt occurs. If cobalt were uniformly distributed over the birnessite surface following treatments at pH 8 and 10 where the Co/Mn atomic ratio is 9 and 6, respectively, the cobalt content would be so great as to obscure photoemission from manganese. Although no direct evidence is available from this study, the large Co/Mn atomic ratios are consistent with the nucleation of a cobalt solid phase at the surface.

Table I. Co(II)/Birnessite XPS Data

pH	Binding Energy Results (eV)			
	Co $2p_{3/2}$	Δ Co $2p*$	Mn $2p_{3/2}$	Co/Mn Atomic Ratio
	(± 0.1 eV)	(± 0.1 eV)	(± 0.1 eV)	
4	781.0	15.0	642.6	0.08
6	781.1	15.1	642.3	0.78
7	781.2	15.1	642.4	2.75
8	781.0	15.8	642.4	9.30
10	781.0	15.9	642.5	6.12
$Co(NO_3)_2 \cdot 6H_2O$	782.2	16.0		
Birnessite			642.5	

$*$ Δ Co $2p$ = BE Co $2p_{1/2}$ - BE Co $2p_{3/2}$

The binding energy results at pH 7 and the Co 2p spectral features are similar to those obtained for CoOOH. If a distinct cobalt-containing phase is formed in the reactions of cobalt(II) with birnessite, the XPS results are more consistent with those for CoOOH when compared with XPS data for other Co(III)-containing oxides, namely Co_2O_3 or Co_3O_4. The corresponding results at pH 8 and 10, namely the presence of satellite features, the Co $2p_{1/2}$ - Co $2p_{3/2}$ splitting, and the Co $2p_{3/2}$ binding energy results, are equal to the results obtained for $Co(OH)_2$. Under the conditions of the experiments at pH 8 and 10 it is reasonable that a $Co(OH)_2$-like surface could be formed via deposition or precipitation of Co(II) on birnessite.

The sorption processes for cobalt complexes can be complicated by hydrolysis reactions of the complex in solution, surface induced ligand loss processes, sorption of hydrolysis products of either amine, protonated amine, or mixed amine/aquo cobalt complexes, and oxidation/reduction processes associated with cobalt. The principal objective of the XPS studies was to evaluate, the chemical state of cobalt and amine ligands, the surface concentration of the respective elements, and the ligand to cobalt ratio as indicated by the surface nitrogen to cobalt atomic ratio.

The XPS results for $Co(NH_3)_6^{3+}$ sorbed on Na-birnessite are summarized in Table II. In the table, core level binding energy results are presented for nitrogen, cobalt, and manganese; cobalt 2p splitting (Δ Co 2p), and manganese 3s splitting (Δ Mn 3s) data are collected; the elemental atomic ratios evaluated from XPS measurements for protonated nitrogen to total nitrogen, for cobalt to manganese, and non-protonated nitrogen to cobalt at the surface are presented. The bulk cobalt to manganese atomic ratio determined from the atomic absorption measurements for digested samples is summarized.

Table II. $Co(NH_3)_6{}^{3+}$/Birnessite XPS Data

pH	Binding Energy Results (eV)					
	N(1) 1s	N(2) 1s	Co $2p_{3/2}$	Δ Co 2p	Mn $2p_{3/2}$	Δ Mn 3s*
	(±0.1 eV)	(±0.1 eV)	(±0.1 eV)	(±0.1 eV)	(±0.1 eV)	(±0.1 eV)
4	401.1	399.2	781.4	15.0	642.1	5.1
6	401.1	399.1	780.9	15.5	642.0	5.1
7	401.1	399.2	780.6	15.6	642.0	5.2
8	401.1	399.3	780.8	15.7	642.0	5.0
10	400.8	399.3	780.6	15.6	642.1	5.1
$Co(NH_3)_6(Cl)_3$	--	399.7	781.8	15.0	--	--
Birnessite	--	--	--	--	642.5	4.7

pH	$\dfrac{N(1)}{N(1)+N(2)}$ Atomic Ratio	Co/Mn Atomic Ratio (Surface)	Co/Mn Atomic Ratio (Bulk)	N(2)/Co Atomic Ratio (Surface)
4	0.31	0.02	.024	2.7
6	0.29	0.04	.033	1.7
7	0.28	0.04	.037	1.3
8	0.27	0.08	.052	1.2
10	0.25	0.16	.078	0.7
$Co(NH_3)_6(Cl)_3$	--	--	--	6.0
Birnessite	--	--	--	--

$N(1) = NH_4{}^+$ - type; $N(2) = NH_3$ - type

* Δ Mn 3s = manganese multiplet splitting energy

Two photopeaks for nitrogen are detected for $Co(NH_3)_6{}^{3+}$-birnessite samples produced at all pHs studied, whereas only one peak is detected for the pure $Co(NH_3)_6{}^{3+}$ complex. The N 1s peak with a binding energy at 401.1 eV is attributed to $NH_4{}^+$ (ammonium ion) (26), while the N 1s peak with a binding energy at 399.2±0.1 eV is attributed to coordinated ammonia (amine). That two N 1s photopeaks are detected suggests that ammonia originally coordinated to cobalt(III) has become uncoordinated and sorbed on Na-birnessite as $NH_4{}^+$. This suggestion is also supported by the N(amine)/Co ratio which decreases from 2.7 to 0.7 as the solution pH is increased from 4 to 10. The N(amine)/Co ratio represents an upper limit for ammonia coordinated to cobalt(III) for the adsorbed complex, since it is likely that ammonia coordinated to cobalt(III) and ammonia sorbed on Na-birnessite would have very similar N 1s binding energies, and thus, could not be easily distinguished. The greater surface concentration of protonated amine compared to non-protonated amine at low pHs indicates that the solution pH plays a role in determining the extent of ammonia protonation. However,

the very slight decrease in the protonated amine to total amine
ratio over the pH range 4 to 10, suggests that the mode of sorption
on Na-birnessite may be a more important factor in determining the
chemical nature of sorbed ammonia. In the current study no attempt
has been made to discover the factors influencing ammonia or
ammonium ion sorption on birnessite.

The XPS results for cobalt at pH 4, particularly the Co 2p
splitting (15 eV) and the absence of shake-up satellite structure,
are indicative of cobalt(III). However, the N(amine)/Co atomic
ratio of 2.7 indicates that some ammonia ligands have been
displaced. Since it is known (27) that hydrolysis rates for
cobalt(III) complexes are very slow, the presence of cobalt with a
low number of coordinated amines, suggests that hydrolysis is
induced via an interaction with the birnessite surface. The cobalt
to manganese ratios for bulk and surface measurements are
equivalent within experimental error, a result which is consistent
with a reaction process occurring primarily at the surface. It is
also noteworthy that the Co/Mn ratio for $Co(NH_3)_6^{3+}$ sorption is a
factor of four lower than that measured in the Co(II)-birnessite
sorption experiments. It is possible that the differences in the
quantity of cobalt sorbed is related to the hydrolysis rate for the
cobalt-containing species at the birnessite surface.

For $Co(NH_3)_6^{3+}$-birnessite samples prepared at pH 6 and 7 the
XPS results are very similar. The Co 2p splitting values 15.5 ± 0.1
eV are equal within experimental error, the N(amine)/Co atomic
ratios are 1.7 and 1.3, respectively, and the individual Co/Mn
surface and bulk ratios are approximately equal at each pH. This
latter result indicates that the sorption process occurs
predominately on the surface at pH 6 and 7. The Co 2p splitting
results are intermediate between values measured for Co(III) and
Co(II)-containing compounds. To account for the Co 2p splitting
result, a cobalt material with such an intermediate splitting or a
mixture of the two cobalt oxidation states must be present. A
survey of representative cobalt-containing materials (19,24,26)
reveals that Co 2p splittings at about 15.5 eV are not common.
This fact taken together with the result that the N(amine)/Co
atomic ratio is significantly below that for the pure complex, and
that some weak shake-up satellite structure is noted in the Co 2p
spectrum for the pH 7 material (Figure 1a), suggest that a mixture
of cobalt oxidation states is present. The suggested presence of
Co(II) is difficult to rationalize, when it was found that aqueous
Co(II) was oxidized to Co(III) during sorption on birnessite.
Although the present results do not permit a definitive
explanation, a possible process could be suggested where a
partially hydrolyzed (and thus labile) Co(III) ammonia complex (27)
is reduced to a Co(II) ammonia/aquo sorbed complex which is not
capable of being oxidized either by manganese in birnessite or
dissolved solution gases. Alternatively, the cobalt could be in
the form of a hydroxy cobalt(II) moiety which resembles a surface
$Co(OH)_2$-like species, which is not easily oxidized to a Co(III)-
containing species. Such a Co(II) hydroxy material could be
similar to the species produced at pH 8 and 10 in the Co(II)-

birnessite sorption experiments discussed earlier. Clearly the XPS results alone are not sufficient to offer a complete interpretation of cobalt(III)-ammonia sorption process at pH 6 and 7.

The surface analysis characterization following sorption at pH 8 and 10, demonstrates further the decrease in sorbed cobalt-ammonia complex species, while the surface Co/Mn atomic ratio increases to values which are larger than the values at lower pH and the corresponding bulk ratios. The Co 2p spectra also exhibit greater shake-up satellite intensity (Figure 1b), a spectral feature which can be associated with the presence of surface Co(II) components. The Co 2p splitting results are equivalent to the values measured for samples prepared at pH 6 and 7, and the results are indicative of a Co(II)-Co(III) mixture on the surface. The Co 2p spectral features indicate the presence of Co(III) probably as a mixture of cobalt-amine complex species and cobalt(III) oxyhydroxide constituents. The cobalt surface chemistry is clearly unlike that produced in the reaction of Co(II) with birnessite at these pHs. That only a Co(II) hydroxide-like surface precipitate is present cannot be correct, since the Co 2p splitting and the shake-up satellite intensity are not equal to those found in $Co(OH)_2$. The surface and bulk Co/Mn ratios are not equal and indicate some process leading to enhanced deposition of cobalt. Further, the Co/Mn surface ratio at pH 8 or 10 is significantly less than the value obtained in the Co(II)-birnessite sorption measurements, suggesting that partial hydrolysis of the adsorbed Co(III) amine complex occurs and yields aquated or hydroxy Co(II) species via reduction (27), or that sorption/precipitation of hydrolyzed and reduced Co(II) hydroxy constituents from solution occurs on the birnessite surface. From the present results it is not possible to give a simple reaction process, or to indicate whether a combination of sorption/deposition reactions occurs in the $Co(NH_3)_6{}^{3+}$ reaction with birnessite at pH 8 and 10.

The surface characterization results after $Co(en)_3{}^{3+}$ interaction with birnessite as a function of pH yield a somewhat different picture for the sorption process. The XPS results for $Co(en)_3{}^{3+}$ sorbed on Na-birnessite (Table III) indicate that the chemical nature of the sorbed species did not change with pH. Only a single nitrogen photopeak was observed and the N 1s binding energy is in the range for coordinated nitrogen. Also the Co $2p_{3/2}$ binding energy and the Co $2p_{1/2}$-Co $2p_{3/2}$ splitting remained constant at 781.2±0.1 eV and 15.1±0.1 eV, respectively. No satellite structure was evident in the Co 2p XPS spectra (Figure 2). The Co 2p XPS data thus indicate the presence of Co(III) on the birnessite surface at all pH values. The N(amine)/Co atomic ratios, values at about 5.0, were invariant over the pH range studied, suggesting that one bidentate ligand was removed giving a 1:1 mole mixture of $Co(en)_3{}^{3+}$ and $Co(en)_2{}^{3+}$ sorbed on Na-birnessite. This finding could be interpreted to mean that a critical number of cobalt-amine bonds must be cleaved before

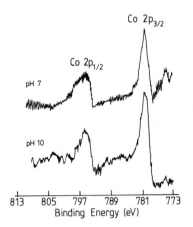

Figure 1. Co 2p XPS spectra for birnessite treated at pH 7 and 10

with $Co(NH_3)_6^{3+}$ (aq.)

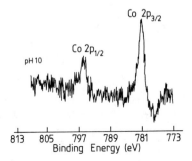

Figure 2. Co 2p XPS spectrum for birnessite treated at pH 10 with

$Co(en)_3^{3+}$ (aq.)

reduction of Co(III) proceeds at a reasonable rate to produce
surface Co(II) species. It is also found that the quantity of
cobalt-containing species sorbed, as indicated by the Co/Mn atomic
surface ratios, is approximately equal at pH values of 4, 6, and 7,
but only slightly less than the ratios at pH values of 8 and 10.
If hydrolysis plays a role in the sorption process, the slight
differences for the Co/Mn ratio at the higher pH could be related
to the relatively lower rates for hydrolysis of the cobalt
ethylenediamine complex at low pH (27).

Table III. $Co(en)_3^{3+}$/Birnessite XPS Data

pH	N 1s (± 0.1 eV)	Co $2p_{3/2}$ (± 0.1 eV)	Δ Co 2p (± 0.1 eV)	Mn $2p_{3/2}$ (± 0.1 eV)	Δ Mn 3s (± 0.1 eV)
			Binding Energy Results (eV)		
4	399.5	781.3	15.0	642.0	5.0
6	399.5	781.2	15.2	642.1	5.2
7	399.4	781.1	15.1	641.9	5.0
8	399.5	781.3	15.2	642.1	5.1
10	399.5	781.2	15.2	642.0	5.1
$Co(en)_3Cl_3 \cdot 3H_2O$	399.0	780.8	15.0	--	--
Birnessite				642.5	4.7

pH	Co/Mn Atomic Ratio (Surface)	Co/Mn Atomic Ratio (Bulk)	N/Co Atomic Ratio (Surface)
4	0.03	.017	5.2
6	0.04	.017	5.0
7	0.04	.022	4.9
8	0.06	.020	5.3
10	0.08	.029	5.2
$Co(en)_3Cl_3 \cdot 3H_2O$	--	--	6.0
Birnessite			

The XPS results for manganese for the cobalt complex-birnessite
samples require discussion. The decrease in Mn $2p_{3/2}$ binding
energy and the greater Mn 3s splitting for the cobalt treated
samples compared to sodium birnessite indicate that reduced
manganese is present (19). Reduced manganese suggests that
manganese(IV) has oxidized a solution or surface species or that
manganese is removed from birnessite and readsorbed as Mn(II). If
cobalt(II) is produced in solution in association with the ligand
loss process, manganese(IV) could behave as the oxidizing agent
yielding surface Co(III) upon cobalt sorption. However, if Co(II)
is produced in solution at all pH values studied, then the chemical
nature of readsorbed cobalt would not be equivalent; Co(III) would
be expected at low pH and Co(II) at higher pH. Also, one would
expect cobalt as cobalt(II) hydroxide to form at pH 8 and 10 via a
process similar to that noted above for Co(II) sorption at pH 8 or

10. The XPS results do not indicate significant cobalt(II)
hydroxide formation following complex sorption at pH 8 or 10 for
$Co(NH_3)_6^{3+}$ or at any pH for the $Co(en)_3^{3+}$ sorption studies. Thus,
surface manganese reactions with reduced cobalt solution species
seem not to occur.

The participation of surface manganese in the reactions of
sorbed cobalt species cannot be evaluated because of the inability
in present experiments to present a complete and unique explanation
for the presence of both Co(II) and Co(III) at the surface. As
noted above, the presence of Co(II) species would require inert
Co(II) in a unique sorbed chemical state.

If reduction of manganese was not due to oxidation of cobalt,
then another possible explanation is that Mn^{2+} ions were released
into solution from the $[MnO_6]$ octahedral layers during the
adsorption process. These Mn^{2+} ions could then be readsorbed onto
the manganese dioxide surface. The presence of surface Mn^{2+} would
lower the composite Mn $2p_{3/2}$ binding energy and increase the Mn 3s
splitting. However, definitive experiments have not been conducted
in this study to test the various possible explanations.

SIMS Analysis

Secondary ion mass spectrometry has become a valuable analytical
method for the analysis of non-volatile materials ([28]). The
formation of secondary ionic species related to the composition of
adsorbates has been demonstrated by several investigators ([29-31]),
and such measurements should be valuable in the study of mineral
surfaces. In the attempt to provide evidence related to the
chemical and molecular composition of sorbed cobalt species,
secondary ion mass spectra were obtained for selected samples.
Spectra were measured for $Co(NH_3)_6Cl_3$, $Co(en)_3Cl_3$, $Co(OH)_2$, CoOOH,
sodium- and cobalt-treated birnessite and for samples obtained
following sorption of $Co(NH_3)_6^{3+}$ and $Co(en)_3^{3+}$ at pH 7 and 10 on
birnessite. Spectra were obtained using equivalent instrumental
conditions for all samples except as noted in the experimental
section.

The SIMS spectrum for sodium saturated birnessite is presented
in Figure 3. The assignment of chemical composition to a
particular mass number is simplified by the fact that elemental
manganese and cobalt are monoisotopic; ^{55}Mn and ^{59}Co. The
principal ions are alkali metal and manganese-containing species.
The dominant ions in the spectrum are the alkali metal ions, Na⁺
m/z 23 and K⁺ m/z 39,41. The presence of intense Na⁺ and K⁺ ion
signals does not necessarily indicate that the surface is dominated
by these ions. It is likely that the large Na⁺ and K⁺ intensity is
due to the greater ionization probability for the alkali metals.
Thus the signals arise from birnessite and trace impurities in the
SIMS instrument. The intensities of these ions are not of interest

in the present discussion, and thus no attempt will be made to evaluate the relationship between alkali metal ions and manganese-containing species. The principal manganese ions are Mn^+, MnO^+, $MnOH^+$, and the dimanganese species Mn_2^+, Mn_2O^+, $Mn_2O_2^+$, $Mn_2O_2H^+$. A series of ions at m/z 104, 121, 138, and 155 can be formulated as $MnO_2(OH)_n^+$ species (n = 1-4). The formation of these species can occur via several processes including direct sputtering and ion-molecule reactions in the selvedge region above the sample surface (28,29).

The SIMS spectra for birnessite saturated at pH 7 and 10 with Co(II) are given in Figure 4. The relative intensity for m/z 59, Co^+, compared to m/z 55, Mn^+ is greater for the pH 10 sample, an expected result based on the XPS measurements. The spectra exhibit manganese-containing ions that are similar to those noted in Na-birnessite. The important cobalt-containing species are Co^+, $CoOH^+$, and $CoOH_2^+$, and combination ions $MnCo^+$, Co_2^+, Co_2O^+, and Co_2OH^+.

The spectrum for the pH 7 Co(II)/birnessite sample (Figure 4) exhibits ionic species similar to those found in CoOOH (Figure 5). The m/z 94, 151, and 152 species at low abundance and the major ions CoO^+, $CoOH^+$, and $CoOH_2^+$, are detected at abundances similar to values from CoOOH (Figure 5). The ion abundances are not in exact agreement, but the presence of ions unique to CoOOH supports the notion that surface cobalt is similar to CoOOH. The non-equivalence of the relative intensities could arise if a mixture of various Co(III) species is present on the surface. Alternatively, the abundances could also be affected by so-called "matrix effects" of the birnessite lattice on ion yields (28,29).

The intensities for the high mass manganese ions for the pH 10 Co(II)-birnessite material are reduced compared to intensities in Na-birnessite, or in the pH 7 Co(II)-birnessite sample, while the cobalt-containing species are the dominant ions. The results indicate a surface coating of cobalt-containing species, perhaps cobalt hydroxide or cobalt oxyhydroxide. A comparison of the pH 10 Co(II)-birnessite results with data for the reference cobalt compounds reveals ions similar to those found in $Co(OH)_2$ and CoOOH (Figure 5). Several features of the pH 10 Co(II)-birnessite spectra are unlike the spectra for CoOOH, in that ions at m/z 94, 151, and 152 are absent or of low abundance, and the $CoOH^+$ and $CoOH_2^+$ ions are not intense. On the other hand, ions detected and the relative intensities of the major ions CoO^+, $CoOH^+$, and $CoOH_2^+$, in the sample and in $Co(OH)_2$ compare favorably. This result suggests that the surface cobalt species at pH 10 is much more like $Co(OH)_2$ than CoOOH.

The comparison of the SIMS spectra and the interpretation of

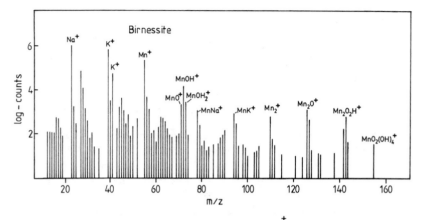

Figure 3.   Positive ion SIMS spectrum for Na$^+$ saturated birnessite.

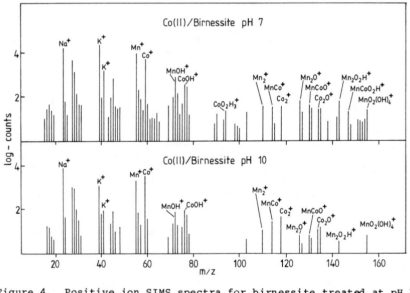

Figure 4.   Positive ion SIMS spectra for birnessite treated at pH 7
and 10 with Co(II) (aq.).

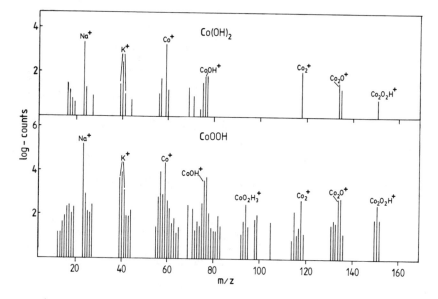

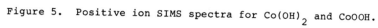

Figure 5.   Positive ion SIMS spectra for Co(OH)$_2$ and CoOOH.

the results support the XPS studies, suggesting cobalt hydroxide and cobalt oxyhydroxide on the surface. However, the SIMS results do not appear to be specific with regard to the distribution of species and the identification of cobalt oxidation states.

The SIMS spectra for birnessite on which $Co(NH_3)_6^{3+}$ was sorbed at pH 7 and 10 are presented in Figure 6. Under the experimental conditions employed, $Co(NH_3)_6^+$, m/z 161, for the adsorbed complex (pH 10) was detected. It is probable that the quantity of complex adsorbed at pH 7 was not sufficient to permit detection of molecular ion type species in that sample. The spectra (Figure 6) also exhibit manganese-containing ions similar to those noted for Co(II)-birnessite samples. The relative amounts of cobalt detected via SIMS, as indicated by the Co^+/Mn^+ ratios, follow a pattern similar to ratios found in the XPS studies (Table IV). The XPS/SIMS relative intensity ratios are reasonably constant for spect a involving a particular sorbate and indicate that SIMS could be used for quantitative detection of metal species sorbed on minerals, if appropriate calibration for ion sputtering processes is taken into account.

Table IV. Elemental Ratios - Surface Analysis - Co/Mn
Cobalt(III) Complexes/Birnessite

| Sorbate | pH | Co/Mn Atomic Ratio | | |
		XPS	SIMS	XPS/SIMS
$Co(NH_3)_6^{3+}$	7	.040	.017	2.4
	8	.080	.036	2.2
	10	.160	.063	2.5
$Co(en)_3^{3+}$	7	.040	.010	4.0
	10	.080	.025	3.2

For the pH 7 and 10 samples, the principal manganese ions include MnO_x^+, $MnOH_n^+$, $Mn_2O_x^+$, $Mn_2O_xH_n^+$ and $MnCOOH^+$ species, x = 1,2; n = 1,2. The most abundant cobalt species occur at m/z 59, Co^+; 76, $CoOH^+$ or $Co(NH_3)^+$; 118, Co_2^+; 144, $Co(NH_3)_5^+$, for the pH 7 sample, while additional important ions at m/z 134, $Co_2(NH_2)^+$ and 135, $Co_2(NH_3)^+$ appear for the sample prepared at pH 10. The intensities of the cobalt-containing ions are greater for the pH 10 sample. This result is expected since a greater quantity of $Co(NH_3)_6^{3+}$ was sorbed at pH 10. The SIMS spectra also demonstrate that sorption of the cobalt complex does not yield the same surface cobalt species as obtained for Co(II)-birnessite samples even though Co(III) is produced upon Co(II) sorption. In particular the ion at m/z 76 is attributed to $Co(NH_3)^+$ rather than $CoOH^+$, according to the following reasoning. It was shown (Figures 4 and 5) that $Co(OH)_2$ and CoOOH surface or bulk components, yield $CoOH^+$ (m/z 76) __and__ $CoOH_2^+$ (m/z 77) ions in the SIMS spectra. The absence of m/z 77 and the presence of $Co_2NH_y^+$ (where y = 2 or 3) type ions for Co complex-birnessite lead to the suggestion that m/z 76 is $Co(NH_3)^+$.

Clearly an unambiguous examination of the chemical nature of sorbed complexes using SIMS in these measurements is complicated by the presence of manganese and manganese-cobalt containing ions in the spectra. The greater relative ion intensities and the intensity distribution differences for the pH 10 sample compared to the pH 7 material, may arise due to the presence of different surface amine species. Alternatively, the difference may be related to different secondary ion formation processes for sorbed species.

The SIMS spectrum for $Co(en)_3^{3+}$ sorbed on birnessite at pH 10 (Figure 7) provides evidence also of cobalt-amine complex species. The SIMS spectrum for pure $Co(en)_3Cl_3$ shows important ions at m/z 76, $Co(NH_3)^+$; 94, $HCo(NH_3)_2^+$; 111, $HCo(NH_3)_3^+$; 144, $Co(NH_3)_5^+$; 153 $Co(en)(NH_3)_2^+$; 197, $HCo(en)_2NH_3^+$; 214, $HCo(en)_2(NH_3)_2^+$; 240, $HCo(en)_3^+$. The abundant ions noted in the $Co(en)_3$-birnessite sample occurred at m/z 76, 94, 111, 144, 153, 197, 213 and 214. It is assumed that ions from the $Co(en)_3$-birnessite sample are identical in composition to the ions detected at these m/z values in the pure complex. The presence of m/z 213, $Co(en)_2(NH_3)_2^+$ in the pH 10 material could be indicative of adsorbed $Co(en)_3^{3+}$.

Ions detected for the treated sample in the range m/z 151-154 could be attributed to $Co(en)_2(NH_2)_2$ (m/z 153) ions, with appropriate addition or loss of protons to give ions above or below m/z 153. Ions in this range and at m/z 134-138 could also correspond to $MnO_n(OH)_m^+$ where n = 1, 2; m = 0-4; so that the ion series 151-155 would be produced by loss of hydrogen from the $MnO_2(OH)_4^+$ (m/z 155) ion. That the intensities of other unique cobalt-containing ions do not increase significantly at high pH, lends support to the assignment of the 151-155 and 134-138 ions as manganese-containing.

Summary

Surface characterization of birnessite following the sorption of Co(II) indicates the presence of oxidized cobalt (CoOOH) at pH values, 4-6, where precipitation of $Co(OH)_2$ does not occur, and a surface coating of $Co(OH)_2$ at pH 8 and 10. XPS and SIMS results

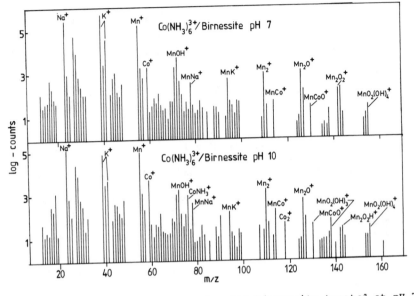

Figure 6. Positive ion SIMS spectra for birnessite treated at pH 7 and 10 with $Co(NH_3)_6^{3+}$ (aq.)

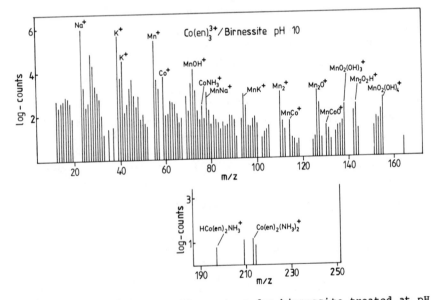

Figure 7. Positive ion SIMS spectrum for birnessite treated at pH 10 with $Co(en)_3^{3+}$ (aq.)

for cobalt amine-birnessite samples suggest the presence of cobalt amine complexes and/or hydrolysis products on the surface of birnessite. The distribution of complex species could not be evaluated from the results. The potential value of SIMS for quantitative analysis of sorbed metal ions was indicated but not explored in detail.

Acknowledgments

Thanks are expressed to the Office of Water Resources and Technology, Va. B-119, for partial support of this work. Funds to purchase the surface analysis instruments were provided by the National Science Foundation and the Commonwealth of Virginia.

Literature Cited

1. Tewari, P. H. "Adsorption from Aqueous Solutions," Plenum Press, New York, 1981.
2. Parfitt, G. D.; Rochester, C. H. "Adsorption from Solution at the Liquid/Solid Interface," Academic Press, New York, 1983.
3. Ottewill, R. H.; Rochester, C. H.; Smith, A. L. "Adsorption from Solution," Academic Press, New York, 1983.
4. Swartzen-Allen, S. L.; Matijevic, E. Chem Rev., 1974, 74, 385.
5. Glasby, G. P. "Marine Manganese Deposits," Elsevier, Amsterdam, 1977.
6. Anderson, M. A.; Rubin, A. J. "Adsorption of Inorganics at Solid-Liquid Interfaces," Ann Arbor Science, Ann Arbor, MI, 1981.
7. Schwab, G.-M. Adv. Catal. 1978, 27, 1.
8. Boudart, M. Adv. Catal. 1969, 20, 153.
9. James, R. O.; Healy, T. W. J. Colloid Interface Sci. 1972, 40, 53, 65.
10. Fripiat, J. J. "Advanced Techniques for Clay Mineral Analysis," Elsevier, Amsterdam, 1981.
11. Stucki, J. W.; Banwart, W. L. "Advanced Chemical Methods for Soil and Clay Mineral Reserach," D. Reidel, Dordrecht, 1980.
12. Perry, D. L. "Application of Surface Techniques to Chemical Bonding Studies of Geologic Materials," Davis, J. A. and Hayes, K. F., eds., ACS Symposium Series, 1986.
13. Cronan, D. S. "Deep Sea Nodules: Distribution and Geochemistry", p. 11; Mero, J. L. "Economic Aspects of Nodule Mining," p. 327; in Marine Manganese Deposits, Glasby, G. P., ed. Elsevier, Amsterdam, 1977.
14. Hem, J. D. Geochim. Cosmochim. Acta. 1983, 47, 2037.
15. Murray, J. W.; Dillard, J. G.; Giovanoli, R.; Moers, H.; Stumm, W. Geochim. Cosmochim. Acta. 1985, 49, 463.
16. Postma, D. Geochim. Cosmochim. Acta. 1985, 49, 1023.
17. Oscarson, D. W.; Huang, D. M.; Defosse, C.; Herbillon, A. Nature 1981, 291, 50.
18. Rapsomanikis, S.; Weber, J. H. Environ. Sci. Technol. 1985, 19, 352.
19. Crowther, D. L.; Dillard, J. G.; Murray, J. W. Geochim. Cosmochim. Acta. 1983, 47, 1399.

20. Giovanoli, R.; Buerki, P.; Giuffredi, M.; Stumm, W. Chimia
    1975, 29, 517.
21. Bjerrum, J.; McReynolds, J. P. Inorg. Syn. 1946, 2, 216.
22. Burstall, F. H.; Nyholm, R. S. J. Chem. Soc. 1952, 3570.
23. Schilt, A. A.; Taylor, R. C. J. Inorg. Nucl. Chem. 1959, 9,
    211.
24. Schenck, C. V.; Dillard, J. G.; Murray, J. W. J. Colloid
    Interface Sci. 1983, 95, 398.
25. Dillard, J. G.; Koppelman, M. H.; Crowther, D. L.; Schenck, C.
    V.; Murray, J. W.; Balistrieri, L. in "Adsorption from Aqueous
    Solutions"; Tweari, P. W., Ed.; Plenum: New York, 1981; p. 227.
26. Wager, C. D.; Riggs, W. M.; Davis, L. E.; Moulder, J. F.;
    Muilenberg, G. E. "Handbook of X-ray Photoelectron
    Spectroscopy"; Perkin Elmer, Phi Division: Eden Prairie, Minn.,
    1978.
27. Basolo, F.; Pearson, R. G. "Mechanisms of Inorganic Reactions,"
    John Wiley, 2nd ed, 1967, p. 124ff.
28. Busch, K. L.; Cooks, R. G. Science 1982, 218, 247.
29. Pierce, J.; Busch, K. L.; Walton, R. A.; Cooks, R. G. J. Am.
    Chem. Soc. 1981, 103, 2583.
30. Unger, S. E.; Cooks, R. G.; Steinmetz, B. J.; Delgass, W. N.
    Surface Sci. 1982, 116, L211.
31. Winograd, N.; Karwacki, E. J. Anal. Chem. 1983, 55, 790.

RECEIVED May 20, 1986

SOLID SOLUTIONS

25

Ionic Solid Solutions in Contact with Aqueous Solutions

Ferdinand C. M. Driessens

Catholic University, Nijmegen, the Netherlands

The knowledge that ionic solutions are mostly
regular, if not ideal (1-3) was used to describe
their solubility behavior in water. It appears that
Roozeboom's class I solid solutions are ideal. In
coprecipitation they follow the Doerner-Hoskins law,
where the distribution coefficient is a simple function of the solubility products of the pure components.
Class II and class III solid solutions were found to
be regular, having a positive and a negative heat of
mixing respectively. Substitutional disorder in ideal
solid solutions gives rise to class II solid solutions,
whereas ordering related to a negative value for the
heat of mixing gives rise to type III solid solutions.

An ionic compound has a fixed composition. It may consist of
several cations (A, B, ...) and several anions (X, Y, ...) so that
its general formula is given by:

$$A_k B_l \ldots X_m Y_n \tag{1}$$

in such a way that k, l, m and n are simple integers.
    An ionic solid solution, on the other hand, is of variable
composition. Its chemical formula can not be written in simple
integer ratio numbers. A theoretical example is that of two
compounds AX and BX which form a solid solution of the general
formula:

$$A_{1-x} B_x X \tag{2}$$

with $0 \leq x \leq 1$, if AX and BX form a continuous series of solid
solutions.
    The equilibrium between an ionic compound like that of formula
(1) and an aqueous solution can be described by a solubility
product defined by:

0097–6156/86/0323–0524$10.25/0

$$(a_A)^k (a_B)^l \ldots (a_X)^m (a_Y)^n = K_{sp} \qquad (3)$$

which is a constant under given temperature and pressure. In Equation (3), a_i represents the activity in the aqueous solution of the ion i. For simplicity, the charge of the ions is omitted in Equation (3) and subsequent expressions. However, equilibrium between a solid solution like that of Formula (2) and an aqueous solution is not characterized by a constant solubility product. In that case the following two Equations apply (4):

$$(a_A) (a_X) = K_{sp}^{AX} a_{AX,s} \qquad (4)$$

and,

$$(a_B) (a_X) = K_{sp}^{BX} a_{BX,s} \qquad (5)$$

where $a_{AX,s}$ and $a_{BX,s}$, respectively, represent the activities of the components AX and BX in the solid solution of Formula (2), whereas K_{sp}^{AX} and K_{sp}^{BX} are the solubility products of pure AX and BX, respectively.

In most cases the study of equilibria between solid solutions and aqueous solutions containing their ions is extremely difficult, since solid state diffusion is virtually absent at ordinary temperatures. Most ionic solid solutions can be made homogeneous only at temperatures above 500°C, where solid state diffusion is relatively fast.

Only in certain cases (a relatively high solubility of both components) is it possible to obtain equilibrium between a solid solution of known composition and an aqueous solution, because the solid solution is homogenized by a relatively fast recrystallization. In other instances, equilibrium develops between the surface of the particles of the solid solution and the aqueous solution.

The present paper is intended to review the most important literature in this field and to extend the theory from the widely accepted ideal solid solutions to the more general models of regular solid solutions (5), with and without ordering (6) or substitutional disorder (2, 3, 7).

The Roozeboom Classification

Roozeboom (8) classified systems of two isomorphous salts, forming solid solutions like those of Formula (2) which vary in respect to only one ion, such that they constitute ternary systems (including water). Three types were distinguished, depending on the relative distribution of the salts between the aqueous and solid phases, as shown schematically in Figure 1a. This diagram, commonly known as a Roozeboom diagram, gives the mole fraction of one of the salts in the aqueous phase (disregarding the water in this phase), e.g.

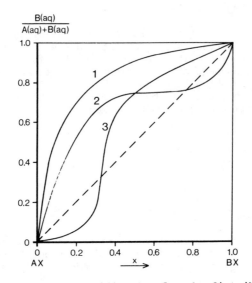

Figure 1a. Roozeboom's classification for the distribution of the ionic compounds AX and BX over the solid phase and the aqueous phase.

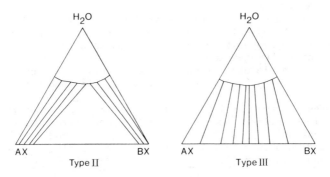

Figure 1b. Representation of type II and type III systems in the usual ternary phase diagram.

$$\frac{BX_{aq}}{AX_{aq} + BX_{aq}} \qquad (6)$$

as a function of its mole fraction x in the solid phase. $(i)_{aq}$ represents the molar concentration of the species i in the aqueous solution.

The meaning of types II and III can also be visualized on a Gibbs triangular diagram (See Figure 1b). The lines of type II would tend to converge on the aqueous solution curve, leading in extreme cases, to the formation of an isothermally and isobaric invariant aqueous phase, in equilibrium with two solid phases, meaning discontinuity in the solid solution. The lines of type III systems would tend to converge on the solid solution curve, leading finally to the formation of a solid compound with a definite composition lying between the two components.

Distribution Laws For Simple Ideal solid solutions. If a solid solution of Formula (2) is in equilibrium with an aqueous phase (aq), the distribution of A and B ions between the aqueous phase and the solid phase (s) can be represented by:

$$AX (s) + B(aq) \gtrless A(aq) + BX (s) \qquad (7)$$

and is described by:

$$\frac{a_{BX,s}}{a_{AX,s}} = D \frac{a_B}{a_A} \qquad (8)$$

provided that the solid phase is homogeneous. The solid solution of Formula (2) is ideal when their heat of mixing is zero and when their entropy of mixing is given by the relation

$$S = 2.303R \left[x \log x + (1-x) \log (1-x) \right] \qquad (9)$$

In that case,

$$a_{AX,s} = 1-x \qquad (10a)$$

and,

$$a_{BX,s} = x \qquad (10b)$$

apply (2, 3). Assuming that the activity coefficients of the A and B ions do not differ significantly, Equation (8) transforms to:

$$\frac{[BX]_s}{[AX]_s} = D \frac{[B]_{aq}}{[A]_{aq}} \qquad (11)$$

where BX_s and AX_s represent the concentration of BX and AX, respectively, in the solid phase. Equation (10) is known as the Berthelot-Nernst distribution law for coprecipitation (9). It represents type I solid solutions according to the classification of Roozeboom. This is illustrated in Figure 2 for two values of D. The corresponding relative amounts of AX and BX coprecipitated are given in Figure 3 which is modified from Gordon (10).

If the equilibrium expressed in Equation (7) is attained only between the crystal surface and the aqueous solution, the equilibrium is described by:

$$\frac{a_{BX,cs}}{a_{AX,cs}} = \frac{a_B}{a_A} \qquad (12)$$

where $a_{BX,cs}$ and $a_{AX,cs}$ represent the activities of components BX and AX, respectively, in the crystal surface layer. In the case of ideal solid solutions, $d\,BX_{cs}$ and $d\,AX_{cs}$, the increments of the components in the precipitated substance in the surface layer, are proportional to their respective solution concentration, i.e.

$$\frac{d\,BX_{cs}}{d\,AX_{cs}} = \frac{b_0 - b}{a_0 - a} \qquad (13)$$

provided that the activity coefficients of the A and B ions in the aqueous solution do not differ significantly. In Equation (13) b_0 and a_0 represent the initial quantities of BX and AX, respectively, in the aqueous solution. The symbols b and a represent the quantities of BX and AX, respectively, which have been deposited in the solid. Integration of Equation (13) yields:

$$\log \frac{B_{aq,i}}{B_{aq,f}} = \log \frac{A_{aq,i}}{A_{aq,f}} \qquad (14)$$

where the subscripts i and f denote the initial and final concentrations in the aqueous solution (10). Equation (14) is known as the Doerner-Hoskins distribution law (12) for coprecipitation, although it was derived first by Kroeker (12). It also represents only Roozeboom's type I systems.

The numerical values of the distribution coefficients λ and D have been derived from experimental data for a large number of systems (e.g. (10, 11, 13). From the constancy of either λ or D values it can be determined whether or not the system yielded homogeneous precipitates. In either case, the numerical value of λ or D should be equal to:

$$\lambda = D = \frac{K_{sp}^{AX}}{K_{sp}^{BX}} \qquad (15)$$

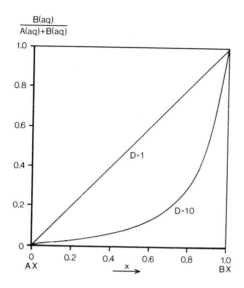

Figure 2. Distribution of the ionic components AX and BX over the solid components AX and BX over the solid phase and the aqueous phase for different values of the distribution parameter D under the assumption that AX and BX form ideal solid solutions and that the solid phase is homogeneous.

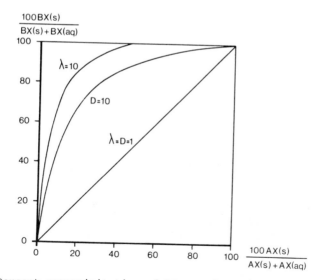

Figure 3. Percent coprecipitation of AX vs. that of BX from an aqueous solution under the assumption that AX and BX form ideal solid solutions. Modified from Gordon et al. (10).

which is easily derived from Equation (4) and (5). By changing the experimental conditions some systems can be made to obey either Equation (11) or Equation (14) (10). In most cases, however, the behavior of a system will be somewhere between that described by Equation (11) or Equation (14).

Distribution Laws For Complex Ideal Solid Solutions. Let A_nX and B_nX be two ionic compounds which form a series of solid solutions of the Formula:

$$A_n(1-x) B_{nx}X \tag{16}$$

In that case the entropy of mixing is:

$$S = 2.303 \; Rn \; \{x \; \log x + (1-x) \; \log (1-x)\} \tag{17}$$

so that:

$$a_{AX,s} = (1-x)^n \tag{18}$$

and,

$$a_{BX,s} = x^n \tag{19}$$

as long as the distribution of A and B ions over their sublattice is random (2). Equations (8) through (14) remain valid under these conditions, so that only type I solid solutions are found among these systems. It is easily shown that in this case

$$= D = \left[\frac{K_{sp}^{AX}}{K_{sp}^{BX}} \right]^{1/n} \tag{15a}$$

which is a more general expression of Equation (15) (14).

Distribution Laws And Regular Solid Solutions. For so-called regular solid solutions (15), Equation (9) still holds but by definition the expression for their enthalpy of mixing is:

$$H_m = x \; (1-x) \; W \tag{20}$$

in which W is the parameter for A - B ion interaction ($\underline{3}$). The activities of the components AX and BX in the solid solution of Formula (2) then becomes:

$$a_{AX,s} = (1-x) \exp \{x^2 W/(2.303 \, RT)\} \qquad (21)$$

and

$$a_{BX,s} = x \exp \{(1-x)^2 W/2.303 \, RT)\} \qquad (22)$$

In that case a plot of log $(a_{AX,s})/(1-x)$ versus x^2 or of log $a_{BX,s}/x$ versus $(1-x)^2$ must yield straight lines with the same slope, from which W can be calculated. Under these conditions Equation (8) transforms to:

$$\frac{BX_s}{AX_s} = D \, \frac{B_{aq}}{A_{aq}} \exp \{-(1-2x) \, W/(2.303 \, RT)\} \qquad (23)$$

so that the apparent distribution coefficient (D in Equation (10), is no longer constant but depends on x. Kirgintsev and Trushnikova ($\underline{16}$) have published a general method to derive $a_{AX,s}$ and $a_{BX,s}$ from experimental distribution data, and they have shown that a number of systems obey Equation (23) in systems with high rates of recrystallization. Figures 4 and 5 give an example of distributions in a system with varying values for W and D. Both type II and type III solid solutions of Roozeboom's classification are found in such systems, depending on whether W has a positive or a negative value, respectively.

The values chosen for $W/(2.303 \, RT)$ in order to construct Figures 4 and 5 are realistic; for most regular ionic solid solutions these values range from 1 to -2 ($\underline{3}$).Due to the differences in solubility products of the components of such solid solutions, however, the value of the distribution coefficient D can deviate several orders of magnitude from unity (see Equation (14a)). By extrapolation from Figures 4 and 5 it can be shown that type II and type III systems are indistinguishable from type I systems when the distribution coefficient D differs by one order of magnitude or more from unity. In those cases, experimental data for the distribution of ions between the solid solution and aqueous solution are not suitable to derive the nature of the solid solutions, as has been proposed by Kirgintsev and Trushnikova ($\underline{16}$). At very small or very large values of D, even miscibility gaps in solid solutions can not be detected by this method.

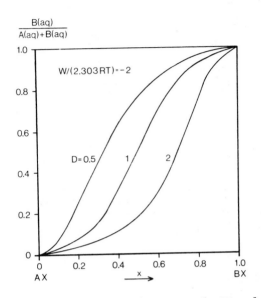

Figure 4. Distribution of the ionic compounds AX and BX over the solid phase and the aqueous phase for different values of the distribution parameter D under the assumption that AX and BX form homogeneous regular solid solutions with a negative value for the interaction parameter W.

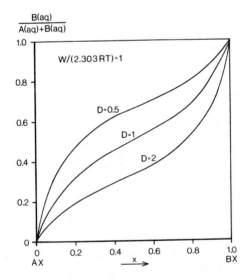

Figure 5. Distribution of the ionic compounds AX and BX over the solid phase and the aqueous phase for different values of the distribution parameter D under the assumption that AX and BX form homogeneous regular solid solutions with a positive value for the interaction parameter W.

DISTRIBUTION LAWS AND SUBSTITUTIONAL DISORDER

Driessens ($\underline{2}$) has discussed the consequences of substitutional disorder on component activities in solid solutions. For example, solid solutions of the Formula:

$$Mn_{2(1-x)} Co_{2x} SiO_4 \qquad (24)$$

with the olivine structure obeyed the relations:

$$0.5 \ln a_{Co_2SiO_4,s} = \ln a_{CoSi_{0.5}O_2,s} = \ln x - 0.2 (1-x)^2 \quad (25)$$

and

$$0.5 \ln a_{Mn_2SiO_4,s} = \ln a_{MnSi_{0.5}O_2,s} = \ln (1-x) - 0.2 x^2 \quad (26)$$

within the limits of experimental error, and thus, these solid solutions appeared to be regular.
However, exchange of Co^{2+} and Mn^{2+} can occur between lattice sites (4a) and (4c), resulting in equilibrium according to the reaction:

$$Co^{2+} (4a) + Mn^{2+} (4c) \stackrel{\rightarrow}{\leftarrow} Co^{2+} (4c) + Mn^{2+} (4a) \qquad (27)$$

Let the disorder parameter be z. Then the structural Formula of these olivines can be written as:

$$Co_{x+z} Mn_{1-x-z} (Co_{x-z} Mn_{1-x+z}) SiO_4 \qquad (28)$$

If the law of mass action applies to Equation (27), the disorder parameter can be estimated to a first approximation by:

$$z = x (1-x) (1-K_{27}) (1+K_{27})^{-1} \qquad (29)$$

where K_{27} is the equilibrium constant for the reaction in Equation (27). Furthermore, the following expressions are obtained for the activities of the components:

$$a_{Co_2SiO_4,s} = (x+z) (x-z) = x^2 - z^2 \qquad (30)$$

and

$$a_{Mn_2SiO_4,s} = (1-x+z) (1-x-z) = (1-x)^2 - z^2 \qquad (31)$$

From these expressions, one can derive

$$0.5 \ln a_{Co_2SiO_4,s} = \ln x + 1/2 \{V(1-x)^2 - 1/2 V^2 (1-x)^4 + ...\} \tag{32}$$

and

$$0.5 \ln a_{Mn_2SiO_4,s} = \ln (1-x)+1/2 \{V x^2-1/2 V^2 x^4 + ...\} \tag{33}$$

where V will be defined below in Equation (34). Equations (32) and (33) agree to a first approximation with the experimental curves given by Formulas (25) and (26).

In this way and by numerical evaluation, Driessens (2) proved that the experimental activities could be explained on the basis of substitutional disorder, according to Equation (27), within the limits of experimental error. It seems, therefore, that measurements of distribution coefficients and the resulting activities calculated by the method of Kirgintsev and Trushnikova (16) do not distinguish between the regular character of solid solutions and the possibility of substitional disorder. However, the latter can be discerned by X-ray or neutron diffraction or by NMR or magnetic measurements. It can be shown that substitutional disorder always results in negative values of the interaction parameter W due to the fact that

$$W(2.303 \ RT)^{-1} = - 1/2 V = - 1/2 (1-K_{27})^2 (1+K_{27})^2 \tag{34}$$

This is also valid for the more complex spinel solid solutions of Fe_3O_4, Mn_3O_4 and CO_3O_4, in which electron exchange occurs in addition to substitutional disorder (2).

Substitutional Disorder In Regular Solid Solutions. Most simple ionic solutions in which substitution occurs in one sublattice only are not ideal, but regular (2, 3). Most complex ionic solid solutions in which substitution occurs in more than one sublattice are not only regular in the sense of Hildebrand's definition (15) but also exhibit substitutional disorder. The Equations describing the activities of the components as a function of the composition of their solid solutions are rather complex (7, 17, 18), and these can be evaluated best for each individual case. Both type II and type III distributions can result from these conditions.

Ordering. New compounds which include the ionic components AX and BX may be formed by ordering of the solid solution (6). In that case, the entropy of mixing may still be given by Equation 17, whereas the enthalpy of mixing is given by:

$$H_m = x^n (1-x)^n W \tag{35}$$

in which W has a negative value. In this expression n is the minimum number of units of AX and BX which unite to form a nucleus of the new compound $A_nB_nX_{2n}$. The solubility behavior of solid solutions of $CaCO_3$ and $MgCO_3$ could be explained with this Equation under the assumption that n = 3 for dolomite, the new compound which forms between $CaCO_3$ and $MgCO_3$. This value of n is in agreement with the content of a lattice cell in the dolomite structure (6).

The appropriate expressions for the activities of the components become:

$$a_{AX,s} = (1-x) \exp\{x^n(1-x)^{n-1}[1-n+(2n-1)x]W\}/(2.303 \, RT) \qquad (36)$$

and,

$$a_{BX,s} = x \exp\{(1-x)^n \, x^{n-1}[n-(2n-1)x]W\}/(2.303 \, RT) \qquad (37)$$

Such systems belong to type III distributions because the value of W is always negative. The system $CaCO_3$ – $MgCO_3$ – H_2O is given as an example in Figure 6.

Comparison With Literature Data

The distribution of components of binary solid solutions over the solid phase and the aqueous phase has been studied for a number of systems. Table I contains a summary of some of these systems with references. This literature review is not complete; more data are available especially for rare earth and actinide compounds, which primarily obey type I Equations to a good approximation. In the following sections, the theory above will be applied to some special systems which are relevant to the fields of analytical chemistry, inorganic chemistry, mineralogy, oceanography and biominerals.

Application In Analytical And Inorganic Chemistry

Knowledge about distribution coefficients is used in analytical chemistry to determine the feasibility of quantitative separation by precipitation. Therefore, D and λ are also called separation factors. In order to precipitate 99.8% or more of the primary substance, λ must be 3.2×10^{-4} or smaller. For larger values of λ more than one precipitation step is necessary, and the number of steps can be calculated when λ is known.

This straightforward application is obvious for type I systems only, for which coprecipitation diagrams like Figure 3 can be calculated and experimentally verified. As can be seen from Figures 4 and 5, the apparent distribution coefficient, λ, for systems of

Table I.

Evaluation of some systems of solid solutions according
to their solubility behavior

AX	BX	Type	D( )*	Reference
NaCl	NaBr	II		Kirgintsev and Trushnikova (16)
KCl	KBr	II		Kirgintsev and Trushnikova
NH$_4$Cl	NH$_4$Br	II		Kirgintsev and Trushnikova
RbCl	RbBr	II		Kirgintsev and Trushnikova
CsCl	CsBr	II		Kirgintsev and Trushnikova
RbBr	KBr	II		Durham et al (19)
KCl	NH$_4$Cl	(II)**		Flatt and Burkhardt (20)
KBr	NH$_4$Br	(II)		Flatt and Burkhardt
AgCl	AgBr	I (?)	104	Yutzy and Kolthoff (21)
TlCl	AgCl	I		Vaslow and Boyd (22)
NH$_4$Al (SO$_4$)$_2$	KAl (SO$_4$)$_2$	I	0.9	Hill et al (13)
NH$_4$Cr (SO$_4$)$_2$	KCr (SO$_4$)$_2$	I	1.6	Hill et al
NH$_4$Al (SO$_4$)$_2$	TlAl (SO$_4$)$_2$	I	2.5	Hill et al
KAl (SO$_4$)$_2$	TlAl (SO$_4$)$_2$	I	2.5	Hill et al
Ca$_5$(PO$_4$)$_3$ OH	Ca$_5$(PO$_4$)$_3$F	(II)		Driessens

AX	BX	Type	D( )*	Reference
Cu(NH$_4$)$_2$(SO$_4$)$_2$	Zn(NH$_4$)$_2$(SO$_4$)$_2$	II		Hill et al (13)
Zu(NH$_4$)$_2$(SO$_4$)$_2$	Ni(NH$_4$)$_2$(SO$_4$)$_2$	I	4	Hill et al
Cu(NH$_4$)$_2$(SO$_4$)$_2$	Ni(NH$_4$)$_2$(SO$_4$)$_2$	I (?)	16	Hill et al
Mg(NH$_4$)$_2$(SO$_4$)$_2$	Cu(NH$_4$)$_2$(SO$_4$)$_2$	II		Hill et al
CuK$_2$(SO$_4$)$_2$	NiK$_2$(SO$_4$)$_2$	II		Hill et al
CoK$_2$(SO$_4$)$_2$	CuK$_2$(SO$_4$)$_2$	II		Hill et al
RaCrO$_4$	BaCrO$_4$	I	5.5	Gordon et al (10,23)
RaSO$_4$	BaSO$_4$	I	1.2	Gordon et al
RaCO$_3$	BaCO$_3$	I	0.18	Gordon et al
SrSO$_4$	BaSO$_4$	I (?)	0.030	Gordon et al
RaBr$_2$	BaBr$_2$	I	9.8	Gordon et al
RaCl$_2$	BaCl$_2$	I	5.0	Gordon et al
Ra(NO$_3$)$_2$	Ba(NO$_3$)$_2$	I	2.0	Gordon et al

Table I. (Continued)

AX	BX	Type	D( )*	Reference
$Pb(NO_3)_2$	$Ra(NO_3)_2$	I	3	Ratner (9)
$BaSO_4$	$PbSO_4$	I (?)	0.08	Kolthoff and Noponen (24)
$PbCrO_4$	$PbMoO_4$	I (?)	250	Kolthoff and Eggertsen (25)
$CaCO_3$	$MgCO_3$	(III)****		Driessens and Verbeeck (6)
$BaHPO_4$	$SrHPO_4$	I (?)	0.31	Spitsyn et al (26)
$MgNH_4PO_4$	$MgNH_4AsO_4$	I	5.7	Kolthoff and Carr (27)
Sm-oxalate	Nd-oxalate	I	1.65	Weaver (22)
Nd-oxalate	Pr-oxalate	I	1.37	Weaver
Sm-oxalate	Gd-oxalate	I	1.66	Weaver
Gd-oxalate	Dy-oxalate	I	2.09	Weaver
Ce(III)-oxalate	Nd-oxalate	I	1.75	Gordon et al (10, 23)
Am-oxalate	La-oxalate	I	5.85	Gordon et al
Yb-oxalate	Nd-oxalate	I	0.69	Gordon et al
$Fe(IO_4)_3$	$Y(IO_4)_3$	I (?)	0.001	Gordon et al
$Th(IO_4)_4$	$La(IO_3)_3$	**	6.5	Gordon et al

*   The case of type I were D is constant rather than   are rare.
**   In this case relation (13) is not valid. A more appropriate
       relation was derived by Gordon et al (10).
*** (II) means that the interaction parameter W is so strongly
       positive that a miscibility gap occurs in the series of solid
       solitions.
**** (III) means that n is so high and W is so negative that a new
       compound is formed under the simultaneous development of two
       miscibility gaps in the series of solid solutions.

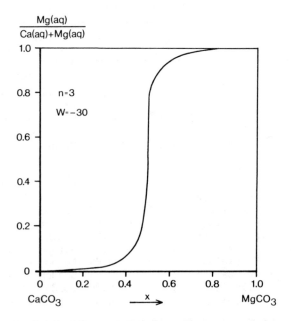

Figure 6. Distribution of the ionic compounds $CaCO_3$ and $MgCO_3$ over the solid phase and the aqueous phase. Ordering occurs in the solid solutions around $x = 0.5$. It is assumed that the solid phase is homogeneous.

regular solid solutions or with substitutional disorder or ordering depends on the initial molar ratio of the components in the aqueous solution. Thus, the calculation of coprecipitation diagrams for type II and type III systems is meaningless, except for theoretical purposes.

There is evidence that the value of λ for type I systems depends on the degree of supersaturation during the precipitation process with a somewhat better separation being reached at lower rates of precipitation, and hence, at lower degrees of supersaturation (29). This may mean that the events at the interface of solid phase and liquid phase are not completely described by Equation (12), e.g. adsorption might also be involved.

The importance of adsorption is especially clear from studies of ion entrapment, a phenomenon whereby occlusion of adsorbed foreign ions occurs by overgrowth of a precipitate (30). Occlusion of chloride in a $BaSO_4$ precipitate can be diminished by adding barium chloride to the sulfate solution rather than the reverse. It is well known that the amount of occlusion generally increases with the speed of formation of a precipitate. However, the rapidly formed crystals produced from relatively concentrated solutions have a higher rate of recrystallization during aging due to their small particle size. Thus, it is advisable in analytical procedures to precipitate rapidly at room temperature followed by aging at slightly higher temperatures (31).

As has been observed by many authors and as seen from Table I, there are only a few type I systems for which the distribution coefficient D is constant. Usually λ is constant which means that precipitates are not homogeneous but contain logarithmic concentration gradients. Taking into account that for type II and III systems the situation is even more complex, one comes to the conclusion that, in general, the precipitate of a solid solution will not be homogeneous, unless the concentrations of the ions in the aqueous solution are held constant during the precipitation process. The gradients in the solid solution will be more pronounced for more extreme values of D and W. As observed by Driessens (2, 3, 7), ideal solid solutions are an exception rather than a rule among ionic solid solutions. Therefore, preparation of homogeneous ionic solid solutions by precipitation from aqueous solutions can in general only be reached by tedious iterative procedures, provided that techniques are developed to keep the concentration of all ions in the aqueous solution constant during the precipitation process. For obvious reasons, the distribution coefficient D must not differ much from unity if one aims to prepare such solid solutions over a wide range of x values. Fortunately, high-temperature techniques, i.e. hydrothermal or solid-state chemical methods, can provide more direct methods to prepare homogeneous ionic solutions for many systems, because they may be operated at temperatures at which diffusion in the solid solutions becomes sufficiently fast.

The System $CaCO_3$ - $MgCO_3$

In the system $CaCO_3$-$MgCO_3$-H_2O several solid phases can
occur. The compound $CaCO_3$ exists in three polymorphs under
atmospheric pressure: calcite, aragonite and vaterite. Vaterite is
the least stable and will not be considered further here. Calcite
is slightly more stable than aragonite (32) at most earth surface
conditions. For $MgCO_3$, magnesite is the stable solid phase,
except at low partial pressures of CO_2 where hydromagnesite
($Mg_4(CO_3)_3(OH)_2 \cdot 3H_2O$) becomes stable (33). Between the
extreme compositions dolomite ($Ca_{0.5}Mg_{0.5}CO_3$) is found as a
stable solid phase (33). The structure of dolomite is that of an
ordered calcite, whereas magnesite is isostructural with calcite
(34). Natural dolomites contain between 40 and 51 mol-% $MgCO_3$
(35-38). In calcite sediments, up to about 6 mol-% $MgCO_3$ is
found, whereas aragonite sediments contain very little $MgCO_3$ (37,
38).

The solubility of dolomite is less than that of either calcite
or magnesite (39). Equilibrium of dolomite with aqueous solutions,
with no added Mg^{2+} ions, leads ultimately to the formation of a
thin calcite layer onto the dolomite particles (40), whereas
equilibration of calcite with aqueous solutions containing Mg^{2+}
ions results in the formation of a thin layer of dolomite on the
calcite particles (41, 42). In both cases these surface layers
become the controlling solid phase in solid-liquid phase equilibria
(43). Under certain conditions calcium-rich dolomite appears to be
more soluble than calcite (44). The most soluble seems to be a
solid solution containing containing between 20 and 30 mol-% $MgCO_3$
(45). Aqueous solutions equilibrated with calcium-rich dolomites
can become supersaturated with aragonite, which can then
precipitate and become the solid phase controlling the solubility
of Ca^{2+} (46).

In precipitation studies (47, 49) it has been shown that,
below a certain Mg/Ca concentration ratio in the aqueous solution,
the rate of nucleation of calcite was faster than that of
aragonite. Above that Mg/Ca ratio the order was reversed. This was
explained by the effect of Mg^{2+} ions on the interfacial tension
between the solution and precipitate, which apparently is larger
for calcite than for aragonite (49). At still higher Mg/Ca ratios
dolomite can be formed (50). Such low temperature precipitates of
dolomite contain ordering defects. The number of defects increases
when precipitation proceeds in a shorter time interval or at lower
temperatures (51).

The solubility, dissolution and precipitation behavior in the
system $CaCO_3$-$MgCO_3$-H_2O can be described by the following
model. Let the general Formula of the Ca-Mg-carbonate be
represented by:

$$Ca_{1-x}Mg_x CO_3 \tag{38}$$

The free energy of mixing of such a solid solution is given by:

$$G^M = G - (1-x)\ G^o\ _{CaCO_3} - x\ G^o\ _{MgCO_3} \qquad (39)$$

in which G and G^o_i represent the free energy of the solid solution and that of the pure component i respectively. On the other hand, the free energy of mixing also equals:

$$G^M = (1-x)\ G^M_{CaCO_3} + x\ G^M_{MgCO_3} \qquad (40)$$

in which G^M_i is the partial free energy of mixing of component i. When a solid solution of Formula (38) is in equilibrium with an aqueous solution of its ions, its solubility behavior is completely described by the reactions:

$$CaCO_3(ss) \overset{\rightarrow}{\leftarrow} Ca^{2+}(aq) + CO_3^{2-}\ (aq) \qquad (41)$$

and

$$CaCO_3(ss) + Mg^{2+}(aq) \overset{\rightarrow}{\leftarrow} MgCO_3(ss) + Ca^{2+}(aq) \qquad (42)$$

where (ss) represents the solid solution. The equilibria (41) and (42) are defined by the respective relations:

$$G^M_{CaCO_3} = 2.303\ RT\ (\log I_{CaCO_3} - \log K_{CaCO_3}) \qquad (43)$$

and

$$\log \frac{a_{Ca}}{a_{Mg}} = \frac{G^M_{CaCO_3} - G^M_{MgCO_3}}{2.303\ RT} + \log \frac{K_{CaCO_3}}{K_{MgCO_3}} \qquad (44)$$

where

I_{CaCO_3} is the ion activity product $(a_{Ca})\ (a_{CO3})$,

$(10^{-8.42})$ K_{CaCO_3} is the solubility product of pure calcite

and $(10^{-8.12})$ K_{MgCO_3} is the solubility product of pure magnesite

Driessens and Verbeeck (6) derived the following analytical expression for G^M:

$$G^M = 2.303 \; RT[(1-x)\log(1-x) + x\log x] + x^3(1-x)^3 \; W \qquad (45)$$

in which the most appropriate value for the interaction parameter W is - 858 kJ mole^{-1}, when the ordering of the Ca and Mg ions in the structure is ideal. This is based on the stability ranges for the solid solutions mentioned above and on the experimental formation energy of well-ordered dolomite.

The graphical form of G^M as a function of x is given in Figure 7. For $0 \leq x < x_1$, calcite is stable, whereas for $x_1 < x < x_2$ calcite is metastable. Solid solutions between x_2 and x_3 are unstable. In the range $x_3 < x < x_4$ and $x_4 < x < (1-x_4)$ dolomites are metastable, and stable, respectively.

This model for the system $CaCO_3$-$MgCO_3$ applies only for ideal ordering of Mg and Ca ions in the dolomite structure. Ideal ordering occurs only in precipitates of dolomite formed at temperatures above about 250°C. Studies in the laboratory (52) show that dolomitization (the development of ordering in the Mg and Ca distribution in the calcite structure) is a very slow process at ordinary temperatures. Therefore, a solid-state chemical model more applicable to precipitated dolomites is:

$$G^M = 2.303 \; RT[(1-x)\log(1-x) + x \log x] + x^3(1-x)^3 \quad \alpha W \qquad (46)$$

where $\quad$ is an ordering parameter which theoretically can vary from 0 to 1. The resulting expression for $\underset{G_{CaCO_3}}{M}$ becomes:

$$G_{CaCO_3}^M = 2.303 \; RT \log (1-x) + x^3(1-x)^2(5 \; x-2) \; \alpha W \qquad (47)$$

whereas

$$G_{CaCO_3}^M - G_{MgCO_3}^M = 2.303 \; RT \log x - \log(1-x) + 3x^2(1-x)^2(1+2x) \; \alpha W \qquad (48)$$

In table II the values of x, through x_4 are given for certain values of the ordering parameter α.

Table II.

Values of x_1 through x_4 as a function of the ordering parameter α for solid solutions of the composition $Ca_{1-x}Mg_xCO_3$ at 25°C

α	1	0.5	0.25
x_1	7.6×10^{-7}	4.3×10^{-4}	1.2×10^{-2}
x_2	0.03	0.06	0.10
x_3	0.28	0.28	0.27
x_4	0.38	0.37	0.31

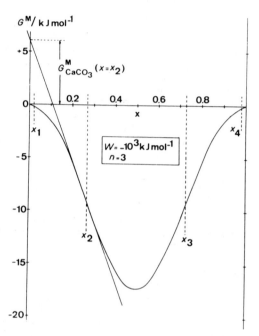

Figure 7. Proposed form for the curve of the free energy of mixing in the system $CaCO_3$ - $MgCO_3$. The curve was calculated with the indicated values for the parameters n and W according to the proposed model of subregular solid solutions.

According to Equation 48 calcite should precipitate from waters
having a Mg/Ca ratio below a certain value, while dolomite should
precipitate from waters having a Mg/Ca ratio above that critical
value. This rule is obeyed under conditions of precipitation from
very slightly supersaturated aqueous solutions like those occurring
in certain areas of the ocean. Ocean water is close to equilibrium
with both calcite and dolomite (53).

   When precipitation occurs under conditions of high supersaturation, kinetic factors become important as well. Then calcite,
aragonite and dolomite can form from solutions having Mg/Ca ratios
in increasing order of magnitude. This is the main reason why not
only calcite and dolomite, but also aragonite is found among
biologically induced carbonatations (53). Interesting in this
respect is that calcium and pH homeostasis in snails (54, 55) and
frogs (56) was shown to reflect a constant ionic product for
calcite, which after proper correction for activity coefficients
was equal to that of the solubility product.

The System Hydroxyapatite - Fluorapatite

Most calcium, containing apatites in nature are heavily carbonated.
The only exception is formed by the mineral in the surface of tooth
enamel which consists mainly of hydroxyapatite $(Ca_5(PO_4)_3OH)$.
Most foods and drinking waters contain enough fluoride to result in
the incorporation of significant amounts of fluoride into this
mineral whereby the solubility decreases. Therefore, the system
hydroxyapatite-fluorapatite is primarily of importance for the
prevention of dental caries. However, in this context its
theoretical treatment is important for geochemists who may be
confronted with so-called subregular solid solutions.

   The logarithm of the solubility product for hydroxyapatite is
-58.6 and that of fluorapatite $(Ca_5(PO_4)_3F)$ is -60.6 (57),
and thus, D = 0.01 in favour of fluoride incorporation into the
solid apatite precipitate. Accordingly, it should be difficult to
prepare solid solutions of these compounds by precipitation from
aqueous solution and if prepared batchwise, they are expected to
contain logarithmic gradients in their internal composition. Yet,
Moreno et al. (58) report linear changes in the lattice parameters
of such solid solutions. They also determined their solubility
behavior.
Given the formula as:

$$Ca_5(PO_4)_3OH_xF_{1-x} \qquad\qquad\qquad (49)$$

their solubility behavior has been evaluated (58) by using the
solubility products

$$K(x) = a_{Ca^{2+}}^5 \cdot a_{PO_4^{3-}}^3 \cdot a_{OH^-}^x \cdot a_{F^-}^{1-x} \cdot \qquad (50)$$

On the other hand, Wier et al (59) have shown that fluoride ions react with the surface of hydroxyapatite particles so that a state of equilibrium is reached as if the aqueous solution is in equilibrium with pure fluorapatite, provided that enough fluoride ions occur in the aqueous solution. Therefore, one should expect, that particles of solid solutions of hydroxyapatite and fluorapatite will react similarly with fluoride ions from an aqueous solution, and that a surface layer is formed which has a composition closer to that of pure fluorapatite than that of the original solid solution.

This solid solution still makes up the bulk of the solid particles after equilibration in an aqueous solution (59), since solid state diffusion is negligible at room temperature in these apatites (60), which have a melting point around 1500°C. These considerations and controversial results justify a thermodynamic analysis of the solubility data obtained by Moreno et al (58). We shall consider below whether the data of Moreno et al (58) is consistent with the required thermodynamic relationships for 1) an ideal solid solution, 2) a regular solid solution, 3) a subregular solid solution and 4) a mixed regular, subregular model for solid solutions.

In that study (58), the average of the logarithms of the solubility products for pure hydroxyapatite ($\log K_{OHA}$) and pure fluorapatite ($\log K_{FA}$) appeared to be - 59.16 and - 60.52 respectively, both with an uncertainty of about $\pm$ 0.30. In the present study the solubility data found for equilibration of solid solutions are expressed as the negative logarithms for the ionic products of hydroxyapatite and fluorapatite, i.e.

$$\log I_{OHA} = 5\log a_{Ca^{2+}} + 3\log a_{PO_4^{3-}} + \log a_{OH^-} \qquad (51)$$

and

$$\log I_{FA} + 5\log a_{Ca^{2+}} + 3\log a_{PO_4^{3-}} + \log a_{F^-} \qquad (52)$$

Subsequently, the apparent activities of the quasibinary components hydroxyapatite OHA and fluorapatite FA were derived as follows:

$$\log a_{OHA} = \log I_{OHA} - \log K_{OHA} \qquad (53)$$

and

$$\log a_{FA} = \log I_{FA} - \log K_{FA} \qquad (54)$$

It is assumed that in this experiment (58), stable or metastable equilibrium had been reached between the aqueous solution and a surface layer of the apatite particles.

The results of the calculations using Equations (51) through (54) are given in Table III, which includes the pH values of the original equilibrations. In addition, mass balance calculations were carried out to see whether the solid particles had accumulated fluoride in their surface layer from the aqueous solutions. The mass balance showed that an accumulation of fluoride had occurred in the equilibration of all solid solutions. This discounts an interpretation of the solubility data as carried out by Moreno et al. (58).

A thermodynamically acceptable explanation for the solubility behavior of solid solutions at x = 0.868 is needed. First, we shall assume that OHA-FA solid solutions are ideal. If the composition of the surface layer of the solid particles is given by Equation (49), then the following equations can be derived (2):

$$\log a_{OHA} = \log x \qquad (55)$$

and

$$\log a_{FA} = \log(1-x) \qquad (56)$$

The data of Table III show that the surface layer of the solid particles is indistinguishable from pure fluorapatite in all equilibrations at x = 0.110, 0.190 and 0.435 and 0.595. However, some equilibrations at x = 0.763 and all at x = 0.868 do deviate significantly from the behavior of pure fluorapatite. A peculiar aspect is that the activity of fluorapatite becomes significantly larger than 1. Simutaneously, the activity of hydroxyapatite approaches unity. This would mean that at all values of x both activities would become smaller than 1, and thus an ideal behavior of the solid solutions would not explain the observed solubility behavior.

Next let us assume that the solid solutions are regular. Then the following relations hold (3, 7):

$$\log a_{OHA} = (1-x)^2 \frac{W}{2.303\ RT} + \log x \qquad (57)$$

and

$$\log a_{FA} = x^2 \frac{W}{2.303\ RT} + \log(1-x) \qquad (58)$$

where W is a parameter for the interaction energy between hydroxyl and fluoride ions within the apatite lattice. For $W/2.303\ RT < 0.88$, the activities as a function of x are similar to those of ideal solid solutions (Figure 8). However, for $W/2.303\ RT \geq 0.88$, a solubility gap occurs which is symmetrical with respect to x = 0.5.

Under such conditions, the free enthalphy curve at a given temperature shows two minima and one maximum as a function of x.

Table III.
Apparent activities of hydroxyapatie (OHA) and fluorapatite (FA)
after equilibration of solid solutions of the formula
$Ca_5(PO_4)_3F_{1-x}OH_x$

x	pH	log a_{OHA}	log a_{FA}	x	pH	log a_{OHA}	log a_{FA}
0.110	3.587	-8.33	0.10	0.595	4.334	-5.04	0.88
	3.604	-8.32	0.06		5.057	-4.25	0.29
	3.960	-7.76	-0.01		5.495	-3.50	0.28
	4.354	-7.08	-0.01		5.956	-2.83	0.81
	4.746	-6.41	-0.01		6.150	-2.86	0.51
	5.181	-5.66	-0.04				
	6.078	-4.27	-0.07	0.763	4.858	-3.38	0.93
					5.305	-2.45	1.35
0.190	3.596	-7.98	0.11		5.676	-2.15	1.11
	3.985	-7.61	0.07		6.276	-2.17	0.60
	4.400	-6.77	0.09				
	4.850	-5.87	0.26	0.868	4.894	-0.84	2.67
	5.261	-5.27	0.10		5.223	-0.78	2.39
	5.746	-4.59	0.17		5.630	-0.66	1.88
	5.823	-4.52	0.15		5.876	-0.57	1.63
0.435	3.637	-7.84	0.26				
	4.202	-7.12	0.11				
	4.433	-5.94	0.17				
	4.838	-5.78	0.21				
	5.257	-5.04	0.28				
	5.750	-4.48	0.06				
	6.062	-4.30	-0.23				

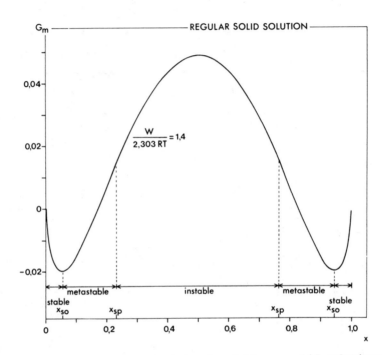

Figure 8. Free enthalpy of mixing G_m of binary solid solutions (regular) $Ca_5(PO_4)_3OH_xF_{1-x}$ as a function of x at W/2.303 RT = 1.4.

An example is illustrated in Figure 8 for $W/2.303\ RT = 1.4$. Two ranges of x values occur near the borders at $x = 0$ and $x = 1$, where the solid solutions are stable. The range of stability end at the x values which apply to the minima in the enthalphy curve, which are known as the limits of solid solubility X_{SO}. The two ranges of x values between the minima and the points of inflection in the enthalphy curve pertain to metastable solid solutions. The points of inflection occur at the so-called spinodal compositions X_{SP} (61). Between the two spinodal compositions any solid solution is unstable and will disproportionate into two solid solutions of the compositions X_{SO}.
Within the two metastable ranges one of the binary components can have an apparent thermodynamic activity larger than 1. The maximum will be reached at $x = x_{SP}$. In this study x_{SP} was derived as a function of $W/2.303\ RT$ by iterative procedures using the relevant equations given by Meyering (61). Subsequently, the thermodynamic activities of the two components were calculated at the extremes which can be reached for variable x_{SP} (Figure 9). Apparently, such high values as $\log a_{FA} = 2$ are reached only for $x > 0.93$. Thus, the assumption of a regular behavior of the solid solutions of OHA and FA does not explain the observed solubility behavior either.

Freund and Knobel (62) have found evidence from infrared studies that complexes of the form F-OH-F are of importance in solid solutions of OHA and FA, which were synthesized by us (63). In that case, the enthalphy of mixing H_m should be of a form typical for subregular behavior such as:

$$H_m = x(1-x)^2 W \tag{59}$$

whereas the following expressions are derived for the activities:

$$\log a_{OHA} = (1-2x)(1-x)^2 \frac{W}{2.303\ RT} + \log x \tag{60}$$

and

$$\log a_{FA} = 2x^2(1-x) \frac{W}{2.303\ RT} + \log(1-x) \tag{61}$$

Calculation of the extreme values of the activities at the spinodal compositions x_{SP} for variable values of $W/2.303\ RT$ results in the data presented in Figure 10. It appears that values as high as $\log a_{FA} = 2$ are reached in the range $x_{SP} > 0.63$. Thus, the assumption of a subregular behavior of the solid solutions of OHA and FA explains the observed solubility behavior qualitatively. It follows further from the calculations that $W/2.303\ RT \geq 8$ so that $W \geq 4.6$ $10^4\ J\ mol^{-1}$.

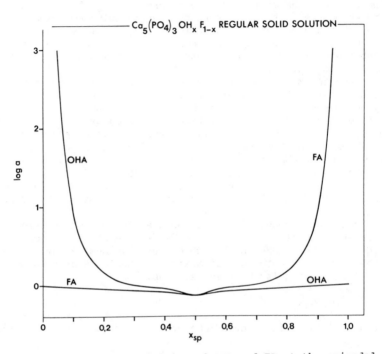

Figure 9. Ultimate activities of OHA and FA at the spinodal compositions x_{SP} in the model of regular solid solutions.

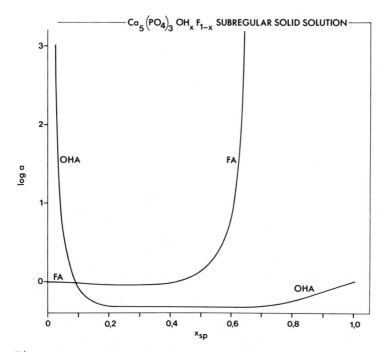

Figure 10. Ultimate activities of OHA and FA at the spinodal compositions x_{SP} in the model of subregular solid solutions.

Freund and Knobel (62) have found that in addition to F-OH-F complexes, F-OH pair interactions are important. Therefore, a mixed type regular and subregular model of the solid solutions should be more appropriate. To a first approximation, the enthalpy of mixing should then have the form:

$$H_m = x(1-x)W_1 + x(1-x)^2 W_2 \qquad (62)$$

If one assumes $W_1 = W_2 = W$ in order to minimise the number of parameters introduced in the model, one obtains the following expressions for the activities:

$$\log a_{OHA} = 2(1-x)^3 \frac{W}{2.303 \ RT} + \log x \qquad (63)$$

and

$$\log a_{FA} = x^2(3-2x) \frac{W}{2.303 \ RT} + \log(1-x) \qquad (64)$$

Calculation of the extreme values of the activities at the spinodal compositions x_{SP} for variable values of $W/2.303 \ RT$ yields the data presented in Figure 11. It appears that values as high as $\log a_{FA} = 2$ are reached in the range $x_{SP} > 0.83$. Accordingly, the value of $\log a_{OHA}$ is about -0.5, which is close to the experimental value at $x = 0.868$. Thus, the assumption of a mixed-type regular and subregular solid solution with $W_1 = W_2 = W$ explains the observed solubility behavior at $x = 0.868$. However, it does not explain the high activities of fluorapatite found in some of the equilibrations at $x = 0.763$. In this model $W/2.303 \ RT \geq 2.0$ so that $W \geq 1.17 \ . \ 10^4$ J mol^{-1}. Further refinement of this model is possible by independent variation of W_1 and W_2.

In the subregular model the absence of a solubility gap at 1000°C would mean $W/2.303 \ RT \leq 0.92$ and thus $W \leq 2.3 \ . \ 10^4$ J mol^{-1} (see Figure 3). On the other hand the solubility data indicate a value of $W \geq 4.6 \ . \ 10^4$ J mol^{-1}. In the mixed-type regular and subregular model with $W_1 = W_2 = W$ the absence of a solubility gap at 1000°C would mean $W/2.303 \ RT \leq 0.46$ so that $W \leq 1.17 \ . \ 10^4$ J mol^{-1}. For that model, the solubility data indicate a value of $W \geq 1.17 \ . \ 10^4$ J mol^{-1}. Therefore , a mixed type regular and subregular solid solutions is the most acceptable model, and the most probable value for the interaction parameter is $W = 1.17 \ . \ 10^4$ J mol^{-1}. Within the scope of this conclusion one should consider the increased activities of fluorapatite at $x = 0.763$ and 0.595 as probably being caused by the fact that their compositions are found beyond the maximum in the free enthalpy curve. Hence, their transformation into fluorapatite may be very slow, unless the concentration of fluoride ions in the aqueous solution is high.

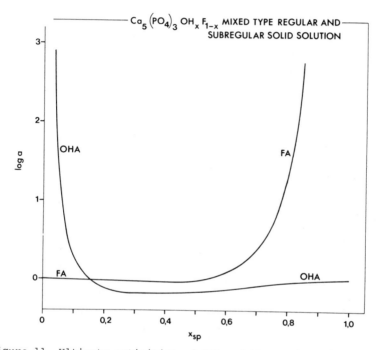

Figure 11. Ultimate activities of OHA and FA at the spinodal compositions x_{SP} in the model of mixed-type regular and subregular solid solutions.

The previous paper (63) also studied the disintegration of solid solutions and for that purpose samples were heated for 300 hours at 250°C, but no signs of disintegration were detected in an X-ray diffractogram. This might be due to the fact that solid state diffusion is still too slow at that temperature. This is supported by the low diffusion coefficient calculated if one extrapolates from the experimental values determined at high temperature (60).

In conclusion, the solubility data indicate that upon precipitation from aqueous solutions which have a F/OH molar ratio less than a certain value, slightly fluoridated hydroxyapatites will be formed (x ≤ 0.15), and above that ratio nearly pure fluor-apatite will be formed. Usually the F/OH ratio varies so that intimate mixtures of hydroxyapatite and fluorapatite will result (64). The effect of fluoride on teeth and bones are discussed elsewhere (53, 57).

The System Calciumhydroxyapatite - Strontiumhydroxyapatite. From a study of the cation distribution over the two cation sublattices in solid solutions of calciumhydroxyapatite and strontiumhydroxy-apatite (65) it was shown that such solid solutions are ideal.

Verbeeck (66) found that the solubility behavior could be explained by assuming ideality; his value for the logarithm of the solubility product of pure strontiumhydroxyapatite was -52.3. Hence, the value of D is 18 in favour of Ca incorporation and against Sr incorporation in mixed precipitates. This seems to be in agreement with discrimination against strontium in the bones and teeth of living organisms.

The System Calciumhydroxyapatite - Leadhydroxyapatite. In this system there is at least one and presumably two miscibility gaps around 1200°C (66). At room temperature there is one large miscibility gap. The solubility product for leadhydroxyapatite (67) is about 10^{-81} so that for this system D is about 30000 in favour of lead incorporation into the apatite. This means that upon precipitation, practically all the lead will precipitate before any calcium coprecipitates.

Calcium Phosphates And Calcified Tissues. Precipitation in the system Ca(OH)$_2$ - H$_3$PO$_4$ - H$_2$O can lead to the formation of several calcium phosphates (shown in Table IV), of which hydroxy-apatite OHA is the most stable above a pH of about 4.1. The relative stabilities are illustrated in Figure 12.

Table IV.

Pertinent calcium phosphates relevant to aqueous systems, their formula, structure and negative logarithm of the solubility product pK

Ca/P	Notation	Formula	Space group	pK	Mineral name
1	DCP	$CaHPO_4$	P1	6.90	monetite
1	DCPD	$CaHPO_4 \cdot 2H_2O$	C2/c	6.59	brushite
1.33	OCP	$Ca_8(HPO_4)_2(PO_4)_4 \cdot 5H_2O$	$P\bar{1}$	68.6[1]	–
1.43	WH	$Ca_{10}(HPO_4)(PO_4)_6$	$R\bar{3}c$	81.7[1]	whitlockite
1.67	OHA	$Ca_{10}(PO_4)_6(OH)_2$	$P6_3/m$	117.2	hydroxy-apatite
1.50	DOHA	$Ca_9(HPO_4)(PO_4)_5(OH)$	$P6_3/m$	85.1[1]	defective hydroxy-apatite

[1]Estimate

Once calcium deficient hydroxyapatite DOHA (between pH 6.8 and 8.2) is formed, a metastable equilibrium is created with the aqueous solution which may last indefinitely at room or body temperature. If carbonate ions are present in addition the apatite preferably formed is

$$Ca_9 (PO_4)_{4.5} (CO_3)_{1.5} (OH)_{1.5} \quad \text{(abbr. HCDOHA)} \quad (65)$$

which has a solubility comparable to that of DOHA. If not only carbonate but also sodium ions are present, the apatite preferably formed is

$$Ca_{8.5} Na_{1.5} (PO_4)_{4.5} (CO_3)_{2.5} \quad \text{(abbr. NCCA)} \quad (66)$$

which also has a solubility comparable to that of DOHA. Magnesium ions give rise to the formation of magnesium whitlockite with the Formula:

$$Ca_9 Mg (HPO_4) (PO_4)_6 \quad \text{(abbr. MWH)} \quad (67)$$

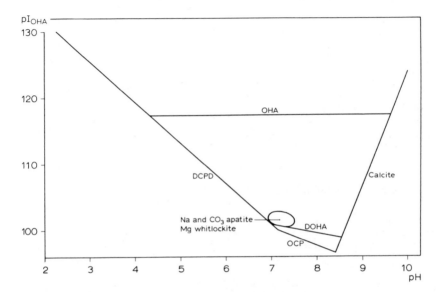

Figure 12. Estimated location of the negative logarithm of the solubility product pK_{OHA} at physiological pH of NCCA and MWH.

which also has a solubility comparable to that of DOHA, whereas magnesium in combination with carbonate and calcium can lead to the formation of dolomite (abbr. DOL), as described previously.

Body fluids have a very high supersaturation with respect to hydroxyapatite, which cannot be explained by the small particle size of bone mineral. In fact, they behave as aqueous solutions which are in metastable equilibrium with DOHA. However, the minerals in bone, dentin, dental enamel and dental calculus contain considerable amounts of Na, Mg and CO_3, in addition to calcium and phosphate which are the major components. Therefore, the phases mentioned above which all show a solubility comparable to that of DOHA, all come into consideration as components of these minerals. A compilation of the phases occurring in the different calcified tissues or calculi is given in Table V. If more than one apatite phase is involved, these generally occur as domains in the same apatitic particles of the mineral and cannot be separated physically (53, 57).

Table V.

Phases occurring in the mineral of calcified tissues and dental calculus (53, 57)

	bone	dentin	enamel	salivary pH	
				≳ 8	< 8
				dental	calculus
OHA			+		
DOHA			+		+
NCCA	+	+	+		+
HCDOHA	+	+			+
MWH	+	+			+
DOL			+	+	
Calcite				+	

These multiphase models were originally based on the observation that in a series of solid solutions the solubility
products differ in general by so much that the value of the separation factor D must be considerably large or very small. For
this reason, the precipitate formed in a quasi-binary system
during relatively slow precipitation, under conditions charac-
terised by small degrees of supersaturation (which resemble the
in vivo conditions), has in general one of the two extreme compositions. In a complex system like that of the body fluids,
several extreme compositions can precipitate simultaneously,
especially because their solubilities in terms of calcium and
phosphate concentration and pH are in fact equal. In this way
these multiphase models lead to the statement that, once bone
is calcified during growth, the calcium and phosphate homeostasis are related to the solubility of the mineral phases in
bone. Thus body fluids and bone mineral are in near equili-
briumn. Similarly, there is equilibrium between saliva and the
mineral in dental enamel. The intervention of crystal growth
inhibitors and nucleation inhibitors, such as organic compounds
(as is assumed by most biochemists) does not truly explain the
vast and enormous degree of supersaturation of body fluids with
respect to hydroxyapatite. The multiphase concept (53, 57) has
replaced this hypothesis by the action of carbonate, sodium and
magnesium ions, which create the metastable states observed between calcium phosphates and aqueous solutions.

Literature Cited

1. Driessens, F.C.M. Ber. Bunsenges. Physik. Chem. 1968, 72,
   754.
2. Driessens, F.C.M. Ber. Bunsenges. Physik. Chem. 1968, 72,
   764.
3. Driessens, F.C.M. Ber. Bunsenges. Physik. Chem. 1968, 72,
   1123.
4. Flood, H. Z. Anorg. Allg. Chem. 1936, 229, 76.
5. Herzfeld K.F.; Heitler, W. Z. Elektrochem. 1925, 31, 536.
6. Driessens, F.C.M.; Verbeeck, R.M.H.Ber. Bunsenges. Physik
   Chem. 1981, 85, 713.
7. Driessens, F.C.M. Ber. Bunsenges. Physik. Chem. 1979, 83,
   583.
8. Roozeboom, P. Z. Physik. Chem. 1891, 8, 504.
9. Ratner, A.P. J. Chem. Phys. 1933, 1, 789.
10. Gordon, L. Precipitation form homogeneous solution, J.
    Wiley and Sons, New York, 1959.
11. Doerner, H.A.; Hoskins, W.M. J. Am. Chem. Soc. 1925, 47,
    662.
12. Kroeker. F. Thesis, Berlin, 1892.
13. Hill, A.E.; Durham, G.S.; Ricci, J.E. J. Am. Chem. Soc.
    1940, 62, 2723.
14. Feibush, A.M.; Rowley K.; Gordon, L. Anal. Chem. 1985, 30,
    1605.

15. Hildebrand, J.H. Solubility of non-electrolytes, Reinhold Publ. Co., New York, 1936.
16. Kirgintsev, A.N.; Trushnikova, L.N. Zh. Neorg. Khim. 1966, 11, 2331.
17. Saxena, S.K.; Ghose, S.G. Am. Mineralogist 1970, 55, 1219.
18. Walker, W.J.; Rodgers, K. Geochim. Cosmochim. Acta 1974, 38, 1521.
19. Durham, G.S.; Rock, E.J.; Frayn, J.S. J. Am. Chem. Soc. 1953, 75, 5792.
20. Flatt, R.; Buckhardt, G. Helv. Chim. Acta 1944, 27, 1605.
21. Yutzy, H.C.; Kolthoff, I.M. J. Am. Chem. Soc. 1937, 59, 916.
22. Vaslow, F.; Boyd, G.E. J. Am. Chem. Soc. 1952, 74, 4691.
23. Gordon, L.; Reimer, C.C.; Burtt, B.P. Anal. Chem. 1954, 26, 842.
24. Kolthoff, I.M.; Noponen, G.E. J. Am. Chem. Soc. 1938, 60, 197.
25. Kolthoff, I.M.; Eggertsen, F.T. J. Phys. Chem. 1942, 46, 616.26. Spitsyn, I.; Mikheev, N.B.; Kherman, A. Akad. Nauk SSSR 1966, 166, 48.
27. Kolthoff, I.M.; Carr, C.W. J. Phys. Chem. 1943, 47, 148.
28. Weaver, B. Anal. Chem. 1954, 26, 479.
29. Hermann, J.A. Ph. D. Thesis, The University of New Mexico, 1955.
30. Kolthoff, I.M. J. Phys. Chem. 1932, 36, 860.
31. Kolthoff, I.M.; Sandell, E.B. J. Phys. Chem. 1933, 37, 443.
32. Jamieson, J.C. J. Chem. Phys. 1953, 21, 1385.
33. Garrels, R.M.; Thompson, M.E.; Seever, R. Am. J. Sci. 1960, 258, 402.34. Graf, D.L. Am. Mineral. 1961, 46, 1283.
35. Goldsmith, J.R.; Graf, D.L. J. Geol. 1958, 66, 678.
36. Barnes, I.; O'Neil, J.R. Geochim. Cosmochim. Acta 1971, 35, 699.
37. Milliman, J.D. Marine Carbonates, Springer Verlag, Berlin, 1974.
38. Chilingar, G.V. et al Eds.; "Developments in Sedimentology, Vol. 9B" Carbonate Rocks, Elsevier Publ. Co.: Amsterdam, 1967.
39. Sveshnikova, B.N. Dokl. Akad. Nauk. SSSR 1960, 85, 357.
40. Brätter, P.; Möller, P.; Rösuk, U. Earth Planet. Sci. Lett. 1972, 14, 50.
41. Möller, P. Z. Anal. Chem. 1974, 268, 28.
42. Sastri, C.S.; Möller, P. Chem. Phys. Lett. 1974, 26, 116.
43. Koss, V.; Möller, P. Z. Anorg. Allg. Chem. 1974, 410, 165.
44. Wigley, T.M.L. Geochim. Cosmochim. Acta 1973, 37, 1397.
45. Chave, K.E.; Deffeyes, K.S.; Weyl, P.K.; Garrels, G.M.; Thompson, M.E. Science 1962, 137, 33.
46. Siebert, R.M.; Hostetler, P.B.; Christ, C.L. J. Res. U.S. Geol. Surv. 1974, 2, 447.
47. Möller, P.; Rajagopalan, G. Z. Phys. Chem. 1975, 94, 297.
48. Caillean, P.; Dragone, D.; Giron, A.; Jacqmin, C.; Rgues, H. Soc. Franc. Mineral. Cristallogr. 1977, 100, 81.
49. Möller, P.; Rajagopalan, G. Z. Phys. Chem. 1976, 99, 187.

50. Möller, P.; Kubanek, F. Neues Jahrb. Mineral. 1976, 126, 199.
51. Schneider, H. Mineral. Mag. 1976, 40, 579.
52. Sureau, J.F. B.U. Soc. Fr. Mineral. Cristallogr. 1974, 57, 300-316.53. Driessens, F.C.M.; Verbeeck, R.M.H. Biominerals, C.R.C. Press, Boca Raton, 1986. 54. Burton, R.F.; Mathie, R.T. Experientia, 1975, 31,543-544.
55. de With, N.D.; van der Wilt, G.J.; van der Schors, R.C. In Westbrook and de Jong Eds.; Biomineralisation and biological
    metal acumulation, D. Reidel Publ., Dordrecht, 1983, p. 149-153.
56. Simkiss, K. Am. J. Physiol. 1968, 214, 627-634.
57. Driessens, F.C.M. Mineral aspects of dentistry, S. Karger, Basel, 1982.
58. Moreno, E.C.; Kresak, M.; Zahradnik, R.T. Caries Res. 1977, 11 (Suppl.) 192.
59. Wier, D.R.; Chien, S.H.; Black, C.A. Soil. Sci. 1972, 36, 285.
60. Rösick, U.; Zimen, K.E. Biophysik. 1973, 9, 120.
61. Meijering, J.L. Philips Techn. Rev. 1965, 26, 52.
62. Freund, F.; Knobel, R.M. J. Chem. Soc. 1977, 1136.
63. Schaeken, H.G.; Verbeeck, R.M.H.; Driessens, F.C.M.; Thun, H.P. Bull. Soc. Chim. Belg. 1975, 84, 881.
64. Driessens, F.C.M. Ber. Bunsenges. Physik. Chem. 1979, 83, 583.
65. Wolf, N.; Gedalia, I.; Yariv, S.; Zuckermann, H. Archs. oral. Biol. 1973, 18, 233.
66. Verbeeck, R.M.H.; Lassuyt, C.J.; Heijligers, H.J.M.; Driessens, F.C.M.; Vrolijk, J.W.G.A. Calcif. Tis. Int. 1981, 33, 243-247.
67. Lassuyt, C.J. Ph.D. Thesis, State University, Gent, 1984.

RECEIVED June 25, 1986

Approach to Equilibrium in Solid Solution-Aqueous Solution Systems: The KCl-KBr-H₂O System at 25 °C

L. Niel Plummer

U.S. Geological Survey, Reston, VA 22092

Thermodynamic calculations based on the
compositional dependence of the equilibrium
constant are applied to solubility data in
the KCl-KBr-H₂O system at 25°C. The experimental distribution coefficient and activity
ratio of Br⁻/Cl⁻ in solution is within a
factor of two of the calculated equilibrium
values for compositions containing 19 to 73
mole percent KBr, but based on an assessment of
uncertainties in the data, the solid solution
system is clearly not at equilibrium after 3-4
weeks of recrystallization. Solid solutions
containing less than 19 and more than 73
mole percent KBr are significantly farther from
equilibrium. As the highly soluble salts are
expected to reach equilibrium most easily,
considerable caution should be exercised before
reaching the conclusion that equilibrium is
established in other low-temperature solid
solution-aqueous solution systems.

Equilibrium between simple salts and aqueous solutions is often
relatively easily demonstrated in the laboratory when the
composition of the solid is invariant, such as occurs in the
KCl-H₂O system. However, when an additional component which
coprecipitates is added to the system, the solid composition
is no longer invariant. Very long times may be required to
reach equilibrium when the reaction path requires shifts in
the composition of both the solution and solid. Equilibrium
is not established until the solid composition is homogeneous
and the chemical potentials of all components between solid
and aqueous phases are equivalent. As a result, equilibrium
is rarely demonstrated with a solid solution series.
      Laboratory studies that have been used to evaluate the
solid component activity coefficients have made the assumption
that equilibrium was established (1-3). Using more recent

thermodynamic data for the end-member compositions, it can be
shown that equilibrium was probably not established over
periods of 8 weeks in the $SrCO_3$-$BaCO_3$-H_2O system studied by
Schmeling (1) at 25°C. $SrCO_3$-$CaCO_3$ (aragonite) solid solutions
did not reach equilibrium by recrystallization in batch
experiments after periods of 3 weeks at 76°C (4). Recrystallization continued for at least several hundred days with alkaline-
earth sulfate solid solutions (5) at room temperature.

The present study examines the approach to equilibrium
in the very soluble salt system KCl-KBr-H_2O. Soluble salt
reactions are known to be relatively rapid and there is greater
likelihood for equilibrium to be established. Solubility in the
KCl-KBr-H_2O system has been well studied at 25°C (6-8) and has
been assumed previously to attain equilibrium (3,8). By examining
the compositional dependence of the experimental distribution
coefficient, Stoessell and Carpenter (9) concluded equilibrium
was not established during coprecipitation of trace Br in KCl.
The present paper uses thermodynamic criteria to test for
equilibrium.

KCl-KBr-H_2O System at 25°C

In this study we have examined recrystallization in the
KCl-KBr-H_2O system at 25°C (8). The solubilities of the end-
members KCl and KBr were determined from both oversaturation and
undersaturation. In studying the solubility of this solid
solution system, solubility experiments were carried out in
pairs (8). For each pair of experiments (A and B), the total
amounts of KCl and KBr in the system were identical. In the
A-type runs (Table I), solid KCl was added to an initial KBr
aqueous solution, and in the B-type runs, solid KBr was added
to an initial aqueous KCl solution. For each pair of runs the
total number of moles of KCl and KBr were identical. The final
composition of the solid and aqueous solution was, therefore,
approached from two different compositional reaction paths.
The experiments were conducted in "solubility tubes" each
containing two glass marbles. The tubes were rotated at
25.00 ± 0.02°C to keep the material finely ground.

For each pair of runs nearly identical solid-aqueous
solution compositions were observed after periods of three
to four weeks, which compared favorably to compositions
previously reported (6,7). Table I summarizes the original
data (8).

Although nearly identical solid-aqueous solution
compositions are observed in recrystallization from two
directions under conditions of total constant composition,
this alone is insufficient proof of the establishment of
equilibrium. In order to test for equilibrium, the solid
solution activity coefficients must be determined and used to
compare observed solid and aqueous solution compositions with
the appropriate values expected at equilibrium.

Table I.   Original solubility data for the system
           KCl-KBr-H₂O at 25°C (8)

No.	Total Composition KBr Wt.%	KCl Wt.%	Liquid Phase KBr Wt.%	KCl Wt.%	Br Br+Cl Moles	Solid Solution KBr Wt.%	Mole Fraction KBr
1	0.00	...	0.00	26.42	0.000	0.00	0.000
2A	10.00	25.00	10.01	20.91	.231	7.89	.051
2B	10.00	25.00	10.28	20.69	.237	6.87	.044
3A	21.00	22.00	20.21	15.31	.453	26.95	.188
3B	21.00	22.00	20.13	15.39	.450	26.99	.188
4A	25.00	21.00	22.85	13.83	.509	37.78	.276
4B	25.00	21.00	22.75	13.92	.506	37.29	.271
5A	34.00	18.00	26.62	11.56	.591	60.0	.485
5B	34.00	18.00	26.42	11.71	.586	59.9	.483
6A	37.00	10.00	30.46	8.74	.686	81.1	.729
6B	37.00	10.00	30.50	8.70	.687	81.0	.728
7A	39.00	5.00	35.09	4.89	.818	92.9	.891
7B	39.00	5.00	35.21	4.75	.823	92.9	.891
8	...	0.00	40.57	0.00	1.000	100.0	1.000

Theory

The final solid solution–aqueous solution compositions of
Table I will fall into one of three catagories: (1) they will
either be at equilibrium, (2) at stoichiometric saturation, or
(3) correspond to some non-equilibrium state.

If the solutions are at equilibrium, the solid component
activities will be related to solution composition by the
equations

$$a_{KCl(s)} K_{KCl} = a_{K^+} a_{Cl^-} \tag{1}$$

and

$$a_{KBr(s)} K_{KBr} = a_{K^+} a_{Br^-} \tag{2}$$

where $a_{KCl(s)}$ and $a_{KBr(s)}$ denote the activities of KCl and
KBr in the solid, a_{K^+}, a_{Cl^-} and a_{Br^-} are the activities of
K^+, Cl^- and Br^- in aqueous solution, and K_{KCl}, K_{KBr} are the
equilibrium constants for the end-member reactions

$$KCl = K^+ + Cl^- \tag{3}$$

and

$$KBr = K^+ + Br^- \tag{4}$$

Furthermore, if the solid solution salt is at equilibrium
with the aqueous solution, the equilibrium constant for the
reaction

$$KBr_xCl_{(1-x)} = K^+ + xBr^- + (1-x) Cl^- \qquad (5)$$

will be defined by

$$K_{(x)} = a_{K^+} a^x_{Br^-} a^{(1-x)}_{Cl^-} \qquad (6)$$

where the activity of the solid $KBr_xCl_{(1-x)}$ is unity by
definition.

Stoichiometric saturation defines equilibrium between an
aqueous solution and homogeneous multi-component solid of
fixed composition (10). At stoichiometric saturation the
composition of the solid remains fixed even though the mineral
is part of a continuous compositional series. Since, in
this case, the composition of the solid is invariant, the
solid may be treated as a one-component phase and Equation 6
is the only equilibrium criteria applicable. Equations 1 and 2
no longer apply at stoichiometric saturation because, owing to
kinetic restrictions, the solid and saturated solution compositions
are not free to change in establishing an equivalence of individual
component chemical potentials between solid and aqueous solution.
The equilibrium constant, $K_{(x)}$, is defined identically for both
equilibrium and stoichiometric saturation.

If the solid is neither at equilibrium nor stoichiometric
saturation, the Equations 1, 2, or 6 are not applicable.

In testing for equilibrium in the KCl-KBr-H$_2$O system (8),
the analysis that follows initially assumes that all solids
are at stoichiometric saturation with the aqueous solution.
This permits provisional calculation of the compositional
dependence of the equilibrium constant and determination of
provisional values of the solid phase activity coefficients
(discussed below). The equilibrium constant and activity
coefficients are termed provisional because it is not possible
to determine if stoichiometric saturation has been established
without independent knowledge of the compositional dependence
of the equilibrium constant, such as would be provided from
independent thermodynamic measurements. Using the provisional
activity coefficient data we may compare the observed solid
solution-aqueous solution compositions with those calculated
at equilibrium. Agreement between the calculated and observed
values confirms, within the experimental data uncertainties,
the establishment of equilibrium. The true solid solution
thermodynamic properties are then defined to be equal to the
provisional values.

If there is no agreement in calculated and observed
solid-solution properties we can only conclude that equilibrium
was not established. The validity of the provisional activity
coefficients depends on the validity of the original assumption
that stoichiometric saturation was established. If independent
data for the standard free energy of formation of the solid

solutions can be introduced from another source, comparison
of calculated and observed equilibrium constants determines
if stoichiometric saturation was established, as required by
Equation 6. Stoessell and Carpenter (9) found close agreement
between the calculated equilibrium distribution coefficient and
that measured in slow growth experiments by assuming
recrystallization of halite containing trace amounts of Br
occurs at stoichiometric saturation.

By examining the compositional dependence of the
equilibrium constant, the provisional thermodynamic properties
of the solid solutions can be determined. Activity coefficients
for solid phase components may be derived from an application of
the Gibbs-Duhem equation to the measured compositional dependence
of the equilibrium constant in binary solid solutions (10).
For the KCl-KBr-H₂O system the following relationships are
valid (10):

$$\log a_{KCl(s)} = -x \frac{\partial}{\partial x}(\log K_{(x)}) + \log K_{(x)} - \log K_{KCl} \qquad (7)$$

and

$$\log a_{KBr(s)} = (1-x) \frac{\partial}{\partial x}(\log K_{(x)}) + \log K_{(x)} - \log K_{KBr} \qquad (8)$$

where x denotes the mole fraction of KBr in the solid, $K_{(x)}$
is defined by Equation 6, and K_{KCl} and K_{KBr} are the end-member
equilibrium constants for Equations 3 and 4. The individual
component activity coefficients, λ_i, are defined by

$$\lambda_{KCl} = \frac{a_{KCl(s)}}{1-x} \qquad (9)$$

and

$$\lambda_{KBr} = \frac{a_{KBr(s)}}{x} \qquad (10)$$

By examining the compositional dependence of the equilibrium
constant, the thermodynamic properties of the solid solution
can be determined if the final solution is either at equilibrium
or stoichiometric saturation. That is, the provisional
activities and activity coefficients will be valid if either
equilibrium or stoichiometric saturation is attained in the
solubility data.

The provisional activities and activity coefficients are
used to calculate the expected equilibrium composition of the
aqueous solution. Two compositional properties to be tested
are the equilibrium distibution coefficient, D_{eq}, and the
equilibrium ratio of the activities of Br⁻ to Cl⁻ in solution.
The distribution coefficient is defined

$$D = \cfrac{\cfrac{x}{1-x}}{\cfrac{m_{KBr}}{m_{KCl}}} \qquad (11)$$

where m denotes molality in solution. At equilibrium, the distribution coefficient is

$$D_{eq} = \frac{K_{KCl}}{K_{KBr}} \cdot \frac{\lambda_{KCl(s)}}{\lambda_{KBr(s)}} \cdot \frac{\gamma_{Br^-}}{\gamma_{Cl^-}} \qquad (12)$$

(11, 12) where γ is the individual ion activity coefficient in the equilibrium solution. The aqueous Br^- to Cl^- activity ratio expected in equilibrium solutions is obtained by combining Equations 1 and 2,

$$\left(\frac{a_{Br^-}}{a_{Cl^-}}\right)_{eq} = \frac{K_{KBr} \; a_{KBr(s)}}{K_{KCl} \; a_{KCl(s)}} \qquad (13)$$

Equilibrium Constants

Equilibrium constants calculated from the composition of saturated solutions are dependent on the accuracy of the thermodynamic model for the aqueous solution. The thermo-dynamics of single salt solutions of KCl or KBr are very well known and have been modeled using the virial approach of Pitzer (13-15). The thermodynamics of aqueous mixtures of KCl and KBr have also been well studied (16-17) and may be reliably modeled using the Pitzer equations. The Pitzer equations used here to calculate the solid phase equilibrium constants from the compositions of saturated aqueous solutions are given elsewhere (13-15, 18, 19). The Pitzer model parameters applicable to KCl-KBr-H$_2$O solutions are summarized in Table II.
As a means of verifying the model parameters of Table II, the osmotic coefficient was calculated from isopiestic vapor pressure measurement data (17) for the KCl-KBr-H$_2$O system at 25°C (Table III).
Using the model parameters of Table II the calculated osmotic coefficient is within 0.15% or better for all solutions investigated. Agreement with the experimental results (17) is within 0.02% or better if $\psi_{Cl,Br,K} = 0.0003$ (Table III) instead of zero (Table II). We may conclude from this comparison that the thermodynamic model of Pitzer (Table II) is very realistic. An uncertainty of 0.0003 in $\psi_{Cl,Br,K}$ leads to uncertainties of less than 0.4% in log $K_{(x)}$. The largest uncertainty in equilibrium constants may thus be attributed to the original analytical data (8).

Table II. Summary of Pitzer model
parameters for the system
KCl-KBr-H₂O at 25°C (14,15).

Parameter	KCl	KBr
β°	.04835	.0569
β^1	.2122	.2212
C^ϕ	-.00084	-.00180

$\theta_{Cl,Br}$ = 0.000
$\psi_{Cl,Br,K}$ = 0.000

Table III. Comparison of calculated and
observed osmotic coefficients in
KCl-KBr-H₂O solutions at 25°C

			Osmotic Coefficient		
			From	Pitzer model	
No.	m_{KBr}	m_{KCl}	(17)	$\psi=0.0$	$\psi=0.0003$
1	0	4.816	.9893	.9893	.9893
2	.935	3.546	.9843	.9833	.9843
3	2.128	2.425	.9919	.9905	.9921
4	2.489	2.153	.9960	.9945	.9961
5	3.048	1.753	1.0026	1.0012	1.0028
6	3.681	1.285	1.0094	1.0082	1.0096
7	4.543	.690	1.0193	1.0185	1.0194
8	5.737	0	1.0354	1.0354	1.0354

Table IV summarizes the final reported compositions of
the solids and aqueous solutions (8) with calculated equilibrium
constants. Values of log $K_{(x)}$ from companion runs at constant
composition differ by no more than 0.003 log K units.
Because there is no obvious reason for selecting equilibrium
constants from runs in which the initial solution contained
either solid KCl or solid KBr (runs A or B), the calculations
that follow are based on average solid compositions and
average equilibrium constants of A and B runs for each total
composition reported (Table IV). The equilibrium constants
for Equation 5 are shown as a function of KBr mole fraction
in Figure 1.

Test for Equilibrium

Values of $\partial\log K_{(x)}/\partial x$ were interpolated from Figure 1
and used to calculate the provisional activities and activity
coefficients of KCl and KBr in the solids using Equations
7-10. Values of log $K_{(x)}$, $\partial\log K_{(x)}/\partial x$, provisional

Table IV. Summary of calculated provisional equilibrium
           constants

No.	Mole Fraction KBr	Molality in Solution KBr	KCl	Log K	Average Values x_{KBr}	log K
1.	.000	.000	4.816	.9037	.000	0.904
2A	.051	.935	3.546	.7071		
2B	.044	.963	3.499	.7039	.048	0.706
3A	.188	2.128	2.425	.5779		
3B	.188	2.118	2.440	.5802	.188	0.579
4A	.276	2.489	2.153	.5685		
4B	.271	2.475	2.169	.5699	.274	0.569
5A	.485	3.048	1.753	.6047		
5B	.483	3.017	1.779	.6047	.484	0.605
6A	.729	3.681	1.285	.7147		
6B	.728	3.688	1.278	.7142	.729	0.714
7A	.891	4.543	.690	.8780		
7B	.891	4.567	.669	.8789	.891	0.878
8	1.000	5.737	.000	1.1288	1.000	1.129

Table V. Summary of provisional activities and
          activity coefficients for $KBr_xCl_{(1-x)}$
          at 25°C

No.	x	log $K_{(x)}$	$\dfrac{\partial \log K_{(x)}}{\partial x}$	λ_{KBr}	λ_{KCl}	$a_{KBr(s)}$	$a_{KCl(s)}$
2	.048	.706	-1.56	.257	.791	.012	.753
3	.188	.579	-.28	.888	.658	.167	.534
4	.274	.569	-.05	.925	.657	.253	.477
5	.484	.605	.29	.863	.708	.421	.363
6	.729	.714	.74	.837	.688	.610	.186
7	.891	.878	1.54	.927	.367	.826	.040

activities and provisional activity coefficients are given in
Table V.
    Table VI summarizes values of the activity coefficient ratio
$\gamma_{Br^-}/\gamma_{Cl^-}$ in the saturated solution for each average solid
composition (as calculated from the model of Table II), the
calculated provisional equilibrium distribution coefficient
(Equation 12) and the provisional equilibrium aqueous solution
activity ratio of Br⁻ to Cl⁻ (Equation 13) based on the data
of Table V.
    Figure 2 compares the experimental data (8) with the
provisional equilibrium compositions on a conventional
Roozeboom diagram. It appears that equilibrium is most
closely approached in the mid-range compositions, but
compositions closer to the end-members KCl and KBr deviate

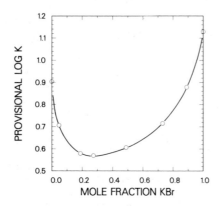

Figure 1. Provisional equilibrium constants of solids KBr$_x$Cl$_{(1-x)}$ at 25°C.

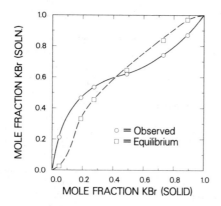

Figure 2. Roozeboom diagram comparing experimental and provisional equilibrium compositions in the KCl-KBr-H₂O system at 25°C.

Table VI. Comparison of experimental and provisional
equilibrium solid/aqueous solution properties

No.	x	$\dfrac{\gamma_{Br^-}}{\gamma_{Cl^-}}$	Experimental		Equilibrium	
			D	a_{Br^-}/a_{Cl^-}	D_{eq}	a_{Br^-}/a_{Cl^-}
2	.048	1.068	.183	.288	1.96	.028
3	.188	1.069	.265	.933	.47	.53
4	.274	1.069	.329	1.228	.45	.89
5	.484	1.071	.554	1.746	.52	1.95
6	.729	1.072	.938	3.082	.53	5.50
7	.891	1.074	1.207	7.200	.25	34.67

more significantly from equilibrium. In the compositional
range $.188 \leq x \leq .730$ the provisional equilibrium distribution
coefficient and Br^- to Cl^- activity ratio are within a
factor of two of the experimental values. Larger deviations
from equilibrium are found outside this compositional range
(Table VI). Figure 3 shows the provisional activities of KCl
and KBr in the solids as a function of KBr mole fraction.
These activities can be verified if it can be shown that the
solubility data of (8) are at stoichiometric saturation,
but as mentioned above, this requires independent thermodynamic
definition of $K_{(x)}$.
    It has already been pointed out that the calculated
equilibrium constants are known better than the analytical
data on which they are based. So we may not attribute the
observed difference in provisional equilibrium values and
experimental values (Table VI) to uncertainties in the aqueous
model. There are, however, uncertainties in estimating
$\partial \log K(x)/\partial x$ from Figure 1. Slopes estimated from Figure 1
are probably known within 20%. Uncertainties of 20% in
$\partial \log K_{(x)}/\partial x$ translate directly to uncertainties of 20%
in solid phase activities and activity coefficients.
Uncertainties of 20% in $\partial \log K_{(x)}/\partial x$ correspond to maximum
uncertainties in D_{eq} and $(a_{Br^-}/a_{Cl^-})_{eq}$ of 40%. This does not
alter the conclusion that many of the final solid solution—aqueous
solution points in the KCl—KBr—H_2O system are out of equilibrium.
    As a further test of the approach to equilibrium, we use the
observed a_{Br^-}/a_{Cl^-} ratio in solution to calculate the expected
slope on plots of log K vs x, if at equilibrium. Equilibrium is
indicated by close agreement in calculated and observed slopes.
The equilibrium slope is defined from the equilibrium aqueous
solution ratio of the activities of Br^- to Cl^- (14)

$$\log\left(\frac{a_{Br^-}}{a_{Cl^-}}\right)_{eq} = \frac{\partial}{\partial x} \log K_{(x)} \tag{14}$$

Using the experimental solution compositions (Table IV) and
the calculated aqueous solution activity coefficient ratio
$\gamma_{Br^-}/\gamma_{Cl^-}$ (Table VI), Figure 4 shows the slopes of log K

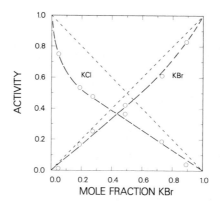

Figure 3. Provisional activities of KCl and KBr in
KCl-KBr solid solutions at 25°C.

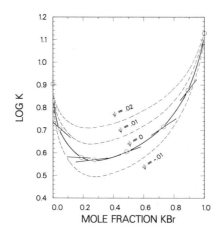

Figure 4. Comparison of slopes $(\partial \log K_{(x)}/\partial x)$ required
if the experimental data (8) correspond to equilibrium (short
line segments through each experimental point) with the
compositional dependence of $\log K_{(x)}$ calculated assuming
$\psi_{Cl,Br,K}$ is -0.01, 0.00, 0.01 and 0.02.

as a function of x which are required if equilibrium is
established. The implied equilibrium slopes deviate as much
as 300% from those estimated from the smoothed log K curve
(Figure 4). The sensitivity of $\log K_{(x)}$ was investigated by
varying the $\psi_{Cl,Br,K}$ parameter of the Pitzer model between
-0.01 and 0.02. Figure 4 shows that if $\Psi_{Cl,Br,K}$ is near
.02, there is close correspondence in slopes of the log K
curve and slopes calculated from the observed Br⁻/Cl⁻

activity ratio using Equation 14. That is, $\Psi_{Cl,Br,K}$ must
be near 0.02 for the solubility data ($\underline{8}$) to correspond to
equilibrium. It was previously shown using isopiestic data
($\underline{17}$) that $\Psi_{Cl,Br,K}$ is near 0.0003. If $\Psi_{Cl,Br,K}$ were 0.02,
the computed osmotic coefficients would deviate by as much as
11% from the observed values ($\underline{17}$). Because the osmotic
coefficient is known to 1 part in 10,000 ($\underline{17}$), it can be
reliably concluded that equilibrium was not established
during recrystallization in the KCl-KBr-H_2O system at 25°C.

Although equilibrium was not established, it was more
closely approached in the KCl-KBr-H_2O system than in carbonate
systems. For example, in a similar analysis of the
strontianite-aragonite solid solution system ($\underline{4}$), it was
found that the experimental distribution coefficient for Sr
substitution from seawater into aragonite is 12 times larger
than the expected equilibrium value. Most of the distribution
coefficients for the KCl-KBr-H_2O system are within a factor
of two of the equilibrium value, but clearly not at equilibrium.
Considerable caution should be exercised before reaching the
conclusion that equilibrium is established at relatively low
temperatures in other solid solution-aqueous solution systems.

Finally, it is not appropriate to derive thermodynamic
properties of solid solutions from experimental distribution
coefficients unless it can be shown independently that
equilibrium has been established. One possible exception
applies to trace substitution where the assumptions of
stoichiometric saturation and unit activity for the predominant
component allow close approximation of equilibrium behavior
for the trace components ($\underline{9}$). The method of Thorstenson and
Plummer ($\underline{10}$) based on the compositional dependence of the
equilibrium constant, as used in this study, is well suited
to testing equilibrium for all solid solution compositions.
However, because equilibrium has not been found, the thermodynamic
properties of the KCl-KBr solid solutions remain provisional
until the observed compositional dependence of the equilibrium
constant can be verified. One means of verification is the
demonstration that recrystallization in the KCl-KBr-H_2O
system occurs at stoichiometric saturation.

Conclusion

Most thermodynamic data for solid solutions derived from
relatively low-temperature solubility (equilibration) studies
have depended on the assumption that equilibrium was
experimentally established. Thorstenson and Plummer ($\underline{10}$)
pointed out that if the experimental data are at equilibrium
they are also at stoichiometric saturation. Therefore,
through an application of the Gibbs-Duhem equation to the
compositional dependence of the equilibrium constant, it is
possible to determine independently if equilibrium has been
established. No other compositional property of experimental
solid solution-aqueous solution equilibria provides an independent
test for equilibrium. If equilibrium is demonstrated, the
thermodynamic properties of the solid solution are also

determined. However, if equilibrium is not attained, the
thermodynamic properties of the solid can be determined from
the solubility data only if the system can be demonstrated
to be at stoichiometric saturation.

In application of this method to solubility data (8)
in the KCl–KBr–H$_2$O system at 25°C, it is found that equilibrium
is in general not attained, though some mid-range compositions
may be near equilibrium. As the highly soluble salts are
expected to reach equilibrium most easily, considerable
caution should be exercised before reaching the conclusion
that equilibrium is established in other low-temperature
solid solution–aqueous solution systems. It is not appropriate
to derive thermodynamic properties of solid solutions from
experimental distribution coefficients unless it can be
demonstrated that equilibrium has been attained.

Acknowledgments

Review comments of E. Busenberg and B. F. Jones are
gratefully acknowledged.

Literature Cited

1. Schmeling, P. Svensk Kem. Tidskr. 1953, 65, 123–34.
2. Crocket, J. H.; Winchester, J. W. Geochim. Cosmochim. Acta
   1966, 30, 1093–1109.
3. Kirgintsev, A. N.; Trushnikova, L. N. Russian J. Inorg. Chem.
   1966, 11, 1250–5.
4. Plummer, L. N.; Busenberg, E., unpublished data.
5. Denis, J.; Michard, G. Bull. Mineral. 1983, 106, 309–19,
6. Amadori, M.; Pampanini, G. Atti acscad. Lineei II 1911, 20, 473.
7. Flatt, R.; Burkhardt, G. Helv. Chem. Acta 1944, 27, 1605–10.
8. Durham, G. S.; Rock, E. J.; Frayn, J. S. J. Am. Chem. Soc.
   1953, 75, 5792–4.
9. Stoessell, R. K.; Carpenter, A. B. Geochim. Cosmochim.
   Acta 1986, 50, in press.
10. Thorstenson, D. C.; Plummer, L. N. Am. J. Sci. 1977, 277,
    1203–23.
11. Vaslow, F.; Boyd, G. E. J. Am. Chem. Soc. 1952, 74, 4691–95.
12. McIntire, W. L. Geochim. Cosmochim. Acta 1963, 27, 1209–64.
13. Pitzer, K. S. J. Phys. Chem. 1973, 77, 268–77.
14. Pitzer, K. S.; Mayorga, G. J. Phys. Chem. 1973, 77, 2300–8.
15. Pitzer, K. S.; Kim, J. J. J. Am. Chem. Soc. 1974, 96–5701–7.
16. McCoy, W. H.; Wallace, W. E. J. Am. Chem. Soc. 1956, 78, 1830–3.
17. Covington, A. K.; Lilley, T. H.; Robinson, R. A. J. Phys.
    Chem. 1968, 72, 2759–63.
18. Harvie, C. E.; Weare, J. H. Geochim. Cosmochim. Acta 1980,
    44, 981–97.
19. Harvie, C. E.; Moller, N.; Weare, J. H. Geochim. Cosmochim.
    Acta 1984, 48, 723–51.

RECEIVED June 25, 1986

27

Modes of Coprecipitation of Ba^{2+} and Sr^{2+} with Calcite

Nicholas E. Pingitore, Jr.

Department of Geological Sciences, University of Texas at El Paso, El Paso, TX 79968-0555

Non-lattice incorporation can play a significant role in the aqueous coprecipitation of large cations with calcite. Coprecipitation experimental results yield a partition coefficient of Ba^{2+} into calcite of 0.04; the partition coefficient is affected by rate of precipitation, presence of other ions, and type of seed used to nucleate growth. Such sensitivity to experimental conditions characterizes non-lattice incorporation, a conclusion suggested by earlier EPR studies of calcites doped simultaneously with Ba^{2+} and Mn^{2+}. Previous coprecipitation experiments have shown that the incorporation of Sr^{2+} into calcite involves both lattice and non-lattice substitution, a finding also consistent with published EPR data. A model of increasing importance of non-lattice incorporation with increasing deviation of the ionic radii of host and trace cations emerges.

The incorporation of trace elements into calcite impacts a number of areas of environmental chemistry and geochemistry due to the wide range of chemical, biological, and geological materials composed of the mineral calcite. These include the shells and tests of many marine and terrestrial invertebrates, the calcareous sediments of the continental shelf and deep ocean, limestone, cave deposits, caliche, human calculi, and hard-water crusts. The trace element compositions of these materials encode, with varying degrees of precision, the compositions of the solutions responsible for their formation or alteration. The analysis and interpretation of such trace element compositions has led to a better understanding of processes of biomineralization, limestone genesis and diagenesis, lithification, speleology, paleo-oceanography, and environmental chemistry (1-10).

Experimental and empirical studies have demonstrated that certain doubly charged cations of ionic radius less than calcium (e.g., Mn^{2+}, Zn^{2+}, Fe^{2+}, Cd^{2+}, and Co^{2+}) can be extensively incorporated into calcite precipitated from aqueous solution at

0097-6156/86/0323-0574$06.00/0

earth surface conditions (11-19). Partitioning experiments, x-ray diffraction, and EPR studies have indicated that many or all of these trace cations substitute for Ca^{2+} in the calcium lattice site in calcite ($CaCO_3$) (20). Such substitution behavior is expected on structural grounds; each of these cations forms a rhombohedral carbonate salt which is isostructural with calcite. Thus these cations form true, or isomorphous, solid solutions with calcite, characterized by complete (Cd^{2+}) or limited miscibility (Mn^{2+}, Zn^{2+}, Fe^{2+}, and Co^{2+}).

The calcium ion is of such a size that it may enter 6-fold coordination to produce the rhombohedral carbonate, calcite, or it may enter 9-fold coordination to form the orthorhombic carbonate, aragonite. Cations larger than Ca^{2+}, e.g., Sr^{2+}, Ba^{2+}, Pb^{2+}, and Ra^{2+} only form orthorhombic carbonates (at earth surface conditions) which are not, of course, isomorphous with calcite. Therefore these cations are incapable of isomorphous substitution in calcite, but may participate in isodimorphous or "forced isomorphous" substitution (21). Isodimorphous substitution occurs when an ion "adapts" to a crystal structure different from its own by occupying the lattice site of the appropriate major ion in that structure. For example, Sr^{2+} may substitute for Ca^{2+} in the rhombohedral lattice of calcite even though $SrCO_3$, strontianite, forms an orthorhombic lattice. Note that the coordination of Sr^{2+} to the carbonate groups in each of these structures is quite different. Very limited miscibility normally characterizes such substitution.

The coprecipitation of Pb^{2+} and Ra^{2+} with calcite has attracted little attention due to the low concentrations of these elements in most natural waters (22). In contrast the partitioning of Sr^{2+} into calcite has been intensively investigated over the past two decades and several studies of Ba^{2+} partitioning also have appeared (3,7, 16,18,19,23-30). Recently Pingitore and Eastman (31) presented evidence that both isodimorphy and non-lattice incorporation (the trapping or bonding of a foreign cation at a locus that is not the site for Ca^{2+} in the calcite lattice) characterize the limited substitution of Sr^{2+} into calcite. The present paper integrates those findings with a new set of Ba^{2+} partitioning experiments which indicates that non-lattice substitution dominates the incorporation of this even larger cation into calcite. Our purpose, then, is to document these modes of cation incorporation, and discuss their recognition and implications for experimental and practical investigations in environmental chemistry and geochemistry. This paper illustrates how the interplay of solid solution formation and adsorption/trapping affects the coprecipitation of large cations with calcite. Familiarity with the range of processes which can occur at the growth surface of a mineral is crucial to the interpretation and application of the results of these and any other coprecipitation experiments. Additional information on characteristics, terminology, and thermodynamic aspects of solid solution formation can be found in Driessens (this volume).

The Incorporation of Sr(II) into Calcite: A Review

At least a dozen papers over the past two decades have discussed the value of the partition coefficient of Sr^{2+} into calcite, k_c^{Sr} (3,7, 16,18,19,23-29). The different experimental systems and their

conditions have yielded some apparently conflicting results. Recently Pingitore and Eastman (31) explored a wide range of coprecipitation conditions with a single experimental technique and from
those results constructed a partitioning model which may explain
previous inconsistencies. The value of k_c^{Sr} was found to fall
between 0.05 and 0.07, but rose to values between 0.1 and 0.2 for
runs with a low Sr^{2+}/Ca^{2+} ratio (below a molar ratio in the 10^{-3}
range in the resultant calcite). Katz et al. (25) reported almost
identical findings at $98°C$ using the aragonite-calcite transformation, but they concluded the data were an experimental artifact.
Pingitore and Eastman (31) interpreted both sets of results to
indicate that some Sr^{2+} partitions favorably into a limited number
of non-lattice sites, probably defects. At low concentrations of
Sr^{2+}, occupancy of the non-lattice sites makes a substantial contribution to the overall calculated partition coefficient. At
higher concentrations the increasing occupancy of lattice sites
dominates the saturated occupancy of the limited number of non-
lattice sites. This concentration effect may account for the
variation in k_c^{Sr} encountered in earlier studies. Addition of small
amounts of Ba^{2+} or large amounts of Na^+ in experiments with low
concentrations of Sr^{2+} decreased the value of k_c^{Sr} to the range
encountered in high Sr^{2+} concentration runs (0.05 to 0.07). This
was interpreted as a competitive cation effect in which these
cations, at their respective concentrations, have greater tendency
than Sr^{2+} to occupy the non-lattice sites. An increase in k_c^{Sr} with
precipitation rate likewise may be attributed to an increase in the
contribution of non-lattice (defect) sites to the partition coefficient due to rapid crystallization.

In summary, Pingitore and Eastman (31) presented a model of
lattice and non-lattice incorporation of Sr^{2+} into calcite, the
contribution of each mode of coprecipitation to the overall calculated value of k_c^{Sr} depending on the specific conditions of the
experimental run.

Experimental Methods for Coprecipitation of Ba(II) with Calcite

For the Ba^{2+} experiments 0.11 g of calcite were dissolved in 200 ml
of water by bubbling with CO_2. Varying amounts of $BaCO_3$ were added
to produce typical starting solution compositions ranging from
7×10^{-8} to $4 \times 10^{-5}M$ Ba^{2+}. NaCl and $SrCO_3$ also were introduced in
selected runs. Upon completion of the dissolution 0.005 g of cal-
cite seed were placed in the flask and the solution allowed to
evolve CO_2 through a constricted opening. The solutions were stirred (magnetic bar) at a rate sufficient to keep the seed suspended
and temperature was maintained at $18°C$. Normal runs lasted from
three days to one week and yielded recoveries of precipitate ranging
from 40 to 85% of the calcite originally dissolved (precipitation
rate approximately 3×10^{-6} g calcite per hour per cc). The calculated partition coefficient was not related to precipitation time or
yield in this series of runs. Speed runs in which decarboxylation
was hastened by bubbling the solutions with nitrogen gas lasted 12
hours (precipitation rate approximately 4×10^{-5} g calcite per hour
per cc). Details of these experimental procedures are available in
Pingitore and Eastman (30,31). Reagents were Johnson Matthey
Chemicals Puratronic $CaCO_3$, Baker Analyzed $SrCO_3$, Fisher Certified

ACS NaCl and BaCO$_3$, and, except as noted, the seed was Alfa Ultrapure CaCO$_3$. The calcite seed typically comprised between 5 and 10%
of the solid recovered at the end of a run.

Solutions and precipitates were analyzed on a Beckman Spectra-
Span VI direct current plasma emission spectrophotometer (DCP).
Precision for the Ca^{2+} analyses was 3% and for the Ba^{2+} 2% except
for the most dilute samples in which it rose as high as 5%. Calcite
mineralogy was determined on a Philips x-ray diffractometer; calcite
was the only phase recorded except in speed runs of under one hour
in duration (not included in this study) which produced vaterite.
Details of analytic procedures are available in Pingitore and
Eastman (30,31).

Because the Ba^{2+}/Ca^{2+} ratio in the solution changed during the
course of an experiment, the Doerner–Hoskins equation (32) was used
to calculate partition coefficients. The Doerner–Hoskins model of
coprecipitation without equilibration of the interior of a crystal
to the evolving composition of the solution has proved appropriate
to low-temperature aqueous coprecipitation.

Ba(II) Experiments: Results and Discussion

In the light of the model presented by Pingitore and Eastman (31), a
new set of Ba^{2+} partitioning experiments was undertaken, covering a
broader range of conditions than those considered in Pingitore and
Eastman (30). These runs were conducted in solutions with widely
varying initial Ba^{2+}/Ca^{2+} ratios, with and without the addition of
SrCO$_3$ or NaCl to the solution, with differing rates of precipitation, and with different types of seed.

Figure 1 demonstrates that $k_c{}^{Ba}$ varies between 0.03 and 0.05
under a wide range of Ba^{2+} concentrations in the solutions and thus
in the resultant calcites (y-axis). A value of 0.04 +/- 0.01 characterizes the partitioning of Ba^{2+} in this system, which covered
three orders of magnitude in the concentration ratio. These results
seem at odds with the concentration effect and absolute value of $k_c{}^{Sr}$
which characterize Sr^{2+} partitioning (31); the extreme misfit of
Ba^{2+} in the calcite lattice would suggest a lower value for $k_c{}^{Ba}$ and
a pronounced dependence at low concentration ratios. The appropriate ionic radii are Ca^{2+}, 1.08 A; Sr^{2+}, 1.21 A; and Ba^{2+}, 1.44 A
(33). An explanation of these results is that little or none of the
Ba^{2+} occupies lattice sites in the calcite. Instead, the Ba^{2+}
adsorbs on the calcite surface, and on various defects, and is
incorporated by physical trapping as the calcite grows. Hahn (34)
has described this process and further differentiated between
internal occlusion (the trace element concentrated along preferred
surfaces) and anomalous mixed crystal formation (the trace element
randomly distributed throughout the crystal).

The scatter of points in Figure 1, with the value of $k_c{}^{Ba}$
ranging from 0.03 to 0.05, may reflect the more random behavior of
coprecipitation by adsorption/trapping as compared to the more
reproducible behavior of lattice substitution. Sensitivity of the
partition coefficient to experimental conditions is, in fact, one of
the tests for distinguishing the former from the latter (21,34).
Attempts to refine the experimental procedure to achieve greater
consistency therefore are not warranted; any resultant more precise
value of the partition coefficient would be applicable only to a
more limited set of conditions.

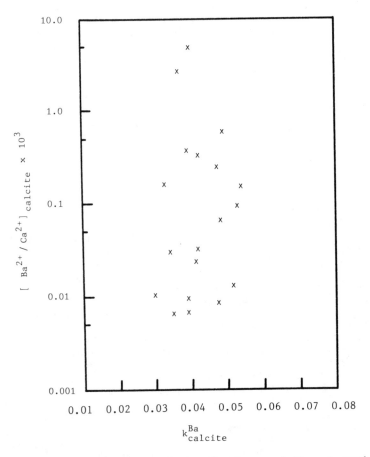

Figure 1.   The partitioning of Ba^{2+} into calcite at various Ba^{2+}/Ca^{2+} solid solution ratios.   Note that $k_c{}^{Ba}$ varies from 0.03 to 0.05 independently of the concentration of Ba^{2+} in the calcite and also, by implication, in the parent solution.

To explore the interpretation that Ba^{2+} does not occupy the lattice sites for Ca^{2+} in these calcites, varying amounts of Sr^{2+} were added to a set of runs (Figure 2), with no apparent effect on k_c^{Ba}. The ratio of Sr^{2+} to Ba^{2+} in the resultant calcites ranged from 0.1 to 100. It appears that Ba^{2+} adsorbs efficiently on the calcite surface and on defects and is not dislodged by Sr^{2+}. Likewise, minor amounts of Ba^{2+} have been shown to displace an equivalent amount of Sr^{2+} from non-lattice sites (31). The more effective adsorption of Ba^{2+} relative to Sr^{2+} is consistent with its larger ionic radius and consequent greater ease of partial dehydration. In view of the fact that Ba^{2+} displaces Sr^{2+} but not vice versa, it is surprising, therefore, to note that the partition coefficient for the non-lattice incorporation of Sr^{2+} into calcite (approximately 0.1 at low concentrations) exceeds that for Ba^{2+} into calcite (0.04). Possible explanations include: 1.) a change in the calcite growth habit due to the presence of Ba^{2+}, yielding growth surfaces with fewer non-lattice sites favorable for Sr^{2+}; 2.) occupancy of the favorable non-lattice sites by Ba^{2+}, but with a low probability of being trapped due to its large ionic radius, yielding a vacancy or unfavorable site after partial encroachment by the growing crystal; 3.) differences in the seed or other experimental conditions in the Ba^{2+} and Sr^{2+} studies rendering a direct comparison of the respective partition coefficients inappropriate.

The addition of 0.48 mols of NaCl per liter of solution significantly depressed k_c^{Ba}; values fell to the 0.02 to 0.03 range (Figure 3). It is not surprising that such an overwhelming quantity of foreign ions can displace Ba^{2+} from adsorption on surface sites. In contrast, NaCl displaced Sr^{2+} from defect sites, but did not significantly lower the lattice value of k_c^{Sr} (31). Sensitivity of a partition coefficient to experimental conditions is a characteristic of internal adsorption and anomalous mixed crystal formation (21,34,35).

Speed runs, in which precipitation was completed in 12 hours, shown in Figure 4 document that higher values of k_c^{Ba} characterize rapid coprecipitation. Possible interpretations include an increase in the number of defect sites due to disorderly crystallization, and/or more efficient capture of adsorbed Ba^{2+} ions due to enhanced probability of physical entrapment with rapid growth.

The possibility that a separate $BaCO_3$ phase precipitated in the speed runs (at levels undetectable by x-ray diffraction) also must be noted. Since calcite is approximately three times as soluble as witherite ($BaCO_3$), the latter can precipitate from a solution at equilibrium with calcite only if the molar ratio of Ba^{2+} to Ca^{2+} exceeds 1:3. Starting compositions for the speed runs had Ba^{2+}/Ca^{2+} ratios from 3.6×10^{-4} to 3.9×10^{-3} and final compositions from 2.6×10^{-3} to 2.3×10^{-2}. To exceed the solubility of witherite during these runs, in which the rapid decarboxylation supersaturates the solution relative to calcite, minimum supersaturations ranging from 14 to 130 with respect to calcite would have been required. Such supersaturations would appear difficult to achieve by decarboxylation, especially in the presence of calcite seed. Viewed in another fashion, the CO_3^{2-} saturation concentration of 0.11 g calcite per 200 ml solution at the start of a run is $9.1 \times 10^{-7}M$. After decarboxylation to atmospheric levels of CO_2 and precipitation of calcite

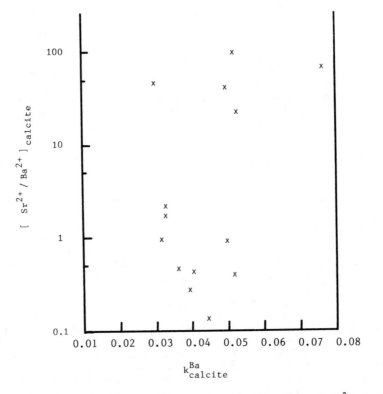

Figure 2. The effect of Sr^{2+} on the partitioning of Ba^{2+} into calcite. The incorporation of up to 100 times as much Sr^{2+} as Ba^{2+} into the calcite does not alter k_c^{Ba}. Note change in vertical axis.

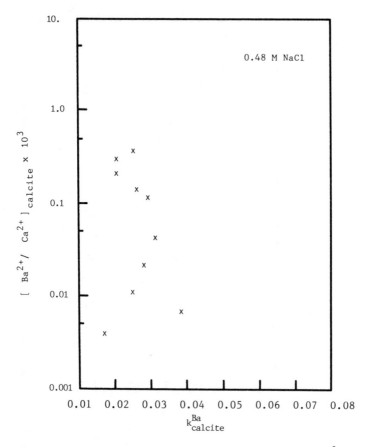

Figure 3.   The effect of NaCl on the partitioning of Ba^{2+} into calcite.   The presence of NaCl at approximately marine concentrations significantly lowers the value of k_c^{Ba}.

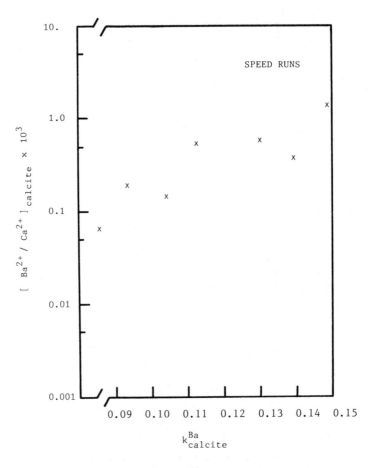

Figure 4. The partitioning of Ba^{2+} into calcite at high rates of precipitation. Reduction of precipitation times from several days to hours yields larger values of $k_c{}^{Ba}$, as compared with Figure 1. Note change of horizontal scale.

the CO_3^{2-} concentration is 1.25 x 10^{-5}. Since the solubility product of witherite is 1.6 x 10^{-9} at low ionic strengths (36) and the maximum Ba^{2+} concentration in the speed runs was 2.9 x 10^{-5}M, the minimum CO_3^{2-} concentration needed for witherite precipitation, 5.5 x 10^{-5}M, apparently was not exceeded.

A final set of experiments demonstrates the importance of the type of seed used to initiate calcite growth. The disparity between the value of k_c^{Ba} reported in Pingitore and Eastman (30), 0.06, and the value encountered in the present work, 0.04, is cause for concern. The slight (7 °C) temperature difference between the two studies is insufficient to explain the substantial difference in k_c^{Ba}. A complete check on procedures showed the seed used in Pingitore and Eastman (30) was produced in experimental runs by rapid decarboxylation by bubbling the solutions with nitrogen gas, whereas Alfa Ultrapure $CaCO_3$ was used in the present study. A fresh batch of laboratory seed yielded the results in Figure 5, which reproduce the findings of Pingitore and Eastman (30). Greater surface area or more surface defects in the seed produced in the laboratory are insufficient to explain the larger value of k_c^{Ba} since many layers of calcite are formed during a run and the partition coefficient was independent of the amount of calcite precipitated. Increases (up to 10 fold) in the amount of seed used in a run also did not affect k_c^{Ba}. Neither seed contained sufficient Ba^{2+} to affect the experiment; the commercial seed contained approximately 3 ppm Ba^{2+} and the laboratory seed was below one ppm. It is nonetheless possible that the rapidly crystallized seed provides a template for defects or consists of polycrystalline masses which propagate during subsequent crystal growth and provide favorable sites for Ba^{2+} incorporation.

In summary, these experiments demonstrate that Ba^{2+} partitioning does not depend on the concentration of Ba^{2+} over the broad range tested and that it is unaffected by moderate amounts of Sr^{2+}. Substantial quantities of NaCl decrease k_c^{Ba} and rapid precipitation increases k_c^{Ba}. These results suggest little or none of the Ba^{2+} is incorporated in Ca^{2+} lattice sites in calcite, but instead adsorbs on surfaces and defects and is trapped by growth of the crystal.

EPR Studies of Sr(II) and Ba(II) in Calcite

The experimental results in Pingitore and Eastman (31) indicate that lattice and non-lattice substitution characterize the coprecipitation of Sr^{2+} with calcite and the results of the present study suggest that Ba^{2+} coprecipitation involves chiefly non-lattice substitution. Electron paramagnetic resonance (EPR) offers an additional probe into conditions in the calcite lattice. Angus et al. (20) overcame the limitation that most cations of interest are not paramagnetic by precipitating calcites with a few hundred ppm Mn^{2+} and measuring the resultant EPR spectra. It had previously been established the Mn^{2+} is incorporated into lattice sites in calcite (37,38). They next produced calcites with varying amounts of Sr^{2+} or Ba^{2+}, along with the small amount of Mn^{2+}. The resultant EPR spectra for the calcites doped with both Sr^{2+} and Mn^{2+} were quite different from the Mn^{2+}-doped calcites produced previously. This indicated a change in the local environment of the paramagnetic Mn^{2+} ions which Angus et al. related to lattice strain associated with

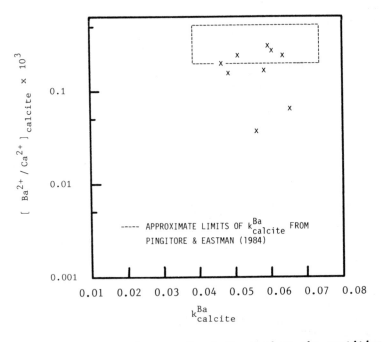

Figure 5. The effect of type of calcite seed on the partition-ing of Ba^{2+} into calcite. Seeding an experiment with calcite nuclei previously produced by rapid precipitation (rather than a commercial reagent) yields values of k_c^{Ba} consistent with the 0.06 reported in Pingitore and Eastman (30).

the incorporation of the large Sr^{2+} ion into Ca^{2+} lattice sites. These effects persisted throughout the 90 to 19,400 ppm range of Sr^{2+} produced in the calcites. In contrast, the incorporation of 240 to 1270 ppm Ba^{2+} in Mn^{2+}-doped calcites produced no significant effect on the Mn^{2+} EPR spectrum. Angus et al. concluded that Ba^{2+} was not incorporated in lattice sites; substitution of such a large cation would yield even more lattice strain than that associated with Sr^{2+}. A similar investigation of Cd^{2+}, with an ionic radius slightly less than that of Ca^{2+}, showed no significant distortion of the EPR spectrum below 20,000 ppm Cd^{2+} in the calcite. The inter-pretation in this case was, of course, that the incorporation of Cd^{2+} into Ca^{2+} lattice sites occurs with minimal strain or distor-tion due to the similarity in atomic radii, a finding consistent with the experimental evidence of Davis et al. (39).

These EPR studies provide independent evidence for at least the partial incorporation of Sr^{2+} in calcite lattice sites and the ex-clusion of Ba^{2+} from lattice sites.

Implications

The range of coprecipitation behavior discussed herein impacts both experimental and practical studies of partitioning of trace elements into calcite. Distinguishing lattice from non-lattice coprecipita-tion emerges as a primary concern. Experiments which explore a wide range of parameters: cation concentration, solution composition, precipitation rate, temperature, etc. provide clues to the nature of coprecipitation.   EPR studies, such as Angus et al. ($\underline{20}$), can provide an obvious aid in this regard.

From a practical standpoint, for the study of carbonate diagenesis it is clear that cations exhibiting lattice substitution exclusively are the most reliable for reconstructing the compositions of diagenetic solutions.   Cations smaller but similar in ionic radius to Ca^{2+} which exhibit exclusively lattice substitution, e.g., Cd^{2+}, are in theory the best diagenetic tracers.   Unfortunately some of these are present in very low concentrations in natural waters and the calcites derived therefrom, making analysis difficult (or impossible by such techniques as electron microprobe) and subject to contamination from the non-carbonate fraction of the rock.   Because of its relative abundance in many carbonate materials and natural waters, Sr^{2+} remains an important diagenetic tool despite the possibility of non-lattice incorporation.   The non-lattice component of the coprecipitation of Sr^{2+} into calcite occurs, in general, under limited and predictable conditions.   The interpretation of the Ba^{2+} contents of calcites is hampered by the potentially large variation of $k_c{}^{Ba}$ with the conditions of precipitation (solution composition, type of carbonate growth surface, precipitation rate).   This suggests that Ba^{2+} may prove inappropriate as a diagenetic tracer.

Acknowledgments

Discussions with my colleagues at UT El Paso in chemistry and geology, especially M.P. Eastman, C.A. Chang, J.D. Hoover, and P.C. Goodell flavored the outcome of this project.   M.P. Eastman generously provided equipment and supplies and C. Podpora, B. Deshler, and O.I. Cilatan provided assistance in the laboratory.   Julieta Aguirre typed the camera-ready copy.   The comments of two anonymous reviewers and the suggestions of editors J.A. Davis and K.F. Hayes improved the manuscript.

Literature Cited

1.   Amiel, A.J.; Friedman, G.M.; Miller, D.A.   Sedimentology 1973, 20, 47–64.
2.   Chave, K.E.   J. Geol. 1954, 62, 266–83.
3.   Kinsman, D.J.J.   J. Sed. Petrology 1969, 39, 486–508.
4.   Pingitore, N.E.   J. Sed. Petrology 1976, 985–1006.
5.   Morrow, D.W.; Mayers, I.R.   Can. J. Earth Sci. 1978, 15, 376–96.
6.   Veizer, J.   In "Stable Isotopes in Sedimentary Geology"; Soc. Econ. Paleont. Mineral.: Tulsa, 1983; Sec. 3, pp. 1–100.
7.   Holland, H.D.; Kirsipu, T.V.; Huebner, J.S.; Oxburgh, U. M. J. Geol. 1964, 72, 36–67.
8.   Lowenstam, H.   J. Geol. 1961, 69, 241–60.
9.   Graham, D.W.; Bender, M.L.; Williams, D.F.; Keigwin, L.D., Jr. Geochim. Cosmochim. Acta 1982, 46, 1281–92.
10.   Dodd, J.R.   J. Geol. 1963, 71, 1–11.
11.   Bodine, M.W.; Holland, H.D.; Borcsik, M.   In "Problems of Postmagmatic Ore Deposition—Proc. Symp.": Prague, 1965; pp. 401–406.

12. Tsusue, A.; Holland, H.D. Geochim. Cosmochim. Acta 1966, 30, 439-54.
13. Crocket, J.H.; Winchester, J.W. Geochim. Cosmochim. Acta 1966, 30, 1093-1109.
14. Dardenne, M. Bull. Bur. Rech. Geol. Min. 1967, 5, 75-110.
15. Michard, G. Comptes Rendus Acad. Sci. Paris, Ser. D 1968, 267, 1685-8.
16. Ichikuni, M. Chem. Geol. 1973, 11, 315-9.
17. Richter, D.K.; Fuchtbauer, H. Sedimentology 1978, 25, 843-860.
18. Lorens, R.B. Ph.D. Thesis, University Rhode Island, Kingston, 1978.
19. Lorens, R.B. Geochim. Cosmochim. Acta 1981, 45, 533-61.
20. Angus, J.G.; Raynor, J.B.; Robson, M. Chem. Geol. 1979, 27, 181-205.
21. Benes, P.; Majer, V. "Trace Chemistry of Aqueous Solutions"; Elsevier: Amsterdam, 1980; p. 136.
22. Langmuir, D.; Riese, A.C. Geochim. Cosmochim. Acta 1985, 49, 1593-1601.
23. Holland, H.D. Final Report, AEC Contract AT(30-1)-2266, 1966, p. 1-53.
24. Kitano, Y.; Kanamori, N.; Oomori, T. Geochem. J. 1971, 4, 183-206.
25. Katz, A.; Sass, E., Starinsky, A.; Holland, H.D. Geochim. Cosmochim. Acta 1972, 36, 481-508.
26. Usdowski, E. Contr. Mineral. Petrology 1973, 38, 177-95.
27. Jacobson, R.L.; Usdowski, E. Contr. Mineral. Petrology 1976, 59, 171-85.
28. Baker, P.A.; Gieskes, J.M.; Elderfield, H.; J. Sediment. Petrology 1982, 52, 71-82.
29. Mucci, A.; Morse, J.W. Geochim. Cosmochim. Acta 1983, 47, 217-33.
30. Pingitore, N.E.; Eastman, M.P. Chem. Geol. 1984, 45, 113-20.
31. Pingitore, N.E.; Eastman, M.P. Geochim. Cosmochim. Acta (in revision).
32. Doerner, H.A.; Hoskins, W.M. J. Am. Chem. Soc. 1925, 47, 662-715.
33. Whittaker, E.J.W.; Muntus, R. Geochim. Cosmochim. Acta 1970, 34, 945-56.
34. Hahn, O. "Applied Radiochemistry"; Cornell Univ. Press: Ithaca, 1936; pp. 98-131.
35. Walton, A.G. "The Formation and Properties of Precipitates"; Interscience: New York, 1967; pp. 102-107.
36. Krauskopf, K.B. "Introduction to Geochemistry"; McGraw-Hill: New York, 1967; p. 82.
37. Fujiwara, S. Anal. Chem. 1964, 36, 2259-61.
38. Wildeman, T.R. Chem. Geol. 1970, 5, 167-77.
39. Davis, J.A.; Fuller, C.C.; Cook, A.D. Geochim. Cosmochim. Acta (submitted).

RECEIVED June 25, 1986

Near-Surface Alkali Diffusion into Glassy and Crystalline Silicates at 25 °C to 100 °C

Art F. White and Andy Yee

Lawrence Berkeley Laboratory, University of California, Berkeley, CA 94720

Alkali uptake by obsidian and feldspar was investigated in batch and recirculating column experiments and by XPS profiling using ion sputtering and variable take-off angles. No measurable alkali diffusion occurred in the feldspars. Diffusion coefficients were determined for Rb, Cs and Sr in glass using a one-dimensional model assuming interdiffusion with Na. Activation energies for Rb, Cs and Sr in glass were 50, 83, and 70kJ respectively which were significantly lower than that measured in previous high temperature experiments and indicate diffusion into a hydrated near-surface matrix.

The immobilization of dissolved chemical species by adsorption and ion exchange onto mineral surfaces is an important process affecting both natural and environmentally perturbed geochemical systems. However, sorption of even chemically simple alkali elements such as Cs and Sr onto common rocks often does not achieve equilibrium nor is experimentally reversible (1). Penetration or diffusion of sorbed species into the underlying matrix has been proposed as a concurrent non-equilibration process (2). However, matrix or solid state diffusion is most often considered extremely slow at ambient temperature based on extrapolated data from high temperature isotopic or tracer diffusion experiments. Only limited attempts have been made to measure matrix diffusion in silicates at lower temperature.

This paper reports results of alkali diffusion experiments on obsidian and feldspars over a temperature range of 25° to 100°C. The immediate significance of this study is on potential retardation of radioactive alkali isotopes such as ^{137}Cs, ^{90}Sr and [226]Ra during transport in potential repository host rocks such as basalt and tuff which contain glass and feldspar phases. The paper hopefully will also shed light on the more general question of the importance of low temperature diffusion in geological environments.

Experimental Results

The three silicates obtained from Wards Scientific Establishment and employed in this study are obsidian from St. Helena, California, microline from Ontario, Canada and albite from South Dakota. The bulk of the samples were crushed and sieved to between 211 and 423 μm. Fines were removed by repeated ultrasonic cleaning and decantation. Measured BET

0097–6156/86/0323–0587$06.00/0

surface areas were low; 0.07, 0.08 and 0.05 m^2 for the K-feldspar, plagioclase feldspar and obsidian, respectively. Individual larger cleavage and fracture fragments were preserved for surface studies using X-ray photoelectron spectroscopy (XPS).

The extent of alkali loss from aqueous solution as a result of sorption and diffusion were determined by both batch and column experiments. Both types of experiments utilized 1x10^{-4} molar solutions of the alkali chloride. Batch experiments consisted of reacting 2.0 L of air-saturated solution with 200 gms of feldspar and glass at 25°C. Batches were stirred once a day over a one-year period. In the column experiments, 650ml of solution stored in an air-saturated reservoir was recirculated by a peristaltic pump through glass-jacketed columns containing 165 gm of obsidian. Experiments were run at 25°, 50° and 75°C for periods up to two months. The pH range during both batch and column experiments was buffered by atmospheric CO$_2$, resulting in a pH range of between 6.5 and 7.5.

Cleavage and fracture fragments of the mineral and glass phases were reacted separately at 25°, 50°, and 100°C. In order to produce levels measurable by XPS, 1x10^{-1} molar alkali chloride solutions were required. After reaction, these samples were washed with deionized water and stored in a vacuum dessicator.

Experimental Results

Mass fluxes of alkali elements transported across the solid-solution interfaces were calculated from measured decreases in solution and from known surface areas and mineral-to-solution weight-to-volume ratios. Relative rates of Cs uptake by feldspar and obsidian in the batch experiments are illustrated in Figure 1. After initial uptake due to surface sorption, little additional Cs is removed from solution in contact with the feldspars. In contrast, parabolic uptake of Cs by obsidian continues throughout the reaction period indicating a lack of sorption equilibrium and the possibility of Cs penetration into the glass surface.

Typical results for the shorter-time column experiments show Rb uptake plotted against the square root of time at several temperatures in Figure 2. No measurable decreases in Li, K or Ba were observed for column experiments conducted at 25°C.

The data of the type shown in Figures 1 and 2 were fitted with a linear regression program to the expression;

$$M_T = M_o + kt^{1/2} \qquad (1)$$

where M_T is the total elemental uptake (moles·cm^{-2}), M_o is the intercept at zero time (moles·cm^{-2}) and k is the parabolic rate constant (moles·cm^{-2}·s$^{-1/2}$).

The M_o term can be used to approximate initial sorption or desorption on the glass surface, and the kt$^{1/2}$ term the longer-term diffusion transport into or out of the surface (3). As shown in Figure 2, the sorption term decreases and the diffusion term increases with temperature for the obsidian experiments. Tabulated values for Equation 1 are presented in Table 1 along with the regression coefficient, r^2, for glass data.

The near-surface alkali-element concentrations in obsidian and feldspar were characterized by their photoelectron peak intensities. An example of the relative Cs $3d_{5/2}$ peak intensities for obsidian and plagioclase plotted against the electron binding energy is shown in Figure 3. The upper solid lines indicate slightly higher levels of Cs in the obsidian. After reaction for 10 minutes with 0.1N HCl, Cs in plagioclase is reduced to a background level indicating reversible ion exchange. In contrast, a

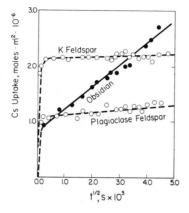

Figure 1. Rate of Cs uptake from batch solutions at 25°C. Straight lines are least square regressions to Equation 1.

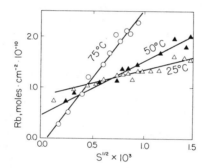

Figure 2. Rate of Rb uptake by obsidian from recirculating column experiments at different temperatures. Straight lines are least square regressions to Equation 1.

Table I. Diffusion rate parameters for Equation 1.
(units are defined in text)

	Temp $C°$	M_o $\times 10^{-10}$	k 10^{-13}	t $\times 10^6$	r^2
Rb	25	0.69	0.61	2.95	.83
	50	0.46	1.03	2.95	.89
	75	0.01	2.66	0.94	.97
Cs	25	1.47	0.28	2.09	.74
	25*	1.97	0.15	2.09	.92
	75	0.23	1.76	0.89	.94
Sr	25	1.06	0.32	1.25	.94
	50	0.64	0.97	1.57	.88

*batch experiment

substantial amount of Cs remains after acid treatment of obsidian demonstrating non-reversibility and the potential for penetration below the glass surface.

Two methods employing XPS analysis, ion sputtering and variable take-off angles, were employed to produce elemental profiles with depth. The sputtering technique, used to create the profiles in Figure 4, involves sequential removal of surface layers by bombardment with a positively charged Ar beam followed by multiplex XPS analysis. The sputter time can be correlated with penetration depth if rates have been calibrated for a given matrix. Sputter rates for obsidian and K-feldspar are not known but are estimated to be 10 Å per minute based on amorphous SiO_2 standards (4) and instrument operating parameters.

The elemental concentrations to the left of the zero time line in Figure 4 are duplicate analyses prior to the onset of sputtering. Cs concentrations in obsidian shown in Figure 4 reach a maximum after sputtering for one minute (~ 10Å) and decrease to a stable background after approximately 4 minutes (~ 40Å). In contrast, the Cs profile for K-feldspar achieves a maximum at the surface and decreases more rapidly with depth indicating less penetration. The high Cs background intensities at greater depths relative to Na and K intensities are related to the greater analytical sensitivity of Cs (5).

Both the Na and K intensities in the K-feldspar profile of Figure 4 are stable with depth indicating a previously documented lack of alkali mobility in the surface layers of feldspars at low temperature (7). In contrast, K increases and Na decreases with depth beneath the obsidian surface demonstrating substantial elemental mobility. The K loss near the surface corresponds to a concentration increase measured in aqueous solution. Sodium profiles in obsidian should exhibit even greater near-surface losses relative to K based on profiles measured by HF leaching (3) and sputter-induced optical emission studies (6).

The anomalous Na decrease with depth in Figure 4 illustrates a problem with ion beam profiling in that some light ions, including Na, are mobile under the beam and can be stripped from the surface or embedded deeper in the matrix. K ions do not appear to be susceptible to this phenomenon and heavier elements such as Rb, Cs and Sr would be expected to be even less affected. Profiles of Si and Al, although not shown in Figure 4, were essentially constant with depth documenting the lack of secondary mineral formation on the surfaces of the obsidian and feldspar.

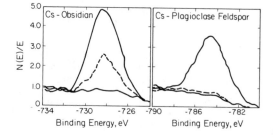

Figure 3. Peak intensity versus electron binding energies for the Cs $3d_{5/2}$ photoelectron peaks. Upper solid lines are after washing with D.I. water, lower dashed lines are after washing with 0.1N HCl, lower solid lines are background Cs concentrations.

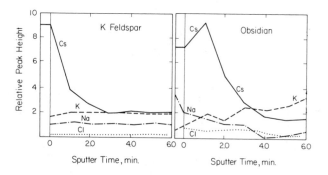

Figure 4. Elemental peak intensities versus duration of Ar^+ sputtering in obsidian.

A second profiling method, termed angle resolved XPS analysis, was used to confirm the penetration of Rb, Cs and Sr. Photoelectrons measured by XPS can travel only a very limited distance through the sample before interacting with interior atoms and losing energy. The inelastic mean free path ($\sim$ 30Å) is generally independent of the trajectory relative to the plane of the sample surface. By providing an axis of rotation in the plane, the electron take-off angle changes. The maximum analytical depth is the product of the inelastic mean free path and the sine of the take-off angle.

A resulting take-off angle profile for Cs in obsidian is shown in Figure 5. In this case, Cs concentrations with depth are plotted as the mole ratio with respect to Si. The use of elemental ratios overcomes the common effect of carbon contamination on the glass surfaces ([8]). The maximum Cs/Si ratio for reaction at 25°C occurs at a depth of approximately 20Å and decreases up to depths of 50Å. The more limited data at 100°C exhibit higher Cs/Si ratios at greater depth. The positions of the Cs maximum and depth of penetration are comparable in the take-off angle and sputtering techniques.

Calculation of Diffusion Parameters

The diffusion coefficients for Rb, Cs and Sr in obsidian can be calculated from the aqueous rate data in Table 1 as well as from the XPS depth profiles. A simple single-component diffusion model ([9]) characterizes one-dimensional transport into a semi-infinite solid where the diffusion coefficient ($cm^2 \cdot s^{-1}$) is defined by;

$$D = (M)^2 \left[\frac{1}{2C_o} \right]^2 \frac{\pi}{t} \qquad (2)$$

The model assumes a constant surface or skin concentration, C_o ($moles \cdot cm^{-3}$). The total mass transferred into the solid, M ($moles \cdot cm^{-2}$), can be calculated from the product of the $kt^{1/2}$ term in Equation 1 and the parameters listed in Table 1.

Application of Equation 2 requires that an estimate of the surface concentration, C_o, be made, and that it is assumed not to change as a function of reaction progress. Conceptually this can be considered equal to concentrations of the sorbed alkali on the glass surface and is estimated from the maximum elemental concentrations measured by XPS analysis. The C_o terms for Rb, Cs, and Sr are 1.3×10^{-3}, 6.5×10^{-4}, and 9.2×10^{-4} moles cm^{-3}, respectively.

An additional assumption is that diffusion is independent of other species in the glass matrix. This cannot be strictly true because interdiffusion of at least one additional species is necessary to maintain charge balance within the glass. Sodium is most likely to be the dominant interdiffusion ion, as has been demonstrated for Sr and Cs diffusion in rhyolite glass at higher temperatures ([10]) and as supported by rapid release rates of Na to solution found in the present study. Codiffusion of hydronium and alkali ions are ignored in the model.

The interdiffusion coefficient, D_i, between Cs and Na for example, can be defined by the Nernst-Plank Equation,

$$D_i = D_{Cs}D_{Na}[N_{Cs} \cdot N_{Na}]/N_{Cs} \cdot D_{Cs} + N_{Na} \cdot D_{Na} \qquad (3)$$

where D_{Na} and D_{Cs} are the self-diffusion coefficients and N_{Na} and N_{Cs} are the mole fractions of each component. If the concentration of one component, as in the case of Cs in obsidian, is much less than that of the other component, Na, then Equation 3 will reduce to $D_i \sim D_{Cs}$ so that the interdiffusion coefficient is approximated by the self-diffusion coefficient of the trace component. This relationship permits the direct calculation of the self-diffusion coefficients of Rb, Cs, and Sr listed in Table II from Equation 2.

Table II. Diffusion Coefficients for Alkali Elements in Obsidian $(cm^2 \cdot s^{-1})$

	T, °C	D		T, °C	D
Rb	25	1.6×10^{-21}	Cs	25	1.2×10^{-21}
	50	4.7×10^{-21}		50	8.1×10^{-21}
	75	3.1×10^{-20}		75	5.6×10^{-20}
			Sr	25	9.7×10^{-22}
				50	8.5×10^{-21}

The diffusion coefficients in Table II can be compared with penetration depths measured by XPS profiling through the relationship (9),

$$\frac{C}{C_o} = \mathrm{erfc} \frac{x}{2\sqrt{Dt}} \tag{4}$$

where C is the concentration $(moles \cdot cm^{-3})$ at some depth x (cm) beneath the surface. Cesium profiles measured at 25°, 50°, and 75°C in obsidian are given in Figure 6. Superimposed on them is the calculated profiles based on Equation 4 with the boundary conditions x=0 when $C=C_o$. As indicated in the figure, the measured and calculated penetration depths are in reasonably good agreement, supporting the above diffusion model.

The temperature dependence of the diffusion coefficients can be described by the conventional form of the Arrhenius expression,

$$\log D = \log D_o - \frac{E_A}{RT} \tag{5}$$

where E_A is the activation energy $(kJ \cdot moles^{-1})$, R is the universal gas constant, T is the absolute temperature (°K), and D_o is the preexponential term $(cm^2 \cdot s^{-1})$. The activation energy for diffusion in alkali glasses consists of two terms (11): (1) the dissociation energy in the breaking of bonds with the non-bridging oxygen ions and (2) the kinetic energy involved in movement or diffusion. The activation energy and D_o correspond to the respective slope and intercept of the plots of log D versus T^{-1} shown in Figure 7 and tabulated in Table III.

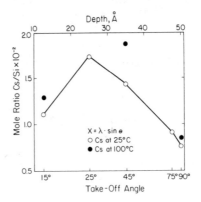

Figure 5. Concentration ratio of Cs to Si in obsidian as a function of analytical XPS take-off angles.

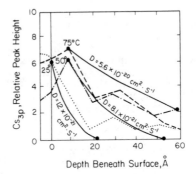

Figure 6. Comparison of XPS depth profiles for Cs in obsidian at various temperatures (broken lines) with predicted penetration depths based on profiles calculated by Equation 4 (solid lines).

Table III. Calculated Activation Energies and Intercepts

	E_A kJ	D_0 $cm^2 \cdot s^{-1}$
Rb	50	10^{-12}
Cs	83	$10^{-6.7}$
Sr	70	$10^{-8.7}$

Comparisons with high temperature diffusion studies

The preceding data, though limited in nature, represent one of the first attempts to measure solid state diffusion rates of alkali elements into the near-surface region of feldspars and natural glasses at low temperature. As such, interesting comparisons can be made with diffusion coefficients and activation energies calculated from numerous high temperature isotope and tracer diffusion studies (11-18).

Such comparisons for diffusion coefficients for Rb, Cs and Sr are shown in Figure 7. The solid lines are high temperature measurements and the dashed lines Arrhenius extrapolations to lower temperatures. Diffusion coefficients for Rb in glass measured in this study fall within the range of extrapolated values shown in Figure 7. Similar results have been found for Na (3). The wide range in extrapolated values is due principally to differing experimental glass compositions, and to the associated mixed alkali effect which influences diffusion rates (11). Extrapolated diffusion rates for Rb in K-feldspar (line 5, Figure 7) at low temperature are extremely small which may explain the lack of observed Rb diffusion in the present study.

The measured Cs and Sr diffusion coefficients shown in Figure 7 are comparable to Rb but are more than an order of magnitude larger than the extrapolated high-temperature data extrapolated to between 25 and 100°C.

The activation energies calculated in the literature (11-22) for high temperature alkali metal diffusion in crystalline and glassy silicates are shown in Figure 8. The wide ranges in apparent energies are in part due to differing compositions and methods of diffusion measurement. In general activation energies increase with ionic radius of an element. For a specific alkali, the activation energy is greater in the crystalline silicates than in the glass.

Discussion

Differences in diffusion rates between alkali glasses and crystalline silicates such as feldspars are not unexpected due to significant structural differences. Alkali metal glasses have been shown to a six-member silicon-oxygen ring structure comparable to tridymite while feldspars exhibit a four-member ring structure (23). This structural difference is in part responsible for the significantly lower density in chemically similar glass and feldspar. For example, albite glass has a reported density of 2.38 compared to an albite feldspar density of 2.62 (24).

The more open structure and lower density in glass decreases the ionic bonding strength as well as the activation energy of alkali diffusion. In addition, the glass structure permits a large number of defect structures, reportedly up to 10^{21} per cm^3 (11), compared to negligible defects in crystalline feldspars. Such defects serve as significant pathways for alkali metal diffusion.

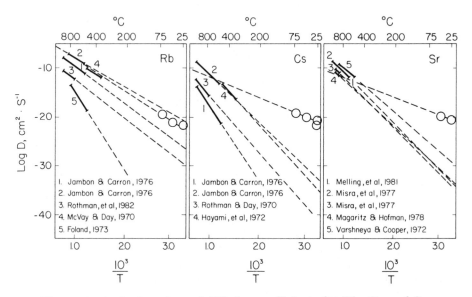

Figure 7. Arrhenius plots of diffusion coefficients for Rb, Cs, and Sr. Solid lines are high temperature data (numbers are literature references). Dashed lines are extrapolated coefficients based on Equation 5. Open circles are from this study.

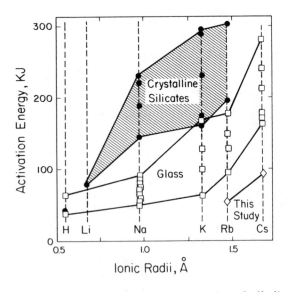

Figure 8. Apparent diffusion activation energies of alkali metals as a function ionic radii. Data are from this study and cited references.

The activation energies calculated for Rb, Cs and Sr in the present study (Table III and Figure 8) are considerably lower than those calculated for high temperature diffusion in both crystalline and glass silicates. This discrepancy in the latter case implies that the glass matrix may be significantly different in high and low temperature diffusion studies.

A surface peak effect has been observed during Rb and Sr diffusion in vitreous silica (13). Such large near-surface concentrations are postulated to result from the exposure at the glass surface of a greater number of interstices or defects over which diffusion can occur. This would lead to steep penetration curves observed in some XPS profiles of glass.

Of potentially greater significance is surface hydration which occurs concurrently with alkali diffusion at relatively low temperature. The average activation energy of water diffusion in obsidian can be estimated at 75kJ between 95° and 245°C (25). A nuclear resonance hydration profile of obsidian at 25°C has yielded a diffusion coefficient of 5×10^{-20} $cm^2 \cdot s^{-1}$ (26). Comparison with alkali diffusion coefficients and activation energies reported in the present study indicates that a hydration front most likely precedes the diffusion of Rb, Cs and Sr into the glass surface.

The incorporation of two to five weight percent of water into glass (27) tends to break oxygen-bridged bonds and produce a volume expansion which increases the number and size of interstices and defects, allowing penetration of large ions such as Cs, Rb, and Sr. The larger activation energies reported from high-temperature studies involving Rb, Cs and Sr tracer or isotopic exchange can be explained by the absence of water in the experiments and the lack of a hydrated glass matrix.

The influence of hydration on alkali metal diffusion rates appears to decrease with ionic ratio as shown by a closer correlation between high- and low-temperature diffusion data (Figure 7) and activation energies (Figure 8) for Rb relative to Cs. Measured Na diffusion coefficients in hydrated obsidian at 25°C can be accurately reproduced by extrapolation of high temperature diffusion rates for nonhydrated obsidian (3) indicating that diffusion rates of smaller ions such as sodium are not affected by the hydration process.

The fact that most natural glasses in contact with water are either partially or totally hydrated implies that larger alkali ions such as Rb, Cs and Sr may penetrate deeper into glass surfaces than previously indicated from extrapolation of higher temperature tracer or isotopic diffusion experiments. Such uptake could have significant implications in predicting the transport and retardation rates of radioactive isotopes of heavy alkali elements in basalt and tuff aquifers which contain abundant glass. The study supports the conclusion that solid state diffusion is not significant in crystalline silicate minerals at ambient temperature.

Acknowledgments

The authors wish to thank Steve Flexser of Lawrence Berkeley Laboratory who prepared samples and William Stickle of the Perkin Elmer Surface Analytical Laboratory who performed the XPS analysis. This work was supported by the U.S. Nuclear Regulatory Commission.

Literature Cited

1. Reyla, R. A.; Serne, R. J. *Battelle Pacific Northwest Lab. Rept.* PNL-SA-7352, 1978, 164-203.
2. Narasimhan, T. N.; Liu, C. W. *Water Resources Res.* (in press).
3. White, A. F. *Geochim. Cosmochim. Acta* 1983, 47, 805-815.
4. Perry, D. L.; Tsao, L.; Gaugler, K. A. *Geochim. Cosmochim. Acta* 1983, 47, 1201-1289.
5. Wagner, C. D; Riggs, W. M.; Davis, L. E.; Mouldeu, S. F.; Milenberg, G. E. *Handbook of X-ray Photoelectron Spectroscopy,* 1979, Perkin Elmer, Praire, Minn. 190p.
6. Tsang, I.S.T. *Science* 1978, 201, 339-344.
7. Holdren, G. R.; Berner, R. A. *Geochim. Cosmochim. Acta 1979* 43, 1146-1171.
8. Bancroft, G. M.; Brown, J. R.; Fyfe, W. S. *Chem. Geology* 1979, 25, 227-235.
9. Crank, J. *The Mathematics of Diffusion,* 1954, Oxford Press, London.
10. Melling, P. J.; Vempati, C. S.; Allnatt, A. R.; Jacobs, P. W. M. *Phys. Chem. Glass* 1981, 22, 49-54.
11. Terai, R. *J. Non-Crystalline Solids,* 1971, 6, 121-135.
12. Jambon, A.; Carron, J. P. *Geochim. Cosmochim. Acta* 1976, 40, 897-903.
13. Rothman, S. J.; Marcuso, T. L. M.; Nowicki, P. M.; Baldo, P. M.; McCormick, A. W. *J. Amer. Ceram. Soc.* 1982, 65, 578-582.
14. McVay, G. L.; Day, D. E. *J. Amer. Ceramic Soc.* 1970, 53, 508-513.
15. Foland, K. A. In *Geochemical Transport and Diffusion.* Holman, A. W.; Giletti, B. J.; Yoder, H. S.; Yund, R. A., 1973, Carnegie Instit. Pub. 634.
16. Hayami, R.; Terai, R. *Phys. Chem. Glasses* 1972, 13, 102-120.
17. Misira, N. K.; Venkatasubramanian, V. S. *Geochim. Cosmochim. Acta* 1978, 41, 837-838.
18. Magaritz, M.; Hofmann, A. W. *Geochim. Cosmochim. Acta* 1978, 42, 595-605.
19. Varshney, A. K.; Cooper, A. R. *J. Amer. Ceramic Soc.* 1972, 55-220-223.
20. Nogami, M.; Tonrozawa, M. *Phys. Chem. Glasses* 1984, 80-85.
21. White, A. F. *Lawrence Berkeley Laboratory Rept.* LBL-11794, 1980, 75p.
22. Jambon, A.; Carron, J. P. *Bull Mineral* 1978, 101, 22-38.
23. Taylor, M.; Brown, G. E. *Geochim. Cosmochim. Acta* 1979, 43, 61-75.
24. Taylor, M.; Brown, G. E. *Geochim. Cosmochim. Acta* 1979, 43, 1467-1473.
25. Friedman, I.; Long, W. *Science* 1976, 191, 347-352.
26. Doremus, R. H.; *Treatise on Material Sci. Tech.,* 1979, 17, 41-69.
27. Ross, C. S.; Smith, R. L. *Amer. Mineral.* 1955, 40, 1071-1089.

RECEIVED August 11, 1986

PRECIPITATION AND DISSOLUTION

29

Mechanisms and Rate Laws in Electrolyte Crystal Growth from Aqueous Solution

Arne E. Nielsen

Medicinsk-Kemisk Institut, Panum Institute, University of Copenhagen, Blegdamsvej 3, DK-2200 Copenhagen N, Denmark

When electrolyte crystals grow in an aqueous solution with a surface controlled rate following a parabolic or an exponential rate law, the rate-determining step is the integration of the cations at kinks in surface steps. The integration rate constant, or frequency, is about one-thousandth of the dehydration frequency of the cations. The factor, 10^{-3}, is assumed to be due to diffusion activation energy. Both the rate laws and the absolute rates observed can be accounted for by calculating the kink density by classical methods, and estimating the adsorption equilibrium constants by means of the ion pair stability constants. The calculated and the observed rates mostly agree within one order of magnitude. The rate-determining mechanism for crystal growth may change between several surface processes and transport processes (diffusion and convection in the liquid phase) when the concentration or the particle size is varied.

In geochemistry, as in chemistry in general, a phenomenon is not considered as completely understood until the essential empirical features of the phenomenon (such as, for instance, its kinetics) are accounted for in terms of a reasonable molecular mechanism, convincingly verified by experimental tests. Geochemistry deals primarily with crystalline bodies, many of which are electrolytes that have crystallized from aqueous solution. The molecular mechanisms of these crystallization processes are therefore of great importance for the understanding of geochemical processes taking place in nature.

When a crystal is growing in a solution two groups of processes are always taking place, <u>transport processes</u> bringing the dissolved growth units (ions or molecules) from the bulk of the solution up to the crystal surface, and <u>surface processes</u> transferring the arriving growth units to the lattice positions (<u>1-3</u>). With some simplifications we may describe the situation in the way that the concentration of the growth units is c in the bulk solution and c' in the solution just outside the crystal surface and any adsorption layer. The total

0097-6156/86/0323-0600$06.00/0
© 1986 American Chemical Society

driving force for crystallization, $(c-c_S)$, where c_S is the solubility, is divided in the driving force $(c-c')$ for the transport processes and $(c'-c_S)$ for the surface processes (1-3). The simplifications made are 1), neglecting that the different ions may have different, perhaps even non-equivalent concentrations in the solution, and 2), neglecting that the concentration very close to a growing crystal may vary along the surface.

Rate Control

If $c_S \approx c' \ll c$ most of the total driving force is used on the transport processes - the surface processes presenting no resistance. In that case the growth rate is transport controlled: A moderate change of the diffusion coefficient or of the liquid flow rate relative to the particles would change the growth rate correspondingly, but a change of the kinetic constants of the surface process would not influence the growth rate perceptibly.

Analogously, if $c_S \ll c' \approx c$ the growth rate is surface reaction controlled. See Figure 1.

In general c and c_S are easy to measure whereas c' is difficult or impossible to determine experimentally. But c' may be eliminated by mathematical methods, or one may conclude that $c' \approx c_S$ or $c' \approx c$. We define the linear growth rate v_g as the linear velocity of displacement of a crystal face relative to some fixed point in the crystal. v_g may be known as a function of c and c', derived from the theory of transport control, and as a function of c' and c_S as well, derived from the theory of surface control. Then c' may be eliminated by equating the two mathematical expressions

$$f_T(c,c') = f_S(c',c_S) \tag{1}$$

solving this equation, and inserting in f_T and f_S (2). This procedure may be relevant in the general case $c_S \ll c' \ll c$, see Figure 1, but in most cases only one of the mechanisms is rate-determining. This may be tested by inserting the total driving force in each of the expressions, which means letting $c' = c_S$ in f_T and $c' = c$ in f_S. If this leads to essentially different values of f_T and f_S the smaller value will be a good approximation to the real growth rate, and the corresponding mechanism (T or S) will be rate-determining, (the "bottle neck" effect).

Transport Controlled Kinetics

We shall mainly discuss the growth of small crystals suspended - if necessary by stirring - in an aqueous solution. It is assumed that the average distance between each particle and the nearest other particle is of the order of several particle diameters, at least, so that it is meaningful to define the bulk concentration as the limit of $c(x)$ for $x \to \infty$, x being the distance from the particle observed. If this condition is not fulfilled the solution of the diffusion problem describing the concentrations as a function of the three space coordinates (x,y,z) and of time (t) is more complicated. But it is always possible to solve the problem even if the particle touch each other, as in a soil. In general $c(x,y,z,t)$ is a solution to

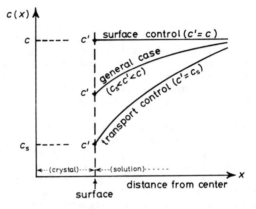

Figure 1. The concentration, c(x) of the solute as a function of the distance, x from the crystal center.

Fick's diffusion laws, depending on the shape of the liquid as determined by the particle surfaces (1,31-33). However, in such systems the rate of depletion of a supersaturated solution by diffusion controlled growth would be very fast, and if the rate is actually slow (measurable) the rate control is likely to be a surface process.

In the following we shall describe the crystals as if they were spheres with the volumes of the actual crystals. This will facilitate the calculation considerably. In this way each particle is ascribed a well-defined radius even if it is not spherical. The approximation of spheres will be sufficient for our purposes unless the particles are needles, thin plates or similar extremely non-spherical objects.

Diffusion. The transport process may consist of two parts, diffusion and convection. When the liquid is stagnant and resting relative to the particle the transport is done by diffusion only. A steady state is quickly established in the solution around the particle (4). (Strictly it is a quasi-steady state since the particle is growing (5)). At the particle surface the concentration gradient becomes equal to $(c-c_s)/r$, which leads to the growth rate

$$v_g \equiv \frac{dr}{dt} = \frac{DV_m(c-c_s)}{r} \equiv k_D(S-1) \tag{2}$$

where r is the (defined) radius of the particle, D is the diffusion coefficient, and V_m the molar volume of the crystalline material. For small ions in water at 25 °C one may estimate $D \approx 10^{-9}m^2/s$. V_m is usually calculated as M/ρ where M = molar mass (molecular weight) and ρ = density of the crystalline material, so that $V_m(c-c_s)$ is dimensionless, c being the molar concentration in the bulk of the solution.

Particles moving sufficiently slowly in a gently stirred solution follow Equation 2. This applies to small particles (r < ca. 5 μm, see next section) falling (sedimenting) due to gravity.

Equation 2 describes the growth of each individual particle when its growth rate is determined by the diffusion of the solute from the bulk of the solution to the particle surface. The total consumption of solute from the solution may be obtained by expressing the growth rate as $dn/dt = V_m^{-1}dV/dt = V_m^{-1} \cdot 4\pi r^2 dr/dt$, and adding the values of dn/dt for all the particles. This is simplest for a monodisperse suspension, but may also be done for a polydisperse system. See Reference (3), page 301, and (34,35).

Convection. Larger particles fall so rapidly that the liquid flow increases the concentration close to the particle, which leads to (6)

$$v_g = \frac{DV_m(c-c_s)}{r} \cdot F \equiv \frac{DV_m(c-c_s)}{\delta} \equiv k_T(S-1) \tag{3}$$

where F is the factor by which the growth rate increases, and $\delta \equiv r/F$ is called the thickness of the diffusion layer or of the unstirred layer (6). F and δ may be estimated as follows (5) for 0 < Pe* < 300, 1 < F < 5

$$Pe^* \equiv 2gr^3\Delta\rho/9D\eta \qquad\qquad (4)$$

$$F = (1+Pe^*)^{0.285} \qquad\qquad (5)$$

$$\delta = r/F \qquad\qquad (6)$$

where g = acceleration of gravity = 9.81 m^2/s, $\Delta\rho$ = density difference particle-liquid, and η the viscosity of the liquid $\approx 8.9\times10^{-4}$ Pa s
at 25 °C and 1.00×10^{-3} Pa s at 20 °C. Particles may be regarded as
small in this respect when $Pe^* < 1$. For $Pe^* = 1$ we have $F = 2^{0.285} =$
1.22 and, for $\Delta\rho$ = 2000 kg/m^3 = 2 g/cm^3, $r = (9D\eta/2g\Delta\rho)^{1/3}$ = 6 µm and
$\delta = v/F$ = 5 µm. Equation (5) is assumed to be valid at least to
$Pe^* = 300$ (5). For this value we have $r = 40$ µm, $F = 5.1$ and $\delta = 8$ µm.
As δ varies very little for $Pe^* > 1$ we may use the rule of thumb,
$\delta \approx r$ for $r \leq 10$ µm; $\delta = 10$ µm for $r > 10$ µm.

Transport - Surface Control Discrimination. When the particles are
smaller than $r \approx 5$ µm, moderate stirring has only a negligible effect
on the growth rate because the particles are carried gently with the
solution. But when $r > 5$ µm, or at intense stirring, the growth rate
will increase with increasing stirring rate. This effect has often
been used for discriminating between "diffusion" (meaning "transport")
control and surface control. For small crystals this is not a relevant criterion. The best way to decide if transport is rate controlling is to calculate the theoretical growth rate, assuming that it
is transport controlled, and compare it with the empirical rate (2).
If the latter is essentially smaller than the former the rate must
be surface controlled, and otherwise it is transport controlled.
There are of course intermediate cases where the rate is of the
order of the transport controlled rate, but somewhat less, and it
is possible to treat these cases more delicately than by choosing
one of the extremes (1-3,7,8).

Surface Controlled Kinetics

The surface processes may comprise adsorption, surface migration
(across terraces or along steps), dehydration of ions, and integration in the growth sites which are assumed to be kinks in surface
steps. Any of these processes may be rate controlling, either alone
or several together (10-11).
    A single adsorbed ion is only bonded in one direction, whereas
an ion in a kink is bonded in three directions. Growth is therefore
likely to proceed through addition of growth units (molecules or
ions) at kinks.
    Moderately and very soluble electrolytes often grow by a linear
rate law (9,10). In some cases the rates agree with the rates calculated for transport control (Equations 2 and 3). In other cases the
rate is significantly smaller than corresponding to transport control,
and one may conclude that the growth is surface controlled. A linear,
surface controlled rate law may be explained by assuming that ion
adsorption is a slow, rate-determining step whereas surface migration
and integration at growth sites are relatively fast steps. We find,
however, that more evidence is needed before this mechanism may be
regarded as sufficiently verified.

Surface Nucleation Control. Before 1949 the only crystal growth mechanism discussed in literature was the surface nucleation model where surface nuclei were assumed to form on a perfect face of the crystal (4,12-14). A face with small groups of adsorbed growth units (subcritical nuclei) presents a higher free energy than a perfect face, due to what may formally be called the free energy of the vertical walls of the group. Depending on the concentration $(c > c_s)$ a certain size of the group is just stable (a critical nucleus) and any larger surface nuclei will grow. On small crystals or at a low degree of supersaturation, each surface nucleus above the critical size will grow until it covers the perfect face right to the edges, and a new surface nucleus must form before the growth can continue (1,15). This mechanism has been observed, but only in very rare case (e.g. silver metal (16)) and it is not typical for electrolyte crystal growth. In probably all cases of surface nucleation control-led growth, many surface nuclei are growing at the same time, and each molecular layer of the crystal is "historically" composed of many intergrown nuclei. All authors agree on the general features of the theory of nucleation controlled growth, although the treatments differ in details. As an example of the kinetic equation for the linear growth rate we may quote (10)

$$v_g = k_e \cdot F(S) \cdot \exp[-K_e/(\ln S)] \qquad (7)$$

where

$$k_e = 2a\nu_{in}(c_{s,ad}V_m)^{4/3}\exp(-\gamma/kT) \qquad (8)$$

$$K_e = \pi\gamma^2/3k^2T^2 \qquad (9)$$

$$F(S) = S^{7/6}(S-1)^{2/3}(\ln S)^{1/6} \qquad (10)$$

a = mean ionic diameter

ν_{in} = integration rate constant (or frequency)

c_{ad} = concentration in the adsorption layer, see Equation 12

Surface Spiral Step Control. Many crystals grow faster at small supersaturation than allowed by Equation 7. This lead Frank (17) to suggest that steps may also originate from the presence of a screw dislocation, and that this kind of steps is not destroyed by spread-ing to the crystal edge, but continues infinitely. The rate law according to this theory is parabolic (7). We shall use the following version of the kinetic equation (10)

$$v_g = \frac{0.1a\nu_{in}c_{s,ad}V_m(S-1)^2}{(\gamma/kT)\exp(\gamma/kT)} = k_S(S-1)^2 \qquad (11)$$

where γ is the (free) energy per growth unit in a step. Using the formality of the surface tension σ in the vertical step surface, $\gamma = a^2\sigma$. In Equation (11) one of the factors $(S-1)$ comes from the

net flux (growth units per unit time) into each kink which varies
linearly with c_{ad} and becomes zero for $S = 1$ (solubility equilibrium).
The other factor $(S-1)$ and the denominator come from the kink density
which is proportional to $(S-1)/[(\gamma/kT)\exp(\gamma/kT)]$.

Surface nucleation and the surface spiral mechanism are parallel
mechanisms. Consequently the growth rate will be equal to the sum of
the rates of these two mechanisms, and in extreme cases approximately
equal to the rate of the faster one among them.

This should be compared with the discussion of Equation 1 where
we concluded that when transport and surface processes are "competing"
for the rate control, the slower one will be rate-controlling.

Estimates of Parameters

In both the equations for surface controlled growth rates we find
three parameters which are normally not known from macroscopic
measurements, namely c_{ad}, γ and ν_{in}.

The Adsorption Layer Concentration. It is not possible to measure
the concentration of the constituent ions in the adsorption layer.
But it has been suggested that c_{ad} may be calculated by means of the
Langmuir equation ($\underline{10},\underline{18},\underline{19}$)

$$c_{ad} = \frac{K_{ad}c}{1+10V_m K_{ad}c} \quad (\approx K_{ad}c \text{ for } 10V_m K_{ad}c \ll 1) \qquad (12)$$

where the adsorption equilibrium constant K_{ad} may be estimated from
the ion pair stability constant K_I,

$$\frac{[XY]}{[X][Y]} = K_I \qquad (13)$$

where X and Y are the cation and the anion, respectively, of the
growing crystal. The result of the discussion is that

$$K_{ad} \approx QK_I \qquad (14)$$

where $Q = 5000$ mol/m^3 for 1,1 electrolytes, 500 for 1,2 and 2,1
electrolytes, and 200 mol/m^3 for 2,2 electrolytes. Ion pair stability
constants are roughly equal for all ion pairs of the same charge type.
The fundamental assumption in the theory of the Q-factors is, that
the force between an electrolyte crystal and an ion adsorbed to it
is the same as the force between the two ions of an ion pair consisting of the adsorbed ion and the ion in the crystal surface at the
adsorption site. The Q-factors account for the different possible
ways of arranging the two ions and their hydration water molecules
in the adsorption situation, and in the ion pair situation ($\underline{18},\underline{19}$).
As average values one may take $K_I = 0.002$, 0.06, and 1 m^3/mol,
respectively. These estimates for Q and K_I give finally ($\underline{18}$)

$$K_{ad} = \left\{ \begin{array}{ll} 10 & \text{for 1,1 electrolytes (as KCl)} \\ 30 & \text{for 1,2 and 2,1 electrolytes (as } K_2SO_4 \text{ and } BaCl_2) \\ 200 & \text{for 2,2 electrolytes (as } BaSO_4) \end{array} \right\} (15)$$

These values of K_{ad} are assumed to be good within factors 0.2 to 5.

<u>Interfacial Tension</u>. The interfacial energy σ between a crystal and an aqueous solution cannot (at least in general) be measured by macroscopic methods. But it may be deduced from homogeneous nucleation data (20-24). For the purpose of determining the edge energy $\gamma \equiv a^2\sigma$ one may either take the individual value determined on the actual substance (<u>if</u> it is determined) or use the general correlation with the solubility c_s, expressed for instance by (10,18)

$$\frac{\gamma}{kT} = 2.82 - 0.272 \ln \frac{c_s}{mol/m^3} \tag{16}$$

The accuracy of this equation was tested on 37 substances and was found to be better than $\pm$ 20% for 57 per cent of the 37 substances. (This test has not been published before, but the data used were published as Figure 4 in Reference (10)).

<u>The Integration Rate Constant</u>. Reich and Kahlweit suggested that ν_{in} is equal to the rate constant (or frequency) ν_{dh} of dissociation of a water molecule from the inner coordination shell of the hydrated cation (25,26). They found only partial verification of this hypothesis because of too few data. It has been shown subsequently that if all the estimates mentioned above are made, a considerable amount of crystal growth data may be explained (18,19) by estimating the integration frequency as

$$\nu_{in} = 10^{-3}\nu_{dh} \tag{17}$$

where ν_{dh} is the dehydration frequency of the cations (27-29). The factor 10^{-3} may be interpreted as the Arrhenius factor $\overline{\exp(-E_D^{\ddagger}/kT)}$ for an ion making a diffusion jump. Values of the dehydration frequencies of all metal ions of interest have been measured in various ways (10).

<u>Prediction of Growth Rates</u>

From the above statements it follows that it should be possible to derive the growth kinetics and calculate the growth rate of uncontaminated electrolyte crystals when the following parameters are known: molecular weight, density, solubility, cation dehydration frequency, ion pair stability coefficient, and the bulk concentration of the solution (or the saturation ratio). If the growth rate is transport controlled, one shall also need the particle size. In table I we have made these calculations for 14 electrolytes of common interest. For the saturation ratio and particle size we have chosen values typical for the range where kinetic experiments have been performed (29,30). The empirical rates are given for comparison.

For the calculation we used

$$V_m = M/\rho \; ; \quad a = (V_m/\nu L)^{1/3} \tag{18;19}$$

$$k_D = DV_m c_s/r \; ; \quad k_T = DV_m c_s/\delta \tag{20;21}$$

$$\delta = r \text{ for } r < 10 \text{ } \mu m \text{ and } \delta = 10 \text{ } \mu m \text{ for } r > 10 \text{ } \mu m$$

Table I. Prediction of Electrolyte Crystal Growth Rates

Electrolyte	M g/mol	ρ g/ml	c_s mol/m³	ν_{dh} log s⁻¹	K_I m³/mol	r μm	s	δ μm	K_{ad}	$\gamma[a]$ kT	k_2^{th} nm/s	k_e^{th} nm/s	K_e^{th}
KNO_3	111.1	2.11	3300	9.0	0.00063	500	1.05	10	3.2	0.616	3250	12300	0.397
$AgCl$	143.3	5.56	0.013	10[e]	1.58	0.1	2	0.1	7900	4.00	3.17	34.6	16.7
K_2SO_4	174.3	2.66	660	9.0	0.0079	500	1.1	10	4.0	1.054	2090	5420	2.37
$Na_2S_2O_3 \cdot 5H_2O$	248.2	1.73	4090	8.6	0.06[e]	500	1.05	10	30	0.558	1670	8456	0.326
CaF_2	78.1	3.18	0.25	8.2	0.06[e]	5	2	5	30	3.20	0.0087	0.0314	10.7
"	=	=	=	=	=	2	42	2	=	=	=	=	=
SrF_2	125.6	4.24	0.95	8.7	0.06[e]	3	2.9	3	30	2.83	0.274	1.16	8.39
"	=	=	=	=	=	0.2	33.3	0.2	=	=	0.174	0.86	=
$Ba(NO_3)_2$	261.4	3.24	380	9.2	0.06[e]	35	1.05	10	30	1.20	1210	12900	1.51
$MgSO_4 \cdot 7H_2O$	246.6	1.68	2240	10[e]	0.17	500	1.01	10	34	0.722	29000	19.1	0.546
$CaSO_4 \cdot 2H_2O$	172.2	2.32	15.1	8.2	0.204	20	1.1	10	41	2.08	11.6	1219	4.53
$BaSO_4$	233.4	4.50	0.010	9.2	0.20	5	10	5	40	4.07	0.00485	0.0108	17.3
"	=	=	=	=	=	1	1600	1	=	=	0.00364	0.0074	=
$MgC_2O_4 \cdot 2H_2O$	148.4	2.45	3.4	5.7	2.7	5	1.3	5	540	2.49	0.000121	0.491	6.49
$CaC_2O_4 \cdot H_2O$	146.1	2.2	0.063	8.2	1.0	2.5	2	2.5	200	3.57	0.0393	0.264	13.3
$CaCO_3$	100.1	2.71	0.063	8.2	1.68	10	2	10	200	3.57	0.0301	0.197	13.3
"	=	=	=	=	=	5	13	5	336	=	=	0.176	=
$AlF_3 \cdot 3H_2O$	138.0	2.0	45	-0.8	?	5	2	5	80[e]	1.78	1.9-08	2.4-07	3.32

Table I. (Continued)

Electrolyte	v_T nm/s	v_P nm/s	v_E nm/s	$\dfrac{v_P}{v_E}$	v_g^{th} nm/s	v_g^{emp} nm/s	v_g^{emp} Ref.	$\dfrac{v_g^{th}}{v_g^{emp}}$	Rate control[b] Theor.	Rate control[b] emp.	Agreement Value[c]	Agreement T/S[d,g]	Agreement P/E[d]
KNO$_3$	850	8.13	0.308	26	8.13	60	9	0.14	P	S[g]	A	A	?
AgCl	3.4	3.17	2.3-09	1.4+09	3.17	0.87	11,19	3.6	P+D	P+D	A	A	A
K$_2$SO$_4$	432	20.9	0.0044	4700	20.9	60	9	0.35	P	P	A	A	A
Na$_2$S$_2$O$_3\cdot$5H$_2$O	2940	4.17	0.912	5	4.17	300	9	0.014	P	P	A	A	B
CaF$_2$	1.2	0.0087	1.27-08	7+05	0.0087	0.0051	29	1.71	P(E)	L	A	A	A
"	504	14.6	2.07	7	14.6	560	29	0.026	P	P	B	A	B
SrF$_2$	17.8	0.78	0.00236	330	0.78	0.001	29	780	P(E)	E	B	A	C
"	4540	181	58.7	3	181	1000	29	0.18	P	E	A	A	B
Ba(NO$_3$)$_2$	154	3.02	4.2-011	7+010	3.02	20	29	0.16	P	L/P	A	A	?
MgSO$_4\cdot$7H$_2$O	329	2.86	6.2-025	5+024	2.86	40	9	0.07	P	L	B	A	B
CaSO$_4\cdot$2H$_2$O	11	0.116	4.6-019	2.5+017	0.116	0.061	19	1.9	P	P	A	A	A
BaSO$_4$	0.93	0.319	0.00042	760	0.319	0.23	11,19	1.4	P	P	A	A	A
"	830	9274	731	13	830	1000	29	0.83	D	D	A	A	A
MgC$_2$O$_4\cdot$2H$_2$O	12	1.1-05[f]	4.3-012	2.6+06	1.1-05	2-04	19	0.055	P	P	B	A	A
CaC$_2$O$_4\cdot$H$_2$O	1.7	0.0393	4.2-09	9+06	0.0393	0.050	19	0.79	P	P	A	A	A
CaCO$_3$	0.23	0.0301	1.8-09	1.7+07	0.0301	0.046	11,19	0.65	P	P	A	A	A
"	5.6	4.00	0.0645	62	4.00	10	11,19	0.40	P+D	D+P	A	A	A
AlF$_3\cdot$3H$_2$O	620	1.9-08	4.2-09	4.5	1.9-08	5-05	30	0.0004	P(E)	P	C	A	A

[a] Estimated, see Equation 16; [b] D = diffusion control; L, P and E, surface control, namely: L = linear rate law, P = parabolic rate law, and E = exponential rate law; [c] value of the quotient v_g^{th}/v_g^{emp}; A: 0.1-10; B: 0.01-0.1 or 10-100; C: <0.01 or >100; [d] A: Correct; B: Correct if one of the values is increased by less than ×10; C: Poorer than B; [e] Estimated, see Reference (10); [f] 1.1-05 means 1.1×10^{-5}. [g] S (surface control) comprises L, P and E; T (transport control) comprises D and convection. Convection is not occurring in the treated cases.

$$v_T \equiv v_g \text{ (transport control) } = k_T(S-1) \tag{22}$$

$$K_{ad} = Q \cdot K_I \tag{23}$$

or, when K_I is unknown, K_{ad} is estimated according to Equation 15;

$$\Phi \equiv \frac{K_{ad}V_m c_s}{1+10 \ K_{ad}V_m c_s S} \tag{24}$$

γ/kT (Equation 16)

$$\Psi \equiv \frac{\gamma}{kT} \exp \frac{\gamma}{kT} \tag{25}$$

$$k_2 = \frac{10^{-4}a\nu_{dh}\Phi}{(\gamma/kT)\exp(\gamma/kT)} = 10^{-4}a\nu_{dh}\Phi/\Psi \tag{26}$$

$$v_P \equiv v_g \text{ (parabolic rate law) } = k_2(S-1)^2 \tag{27}$$

$$k_e = \frac{0.002 \cdot a\nu_{dh}\Phi^{4/3}}{\exp(\gamma/kT)} \tag{28}$$

$$E(S) \equiv \exp\left(-\frac{\pi}{3}\left(\frac{\gamma}{kT}\right)^2/\ln S\right) \tag{29}$$

$$F(S) = S^{7/6}(S-1)^{2/3}(\ln S)^{1/6} \quad \text{(Equation 10)}$$

$$v_E \equiv v_g \text{ (exponential rate law) } = k_e E(S)F(S) \tag{30}$$

Then let v_S = the larger of v_P and v_E, and let the resulting growth rate be $v_g \equiv$ the smaller of v_S and v_T.

Discussion

The set of equations resulting from the theory has been tested on 14 electrolytes for which experimental growth rates had been measured. In Table I theoretical rates have been calculated for concentrations in the experimental range, one concentration for each of ten of the electrolytes and two for each of four where special effects had been observed. In all of the 18 cases the theory lead to the correct conclusion whether the growth rate was controlled by transport (1 case) surface processes (15) or both together (2 cases). In 16 cases the rate calculated was within 0.01 and 100 times the measured rate, and in 14 cases between 0,1 and 10 times the empirical rate. This result should be viewed with respect to the variation of the empirical rate from 5×10^{-5} to 10^3 (a range of 2×10^7). The only serious shortcomings of the theory were the prediction of a parabolic rate law instead of an exponential one for SrF_2 and CaF_2, and a deviation by the factor 0.0004 for $AlF_3 \cdot 3H_2O$. The appearance of an exponential (surface nucleation) rate law instead of a parabolic (surface spiral) law may be a consequence of a rather high degree of perfection in the crystals. And the relatively large disagreement in the case of

$AlF_3 \cdot 3H_2O$ may perhaps be a consequence of the rather special (not typically "electrolyte") structure of the crystals (30). In Figure 2 we have plotted (logarithmically) the theoretical values of the rate constants as a function of the empirical values, for the ten substances of Table I which follow a parabolic rate law. Considering the large range covered (1 to 10^9) the agreement is satisfactory.

Conclusion

We have reviewed today's knowledge of the mechanisms for growth of electrolyte crystals from aqueous solution: Convection, diffusion, and adsorption (?) mechanisms leading to linear rate laws, as well as the surface spiral mechanism (parabolic rate law) and surface nucleation (exponential rate law). All of these mechanisms may be of geochemical importance in different situations.

It is possible to predict, mostly within an order of magnitude, the growth rate of an electrolyte crystal growing from aqueous solution at 25 °C, and it is possible with a corresponding accuracy to predict whether the growth will be controlled by the transport in the surrounding liquid or by processes in the surface of the crystal. The theory is not (yet) precise enough for predicting whether a given surface process follows a parabolic or an exponential rate law.

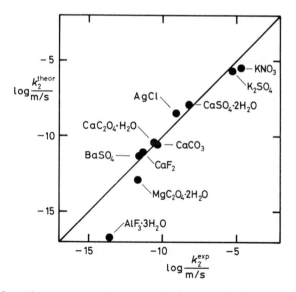

Figure 2. The rate constants k_2 for some electrolytes following the parabolic rate law, $v_g = k_2(S-1)^2$. The theoretical values of k_2 are plotted as a function of the experimental values, in a logarithmic diagram.

Legend of Symbols

a   mean ionic diameter = $(V_m/Lv)^{1/3}$

c   concentration (bulk)

c(x)   concentration in the distance, x from the particle center (Figure 1)

c'   concentration close to crystal surface

c_s solubility

D   diffusion constant

f   function, mathematical expression

F   rate increase factor due to convection

F(S)   function of S

g   acceleration of gravity (+ centrifugation, if any)

k   Boltzmann constant

k_{index} rate constant (SI unit m/s)

K_e constant

K_{ad} adsorption equilibrium constant, $\lim(c_{ad}/c)$ for c → 0

K_I ion pair stability constant

L   Avogadro constant

M   molar mass ("molecular weight")

Pe*   Peclet number of mass transfer

Q   constant

r   defined radius of particle = $(3v/4\pi)^{1/3}$

S   saturation ratio = c/c_s

t   time

T   temperature

v   particle volume

v_g growth rate ≈ dr/dt

V_m molar volume = M/ρ

γ edge energy = $a^2\sigma$

δ thickness of diffusion layer ("unstirred layer")

η viscosity of solution

ν number of ions in a formula unit

ν_{in} integration rate constant, or frequency (SI unit, s^{-1})

ν_{dh} dehydration rate constant (or frequency) for cation (SI unit, s^{-1})

ρ density; $\Delta\rho$ = density difference crystal-liquid

σ surface tension

Indices, etc.

ad   adsorption layer

dh   dehydration

D   diffusion control

e,E   exponential rate law (polynuclear surface nucleation)

in   integrating jump

L   linear rate law

m   molar

P   parabolic rate law

s   solubility equilibrium, saturated

S   surface control

T   transport control

2   parabolic rate law (surface spiral step)

Acknowledgments

This work has been supported by travel grants from the Danish Natural Science Research Council, and by the Petroleum Research Fund of the American Chemical Society.

Literature Cited

1. Nielsen, A. E.   "Kinetics of Precipitation"; Pergamon Press: Oxford etc., 1964.
2. Nielsen, A. E.   Croatica Chem. Acta 1980, 53, 255-79.
3. Nielsen, A. E.   In "Treatise on Analytical Chemistry"; Kolthoff, I. M.; Elving, P. J., Eds.; Wiley: New York, 1983; Part 1, Vol. 3, Chap. 27.
4. Volmer, M.   "Kinetik der Phasenbildung"; Steinkopff: Dresden and Leipzig, 1939.
5. Nielsen, A. E.   J. Phys. Chem. 1961, 65, 46-9.
6. Nernst, W.   Z. Physik. Chem. 1904, 47, 52-5.
7. Burton, W. K.; Cabrera, N.; Frank, F. C.   Phil. Trans. Roy. Soc. London 1951, A243, 299-358.
8. Nielsen, A. E.   Acta Chem. Scand. 1959, 13, 1680-6.
9. Mullin, J. W.   "Crystallisation"; Butterworths: London, 1972.
10. Nielsen, A. E.   J. Crystal Growth 1984, 67, 289-310.
11. Nielsen, A. E.   In "Industrial Crystallization 78"; de Jong, E. J.; Jančić, S. J., Eds. North Holland Publishing Company: Amsterdam, 1979, pp. 159-68.
12. Stranski, I. N.   Z. physik. Chemie 1928, 136, 259-78.
13. Kaischev, R.; Stranski, I. N.   Z. physik. Chem. 1934, B26, 317-26.
14. Becker, R.; Döring, W.   Ann. Physik 1935, 24, 719-52.
15. Cabrera, N.; Burton, W. K.   Disc. Faraday Soc. 1949, 5, 40-8.
16. Bostanov, V.   J. Crystal Growth 1977, 42, 194-200.
17. Frank, F.C.   Disc. Faraday Soc. 1949, 5, 48-54.
18. Nielsen, A. E.   Pure and Appl. Chem. 1981, 53, 2025-39.

19. Nielsen, A. E.   In "Industrial Crystallization 81"; Jančić, S. J.; de Jong, E. J., Eds. North Holland Publishing Company: Amsterdam, 1982, pp. 35-44.

20. Nielsen, A. E.   Acta Chem. Scand. 1961, 15, 441-2.

21. Nielsen, A. E.   In "Crystal Growth"; Peiser, H. S., Ed.; Supplement to the Journal of Physics and Chemistry of Solids, Pergamon Press: Oxford and New York, 1967, D14, 419-26.

22. Nielsen, A. E.   Kristall und Technik 1969, 4, 17-38.

23. Nielsen, A. E.; Sarig, S.   J. Crystal Growth 1971, 8, 1-7.

24. Nielsen, A. E.; Söhnel, O.   J. Crystal Growth 1971, 11, 233-42.

25. Reich, R.   "Zur Kinetik des Kristallwachstums in Wässerigen Lösungen", Dissertation, Göttingen, 1965.

26. Reich, R.; Kahlweit, M.   Ber. Bunsenges. 1968, 72, 66-74.

27. Eigen, M.; Maass, G.   Z. Physik, Chem. N. F. 1966, 49, 163-177.

28. Burgess, J.   "Metal Ions in Solution"; Horwood, Chichester/Wiley: New York, 1978; Chap. 11.

29. Nielsen, A. E.; Toft, J. M.   J. Crystal Growth 1984, 67, 278-88.

30. Nielsen, A. E.; Altintas, N. D.   J. Crystal Growth 1984, 69, 213-30.

31. Crank, J.   "The Mathematics of Diffusion"; Clarendon Press: Oxford, 1957.

32. Jost, W.   "Diffusion in Solids, Liquids, Gases; Academic Press: New York, 1960.

33. Sissom, L. E.; Pitts, D. R.   "Elements of Transport Phenomena"; McGraw-Hill: New York, 1972, Chap. 7.

34. Randolph, A. D.; Larson, M. A.   "Theory of Particulate Processes"; Academic Press: New York, 1971.

35. Nývlt, J.; Söhnel, O.; Matuchová, M.; Broul, M.   "The Kinetics of Industrial Crystallization"; Elsevier: Amsterdam, 1985.

RECEIVED June 25, 1986

Influence of Surface Area, Surface Characteristics, and Solution Composition on Feldspar Weathering Rates

Michael Anthony Velbel

Department of Geological Sciences, Michigan State University, East Lansing, MI 48824-1115

Current best estimates for natural plagioclase weathering rates are one to three orders of magnitude lower than laboratory rates. Surface characteristics which may play a role in determining rates and mechanisms of feldspar dissolution (including non-stoichiometric dissolution and parabolic kinetics) in the laboratory include adhered particles, strained surfaces, defect and dislocation outcrops, and surface layers. The narrow range of rates from experiments with and without pretreatments indicates that these surface characteristics alone cannot account for the disparity between artificial and natural rates. Either the intrinsic surface area and characteristics of natural materials are significantly different from those used in laboratory experiments, or other non-mineralogical factors must account for the disparity between laboratory- and field-determined rates of feldspar weathering.

Predicting the rates of natural water-rock interactions on the basis of theoretical models and extrapolation from laboratory studies is of paramount importance in understanding pressing environmental and geochemical issues such as the stability of nuclear wasteforms, the chemical susceptibility of landscapes to acid deposition, and the distribution of porosity and permeability in hydrocarbon reservoir rocks. The necessity of solving these problems has rejuvenated the study of geochemical kinetics in water-rock interactions, and made it imperative to evaluate the existing data on reaction rates and their applicability to natural systems. The purpose of this paper is to review and evaluate the recent literature on the mechanisms and rates of feldspar weathering, and to suggest, on the basis of the literature review, some areas in which there is urgent need for further research.

0097–6156/86/0323–0615$06.00/0
© 1986 American Chemical Society

Weathering of Feldspar

Feldspars are the most abundant minerals of igneous and metamorphic
rocks (1,2). Being the most abundant rock-forming minerals of the
earth's crust, they have received a proportionately large share of
attention from students of weathering, and most of the major
analytical advances and conceptual models which have been applied
to other mineral groups have also been applied to feldspars.
   Feldspar is a framework silicate in which tetrahedra containing
Al or Si are linked to one another by shared oxygens in all
directions rather than in chains or sheets (3). The plagioclase
solid-solution series

$$Ca_x Na_{1-x} Al_{1+x} Si_{3-x} O_8, \text{ where } 0 \leq x \leq 1$$

is the commonest rock-forming series (2). Plagioclase feldspars
vary widely in their susceptibility to weathering; Goldich (4)
observed that calcic plagioclase (anorthite) is extremely
susceptible whereas successively more sodic plagioclases are
progressively less weatherable. This compositional effect on
natural weathering of plagioclase has been confirmed by numerous
other observational studies.
   Alkali feldspars $\{(K,Na)AlSi_3O_8\}$ form a continuous solid
solution series at high temperatures, but tend to exsolve upon
cooling into a potassium-rich phase and a sodium-rich phase,
resulting in perthitic intergrowths (2). Sodic plagioclases and
potassium-rich feldspars tend to be studied together, in part
because of similarities in structure, composition, occurrence, and
weatherability (e.g., 4).

Parabolic Kinetics

"Parabolic kinetics" refers to the observation that the
concentration of a species released to an aqueous solution by
alteration of a primary mineral plots as a linear function of the
square root of time. Figure 1 (data from 5) illustrates this.
Numerous workers (e.g., 5-11) have observed this behavior.
   The importance of "parabolic kinetics" in laboratory studies of
mineral dissolution has varied as interpretations of the underlying
rate-controlling mechanism have changed. Much of the research on
silicate mineral weathering undertaken in the past decade or so
served to test various hypotheses for the origin of parabolic
kinetics.
   At least four different explanations have been proposed to
account for parabolic kinetics. The oldest and best established is
the "protective-surface-layer" hypothesis. Correns and von
Englehardt (6) proposed that diffusion of dissolved products
through a surface layer which thickens with time explains the
observed parabolic behavior. Garrels (12,13) proposed that this
protective surface consists of hydrogen feldspar, feldspar in which
hydrogen had replaced alkali and alkaline earth cations. Wollast
(5) suggested that it consists of a secondary aluminous or
alumino-silicate precipitate. In either case, a protective surface
layer explains parabolic kinetics as follows: If the concentration
of any dissolved product at the boundary between the fresh feldspar

and the protective surface layer is fixed by either thermodynamic
equilibrium or the rate of transformation of the primary mineral,
and the concentration at the outer surface is that of the ambient
solution, Fick's first law of diffusion

$$J = -D(dc/dx) \qquad (1)$$

states that the flux of dissolved products J is proportional to the
product of the diffusion coefficient D and the concentration
gradient across the protective surface layer. As the surface layer
thickens, either by encroachment of the hydrogen feldspar "front"
into the fresh feldspar, or by precipitation of products from
solution, the concentration gradient across the protective surface
layer decreases with time, and the rate of material transport out
of the surface layer decreases with time. Wollast (5) derived
diffusion equations to model this phenomenon, and found that the
concentration vs. time curves calculated from the model closely
matched his experimentally determined data for dissolved silica
well. Helgeson (14,15) constructed a mass-transfer model on the
basis of progressively thickening layers of secondary precipitates,
with similar favorable results.
    The second explanation proposed for parabolic kinetics is that
the early, high-dissolution-rate stage of the reaction is due to
the presence of a structurally distorted or strained outer layer of
the mineral, or the presence of large quantities of hyperfine
particles adhering electrostatically to the outer surface of the
mineral. In either case, this disturbed outer layer is made up of
material with higher-than-normal energy; the large abundance of
strained or broken bonds, exposed edges of fine particles, etc.,
cause this disturbed layer to be less stable with respect to
dissolution than the undisturbed bulk mineral material (e.g., 11,
Appendix A). The disturbed layer therefore goes into solution more
readily; as the disturbed layer and/or fine particles are
destroyed, the dissolution rate tapers off to the linear rate of
destruction of the bulk mineral. The disturbed surface layer is
created by crushing and grinding of the mineral in preparation for
laboratory dissolution experiments; Huang and Kiang (7) and Nickel
(8) (among others) suggested that the high initial dissolution
rates might be attributed to such disturbed surface layers.
    Holdren and Berner (11) demonstrated clearly that adhered fine
particles could, in fact, be responsible for the pseudo-parabolic
initial stage of laboratory dissolution experiments. Figure 2
(modified from 11) shows the results of three experimental runs.
Curve "a" represents freshly ground albite, which has not been
washed or pretreated in any way. Scanning electron
photomicrographs revealed that this "unwashed" albite was covered
with adhered fine particles (submicron to ten microns) of albite.
Curve "b" shows the dissolution of "washed" albite, which had been
ultrasonically cleaned in acetone to remove as much fine adhered
material as possible. SEM photos revealed that much adhered
particulate matter had been removed, but that much very fine
(submicron) material had withstood attempts to remove it
ultrasonically. Note on Figure 2 that the removal of the coarser
adhered particles (~0.5-10 microns) greatly reduced the initial
high-dissolution-rate. Curve "c" represents the dissolution curve

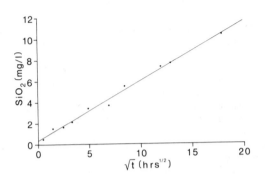

Figure 1   Example of parabolic kinetics showing linear
behavior of silica concentration vs. the square root of time.   Data
from Ref. 5.

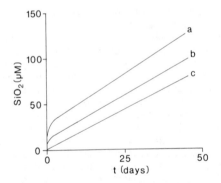

Figure 2   Effect of sample pretreatment on feldspar
dissolution kinetics (modified from Ref. 11).   See text for
discussion.

for ground and washed feldspar from which all adhered fine
particles had been removed by treating the grains with
$HF-H_2SO_4$. No pseudo-parabolic initial stage occurs; the
dissolution of albite in the absence of fine adhered particles is
linear. Holdren and Berner (11) suggested that much "parabolic"
behavior observed in previous dissolution studies was due not to
diffusion through a protective surface layer, but to dissolution of
fine particles generated during sample preparation.

The third proposed explanation for parabolic kinetics is that
dissolved products may be released from the mineral surface
linearly, but that non-linear precipitation of secondary minerals
from solution accounts for the non-linear concentration vs. time
behavior (16).

The fourth explanation for non-linear kinetics differs from the
previous three in that it concerns the composition of the solution
rather than any intrinsic property of the solid reactants or
products. Changing solution composition can produce apparent or
true parabolic dissolution kinetics either through the influence of
changing pH and CO_2 equilibria, or through the effect of chemical
affinity and the reverse reaction rate. These phenomena have been
discussed in detail by Helgeson and Murphy (17) and Helgeson and
others (18).

One additional aspect of laboratory dissolution experiments is
the question of stoichiometric vs. non-stoichiometric dissolution.
Many of the studies cited above analyzed only a few of the elements
released by feldspar; that is, although alkalis, alkaline earths,
silica, and aluminum may be released during dissolution of
feldspar, few studies report analyses for all elements. Often,
only silica was analyzed. Where multiple elements are analyzed,
they are often released to the solution in proportions which do not
correspond to the bulk stoichiometry of the feldspar (19).
Usually, alkalis and alkaline earths are released in excess of
silica, and dissolved aluminum is the least abundant. This
observed non-stoichiometry suggested that silica and aluminum were
being preferentially retained in some solid phase relative to
alkalis and alkaline earths. Such preferential retention,
manifested as non-stoichiometric dissolution, was long thought to
be consistent with the concept of some kind of residual surface
layer.

By the mid-1970's the diffusion-inhibiting surface was widely
accepted as the rate-controlling factor in the dissolution not only
of feldspars, but of magnesium silicates (20) and, ultimately, of
almost all non-carbonate and non-sulfur-bearing rock-forming
minerals (8). As shown by Berner and Holdren (21,22), however, the
protective-surface-layer hypothesis proved to have numerous
difficulties, which became increasingly apparent as petrographers
trained their optical and electron microscopes upon silicate
minerals (see below). The newfound electron-microscopic
observations led to wider acceptance of kinetic models for feldspar
weathering based on surface- (or interface-) controlled reactions
and their associated linear kinetics (see below). As a result of
this challenge to the protective-surface-layer hypothesis, two
avenues of investigation dominate the feldspar weathering
literature of the 1980's. Studies of surface morphology and
surface chemistry seek evidence of reaction mechanisms besides that

provided by dissolution studies (see below), and further laboratory
studies of dissolution kinetics have been undertaken to resolve
apparent ambiguities between earlier laboratory studies and the
newer surface morphology and surface chemistry results discussed
below.

Chou and Wollast (23,24) performed experiments on feldspar
dissolution using a fluidized-bed reactor, which allowed them to
vary the composition of solutions during the course of the
experiment (for instance, after the linear, steady-state stage of
the reaction had been attained) and monitor the effects of the
solution perturbations on the dissolution kinetics. They also
analyzed their solutions for Na, Al, and Si, rather than for any
one element alone, allowing them to evaluate not only the behavior
of individual elements with time, but also the stoichiometric or
non-stoichiometric behavior of the reactions. The experimental
results presented by Chou and Wollast (23,24) show
non-stoichiometric dissolution in early stages of the reaction,
resulting in the formation of a residual surface layer determined
by material balance to be "a few tens of Angstroms thick", the
composition of which is strongly pH-dependent. According to Chou
and Wollast, diffusion is rate-limiting with respect to alkali
release at this stage of the reaction. Linear kinetics and
stoichiometric dissolution prevail in later stages of the reaction
(at any given pH), suggesting that the layer reaches a
quasi-steady-state thickness maintained by destruction of the outer
surface of the residual layer at the same rate at which fresh
feldspar is transformed at the interface between it and the
residual layer. More interesting, however, is their observation
that instantaneous modification of solution pH results in a new
episode of initially non-linear kinetics, and release of elements
in different proportions than at other pH's, implying the formation
of a residual surface layer with a different (pH-dependent)
composition. They argue that only in the first portion of any
experimental run could parabolic behavior of solutes be explained
by fine particles: parabolic kinetics observed upon changing the
solution pH must be due to the formation of a new surface layer,
the formation of which, they argue, must be diffusion-controlled.
Chou and Wollast (23) suggested that their own results clearly
indicated the formation of a sodium-depleted surface layer.
Wollast and Chou (25) also apply and critique Helgeson and others'
(18) formulation of activated complex theory, and make suggestions
for future refinements to facilitate its application to feldspar
weathering studies.

The recent work of Holdren and Speyer (26,27) has suggested
that at least some studies are strongly affected by the abundance
and distribution of defects and dislocation outcrops in ways that
are not simply related to surface area.

Petrographic Observations

Wilson (28) noted the presence of etch pits (crystallographically
controlled voids or features of negative relief, or "negative
crystals") on some soil feldspars, and reviewed similar
observations from earlier studies. Some examples of etch pits on
naturally weathered feldspars are shown in Figure 3. Etch pits

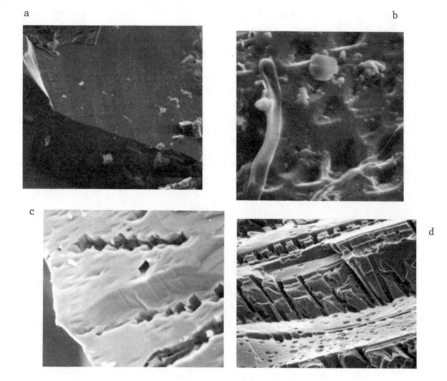

Figure 3   Scanning electron photomicrographs of feldspar
surfaces in various stages of weathering. a) Fresh surface. b)
Incipient formation of shallow almond-shaped etch pits. c) Moderate
development of prismatic etch pits. d) Extensive penetration of
prismatic etch pits into feldspar interiors. Photographs b-d are
from naturally weathered materials. All photomicrographs by Alan
S. Pooley and the author.

reflect the presence of dislocations (linear arrays of crystal
defects, mismatched bonds, etc.) in a crystal. Dislocations,
because they reflect imperfections in a crystal, are sites of
enhanced energy, and are readily susceptible to selective attack.
Until the advent of sophisticated electron microscopic techniques,
the correspondence between dislocations and etch pits was
tautological. Metallurgists and materials scientists determined
distributions and densities of dislocations by attacking the
crystal with an etchant and then counting the resultant etch pits
under a microscope. Site-selective attack (that is,
surface-reaction control), however, contradicts the notion of a
protective surface layer (28-30). A mineral dissolved by diffusion
control should possess smooth, rounded surfaces. As dissolved ions
are released by parent-mineral dissolution, they "pile up" in the
diffusion-inhibiting layer, the local concentration goes up, and
the local dissolution rate slows down. This is because the
dissolution rate is a function of the degree of undersaturation
(or, strictly speaking, the thermodynamic chemical affinity) of the
reaction (e.g., 31): e.g., in general

$$R = k(c_s - c)^n$$

where R is the dissolution rate, k is the rate constant, c_s is
the equilibrium concentration, c is the dissolved concentration in
the ambient solution and n is rational, usually between 0 and 2. As
diffusion-inhibition causes local concentration buildups around
sites of enhanced crystal energy (dislocations, edges and corners,
etc.) local dissolution rates around these sites should decrease,
allowing dissolution of the surrounding bulk surface to catch up.
The result is that transport-controlled dissolution attacks mineral
surfaces uniformly, preventing the development of site-selective
attack and any surface features (e.g., etch pits) which would
indicate site-selective attack. Wilson (28) correctly inferred
that the presence of etch pits militates against the
protective-surface-layer hypothesis. Numerous workers have also
observed etch pits on both naturally and artificially weathered
feldspar (21,22,27,32-40), and electron microscopic examination of
altered mineral surfaces for etch pits has become a routine test
for surface-reaction control.

Electron microscopy makes possible another important kind of
observation; Berner and Holdren (21,22) observed that clay coatings
on weathered feldspar are patchy, discontinuous, and sufficiently
hydrous that they form "micro-mudcracks" and are detached from the
feldspar surface upon dessication during sample storage and
preparation. The "protective-surface-layer", if it were to be
diffusion-inhibiting as its proponents propose, should have a
diffusion coefficient (D in equation 1) on the order of 10^{-20}
cm^2/sec, equivalent to diffusion through a crystalline solid.
(Diffusion through a porous layer could be rate limiting in
principle even if D were much greater then 10^{-20}, provided that
diffusion resulted in a rate that is slower than the surface
controlled rate.) The coating observed by Berner and Holdren
(21,22) on naturally weathered feldspars behaves much more like mud
(which has a diffusion coefficient of approximately 10^{-6}
cm^2/sec) than like a crystalline solid and therefore is not

consistent with the protective-surface-layer hypothesis of feldspar weathering. Petrovic (41) had reached a similar conclusion on theoretical grounds and Fung and Sanipelli (42) confirmed the earlier findings. The rapidly growing body of micromorphologic evidence militated against the protective-surface-layer hypothesis.

Helgeson and others (18) suggested that etch-pitting of minerals increases the surface roughness of the grains such that total mineral surface area in the system can increase to a maximum value more than an order of magnitude higher than initial values as the reaction progresses. (This phenomenon was demonstrated experimentally for olivine by Grandstaff, 43). They note that these theoretical findings are in at least qualitative accord with observations of surface morphology using the SEM. Depending on the precise functionality of the roughness factor, surface area may increase appreciably after as little as 0.1% of the starting material has been dissolved. Generally, however, the surface area increases by one order of magnitude or more (the reason for choosing one order of magnitude as the critical value for discussion will become apparent in a later section) only after some 10% of the starting material has been dissolved, and surface area reaches its maximum when about 55% of the starting material has been consumed. As a consequence, surface area increases do not affect room-temperature laboratory studies, which are generally of such short duration that only a small fraction of the starting material has reacted.

Surface Composition

The morphology of weathered feldspar surfaces, and the nature of the clay products, contradicts the protective-surface-layer hypothesis. The presence of etch pits implies a surface-controlled reaction, rather than a diffusion (transport) controlled reaction. Furthermore, the clay coating could not be "protective" in the sense of limiting diffusion. Finally, Holdren and Berner (11) demonstrated that so-called "parabolic kinetics" of feldspar dissolution were largely due to enhanced dissolution of fine particles. None of these findings, however, addressed the question of the apparent non-stoichiometric release of alkalis, alkaline earths, silica, and aluminum. This question has been approached both directly (e.g., XPS) and indirectly (e.g., material balance from solution data).

Nickel (8) calculated the thickness of the proposed "residual layer" on albite from the mass of dissolved alkalis and alkaline earths released during laboratory weathering and the measured surface area, and determined that the thickness ranges from 0.8 to 8.0 nm in the pH range of natural surface waters. Although he interpreted his results differently, they anticipate later findings on the pH dependence of residual layer compositions (see below).

Petrovic and others (44) used X-ray photoelectron spectroscopy (XPS) to analyze the K, Al, and Si contents of experimentally altered K-feldspar grains, and found that alkali depletion, if it existed at all, could not extend to greater than 1.7 nm depth; i.e., the "leached layer", if it exists at all, is less than 1.7 nm thick. Holdren and Berner (11) using XPS observed a slight decrease in the Na/Si ratio of experimentally weathered alkali feldspar surfaces, and suggested that some exchange of Na^+ for H^+ (or K^+ from the KOH buffer solution they used) like that proposed by Garrels and Howard (13) might be taking place. Otherwise, they were able to document no other change in the outer 2.0-4.0 nm of the weathered surfaces relative to the fresh mineral interior. Perry and others (45) observed some loss of K (presumably due to exchange with hydrogen) in the outer portion of a K-feldspar treated with HF/H_2SO_4; they estimate the sampled thickness to be about 6.0 nm. Finally, Della Mea and others (46) using ion beam techniques observed that sodium losses from experimentally leached albite are negligible at distances greater than 30-40 nm from the mineral surface; more recent experiments (Petit, personal communication) indicate that there is no detectable hydrogen below this same depth, and that very little occurs even within the layer of sodium depletion.

Chou and Wollast (23) challenged XPS studies which indicate that incongruent surface layers thicker than several Angstroms do not exist. They argue that material balance calculations require some sort of altered layer in order to account for the observed incongruency between alkalis, silica, and aluminum. Their material balance calculations suggest that the layer thickness must be on the order of only tens of nanometers, which, despite their arguments to the contrary, is not inconsistent with the surface chemistry observations (e.g., XPS) they seek to refute.

Berner and others (47) suggest that the contradiction between dissolution studies and surface chemistry studies "may be illusory". Noting that limited incongruency could be accomodated with the XPS data sets Chou and Wollast (23) challenge, Berner and others (47) suggest that spatially inhomogeneous incongruent dissolution along "tubes" (for instance, etch pits which penetrate deep into the interiors of mineral grains) could reconcile XPS measurements (to which the interior surfaces of "tubes" would be inaccessible) with the incongruency and material balance requirements of laboratory studies, while still allowing surface-reaction control. Chou and Wollast (48) agree, noting that in the quasi-steady-state model outlined above, diffusion limits dissolution only until the residual surface layer attains its quasi-steady-state thickness - after that, no further encroachment of the inner weathering "microfront" can occur until the outer portion of the layer is removed, which may well be a surface-controlled reaction. Chou and Wollast (48) agree with Berner and others (47) regarding the possible role of dissolution-void interiors in reconciling results of previous studies.

Other aspects of the inhomogeneous distribution of "active sites" are discussed by Wollast and Chou (25), who suggest that their earlier estimates of residual layer thickness are minima, because restricting the residual layer to the vicinity of active sites would mean that more of the reaction takes place over a smaller area, increasing the thickness of the residual layer there. "If a residual layer exists, then it would probably cover less than 10% of the total surface and accordingly the value for the calculated thickness would be increased proportionally" (p. 88). However, their discussion is inconsistent with the model of

weathered surfaces proposed by Berner and others (47). Etching
during dissolution may create large areas of new surface in the
form of the walls of "deep cracks, tubes, holes, etc." (47). If
preferential attack of the active sites results in penetration of
long "tubes" deep into the weathering grains of feldspar, the total
surface area may increase significantly (47; see Figure 3).
Depending on how much deeper the holes are than the XPS sample
depth, the actual (local) layer could be thicker than the bulk
average calculated by Chou and Wollast (23,24) (if the holes are
only slightly deeper than the XPS sample depth), as Wollast and
Chou (25, p. 88) suggest, equally thick (if the holes are deep
enough that their surface area equals the bulk surface area), or
thinner, if the holes have an internal surface area greater than
that of the bulk mineral surface.

SEM observations of weathered feldspar surfaces (e.g.,
21,22,37; Figure 3) suggest that etch-pit weathering greatly
increases surface area (see also 43). If so, material balance
requires that the experimentally measured mass loss be divided not
over a smaller area (only near active sites), resulting in an
increase in the calculated local layer thickness as suggested by
Wollast and Chou (25), but over a greater surface area, meaning
that the residual layer would actually be thinner than that
calculated by previous workers (8,9,10,19,23,25). This would then
bring the calculations of mass loss during dissolution experiments
(e.g., 23,24) into direct agreement with XPS measurements (e.g.,
47) showing an undetectably thin residual layer, and make both
consistent with observations of the surface morphology of weathered
feldspar.

Much remains to be done regarding the surface composition of
weathered feldspars. In addition to actual analytical work on
feldspar surfaces, work is required in comparing the results of
surface analysis (determining the amounts of elements removed from
the mineral) with results from dissolution experiments (which tell
us how much material was released to the solution). Nickel's (8)
work, for instance, calculated "residual layer" thicknesses from
dissolved concentration, and later work on feldspar chemistry has
been consistent with his results, suggesting that his results are
of the correct order of magnitude, i.e., if a hydrogen-exchange
cation-depleted surface layer does exist, it is at most several
tens of nanometers thick, and is probably far too thin to be
diffusion inhibiting.

Laboratory Studies of Feldspar Weathering - A Summary

Taken as a whole, a synthesis of all the above observations
suggests that the weathering of feldspar proceeds as follows:
1. Hydrogen-ion exchange for alkalis and alkaline earths
creates a thin layer of hydrolyzed aluminosilicate, the composition
of which is pH dependent. Gardner (19) has calculated, using the
data of Busenberg and Clemency (9,10), that the thickness of the
"leached layer" produced in their experiments was about two unit
cells (about 1.5-2.5 nm) thick. Similar calculations and empirical
observations by other workers (8,22,43,45,46) all suggest that the
altered residual layer is no more than several to a few tens of
nanometers thick. This initial exchange accounts for the initial

non-stoichiometric release of alkalis and alkaline earths relative
to silica and aluminum (as per 13). Further non-stoichiometry
(e.g., between Na and K) may be apparent only. Gardner (19)
postulated that such non-stoichiometry may be due to differential
dissolution of two distinct phases (recall that alkali feldspars
usually consist of microperthitic exsolution features, consisting
of discrete sodic plagioclase lamellae in discrete potassium
feldspar). Congruent dissolution of two distinct but intimately
intergrown phases at different rates may appear to be incongruent
dissolution of the bulk phase. Wilson and McHardy (49) have
documented differential attack on such perthitic intergrowths;
results of Holdren and Speyer (27) may be related to this
phenomenon. However, in the initial stage of dissolution,
non-stoichiometric release (e.g., Si and Al) is real, reflecting
the pH-dependent composition of the residual layer (23-25).
 2. Continued dissolution removes whatever hyperfine particles
may have been created during sample preparation for laboratory
dissolution. Where fines were removed, or were never present (as
in most natural materials), further dissolution breaks down the
outer surface of the residual layer at the same rate that alkalis
are replaced by hydrogen at the interface between the fresh
feldspar and the residual layer, releasing all constituents to
solution. Element release during this stage of the reaction is
stoichiometric (based on observations of surface morphology and
surface chemistry (e.g., 11,22) and solution chemistry (e.g., 25)),
but further work on the surface chemistry of naturally weathered
feldspar is required to determine this with certainty. This stage
of the reaction is also linear with time (11,25); the residual
surface layer is not thick enough to inhibit transport (30), it
attains a quasi-steady state thickness (provided solution
composition remains constant; 23-25), and no continuous, tenaceous
layer of secondary precipitates forms (21,22,41). The
rate-limiting step is the detachment of species from the
mineral-solution interface, i.e., a surface-controlled reaction
(21,22).

Rates of Feldspar Weathering in the Laboratory

Numerous surface characteristics have been postulated to play a
role determining mechanisms, rate-limiting steps, and rates of
feldspar dissolution during weathering (as discussed above). These
include:
  1. Adhered fine particles (e.g., 11).
  2. Defects and dislocation outcrops (e.g., 28,11).
  3. Artificially strained surfaces (e.g., 7,8).
  4. Residual surface layers of altered composition (e.g., 13).
  5. Surface layers consisting of secondary precipitates (e.g.,
     13,14).
  6. Surface layers whose composition has been altered
     chemically during sample preparation (e.g., 45).

Table I.   Silica Release Rates from Sodic Feldspars

pH	System	Treatment	Laboratory Studies Rate x 10^{12} $(moles/m^2/sec)$	Reference
5.6	Flowthrough	None	3.1	(8)
4.5-5.1	Batch	None	4.4	(9,10)
6	Batch	Ultrasonic & HF	13	(11)
5.1	Flowthrough	None	15	(23)
5	Batch	Ultrasonic	22	(26)
		Field Studies		
			0.024	(56)
			0.89	(57)

Numerous laboratory dissolution studies have tested for the effects of many of the aforementioned surface characteristics. Table I shows the results of several such studies. The pH's of the specific experimental runs used in Table I were chosen to facilitate comparison with natural (fresh) surface waters. Rates were taken directly from reported results, or were determined from the linear portions of tabulated or graphed data. Rates were normalized using either reported surface areas, or estimates of surface area based on reported particle size data. Results in Table 1 compare favorably with a similar compilation by Lasaga (50). Table I shows that feldspar dissolution rates in slightly acidic solutions vary by less than one order of magnitude, despite different experimental conditions and sample pretreatments. Helgeson and others (18) come to an identical conclusion on the basis of a more rigorous analysis of numerous published experimental studies, including many (but not all) of those included in this compilation. They conclude that the greatest ambiguity attending interpretation of experimental rates of silicate hydrolysis arises from several factors, which include uncertainty about the effective surface area, and the extent to which surface area and surface roughness change with reaction progress.

Rates of Feldspar Weathering in Natural Weathering Profiles from
Geochemical Mass Balance of Small Watersheds

Geochemical mass balance studies (also known as input-output budgets) invoke a simple conservation-of-mass principle. If the flux of any element leaving a watershed (e.g., via streams), and the flux of that element into the watershed (e.g., via atmospheric precipitation) are known, the difference between the two can be calculated, and this difference must be due to the sum of all reactions and transformations involving that element which took place within the watershed. Pioneering mass balance studies on weathering profiles and/or small watersheds include those of Garrels and Mackenzie (51,52) and Cleaves and Bricker and their coworkers (53,54). Geochemical mass balance studies are widely recognized as the most reliable means of estimating mineral weathering rates in nature (55).

Many mass balance studies which report weathering rates as a function of unit area of landscape surface do not permit comparison of those rates with laboratory dissolution rates, and cannot, therefore, contribute to the objectives of this paper. Only two published studies have thus far attempted to renormalize such calculated rates to mineral surface area. Discussion of these studies therefore forms the basis for comparisons of laboratory rates with natural weathering rates.

Paĉes (56) estimated the rate constant of oligoclase dissolution from the mass balance of sodium in European drainage basins. Several orders of magnitude of variability are associated with his estimate, due to uncertainties in estimating the area of mineral surface in contact with percolating fluids. However, despite this uncertainty, the total range of estimated values falls outside the range of laboratory rate constants; natural feldspar weathering rates are, according to Paĉes (56), one to three orders of magnitude slower than his favored laboratory rate. Paĉes (56) attributes this discrepancy to the difference in the character of feldspar surfaces in nature as opposed to those used in laboratory experiments. He suggests that experimentally dissolved feldspars are fresh, rough, characterized by kinks, ledges and terraces which would result in higher dissolution rates than old, "smooth, rounded" surfaces of naturally weathered feldspar which may also be "partly covered with weathering products which may act as inhibitors of dissolution" (p. 1861). His explanation for the discrepancy between laboratory and natural rates is at variance with many observations and theoretical inferences; for instance, it is widely believed that clay coatings could not inhibit feldspar dissolution (e.g., 41,21,22). It is also well established that the surfaces of naturally weathered feldspar grains are not smooth and rounded, but are instead rough and deeply etched (e.g., Figure 3); furthermore, both observations (e.g., 43 and Figure 3) and theoretical considerations (18) indicate that surface roughness (and surface area) may actually increase with reaction progress. Despite the inadequacy of Paĉes' (56) suggested explanation for the discrepancy between his field-determined rate constants and laboratory rate constants, the quantitative results themselves are unaffected by whatever argument is invoked to explain them.

Velbel (57) calculated rates of weathering of individual minerals from watershed geochemical mass balances for seven small forested control (unmanipulated) watersheds in the southern Blue Ridge of North Carolina. Using a system of linear equations formalized by Plummer and Back (58) combined with volumetric estimates of mineral abundance (based on petrographic data), estimates of mineral grain size and geometry, and the average thickness of the weathering profile, Velbel (57) transformed the results of the mass balance calculations for one of these watersheds (in moles of reaction/ha of watershed/yr) into moles/m^2 of mineral surface/sec, the units in which laboratory experimental rate data are often reported. The results (57) for rates of plagioclase weathering are approximately one order of magnitude slower than rates determined in laboratory experiments under similar hydrogeochemical conditions. Velbel (57) suggested that two major sources of the remaining rate discrepancy are (a) the character of artificially treated mineral surfaces in laboratory experiments, which renders them more reactive than their natural counterparts (e.g., 11) and/or (b) difficulties in estimating the reactive mineral surface area in natural systems. These results are discussed at length in Velbel (57).

Rates estimated in the above studies are shown in Table I. Watershed-scale geochemical mass balance studies yield calculated feldspar weathering rates one to three orders of magnitude slower than rates determined in laboratory experiments.

Summary and Evaluation

Studies of weathering rates in nature have identified two major sources of uncertainty associated with calculation of natural weathering rates: 1) differences in surface characteristics between natural and artificial feldspar surfaces, and 2) uncertainties in the effective surface area in natural systems. As shown in this review, however, the surface characteristics which play major roles in influencing feldspar dissolution rates have either not been identified in natural materials (see discussion of 56 above), or do not result in sufficient variability in (laboratory determined) rates to account for the difference between laboratory- and field-determined rates (as was optimistically suggested by Velbel, 57). In other words, the range of variability in feldspar weathering rates due to fine particles, strained surfaces, defects and dislocation outcrops, and surface layers is too small to account for the difference between laboratory and field-estimated rates of feldspar weathering. Therefore, either the surface characteristics of natural materials fall well outside the range which includes all laboratory experiments, or factors other than those hitherto investigated must account for the disparity between laboratory- and field-determined rates of mineral weathering. Furthermore, the sense and magnitude of the disparity suggests that either 1) the volume (modal abundance) and corresponding surface area have been vastly overestimated in natural systems, or, 2) the fraction of the natural mineral surface area which actually reacts with solutions is much smaller than that in laboratory experiments. The latter could still be due to some property of the mineral surfaces themselves. Alternatively, hydrological factors controlling the distribution of water circulation in weathering profiles could exert a significant influence on how much of the available mineral surface will actually participate in mineral-solution interactions. In other words, the role of surface area and surface characteristics is not well understood, because 1) estimates of total surface area in natural systems may be fundamentally flawed; 2) the ratio of effective surface area to total surface in natural systems may be significantly lower than in laboratory experiments; or 3) there may be some additional hydrological factor which must be introduced in order bring natural weathering rates into quantitative agreement with laboratory rate data.

One additional factor which plays a significant role (at least in principle) in determining rates of mineral weathering is the chemical affinity of the weathering reaction - i.e., how far the system is from thermodynamic equilibrium. As shown by Aagaard and Helgeson (59) and Helgeson and others (18), the effect of affinity on rates of feldspar dissolution in most laboratory experiments is negligibly small. This is because the experiments are usually of such short duration that concentrations of dissolution products have not built up to levels sufficient to significantly reduce the affinity. Mineral-solution contact times in natural systems are generally much longer than in laboratory systems; saturation with respect to secondary mineral products is obviously widely attained, for such secondary minerals (e.g., gibbsite, kaolinite, etc.) abound in natural weathering profiles, and their presence can often be related directly to the composition of coexisting solutions (e.g., 51,52,60-73). Under these conditions, mineral-solution interaction results in solutions highly evolved toward equilibrium. The chemical affinity and, consequently, the rates of the weathering reactions, will decrease (17). Natural weathering rates are, therefore, expected to be slower than laboratory rates; the magnitude of the difference in rates depends on the difference between the chemical affinities of natural vs. laboratory systems. Some effects of solution composition on reaction rates in laboratory settings have recently been investigated by Chou and Wollast (24). Unfortunately, aqueous geochemical data for natural weathering systems are generally inadequate to permit definitve quantitative evaluation of this hypothesis.

One suitable compilation of data for performing a preliminary evaluation of this hypothesis does exist, that of Pačes (64). Paces determined a disequilibrium index,

$$I = \log_{10}(Q/K)$$

where Q is the reaction quotient for the reaction and K is the equilibrium constant. According to Aagaard and Helgeson (59) the affinity term,

$$A = RTln(K/Q)$$

has a negligible effect on reaction rate (that is, the actual rate is 95% or more of its maximum possible value) if $A/\sigma RT > 3$. If σ is assumed to be 1 (59, p. 259), then Pačes' I is related to the affinity term in rate expressions based on transition state theory such that $(-2.3I = A/RT)$. Rearranging and relating the result to the inequality from Aagaard and Helgeson (59) reveals that solution composition has a negligible effect on reaction rate if $(-I \geq 1.3;$ or, $I \leq 1.3)$. The compilation of Pačes (64) suggests that many natural solutions are sufficiently close to thermodynamic equilibrium with potassium feldspar and clays to require the consideration of the affinity term in rate expressions. However, many natural solutions are sufficiently far from equilibrium with respect to albite and clays to render the affinity term (and any consequent rate variation) negligible, and virtually all of the solutions in Pačes' (64) compilation are far enough from equilibrium with respect to anorthite and clays so that affinity plays no role in affecting natural weathering rates of anorthite.

Furthermore, many of the solutions which are close to equilibrium are geothermal waters; virtually all solutions less than 25°C are far from equilibrium. Apparently, many natural systems are sufficiently far from equilibrium with plagioclase feldspars that the chemical affinity in these systems will not significantly slow reaction rates relative to those in laboratory systems. Nevertheless, the disequilibrium indices of Paĉes (64) do not cover the full range of possible situations. For example, Paĉes (64) did not calculate indices of disequilibrium for stoichiometric dissolution of feldspars <u>unaccompanied</u> by formation of clay minerals. Preliminary calculations, using the same data as were used by Velbel (57) to calculate natural weathering rates, suggest that solutions in the North Carolina watershed are <u>probably</u> far enough from equilibrium to render the affinity term in rate expressions negligible. Confirmation requires both more detailed examination of existing data (especially pH) and, especially, much better constraints on dissolved aluminum than are presently available. Finally, of course, the chemical affinity for aluminosilicate mineral hydrolysis in any specific natural system must be determined for each individual case; existing data do not permit generalizations regarding the effect of affinity on natural weathering rates. Nevertheless, preliminary examination of existing data suggests that affinity effects offer a promising avenue of future research in relating laboratory weathering rate to their natural counterparts.

Many of the same factors which complicate the interpretation of laboratory kinetic studies are among the most important limitations on the application of laboratory dissolution rate data to natural systems. These include uncertainty about 1) the effective surface area in natural systems (56,57); 2) the extent to which surface area and surface roughness change with reaction progress (18); and 3) the magnitude of solution composition effects on rates in natural systems.

Conclusions

1. Numerous surface characteristics may play a role in determining mechanisms, rate-limiting steps, and rates of feldspar dissolution during weathering (as discussed above). These include: A. Adhered fine particles; B. Defects and dislocation outcrops; C. Artificially strained surfaces; D. Residual surface layers of altered composition; E. Surface layers consisting of secondary precipitates; F. Surface layers whose composition has been altered chemically during sample preparation.

2. Sample pretreatment and varied experimental procedures designed to control or eliminate the effects of one or more of the aforementioned characteristics do not result in more than one order of magnitude change in the experimental dissolution rates.

3. Watershed-scale geochemical mass balance studies yield calculated feldspar weathering rates one to three orders of magnitude slower than rates determined in laboratory experiments.

4. The range of variability in feldspar weathering rates due to fine particles, strained surfaces, defects and dislocation outcrops, and surface layers is too small to account for the

difference between laboratory and field-estimated rates of feldspar weathering.
5. Either the intrinsic surface characteristics (e.g., ratio of effective-to-total surface area) of natural materials are significantly different from those used in laboratory experiments, or other factors must account for the disparity between laboratory- and field-determined rates of feldspar weathering. Possible non-mineralogical factors include inhomogenous fluid migration through the weathering profile. Differences in solution composition between laboratory and natural systems may also play a significant role, which remains to be evaluated.

Future improvements in the application of laboratory dissolution data to natural systems will come not (only) from additional work on laboratory kinetics, but will also depend heavily on much more comprehensive studies of surface area distribution, evolution, and accessibility to attack by fluids in natural systems, and by improved understanding of thermodynamic properties of natural fluids. Only in this way will laboratory kinetic data contribute to solving environmental problems such as nuclear waste disposal and evaluating the impact of acid deposition.

Acknowledgments

I appreciate the assistance of the editors, an anonymous reviewer, and, especially, D. Brandt Velbel and William M. Murphy, for their comments and criticisms. Dr. Alan S. Pooley of the Yale Peabody Museum assisted in taking the scanning electron photomicrographs of Figure 3. Preparation of this review was supported by NSF grant BSR-8514328.

Literature Cited

1. Hyndman, D.W. "Petrology of Igneous and Metamorphic Rocks"; McGraw-Hill: New York, 1972.
2. Deer, W.A.; Howie, R.A.; Zussman, J. "Rock-Forming Minerals, Volume 4 - Framework Silicates"; Longmans: London, 1963.
3. Deer, W.A.; Howie, R.A.; Zussman, J. "An Introduction to the Rock-Forming Minerals"; Longmans: London, 1966.
4. Goldich, S.S. J. Geology 1938, 46, 17-38.
5. Wollast, R. Geochim. Cosmochim. Acta 1967, 31, 635-48.
6. Correns, C.W.; von Engelhardt, W. Chimie der Erde 1938, 12, 1-22.
7. Huang, W.H.; Kiang, W.C. Amer. Mineralogist 1972, 57, 1849-59.
8. Nickel, E. Contributions to Sedimentology 1973, 1, 1-68.
9. Busenberg, E.; Clemency, C.V. Proc. Int'l. Symp. Water-Rock Interactions, 1976, pp. 388-94.
10. Busenberg, E.; Clemency, C.V. Geochim. Cosmochim. Acta 1976, 40, 41-9.
11. Holdren, G.R., Jr.; Berner, R.A. Geochim. Cosmochim. Acta 1979, 43, 1161-71.
12. Garrels, R.M. In "Research in Geochemistry"; Abelson, P.H., Ed.; Wiley, 1959, pp. 25-37.

13. Garrels, R.M.; Howard, P. Clays and Clay Minerals, Proc. 6th
     Natl. Conf., Berkeley, California, 1957; Swineford, A., Ed.;
     Pergamon, New York, 1959, pp. 68-88.
14. Helgeson, H.C. Geochim. Cosmochim. Acta 1971, 35, 421-69.
15. Helgeson, H.C. Geochim. Cosmochim. Acta 1972, 36, 1067-70.
16. Holdren, G.R., Jr.; Adams, J.E. Geology 1982, 10, 186-90.
17. Helgeson, H.C.; Murphy, W.M. Math. Geology 1983, 15, 109-30.
18. Helgeson, H.C.; Murphy, W.M.; Aagaard, P. Geochim. Cosmochim.
     Acta 1984, 48, 2405-32.
19. Gardner, L.R. Geology 1983, 11, 418-21.
20. Luce. R.W.; Bartlett, R.W.; Parks, G.A. Geochim. Cosmochim.
     Acta 1972, 36, 35-50.
21. Berner, R.A.; Holdren, G.R., Jr. Geology 1977, 5, 369-72.
22. Berner, R.A.; Holdren, G.R., Jr. Geochim. Cosmochim. Acta
     1979, 43, 1173-86.
23. Chou, L.; Wollast, R. Geochim. Cosmochim. Acta 1984, 48,
     2205-17.
24. Chou, L.; Wollast, R. Amer. J. Science 1985, 285, 963-93.
25. Wollast, R.; Chou, L. In "The Chemistry of Weathering";
     Drever, J.I., Ed.; NATO ASI Series No. C149, D. Reidel:
     Dordrecht, Holland, 1985; pp. 75-96.
26. Holdren, G.R., Jr.; Speyer, P.M. Geochim. Cosmochim. Acta
     1985, 49, 675-81.
27. Holdren, G.R., Jr.; Speyer, P.M. Amer. J. Science 1985, 285,
     994-1026.
28. Wilson, M.J. Soil Science 1975, 119, 349-55.
29. Berner, R.A. Amer. J. Science 1978, 278, 1235-52.
30. Berner, R.A. In "Kinetics of Geochemical Processes"; Lasaga,
     A.C.; Kirkpatrick, R.J., Eds.; Reviews in Mineralogy Vol. 8,
     Mineralogical Society of America: Washington, D.C., 1981; pp.
     111-34.
31. Lasaga, A.C. In "Kinetics of Geochemical Processes"; Lasaga,
     A.C.; Kirkpatrick, R.J., Eds.; Reviews in Mineralogy Vol. 8,
     Mineralogical Society of America: Washington, D.C., 1981; pp.
     1-68.
32. Schwaighofer, B. Geoderma 1976, 16, 285-315.
33. Tazaki, K. Papers Inst. Thermal Springs Research, Okayama
     Univ., 1976, No. 45, pp. 11-24.
34. Tazaki, K. Chikyu Kagaku (Earth Science) 1978, 32, 58-62.
35. Dearman, W.R.; Baynes, F.J. Proc. Ussher Soc. 1979, 4,
     390-401.
36. Gilkes, R.J.; Suddhiprakarn, A.; Armitage, T.M. Clays and
     Clay Minerals 1980, 28, 29-34.
37. Velbel, M.A. In "Pétrologie des Altérations et des Sols",
     Volume I; Nahon, D.; Noack, Y., Eds.; Sciences Géologiques,
     Mémoires (Strasbourg), 71, pp. 139-47.
38. Knauss, K.G.; Wolery, T.J. Geol. Soc. Amer., Abstr. with
     Programs 1983, 15, 616.
39. Anand, R.R.; Gilkes, R.J. Geoderma 1984, 34, 261-80.
40. Anand, R.R.; Gilkes, R.J.; Armitage, T.M.; Hillyer, J.W.
     Clays and Clay Minerals 1985, 33, 31-43.
41. Petrovic, R. Geochim. Cosmochim. Acta 1976, 40, 1509-21.
42. Fung, P.C.; Sanipelli, G.G. Geochim. Cosmochim. Acta 1982,
     46, 503-12.
43. Grandstaff, D.E. Geochim. Cosmochim. Acta 1978, 42, 1899-901.

44. Petrovic, R.; Berner, R.A.; Goldhaber, M.B. Geochim.
    Cosmochim. Acta 1976, 40, 537-48.
45. Perry, D.L.; Tsao, L.; Gaugler, K.A. Geochim. Cosmochim. Acta
    1983, 47, 1289-91.
46. Della Mea, G.; Dran, J.-C.; Petit, J.-C.; Bezzon, G.;
    Rossi-Alvarez, C., paper presented at 6th Ion Beam Analysis
    Meeting, Tempe, Arizona, 1983.
47. Berner, R.A.; Holdren, G.R., Jr.; Schott, J. Geochim.
    Cosmochim. Acta 1985, 49, 1657-8.
48. Chou, L.; Wollast, R. Geochim. Cosmochim. Acta 1985, 49,
    1659-60.
49. Wilson, M.J.; McHardy, W.J. J. Microscopy 1980, 120, 291-302.
50. Lasaga, A.C. J. Geophys. Research 1984, 89, 4009-25.
51. Garrels, R.M. In " Researches in Geochemistry", Volume 2;
    Abelson, P.H., Ed.; Wiley, pp. 405-20.
52. Garrels, R.M.; Mackenzie, F.T. In "Equilibrium Concepts in
    Natural Water Systems"; Stumm, W., Ed.; ADVANCES IN CHEMISTRY
    SERIES No. 67, American Chemical Society: Washington, D.C.,
    1967; pp. 222-42.
53. Cleaves, E.T.; Godfrey, A.E.; Bricker, O.P. Geol. Soc. Amer.
    Bull. 1970, 81, 3015-32.
54. Cleaves, E.T.; Fisher, D.W.; Bricker, O.P. Geol. Soc. Amer.
    Bull. 1974, 85, 437-44.
55. Clayton, J.L. In "Impact of Intensive Harvesting on Forest
    Nutrient Cycling"; College of Environmental Science and
    Forestry, State University of New York: Syracuse, New York,
    1979; pp. 75-96.
56. Pačes, T. Geochim. Cosmochim. Acta 1983, 47, 1855-63.
57. Velbel, M.A. Amer. J. Science 1985, 285, 904-30.
58. Plummer, L.N.; Back, W. Amer. J. Science 1980, 280, 130-42.
59. Aagaard, P.; Helgeson, H.C. Amer. J. Science 1982, 282,
    237-85.
60. Feth, J.H; Roberson, C.E.; Polzer, W.L. U.S. Geol. Survey
    Water-Supply Paper 1535-I, 1964.
61. Drever, J.I. J. Sedimentary Petrology 1971, 41, 951-61.
62. Tardy, Y. Chem. Geology 1971, 7, 253-71.
63. Reynolds, R.C.; Johnson, N.M. Geochim. Cosmochim. Acta 1972,
    36, 537-53.
64. Pačes, T. Geochim. Cosmochim. Acta 1972, 36, 217-40.
65. Pačes, T. Geochim. Cosmochim. Acta 1973, 37, 2641-63.
66. Marchand, D.E. U.S. Geol. Survey Professional Paper 352-J,
    1974.
67. Norton, D. Geochim. Cosmochim. Acta 1974, 38, 267-77.
68. Smith, T.R.; Dunne, T. Earth Surface Processes 1977, 2,
    421-425.
69. Verstraten, J.M. Earth Surface Processes 1977, 2, 175-84.
70. Pačes, T. Geochim. Cosmochim. Acta 1978, 42, 1487-93.
71. Bouchard, M. Rev. geol. dyn. geo. physique 1983, 24, 363-79.
72. Velbel, M.A. In "The Chemistry of Weathering"; Drever, J.I.,
    Ed.; NATO ASI Series No. C149, D. Reidel: Dordrecht, Holland,
    1985; pp. 231-47.
73. Katz, B.G.; Bricker, O.P.; Kennedy, M.M. Amer. J. Science
    1985, 285, 931-62.

RECEIVED May 20, 1986

Dislocation Etch Pits in Quartz

S. L. Brantley, S. R. Crane, D. A. Crerar, R. Hellmann, and R. Stallard

Department of Geological and Geophysical Sciences, Princeton University, Princeton, NJ 08544

Quartz samples were etched hydrothermally at 300°C in etchants of controlled Si concentration to measure the concentration above which dislocation etch pits would not nucleate. The C_{crit} for 300°C was predicted to be $0.6C_o$ and the measured C_{crit} was $0.75C_o \pm .15$ (C_o = equilibrium concentration). Our observations suggest that for $C > C_{crit}$, dissolution occurs at edges and kinks on the surface; while for $C < C_{crit}$, dislocation etch pits form rapidly, contributing to the overall dissolution rate. Analysis of quartz particles from a soil profile revealed a transition from angularly-pitted grain surfaces at the top to rounded surfaces at the bottom, suggesting that downward permeating fluids pass through the critical Si concentration. The theory of etch pit formation may be useful in interpreting the chemical conditions of low temperature mineral-water interactions.

Dissolution of a crystal surface is initiated at sites of high surface energy: edges, corners, cracks, scratches, and holes are favorable sites for fast dissolution. At the micro-scale, trapped impurities, point defects, twin boundaries, and dislocations can also cause enhanced dissolution. The formation of etch pits by dissolution at dislocations has been of particular interest to experimentalists interested in testing and developing theories of dissolution, lattice strain, and material deformation. Frank (1), Cabrera, et al. (2), and Cabrera and Levine (3) developed the first theory of etch pit formation based on dislocation lattice strain. Experiments by Sears (4), Gilman, Johnston and Sears (5), and Ives and Hirth (6) showed that these simple theories appeared consistent with dissolution of LiF. Lasaga (7) recently pointed out implications for interpretation of dissolved mineral surfaces and paleo-fluid histories. Several types of experiments are suggested by these theories which apply to geochemical processes such as hydrothermal alteration, weathering, and other dissolution reactions.

We describe here an experiment which indicates that the dislocation etch pit theory is a useful tool in interpreting formation of

0097-6156/86/0323-0635$06.00/0

etch pits in quartz by hydrothermal dissolution. We have also tested
the theory by documenting the incidence of etch pits on surfaces of
quartz grains sampled from a soil profile. Finally, we discuss other
approaches suggested by our experiments and by experiments in the
literature which would provide useful geochemical information about
the rates and mechanisms of natural dissolution processes.

Theory of Etch Pit Formation

Pit formation. If we consider a dissolution nucleus at a screw
dislocation intersecting the surface which consists of a cylindrical
hole of radius r, one atom layer deep (a), then the free energy of
formation of this nucleus will be composed of a volume energy,
surface energy, and elastic strain energy term, respectively, as
follows:

$$\Delta G = \pi r^2 a g + 2\pi \, r a \gamma - a \tau b^2 (\ln(r/r_0))/4\pi \qquad (1)$$

where τ is the shear modulus, b is the Burger's vector, r_0 is the
dislocation core radius, γ is the surface energy, and g is the free
energy of dissolution per unit volume (2,3). Equation 1 shows that
opening of a pit on a crystal surface is a competition between terms
which decrease the free energy (dissolution of a volume of crystal
into an undersaturated medium and release of dislocation strain
energy) and a term which increases the free energy (creation of
additional surface area). The cylindrical hole geometry is chosen
for simplicity. To predict whether an etch pit will form at a
dislocation, we want to determine the variation of ΔG with radius,
r, and free energy of dissolution, g, defining g as the chemical
affinity per unit volume and neglecting activity corrections:

$$g = RT \ln(C/C_0) / V \qquad (2)$$

where C = concentration of dissolving species, C_0 = equilibrium
solubility of the species, V is the molar volume, R is the gas
constant and T is the absolute temperature. Equation 1 expresses
the free energy in enlarging the pit from a radius r_0 to a radius r.
The core radius, r_0, is the radius of the central volume of the
dislocation where the continuum approximation breaks down and electronic energies become important. For $r < r_0$, the elastic strain
term of Equation 1 cannot adequately predict the dislocation energy.
Because the nature and energetics of the dislocation core are not
well understood, the value of r_0 is undetermined, and we have chosen
to set $r_0 = b$, following other workers (8). For quartz at 300°C,
using a shear modulus of 0.48×10^{11} Pa (9), a Burger's vector of
7.3 A (9), a surface energy of 360 mJm^{-2} (10), and a molar volume of
22.688 cm^3, we can calculate ΔG as a function of r for different
values of the saturation index, C/C_0. In Figure 1 we have plotted
calculated values of ΔG as a function of r ($> r_0$) and C for a pit
one molecule deep in quartz at 300°C, a temperature where quartz
readily dissolves in water. Lettered curves correspond to the
following concentrations: A ($0.04C_0$), B ($0.36C_0$), C ($0.51 C_0$), D
($0.64C_0$), E ($0.73C_0$), F ($0.82C_0$), G ($0.89C_0$). Concentrations were
chosen to correspond to selected experimental run conditions. Note
that all calculations have used Equation 1, which is strictly valid

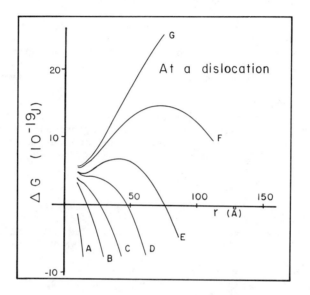

Figure 1.   Calculated values of ΔG, free energy of formation of a pit at a dislocation on a quartz surface at 300°C, plotted vs. pit radius, r. Labels defined in text. Adapted with permission from Ref. 16. Copyright 1986 Pergamon Press.

only for screw dislocations. For edge dislocations, the strain energy term is modified by the factor $(1/(1-\nu))$ where ν is Poisson's ratio. For quartz, $\nu = 0.077$, and the correction is very small.

As Figure 1 shows, all the calculated ΔG curves above curve D show a minimum (very near to $r = 10$ Å) and a maximum (at large values of r_o) in the free energy. The critical concentration where the minimum and maximum in the ΔG curve disappear can be determined by maximizing ΔG with respect to r and solving for r (2,3):

$$r = (-\gamma/2g)[1 \pm (1 - \tau b^2 g/2\pi^2 \gamma^2)^{1/2}] \tag{3}$$

Values of r satisfying Equation 3 (corresponding to the minimum and maximum points in ΔG) will yield steady state solutions where a pit radius should remain constant, while the rest of the crystal grows or dissolves depending on the chemical affinity (Equation 2). If the term $\tau b^2 g /2\pi^2\gamma^2 > 1$, there are no real solutions to Equation 3 and there is no steady state value of r, which indicates that a small pit nucleated at a dislocation core should spontaneously open up to form a macroscopic etch pit. The critical concentration at which this occurs (setting the above term equal to one) is :

$$C_{crit} = C_o \exp(-2\pi^2\gamma^2 V/RT\tau b^2) \tag{4}$$

For $C = C_{crit}$, there is a double root to the maximization equation, and there is an inflection point in the ΔG function (curve D on Figure 1). Since there is no activation barrier to opening up the etch pit, any pit nucleated at a dislocation should open up into a macroscopic etch pit. Similarly, for $C < C_{crit}$, there are no real solutions and no maxima and minima in the ΔG function, and nucleated pits open up into etch pits. At 300°C, the calculated C_{crit} for quartz equals $0.6C_o$.

Above C_{crit} (i.e. E or F in Figure 1), there are two real roots to the equation, so there is a minimum and a maximum in the ΔG function. If a pit is nucleated at the core, the pit should spontaneously open until its radius fulfills the condition that ΔG is at a minimum (~10 Å). There is then an activation barrier ΔG^* ($=\Delta G_{maximum} - \Delta G_{minimum}$) toward further opening of the pit into a macroscopic etch pit. Monte Carlo simulations of etch pit formation have shown that such hollow tubes should be stable for some materials, including quartz (27). Above C_{crit}, the height of the activation barrier (ΔG^*) will determine the rate of formation of etch pits. If metastable equilibrium is assumed for the pit nuclei size distribution, the rate of formation of pits per unit area, J, for concentrations above critical should have the form:

$$J = X_d A \exp(-\Delta G^*/RT) \tag{5}$$

where X_d is the fraction of surface sites intersected by dislocations, and A is a frequency factor (11). If the core energy is included in the ΔG calculation, a small activation barrier exists even for pit nucleation below C_{crit}. Pit nucleation in highly undersaturated solutions should then also show a rate dependence as in Equation 5, with a substantially smaller activation energy.

Dissolution kinetics at etch pits. If an etch pit opens up at a dislocation, the slope of the sides of the pit, determined by the ratio of downward dissolution rate v_n to outward dissolution rate v_s, must be high in order for the etch pit to be microscopically observable. If the crystal surface is close-packed, then holes nucleated in the close-packed surface will consist of high-index faces composed of stepped ledges. Anisotropy of v_s will cause etch pits to have crystallographically-controlled non-cylindrical geometries. The rate v_n is the rate of formation of holes at a dislocation, and the rate v_s is the ledge velocity or the rate of recession of ledges away from the nucleated holes. Kinks in the ledges serve as sites of easiest transfer of molecules from the surface into solution. Johnston (12) suggested that the ratio v_n/v_s > 0.1 for an etch pit to be microscopically visible.

When v_n is small compared to v_s, ledge spacing at the apex of the nucleated pit will be large, and as the ledges recede, the walls of the pit will be maintained at a very shallow angle, making the pit unidentifiable. Sears and co-workers (4,5) pointed out that the adsorption of poisons onto the dissolving pit surface can decrease v_s, producing a more visible pit. Based on observed etching on etching in LiF, they concluded that poisons were essential for the formation of etch pits on crystal surfaces. The importance of poisons in the formation of etch pits in other substances needs to be investigated.

Etch Pits in Quartz: A Test of the Theory

Hydrothermal etching. To test the prediction of a critical concentration in etch pit formation, we investigated the hydrothermal dissolution of quartz. Previous workers (13,14) noted two types of triangular etch features, deep pyramidal pits and shallow flat pits, produced by hydrothermal etching with distilled water on the rhombohedral face. These workers argued that the shallow flat pits correspond to surface defects while the deeper pits correspond to dislocations. Because of the background work completed on dislocation etch pits on rhombohedral faces of quartz, we decided to investigate etching on this surface.

We ran two different types of dissolution experiments: closed and flow. For the closed experiments, we placed cut pieces of Arkansas quartz with 50 ml of silica solution (prepared according to the method of Crerar et al., (15) and neutralized to pH 7 with NaOH) in standard sealed autoclaves in a temperature-controlled oil bath. In each experiment, all pieces were cut from one face of a large crystal. Different crystals, all of the coarse-crystallized variety from Hot Springs, Arkansas, were used for each run. Dissolution was allowed to proceed for 6.5 hours at 300°C at saturated vapor pressure. Heat up time was from 3-5 hours, and silica concentration during the run usually increased to supersaturated conditions. In the flow experiments, quartz was placed in a flow-through chemical reactor at 300°C (heat up time approximately 30 minutes), and a Si solution of known concentration was pumped through at a rate of 0.7 ml/min at saturated vapor pressure. Solution chemistry was monitored throughout the run. After the experiment, etching on the crystal face was analyzed by SEM.

Specific experimental run conditions and observations are described in Brantley et al. (16). In order to quantify the presence of

etch pits, pit density counts were made for most samples. Fifteen
to forty surface samples were randomly selected and imaged at 1000x
under SEM, and deep pyramidal pits were counted. Pit densities are
plotted in Figure 2, along with the calculated critical concentration. Error bars (see 16) are liberal estimates of statistical
counting error and the systematic error involved in distinguishing
etch pits. A least squares linear fit to the data from each individual crystal face is plotted in order to estimate the C_{crit} where
pit densities reach background levels.

For all runs, there is a clear decrease in measured pit density
with increasing C/C_0 ratio. In the closed experiments, pit densities reach background levels ($<3 \times 10^3$ cm^{-2}) at $C/C_0 = 0.75$. In the
flow experiments, samples from crystal R5 (etched for 6.5 hours) and
R5SE (crystal R5 after etching 6.5 hours cleaned and re-etched for
25 more hours) show background levels (1×10^4 cm^{-2}) above $C/C_0 =$
0.8. Figure 3 shows R5SE surfaces etched above and below 0.8 C_0.

Crystal R9, whose unreacted surface was rougher and more
disturbed than the R5 surface, shows a significant decrease in pit
density between $C/C_0 = 0.75$ and 0.89. Extrapolating the limited R9
data to background pit density ($\sim 2 \times 10^3$ cm^{-2}) predicts $C_{crit}/C_0 =$
0.9. Although formation of etch pits decreased markedly above
$0.75C_0$ for R9, some etch pits still formed at $C/C_0 = 0.9$. In addition, crystal R9 showed considerable "arcuate etching", a general
term which describes a variety of unusual etch features common to
quartz: curved etch lines, elongated etch triangles, and linear
arrays of etch lines and etch triangles, usually with curved edges
(16,17). These features could be associated with surface scratches,
high impurity content, inclusions, or other flaws disturbing the
surface. TEM analysis of samples R5 and R9 indicated that dislocation densities were similar in the two crystals; therefore,
heightened etching of R9 might be attributed to segregated impurities which perturb the surface energy of the R5 or R9 dislocations.
Alternatively, increased etching might result from differences in
Burger's vectors or the screw or edge character of the two crystals'
grown-in dislocations. The presence of disturbed surface layers on
some crystals of natural quartz was also noticed in the etching
experiments of Hicks (17).

Our best estimate from the experimental data in Figure 2 for
C_{crit} at 300°C is .0078 $\pm$ .0007 m Si which corresponds to 0.8 ($\pm$
0.07) C_0. However, we have used Walther and Helgeson's (18) value
of 0.0097 m Si as the best value for the equilibrium concentration,
C_0. If we use Fournier and Potter's (19) value of 0.011m Si for C_0,
then our experimental $C_{crit} = 0.7C_0$, which is closer to the theoretically predicted C_{crit} of $0.6C_0$. Because of the uncertainty in C_0
and the uncertainty in our data, our best experimental estimate is
$C_{crit} = 0.75 \pm .15 \ C_0$ at 300°C.

A very accurate measurement of C_{crit} would allow back-calcula-
tion of the surface energy for a given crystal. Because C_{crit} is
dependent on the square of γ, such a measurement could be a very
sensitive method of measuring interfacial energy at dislocation
outcrops. The calculated interfacial energy from our experiments is
280$\pm$ 90 mJm^{-2} for the rhombohedral face of quartz at 300°C. Parks
(10) estimated 25°C value of 360 $\pm$ 30 mJm^{-2} is well within the
experimental error of our measurement. The best way to determine
the value of C_{crit} would be to measure etch pit nucleation rate on

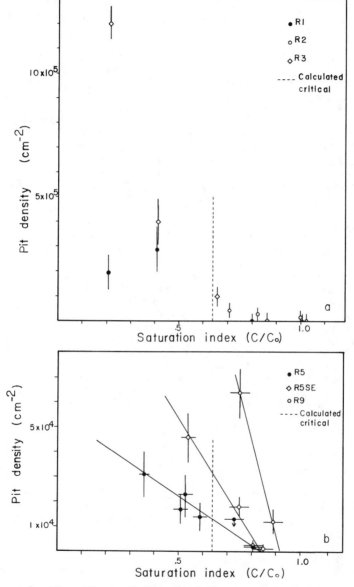

Figure 2. The effect of dissolved Si concentration on etch pit density on quartz surfaces etched: a) in sealed autoclaves for 6.5 hours, b) in a flow reactor for 6.5 hours (R5), 31.5 hours (R5SE), and 25-28 hours (R9). Reproduced with permission from Ref. 16. Copyright 1986 Pergamon Press.

one surface under different silica concentrations. By plotting log
(rate) vs. log $(C/C_0)^{-1}$ as suggested by Equation 5, a break in slope
should occur at C_{crit} regardless of the dislocation density of the
starting material.

Low temperature etching. Our data suggests that, under hydrothermal
conditions the rate of pit formation is dramatically reduced, although perhaps not completely stopped, at $C = C_{crit}$. Etch pits on a
natural, hydrothermally-etched quartz surface therefore indicate extended dissolution times, but not necessarily etching at $C < C_{crit}$.
This is because the rate of etch pit formation even above C_{crit} can
be significant at elevated temperatures (as shown by crystal R9).
However, at low temperatures, formation of etch pits when $C > C_{crit}$
would be less likely, and natural surfaces etched at low temperature
should record the saturation state of the etching fluid.

In order to test this hypothesis, Crane (20) analyzed the
surfaces of quartz grain samples from a 90 cm deep soil profile
developed in situ on the Parguaza granite, Venezuela (21). Figure 4
shows characteristic surface morphologies from sand grains from just
above granite bedrock (90 cm deep) and from 50 cm above bedrock (40
cm deep). As suggested by this figure, a transition occurs at a
depth between 60 and 80 cm from angularly-pitted surfaces to rounded
surfaces, suggesting that the critical concentration is reached at
that point. These observations suggest that, at 25°C, rates are
slow enough that for $C > C_{crit}$, no etch pit formation occurs.

Based on predicted weathering and erosion rates of the region,
we estimate the profile to be several million years old. Because
the soil has developed in situ, the topmost grains have reacted with
water for the greatest extent of time. With depth, the total "life-
time" of the particles as soil decreases. This implies that the
topmost quartz surfaces should be "reactively mature" (all fines
removed, deep grown-together etch pits) and the bottom-most quartz
surfaces should be "reactively young" (plentiful fines, fresh surfaces).

Reactively young surfaces in contact with undersaturated solutions should show high rates of pitting. The lack of pitting in
bottom samples suggests that the critical Si concentration in the
permeating fluid has been exceeded and etch pits are not forming at
significant rates. Modeling (21), corroborated by the observation of
rounded grain shapes observed in bottom-most layers, indicates on-
going quartz dissolution. Apparently, dissolution of quartz continues in this zone by dissolution of fines, edges, cracks, etc., but
without the formation of etch pits. At a slightly higher zone of
the profile, reactively young surfaces are exposed to solutions with
$C < C_{crit}$. At this point in the profile, "reactively young" quartz
meets very reactive solution, and aggressive quartz dissolution and
pitting occurs (Figure 4b). Topmost quartz grains show deep, angular, grown-together pits as expected (16).

Rates of Pit Growth. Joshi and Vag (14) measured the rate of etch
pit growth in quartz at several temperatures in a concentrated NaOH
solution. They report that the rate of dissolution normal to the
surface, v_n, is 0.09 micron/min at 200°C, 0.10 micron/min at 250°C,
and 0.12 micron/min at 275°C, and 0.166 micron/min at 300°C.   By
assuming as a first approximation that precipitation into a disloca-

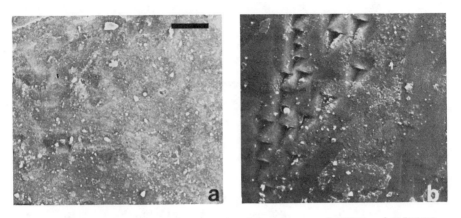

Figure 3. SEM photomicrograph of surfaces of R5SE: a) R5S1SE etched 31.5 hours at 0.008 m Si, b) R5S3SE etched 31.5 hours at 0.006–0.007 m Si (Scale bar = 10 microns).

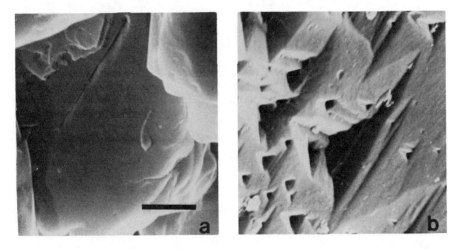

Figure 4. SEM photomicrograph of characteristic surfaces of sand grains from a Venezuelan soil profile. Samples from: a) 90 cm deep, b) 40 cm deep. (Scale bar = 2.5 microns).

tion pit is zero throughout the experiment, we can assume that v_n is directly proportional to the rate constant for dissolution at a dislocation. By regressing ln (v_n) vs. $(1/T)$ for this data, we can estimate an activation energy for this dissolution process at high pH: E_a = 13 kJ/mol. As expected, this activation energy is smaller than the activation energy for dissolution of bulk quartz in water determined for 0°C to 300°C (E_a = ~70kJ/mol, 22) at near-neutral pH. It is interesting to note that this former activation energy is of the same order of magnitude as that measured by Rimstidt and Barnes. Since dissolution at a dislocation has a smaller activation energy than that measured for bulk quartz, pit formation is probably not the rate-limiting step for hydrothermal quartz dissolution. Further experiments on etch pit kinetics are clearly necessary in order to conclusively interpret bulk dissolution data.

Although we did not measure the rate of pit deepening, we did measure the width of etch pits produced on the quartz etched in the flow systems. Pits were imaged by SEM and the small dimension of the etch triangle was measured. By comparing pit widths from crystal R5 etched for 31.5 hours in different Si concentrations, we can compare pit growth rates for different C/C_o values. Pit widths for crystal R5, along with Si concentration of flowing fluids, are tabulated in Table I. We observed a fairly broad variation in pit size for each sample. In particular, we observed that some pits were asymmetric (the deepest etch point off-center from the broader triangle), while others were symmetric; the asymmetric pits were generally smaller than the symmetric pits. Pit symmetry is related

Table I. Crystal R5: Etch Pit Widths

Sample	Si Concen.	Average pit width	Etch time
R5S1SE	0.008 m	1.5 + .5 microns	31.5 hours
R5S2SE	0.007-.008 m	4.4 + .9	31.5 hours
R5S3SE	0.006-.007 m	5 + 2	31.5 hours
R5S5SE	0.005 m	6 + 2	31.5 hours

to the angle between the dislocation line and the surface as well as the anisotropy of dissolution rate. Despite the uncertainty in the data, there is a noticeable decrease in rate with increasing C/C_o. In addition, the largest change in growth rate occurs above C = C_{crit} (~0.8 C_o). Etch pits on crystal R9 were larger than pits observed on R5. Apparently, whatever surface feature of R9 causes enhanced formation of pits also causes increased rates of pit widening.

Ives and Hirth (6) report similar data for etch pits in LiF, in which the rate of pit widening drops to 0 at $0.22C_o$ and 32°C in dilute ferric fluoride solution, where the 2.5 ppm Fe^{3+} acts as a dissolution poison. They argue that, when no poison is present, the rate should decrease linearly with increasing C/C_o, until the rate equals 0 at C = C_o. In the presence of a poison adsorbed to a kink, however, dissociation of a molecule should be slowed, reducing the local equilibrium concentration to a value of $C' < C_o$. If poison molecules completely cover the kink sites, the rate of pit widening could reach 0 at C' rather than C_o. The contribution of poisons to formation of etch pits in quartz is unclear. In our experiments, we observed that etch pits formed during closed runs were better defined than those of the flow experiments, which could be related either to the higher Na content of the closed run solutions or to the flow conditions. We also tested for Fe content in the flow samples and found less than 1 ppm Fe.

Etch Tubes and Terraced Pits. One interesting feature we observed in some long quartz dissolution experiments was the development of very deep etch holes centered on triangular pits. Hicks (17) has also reported these features. Figure 5 shows an example from sample R5S5SE etched 31.5 hours at 0.5 C_o. Notice that the shape of the etch hole is distinctly different than the larger, triangular etch figure. We believe these features are examples of etch tubes, first noted by Nielsen and Foster (23) in both synthetic and natural quartz etched in 48% HF. These features document conditions such that v_n, rate of pit deepening, is very large. They are thought to be characteristic of quartz dislocations with segregated impurities. Apparently, v_n can be changed quite dramatically by dislocation impurity content. The change in etch hole shape could also be due to impurities in the quartz. Formation of extremely large, widely-spaced etch holes as observed in feldspar (25), and in quartz (26), may also be related to impurity or inclusion content of the crystal. Figure 5 also shows another feature previously described for etched amethyst crystals (24): terraced etch pits. Terraced etch pits are those in which continued dissolution reveals that the dislocation line has a stepped configuration. As the pit deepens, the dislocation line jogs or branches, which causes the center of dissolution to move from the center of the etch triangle. Several examples of etching along apparently stepped or branching dislocations are shown in Figure 5.

In general, the shape and character of etch pits may reveal information about the impurity content of the crystal. "Beaked pits" (pits with curved apexes, see 12) can indicate impurity haloes. Some forms of the arcuate etching we observed in quartz (16) may be examples of beaking. Very shallow pits can form at aged dislocations while very deep pits form at new dislocations. "Aging" may be related to impurity diffusion in the crystal lattice.

Conclusions and Implications for Future Research

We have described a set of experiments which suggests that the etch pit formation theory developed by Cabrera and Levine (2,3) and Frank (1) works well in predicting hydrothermal etching of quartz. Our work shows that at 300°C there is a critical concentration (0.75 ± 0.15 C_o) above which the rate of etch pit formation slows dramatically. We also reported qualitative observations of surface features of quartz sand grains from a 90 cm soil profile which show a systematic surface morphology variation with depth. In the profile, we observed a transition from angularly-pitted surfaces to rounded surfaces at a depth between 60 and 80 cm, suggesting that the critical Si concentration is reached in the permeating fluids at that point. We suggest that the observation of etch features on these low- temperature etched grains indicates that our high temperature work does have relevance to dissolution features formed under

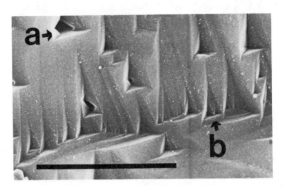

Figure 5. SEM photomicrograph of etch tubes (a) and terraced
etch pits (b) on sample R5S5SE etched 31.5 hours at 0.53 C_o.
(Scale bar = 10 microns.)

different temperature conditions. The impurity content of the
quartz and the solution may also affect etching. If so, natural
etching may give clues to the saturation index and the impurity
content of paleo-fluids.

Several refinements of our experiments could test these theories further. By measuring etch pit densities as well as pit dimensions on sequentially-etched crystals, nucleation rate data and pit
growth data could be collected, yielding information about the rate-
limiting steps and mechanisms of dissolution. In addition, since the
critical concentration is extremely dependent on surface energy of
the crystal-water interface (Equation 4), careful measurement of
C_{crit} yields a precise measurement of γ. Our data indicates an
interfacial energy of 280 ± 90 mJm^{-2} for Arkansas quartz at 300°C,
which compares well with Parks' value of 360 mJm^{-2} for 25°C (10).
Similar experiments on other minerals could provide essential surface energy data.

Whether etch pit formation is important in determining dissolution kinetics and controlling the general reactivity of a crystal
will depend on the nature of different crystals, as well as the
overall dislocation density. Normally, crystal edges are a ready
source of ledges for crystal dissolution. At low concentrations (C <
C_{crit}), crystals will dissolve at edges, steps, cracks, etc., as
well as at nucleated pits in both perfect and imperfect crystal.
The fastest of these parallel processes should be rate-determining.
If dislocation density is large enough, or if dissolution from edges
is slow enough, etch pit dissolution may make a major contribution
to the rate of bulk dissolution and be rate-determining. In this
case, these minerals would show a change in rate-limiting step as C
rises above C_{crit} and dissolution at dislocations slows.

Wintsch and Dunning (8) calculated that the solubility of
plasticly deformed quartz in water should not be significantly
higher than the ideal equilibrium value. However, enhanced dissolution at dislocations could significantly increase the dissolution

rate. This enhanced reactivity would be important in determining
rates of mass transfer in zones of intense deformation. We are
currently conducting dissolution kinetics experiments with such
deformed material.

Formation of etch pits on naturally-weathered mineral surfaces
has been noted by many workers (28-30). A theory based on laboratory etching experiments (31,32) has suggested that, for feldspars,
surface layer buildup in etching holes may explain non-congruent
dissolution (33). Recent evidence from Holdren and Speyer (34) has
also shown that the dissolution rate of feldspar is not always
directly proportional to surface area, but may instead be proportional to active site concentration. These results seem to imply
that etch pit formation may be a rate-controlling step in feldspar
dissolution under some conditions. In order to generalize laboratory dissolution rates for these materials to natural weathering
processes, it may be necessary to measure natural defect densities.
However, the reproducibility of dissolution rates for fluorite and
calcite in our laboratory (35 and unpublished data) and for quartz
(22) indicates that presumed differences in dislocation etch pit
density does not affect the dissolution kinetics of these minerals
within a factor of +10%. This could be explained by assuming that
laboratory samples have dislocation densities which are equal to
within $\pm$ 10%, or that laboratory preparation techniques produce
equivalent dislocation densities. Alternatively, these observations
could imply that etch pit contribution to overall bulk dissolution
rate for these minerals is minimal.

To further understand and model bulk dissolution of minerals,
careful experiments such as we have discussed above will be necessary. Dislocation etch pits, as sources of ledges on a surface,
provide good control in measuring ledge or kink velocities. Work in
this area, applied to geologically important minerals, will extend
our understanding of the rates and mechanisms of alteration reactions. In addition, our soil profile work suggests that there may
be information recorded on mineral surfaces which will be useful in
reconstructing flow, compositional, and temperature histories of
paleo-fluids.

Legend of Symbols

a	thickness of one molecular layer of quartz
A	frequency factor
b	Burger's vector of dislocation
c	concentration
c_{crit}	critical concentration
c_o	equilibrium concentration
g	chemical affinity per unit volume
ΔG	free energy of formation of a pit at a dislocation
ΔG^*	activation barrier toward formation of pit of cri-
	tical radius on surface
J	pit nucleation rate
r	radius of pit
r_o	dislocation core radius
R	gas constant
T	temperature

v_n dissolution rate in a direction normal to the surface

v_s dissolution rate in a direction parallel to the surface

V molar volume

X_d fraction of surface sites intersected by dislocations

γ interfacial energy

τ shear modulus

ν Poisson's ratio

Acknowledgments

This work was funded by NSF grant #EAR-82-18726 and #EAR-84-19421. The soil profile analysis and much of the hydrothermal work was completed by S.R. Crane as her senior thesis at Princeton University. Maria Borscik, Elaine Lenk, and Laurel Pringle-Goodell helped with chemical, SEM and TEM analyses. D.A.C. gratefully acknowledges support from the Shell Companies Foundation and R.F.S. acknowledges support from the Venezuelan Ministerio del Ambiente y Recursos Naturales Renobales, a Dusenbery Preceptorship, and N.S.F. grant #EAR 84-07651.

Literature Cited

1. Frank, F.C. Acta Cryst. 1951, 4, 497.
2. Cabrera, N.; Levine, M.M.; Plaskett, J.S. Phys.Rev. 1954, 96, 1153.
3. Cabrera, N.; Levine, M.M. Phil. Mag. 1956, 1, 450.
4. Sears, G.W. J. Chem. Physics 1959, 32, 1317.
5. Gilman, J.J.; Johnston, W.G.; Sears, G.W. J. Appl. Physics 1958, 29, 747.
6. Ives, M.B.; Hirth, J.P. J. Chem. Phys. 1960, 33, 517.
7. Lasaga,A. 4th Int. Symp. Water-Rock Inter. Ext. Abstr.,1983, p.269.
8. Wintsch, R.P.; Dunning, J. J. Geophys. Res. 1985, 90, 3649.
9. Heinisch, H.L. Jr.; Sines, G.; Goodman, J.W.; Kirby, S.H. J. Geophys. Res. 1975, 80, 1885.
10. Parks, G.A. J. Geophys. Res. 1984, 89, 3997.
11. Hirth, J.P.;Pound, G.M. "Condensation and Evaporation: Nucleation and Growth Kinetics", MacMillan Co.: New York, 1963.
12. Johnston, W.G. Prog. Ceram. Sci. 1962, 2, 3.
13. Joshi, M.S.; Vag, A.S. Sov. Phys.Cryst. 1968, 12, 573.
14. Joshi, M.S.; Kotru, P.N.; Ittyakhen, M.A. Sov. Phys. Cryst. 1960, 15, 83.
15. Crerar, D.A.; Axtmann, E.V.,; Axtmann, R.C. Geochim. Cosmochim. Acta 1981, 45, 1259.
16. Brantley, S.L.; Crane, S.R.; Crerar, D.A.; Hellmann, R.; Stallard, R. Geochim. Cosmochim. Acta, in press.
17. Hicks, B.D. Masters Thesis, University of Missouri-Columbia, Missouri, 1985.
18. Walther, J.V.; Helgeson, H.C. Amer. Jour. Science 1977, 277, 1315.
19. Fournier, R.O.; Potter, R.W. Geochim. Cosmochim.Acta 1982, 46, 1969.

20. Crane, S.R. Senior Thesis, Princeton University, Princeton, New Jersey, 1985.

21. Stallard, R.F. In "The Chemistry of Weathering"; Drever, J.I., Ed.; NATO ASI SERIES vol. 149, D. Reidel Publishing Co.: Dordrecht, 1984, pp.293-316.

22. Rimstidt, J.D.; Barnes, H.L. Geochim.Cosmochim.Acta 1980,44, 1683.

23. Nielsen, J.W.; Foster, F.G. Am. Mineral. 1960, 45, 299.

24. Joshi, M.S.; Kotru, P.N.; Ittyachen, M.A. Am. Mineral. 1978, 63, 744.

25. Holdren, G.R.; Speyer, P.M. Geochim. Cosmochim. Acta 1985, 49,675.

26. Stein, C.L. Chem. Geology, in press.

27. Lasaga, A.C.; Blum, A.E. Geochim. Cosmochim. Acta, in press.

28. Wilson, M.J. Soil Sci. 1975, 119, 349.

29. Berner, R.A. Am. Jour. Sci., 1978, 278, 1235.

30. Velbel, M.A. In "Environmental Geochemistry"; Fleet, M.E., Ed.; MAC SHORT COURSE HANDBOOK Vol. 10, Mineralogical Society of Canada: Toronto, 1984, pp. 67-111.

31. Chou, L.; Wollast, R. Geochim. Cosmochim. Acta, 1984, 48,2205.

32. Holdren,G.R.; Speyer,P.M. Am. Jour. Sci., 1985, 285, 994.

33. Berner, R.A.; Holdren, G.R.; Schott, J. Geochim. Cosmochim. Acta 1985, 49,1657.

34. Holdren, G.R.; Speyer, P.M. Geochim. Cosmochim. Acta 1985, 49, 675.

35. Posey-Dowty, J., Crerar, D.A., Hellmann, R., and Chang, C.D. Am. Mineral. 1986, 71, 85.

RECEIVED August 4, 1986

32

The Growth of Calcium Phosphates

S. J. Zawacki[1], P. B. Koutsoukos[2], M. H. Salimi[1,3], and G. H. Nancollas[1]

[1]State University of New York at Buffalo, Department of Chemistry, Buffalo, NY 14214
[2]University of Patras, Department of Chemistry, Physical Chemistry Laboratory, Patras, Greece

The interactions of ions with growing calcium phosphate interfaces were investigated. The effect of various background electrolytes such as NaCl, KCl, and KNO_3 on the growth rate of hydroxyapatite HAP was measured, using a constant solution composition method. Parallel electrophoretic mobility measurements have also been made. The growth rate of HAP is markedly inhibited in the presence of magnesium ions which also induce a reversal of surface charge. The rate of octacalcium phosphate (OCP) crystallization is reduced in the presence of magnesium ion, although to a lesser extent than HAP. In contrast, dicalcium phosphate dihydrate (DCPD) crystallization is uneffected by magnesium. Strontium ions reduce the growth of both HAP and DCPD, and, unlike magnesium, are incorporated into these grown phases.

The growth of calcium phosphate salts is of importance both in the environment and in biological mineralization. In recent years there has been a resurgence of interest in these minerals because of their involvement in areas such as the removal of phosphate from waste water, the fate of elements such as aluminum, iron, and other heavy metals in the formation of lake and ocean sediments, and in industry where the production of scale on metal surfaces is a continuing problem. The increase in phosphate concentrations in lakes and rivers near heavily populated areas is another reason why the elucidation of the mechanism of precipitation and the nature of the phases which form are problems of considerable importance (1,2). In the environment, the adsorption of metal ions on the surface of calcium phosphate salts may serve to immobilize them in natural waters. In this process, changes in morphology and stoichiometry of the calcium phosphate crystals may prevent them from forming hard scale deposits when these waters are used industrially in applications such as cooling towers. Since both calcium and phosphate concentrations may be relatively high, perhaps through the use of lime additions for the removal of phosphates from sewage, calcium phosphate precipitation may be of particular importance. Higher phosphate levels are also being encountered in cooling waters due to increased water re-use, the use of lower quality sewage plant effluent and corrosion inhibitors which are degraded to orthophosphate.

[3]Current address: Department of Environmental Science and Engineering, Rice University, PO Box 1892, Houston, TX 77251.

0097-6156/86/0323-0650$06.00/0
© 1986 American Chemical Society

Calcium phosphate precipitation may also be involved in the fixation of phosphate fertilizer in soils.   Studies of the uptake of phosphate on calcium carbonate surfaces at low phosphate concentrations typical of those in soils, reveal that the threshold concentration for the precipitation of the calcium phosphate phases from solution is considerably increased in the pH range 8.5 - 9.0 (3).   It was concluded that the presence of carbonate ion from the calcite inhibits the nucleation of calcium phosphate phases under these conditions.   A recent study of the seeded crystal growth of calcite from metastable supersaturated solutions of calcium carbonate, has shown that the presence of orthophosphate ion at a concentration as low as 10^{-6} mol L^{-1} and a pH of 8.5 has a remarkable inhibiting influence on the rate of crystallization (4).   A seeded growth study of the influence of carbonate on hydroxyapatite crystallization has also shown an appreciable inhibiting influence of carbonate ion.(5).

Despite the importance of the precipitation of calcium phosphates, there is still considerable uncertainty as to the nature of the phases formed in the early stages of the precipitation reactions under differing conditions of supersaturation,   pH,   and   temperature.   Although   thermodynamic considerations yield the driving force for the precipitation, the course of the reaction is frequently mediated by kinetic factors.   Whether dicalcium phosphate   dihydrate   ($CaHPO_4.2H_2O$,   DCPD),   octacalcium   phosphate ($Ca_4H(PO_4)_3$, 2.5 H_2O, OCP), hydroxyapatite ($Ca_5(PO_4)_3(OH)$, HAP), amorphous calcium phosphate (ACP), or a defect apatite form from aqueous solution depends both upon the driving force for the precipitation and upon the initiating surface phase.   Thermodynamically, the relative supersaturation, σ, is given by

$$\sigma = (IP^{1/v} - K_{so}^{1/v})/K_{so}^{1/v}$$

where v represents the number of ions in a formula unit of a calcium phosphate. The ionic activity product, IP, of the solution with respect to the three most dominant calcium phosphate phases are:

DCPD : IP = $(Ca^{2+})(HPO_4^{2-})$, v = 2
OCP : IP = $(Ca^{2+})^4(HPO_4^{2-})(PO_4^{3-})^2$, v = 7
HAP : IP = $(Ca^{2+})^5(PO_4^{3-})^3(OH)$, v = 9

K_{so}, the thermodynamic solubility values at 37°C are, 1.87 x $10^{-7}(mol\ L^{-1})^2$ for DCPD, 5.0 x $10^{-50}(mol\ L^{-1})^7$ for OCP, and 2.35 x $10^{-59}\ (mol\ L^{-1})^9$ for apatite. The driving force for crystallization is expressed as a free energy of transfer, ΔG, of an average ion of the calcium phosphate from supersaturated to a saturated solution:

$$\Delta G = -RT\ \ln(IP/K_{so})^{1/v}$$

The experimental conditions, free energies, and the observed rates of growth of the different calcium phosphates at pH 6.0 and at 37° C are summarized in Table 1.   Although thermodynamically, HAP may be the preferred phase, kinetically it has a slow growth rate even though it has the highest thermodynamic driving force and it is quite sensitive to the presence of other ions in the supernatant solution.   Moreover, more acidic precursor phases may persist for long periods, especially as surface components, without conversion to the thermodynamically most stable phase.

Table 1

Experimental rates of growth, pH 6.00, 37°C,
0.100 mol L⁻¹ KNO₃ background electrolyte

Phase	Solution Concentration /10⁻³ mol L⁻¹ Calcium	Phosphate	σ phase	ΔG(DCPD) /KJ mol⁻¹	ΔG(OCP) /KJ mol⁻¹	ΔG(HAP) /KJ mol⁻¹	Growth Rate /mol min⁻¹ m⁻²	Ref
DCPD	5.20	5.20	0.39	-0.85	-1.98	-5.03	3.32×10^{-4}	6
DCPD	4.60	4.60	0.24	-0.56	-1.73	-4.77	0.93×10^{-4}	6
DCPD	4.30	4.30	0.17	-0.46	-1.59	-4.62	0.61×10^{-4}	6
OCP	4.50	3.38	0.74	-0.16	-1.43	-4.50	1.94×10^{-6}	6
OCP	4.10	3.08	0.61	0.62	-1.23	-4.30	0.63×10^{-6}	6
OCP	3.60	2.70	0.45	0.38	-0.96	-4.02	0.23×10^{-6}	6
HAP	5.05	3.03	5.00	-0.42	-1.71	-4.62	2.68×10^{-7}	26
HAP	3.12	1.87	3.00	1.93	-0.54	-3.58	5.67×10^{-8}	26
HAP	1.40	0.84	1.00	5.93	+1.46	-1.78	2.43×10^{-9}	26

Modifications of surface layers due to lattice substitution or adsorption of other ions present in solution may change the course of the reactions taking place at the solid/liquid interface even though the uptake may be undetectable by normal solution analytical techniques.   Thus it has been shown by electrophoretic mobility measurements, (6,7) that suspension of synthetic HAP in a solution saturated with respect to calcite displaces the isoelectric point almost 3 pH units to the value (pH = 10) found for calcite crystallites.   In practice, therefore, the presence of "inert" ions may markedly influence the behavior of precipitated minerals with respect to their rates of crystallization, adsorption of foreign ions, and electrokinetic properties.

In the environment, the presence of other alkaline earth cations such as magnesium and strontium may also markedly influence the course of the calcium phosphate precipitation.   In natural water systems, magnesium ion concentrations may be as high as 5×10^{-2} mol L^{-1} (1), while in biological calcification, magnesium concentrations ranging from 0.5% in outer tooth enamel layers to 2% in the innermost dentine are likely to have important consequences on the rate of remineralization of carious enamel (8).   It has been suggested that magnesium ions kinetically hinder the nucleation and subsequent growth of HAP by competing for lattice sites with the chemically similar but larger calcium ions (9).   It was shown that in the presence of magnesium ions, magnesium-containing tricalcium phosphate was formed (10,11).   In contrast to the influence of magnesium ion which is virtually excluded from the growing calcium phosphates, strontium is readily incorporated into HAP lattices because of the similarity of its ionic radius with that of calcium (12,13).   Thus a complete series of solid solutions can be prepared with lattice parameters linearly dependent upon the extent of strontium incorporation into the apatite lattice (14,15).   The interest in the incorporation of strontium into calcium phosphate stems from the concern about the ^{90}Sr content of bones and teeth.   Moreover, it has been suggested that the lack of strontium in the diet causes a high incidence of dental caries and poor growth conditions (16).

In the present work, a constant composition method has been used to investigate the growth of HAP from solutions of low supersaturation and in the presence of different background electrolytes.   The influence of magnesium and strontium ions both on the rate of crystallization and upon the electrokinetic properties of the crystallite surfaces has also been investigated.

Experimental

Experiments were made in a nitrogen atmosphere using reagent grade chemicals and triply distilled carbon dioxide-free water.   Standard solutions of phosphate were prepared from potassium dihydrogen phosphate (J.T. Baker Co., Ultrex grade), after drying at 105°C.   Ultrapure calcium nitrate tetrahydrate (Alfa Products) and reagent grade magnesium and strontium nitrates (J.T. Baker Co.) were used to prepare standard solutions.   Alkaline earths were determined by atomic absorption spectrophotometry (Perkin Elmer Model 503); phosphate was determined spectrophotometrically as the phosophovanodomolybdate complex as described previously (17).   All standard solutions were prepared using the same electrolyte composition as the growth media.   Specific surface area (SSA) was measured by BET nitrogen adsorption (30/70 nitrogen/helium mixture, Quantasorb II, Quantachrome, Greenvale, N.Y.).   HAP seed crystals were prepared by the method of Nancollas and Mohan (18) using calcium nitrate and potassium dihydrogen phosphate with potassium hydroxide for the control of pH.   Crystal composition was verified by infrared spectroscopy (Perkin Elmer grating infrared spectrometer

Model 467), x-ray powder diffracton (Philips XRG-3000, x-ray diffractometer, CuKα radiation Ni filter), and scanning electron microscopy, SEM (ISI scanning electron microscope, Model II).   Infrared spectra and powder diffraction data were in agreement with the published values for apatite (19,20).   Chemical analysis of the solid gave a molar ratio of Ca/P = 1.64 ±0.01;   the SSA was 21.5 m^2g^{-1}.

Crystallization experiments were made at 37°C in a water thermostatted double-walled vessel in metastable supersaturated solutions using the constant composition method.   Presaturated nitrogen gas, at 37°C was bubbled through the solutions which were stirred with Teflon coated magnetic stirring bars. The stability of supersaturated solutions prepared by mixing calcium nitrate, potassium phosphate, and potassium hydroxide solutions was verified by the constancy of pH for a period of at least 4h.   Following the introduction of seed crystals, crystallization started immediately, and two titrant solutions consisting of (i) calcium and potassium nitrate, and (ii) potassium phosphate and potassium hydroxide were added automatically from mechanically coupled burets (Metrohm Herisau, Model 3D Combititrator) in order to maintain the pH constant.   The pH was measured by means of a glass electrode together with a silver/silver chloride reference electrode separated from the cell solution by means of an intermediate salt-bridge consisting of 0.1 mol L^{-1} potassium nitrate.   Crystallization experiments in the presence of magnesium and strontium ions required a third titrant containing these ions in order to compensate for dilution effects during the crystallization reactions.   Titrant addition was continuously monitored and aliquots were periodically withdrawn from the precipitation cell, filtered (0.2 μm filters, Millipore, Bedford, Ma.), and the solution phase was analyzed for divalent metal and phosphate ions, to verify constancy of composition.   Typical constant composition rate curves for the dominant calcium phosphate phases at pH = 6.0 are shown in Figure 1.

The adsorption of magnesium by the HAP substrates was investigated by equilibrating samples of the solid with solutions of calcium phosphate in 0.1 mol L^{-1} potassium nitrate calculated to be saturated with respect to HAP and with pH adjusted to 8.0 ±0.1.   The polycarbonate equilibration vials were gently rotated end over end for 24h at 37°C.   Following this period, solid and liquid phases were separated by centrifugation and the supernatant analyzed for magnesium.   The electrophoretic mobilities of the HAP particles in the presence and absence of adsorbed magnesium were measured using a Rank Microelectrophoresis Mark II instrument with a four electrode cylindrical cell. The HAP solutions containing pre-calculated saturation concentrations of calcium and phosphate were equilibrated with the solid phases overnight and the pH was adjusted to the required values by the addition of potassium hydroxide or nitric acid.   Mobility measurements were made on at least forty particles.

Results and Discussions

In order to prepare calcium phosphate solutions of known supersaturation with respect to each of the phases, it was necessary to calculate the activities of the free ionic species by successive approximation for the ionic strength, from phosphate protonation, calcium phosphate, magnesium phosphate, and strontium phosphate ion-pair equilibrium constants together with mass balance and electroneutrality expressions as described previously (8,21). In the experiments containing magnesium or strontium ions, it was important to maintain the ionic strength constant by adjusting the concentration of added background electrolyte (potassium nitrate).

In a conventional study of the seeded growth of HAP crystals, in which the calcium and phosphate concentrations were allowed to decrease during the reactions, it was shown that the rate of crystallization varied depending upon the nature of the background electrolyte (22). Towards the end of the growth reaction (1,000 min) the extent of crystal growth increased in the order KCl < CsCl < NH_4Cl < LiCl < NaCl at pH = 7.4. Moreover, it was shown that the precipitated solids at these extended times of reaction contained from 5.5 to 6.0 mol % of sodium in the presence of sodium chloride but almost no potassium in the presence of potassium chloride. These results were in general agreement with those of Newman and co-workers (23), who found that sodium could replace calcium in the calcium phosphate solid whereas potassium was reversibly adsorbed on the surface.

A disadvantage of the conventional precipitation method in which the supersaturation was allowed to decrease during the reactions, was that different calcium phosphate phases could form and subsequently dissolve during the course of the reactions. In the present work, the constant composition method was used to investigate the influence of sodium chloride, potassium chloride, and potassium nitrate, as background electrolyte upon the rate of crystallization of HAP in solutions supersaturated only with respect to this phase. These experiments were made in solutions containing total concentrations of calcium, T_{Ca}, and phosphate, T_p, of 1.0×10^{-3} and 0.6×10^{-3} mol L^{-1}, respectively, pH = 7.0, 37°C, with ionic strength made up to 0.10 mol L^{-1} using background electrolyte and 50.5 mg of inoculating apatite seed (21.5 m^2g^{-1}). The rates of crystallization (±5%) in NaCl, KCl and KNO_3, background electrolytes were 2.34×10^{-8}, 2.09×10^{-8} and 1.92×10^{-8} mol apatite $m^{-2}min^{-1}$, respectively. To interpret the order of the crystallization rate NaCl > KCl > KNO_3, electrophoretic mobility measurements were made on HAP particles suspended in solutions of compositions similar to those of the crystallization experiments. In addition, potentiometric titrations were made in the presence of these electrolytes (24) in order to determine the point of zero charge, pzc. Electrophoretic mobility results as a function of pH, shown in figure 2, reveal that in the case of potassium nitrate no specific interaction of the electrolyte with HAP takes place since the isoelectric point (iep), pH = 6.7 in figure 2,coincides with that corresponding to the pzc. In potassium nitrate solutions, the surface charge of HAP as determined by microelectrophoresis did not show any change for extents of growth up to 60% of new material deposited on the inoculating seed. In contrast, in the presence of potassium chloride, the iep (pH = 6.4 in figure 2) is markedly different from the pzc value (at pH = 8.5) indicating a stronger interaction of Cl^- compared to NO_3^- ion. In the case of sodium chloride (figure 2) the iep is shifted to pH = 5.20 reflecting the combined effect of both sodium and chloride ions. Under the conditions of the crystallization experiments, it can be seen from figure 2 that at pH = 7, the HAP surface carries the most negative charge in the presence of sodium chloride. The observed decreasing negative charge, NaCl > KCl > KNO_3 follows the same trend as the corresponding rates of crystallization.

An advantage of the constant composition technique is that relatively large extents of growth and enhanced crystallinity can be achieved at low supersaturations. Improved crystallinity of the particles during crystallization is reflected in lower specific surface areas of the solid phases; x-ray powder diffractograms of the solid phases removed from the crystallization cell also show increases in sharpness. Experiments in which crystal growth was allowed to proceed until five or six times the amount of

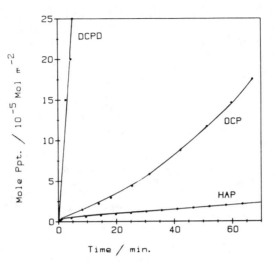

Figure 1.   The rate of growth curves for the three main calcium phosphate phases. DCPD, σ = 0.17;   OCP, σ = 0.74;   and HAP, σ = 5.0 as given in Table 1.

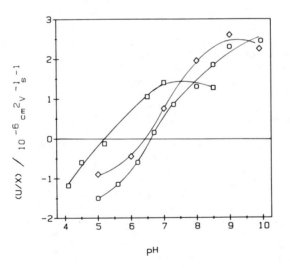

Figure 2.   The electrophoretic mobility of HAP plotted against pH, in different background electrolytes (0.01 mol L^{-1}) at 37°C; $\square$, NaCl; $\diamondsuit$, KCl; $\bigcirc$, KNO$_3$.

original seed was deposited, showed striking changes in morphology in the presence of chloride ion when added as potassium chloride (25) or lithium chloride (26). In contrast to the needle-like HAP morphology grown in the absence of chloride ion, platelike crystals were formed in its presence and measurements of the unit cell lattice parameters revealed a slight increase in the a axis and decrease in the c axis during crystallization. Such changes may reflect the presence of chloride ion in the apatite lattice and this suggestion is supported by the electrophoretic mobility results given above. In experiments conducted in the absence of chloride ion in which the supersaturated calcium phosphate solutions and titrant solutions were prepared using calcium hydroxide and phosphoric acid, characteristic needle-like HAP crystallites with the required hexagonal unit cell lattice parameters were obtained (25).

Uptake of magnesium ions by HAP surfaces at pH = 8.0 is illustrated in the adsorption isotherm shown in figure 3. The adsorption markedly exceeds that corresponding to a monolayer coverage (approximately $1.5 \, \mu mol \, m^{-2}$, calculated assuming that the hydrated ion radius is equal to its magnesium crystal radius plus the diameter of a bound water molecule). The sharply rising adsorption at low magnesium concentrations is indicative of the high affinity between substrate and adsorbent. No plateau was observed in the isotherm and further increase of magnesium concentration in the equilibrium solution leads to magnesium uptake levels as high as $27 \, \mu mol \, m^{-2}$ (26). Studies of the influence of magnesium ion upon the crystallization of calcium phosphate phases, showed little or no evidence for incorporation of this ion into the lattice. It has been suggested that the inhibition was due to adsorption of the added metal ion at the surface of the crystals (8).

The results of the electrophoretic mobility measurements on HAP particles having surface concentrations of magnesium ion of 2.5 and 25×10^{-3} $\mu mol \, m^{-2}$ are shown in figure 4. It can be seen that adsorption of magnesium ions leads to marked changes in zeta potential. It appears that magnesium uptake on HAP surfaces may not be a simple adsorption process since the isoelectric point is shifted in a direction opposite to that expected for cation uptake. An apparent surface concentration greater than that corresponding to a monolayer of hydrated magnesium ions may be attributed to partial dehydration of the adsorbed ions. This may suggest the formation of a "surface phase" of calcium-magnesium-phosphate. The electrophoretic mobility profile as a function of pH of the HAP samples with increasing concentration of magnesium ions at the surface, approaches that exhibited by Whitlockite (26).

The influence of magnesium and strontium ions upon the crystallization rates of calcium phosphate phases are summarized in figures 5 and 6. Previous work has shown that while having no detectable effect on the growth of DCPD, magnesium ions appreciably retard the rates of OCP and HAP crystallization (8). In figure 5 it can be seen that the moderate retardation of OCP crystallization by magnesium ion at pH = 6.0 contrasts the much greater inhibition of HAP growth at pH 7.4 and pH 8.5 (28).

Unlike magnesium, the strontium ion is readily incorporated into the growing calcium phosphate crystal lattices (21). It can be seen in figure 4 that at pH = 7.40 and ionic strength of $0.01 \, mol \, L^{-1}$, the inhibiting influence of magnesium ions on the crystallization of HAP is considerably greater than that observed in the presence of strontium ion. In contrast to the insensitivity of DCPD growth to the presence of magnesium ion, small decreases in crystallization rate of 6% relative to results in the absence of strontium accompany the incorporation of 3 mole per cent strontium into the DCPD lattice as shown in figure 6. Incorporation of 2% strontium resulted in a 40%

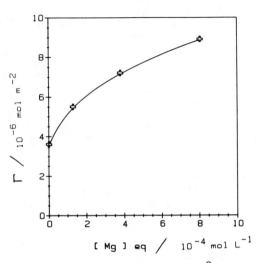

Figure 3.   Plot of magnesium adsorbed per m² of HAP against the equilibrium magnesium concentrations, remaining in solution at pH 8.0, 0.10 mol L⁻¹   KNO₃ background electrolyte at 37°C.

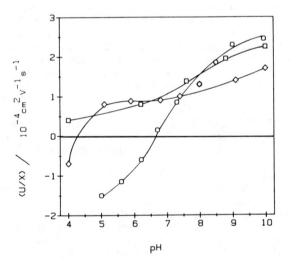

Figure 4.   Influence of magnesium adsorption on the electrophoretic mobility of HAP at 37°C, 0.01 mol L⁻¹   KNO₃ background electrolyte; O , HAP surface, no adsorption; □ , 2.5 μ mol m⁻²; ◇ , 25 μ mol m⁻², Mg²⁺ adsorbed.

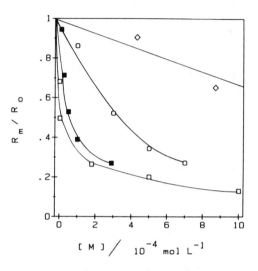

Figure 5. Influence of magnesium and strontium on the rate of precipitation of apatites at 37°C. R_M and R_0 are the rates in the presence and absence of metal ions, respectively. □ , Mg^{2+} on HAP, pH 8.50, 0.01 mol L^{-1} KCl, T_{Ca} = 3.0 x 10^{-4} mol L^{-1}, T_p = 1.8 x 10^{-4} mol L^{-1}; ■ , Mg^{2+} on HAP, pH 7.40, 0.01 mol L^{-1} KCl, T_{Ca}=5.0 x 10^{-4} mol L^{-1}, T_p = 3.0 x 10^{-4} mol L^{-1}; ○ , Sr^{2+} on HAP, pH 7.40, 0.01 mol L^{-1} KCl, T_{Ca} = 5.5 x 10^{-4} mol L^{-1}, T_p = 3.3 x 10^{-4} mol L^{-1}; ◇ , Mg^{2+} on OCP; pH 6.00, 0.10 mol L^{-1} KNO_3, T_{Ca} = 3.7 x 10^{-3} mol L^{-1}, T_p = 2.98 x 10^{-3} mol L^{-1}.

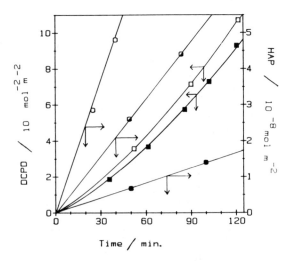

Figure 6.   Influence of strontium ion on the initial precipitation
of DCPD and HAP. □ , DCPD, pH = 5.60, 8 x 10^{-3} mol L^{-1} Ca(NO$_3$)$_2$
and 8 x 10^{-3} mol L^{-1} KH$_2$PO$_4$, 0.078 mol L^{-1} KNO$_3$; ■, DCPD with 3% Sr
incorporation, 6.6 x 10^{-3} mol L^{-1} CaCl$_2$ + 1.4 x 10^{-3} mol L^{-1} SrCl$_2$
and 8 x 10^{-3} mol L^{-1} KH$_2$PO$_4$, 0.078 mol L^{-1} KCl; ○ , HAP, pH = 7.40,
0.55 x 10^{-3} mol L^{-1} CaCl$_2$ and 0.33 mol L^{-1} KH$_2$PO$_4$ 0.01 mol L^{-1} KCl;
 ◑ , HAP, 2% Sr incorporation; ● , HAP, pH = 7.40, 12% Sr
incorporation.

retardation whereas a 12% strontium uptake reduced the growth rate by 57% (figure 5). The incorporation of strontium into the DCPD lattice was accompanied by small but significant expansions of the a and c lattice parameters (28) as was found for the HAP system (21). In contrast to the growth experiments in the presence of magnesium ion, the addition of strontium markedly reduces the average size of the crystallites formed (21).

In conclusion, it has been found that ions frequently found in the environment may have very different effects on the growth and surface properties of calcium phosphate phases. Chloride ion appears to accelerate the reactions either due to its electrostatic interactions with the apatite surface or as a result of substitution for hydroxyl ions. In contrast, sodium ions reduce the rate and may be incorporated into the growing phase (22). Strontium and magnesium ions appear to influence the rate by very different mechanisms. The small magnesium ion has no measurable effect on the least thermodynamically stable and fastest growing DCPD, it has a moderate inhibiting effect on OCP and markedly inhibits the slow growing apatite, even at micromolar concentrations. Since there is little evidence for appreciable incorporation of magnesium into any of these growing crystallites, this rate reduction probably results from surface interactions. The strontium ion, being similar in size to calcium, not only slows down the growth of DCPD and HAP, but is readily incorporated into the growing crystals.

Acknowledgments

We thank the National Institute of Dental Research of the National Institute of Health for a grant (Number DE03223) in support of this work. We also acknowledge a grant by NATO (Number 614-83) to Peter G. Koutsoukos and George H. Nancollas.

Literature Cited

1. Lindsay, W.L. "Chemical Equilibria in Soils"; Wiley: New York, 1979.
2. Brown, W.E. In "Environmental Phosphorus Handbook"; Griffith, E.J.; Beeton, A; Spencer, J.M. and Mitchel, D.T., Eds.; Wiley and Sons: New York, 1973, p.203.
3. Boischot, P.; Coppenet, M.; Herbert, J. Ann. Agron. 1949, 19, 103.
4. Kazmierczak, T.F. Ph.D. Thesis, SUNY, Buffalo, 1978.
5. Koutsoukos, P.G. Ph.D. Thesis, SUNY, Buffalo, 1984.
6. Hassan, K.A.R. MS Thesis, SUNY, Buffalo, 1984.
7. Somasundaran, P.; Wang, Y.H.C. In "Adsorption and Surface Chemistry of Hydroxyapatite"; Misra, D.N. Ed.; Plenum Press: New York, 1980, p.129.
8. Salimi, M.H.; Heughebaert, J.C.; Nancollas, G.H. Langmuir 1985, 1, 119.
9. Ferguson, J.; McCarty, P.L. Environ. Sci. Technol. 1971, 5, 534.
10. Hayek, E.; Newesely, H. Monatsch. Chem. 1958, 89, 88.
11. Rowles, S.L. Bull. Soc. Chim. Fr. 1958, 1797.
12. Elliot, J.C. Clin. Orthop. 1973, 93, 313.
13. Young, R.A. Colloq. Int. C.N.R.S. No. 230 1973m 21.
14. Colin, R.L. J. Am. Chem. Soc. 1959, 81, 5275.
15. Hayek, E.; Petter, H. Monatsch. Chem. 1960, 91, 356.
16. Shield, C.P.; Curzon, M.E.J.; Featherstone, J.D.B. Caries. Res. 1984, 18, 495.
17. Tomson, M.B.; Barone, J.P.; Nancollas, G.H. At. Absorpt. Newsl. 1977, 16, 117.

18.   Nancollas, G.H.; Mohan, M.S. Arch. Oral Biol. 1970, 15, 731.
19.   Baddiel, K.B.; Berry, E.E. Spectrochim. Acta. 1966, 22, 1407.
20.   ASTM, "X-ray Powder Diffraction File" No. 9-77.
21.   Koutsoukos, P.G.; Nancollas, G.H. J. Phys. Chem. 1981, 85, 2403.
22.   Nancollas, G.H.; Tomazic, B. J. Phys. Chem. 1974, 78, 2218.
23.   Newman, W.F.; Toribara, T.Y.; Mulryan, B.J. Arch. Biochem. Biophys. 1962, 98, 384.
24.   Koutsoukos, P.G.; unpublished data.
25.   Koutsoukos, P.G.; Nancollas, G.H. J. Cryst. Growth 1981, 55, 369.
26.   Koutsoukos, P.G.; Zawacki, S.J.; Nancollas, G.H., in preparation.
27.   Amjad, Z.; Koutsoukos, P.G.; Nancollas, G.H. J. Coll. Int. Sci. 1984, 101, 250.
28.   Shyu, L.J. Ph.D. Thesis, SUNY, Buffalo, 1982.

RECEIVED June 3, 1986

Author Index

Subject Index

Recent ACS Books